Virtual Synthesis of
Nanosystems by Design

Virtual Synthesis of Nanosystems by Design

From First Principles to Applications

by

Liudmila A. Pozhar

AMSTERDAM • BOSTON • HEIDELBERG • LONDON
NEW YORK • OXFORD • PARIS • SAN DIEGO
SAN FRANCISCO • SINGAPORE • SYDNEY • TOKYO

Acquiring Editor: Christina Gifford
Development Editor: Jeff Freeland
Project Manager: Surya Narayanan Jayachandran
Designer: Greg Harris

Elsevier
225 Wyman Street, Waltham, MA 02451, USA
The Boulevard, Langford Lane, Kidlington, Oxford OX5 1GB, UK

Copyright © 2015 Elsevier Inc. All rights reserved.

No part of this publication may be reproduced or transmitted in any form or by any means, electronic or mechanical, including photocopying, recording, or any information storage and retrieval system, without permission in writing from the publisher. Details on how to seek permission, further information about the Publisher's permissions policies and our arrangements with organizations such as the Copyright Clearance Center and the Copyright Licensing Agency, can be found at our website: www.elsevier.com/permissions.

This book and the individual contributions contained in it are protected under copyright by the Publisher (other than as may be noted herein).

Notices
Knowledge and best practice in this field are constantly changing. As new research and experience broaden our understanding, changes in research methods, professional practices, or medical treatment may become necessary.

Practitioners and researchers must always rely on their own experience and knowledge in evaluating and using any information, methods, compounds, or experiments described herein. In using such information or methods, they should be mindful of their own safety and the safety of others, including parties for whom they have a professional responsibility.

To the fullest extent of the law, neither the Publisher nor the authors, contributors, or editors, assume any liability for any injury and/or damage to persons or property as a matter of products liability, negligence or otherwise, or from any use or operation of any methods, products, instructions, or ideas contained in the material herein.

British Library Cataloguing-in-Publication Data
A catalogue record for this book is available from the British Library

Library of Congress Cataloging-in-Publication Data
A catalog record for this book is available from the Library of Congress

ISBN: 978-0-12-396984-2

For information on all Elsevier publications
visit our website at http://store.elsevier.com/

Contents

Preface	ix

Part One Quantum Statistical Mechanics Fundamentals — 1

1 Transport Properties of Spatially Inhomogeneous Quantum Systems From the First Principles — 3

- 1.1 Introduction — 3
- 1.2 Charge and spin transport in spatially inhomogeneous quantum systems — 7
  - 1.2.1 Expectation values of the charge and current densities — 9
  - 1.2.2 Space-time Fourier transforms of the expectation values of the charge and current densities — 10
  - 1.2.3 Space-time Fourier transforms of the generalized susceptibility and microcurrent-microcurrent EGFs — 11
  - 1.2.4 Generalized longitudinal sum rule — 13
  - 1.2.5 Dielectric permittivity of a spatially inhomogeneous quantum system in a weak external electromagnetic field — 14
  - 1.2.6 Generalized susceptibility of a spatially inhomogeneous quantum system in a weak external electromagnetic field — 18
  - 1.2.7 Longitudinal quantum conductivity of a spatially inhomogeneous system in a weak external electromagnetic field — 23
  - 1.2.8 Transversal conductivity of a spatially inhomogeneous quantum system in a weak external electromagnetic field — 24
- 1.3 Optical properties: the tensor of refractive indices — 27
- 1.4 Calculation of equilibrium Green's functions — 29
- 1.5 Zubarev-Tserkovnikov's pojection operator method — 33
  - 1.5.1 Definitions and the major properties of two-time temperature GFs used in statistical physics — 33
  - 1.5.2 ZT projection operator method: energy-dependent representation — 42
  - 1.5.3 ZT projection operator method: time-dependent representation — 50
  - 1.5.4 Advantages and shortcomings of the ZT projection operator method — 51
  - 1.5.5 Prospects of applications of the ZT projection operator method to include finite and/or spatially inhomogeneous quantum systems — 53
- References — 54

2 Quantum Properties of Small Systems at Equilibrium: First Principle Calculations — 61

- 2.1 Introduction — 61
- 2.2 Variational methods — 62
  - 2.2.1 The variation theorem and extended variation method — 62
  - 2.2.2 Non-degenerate perturbation theory and the variation-perturbation method — 64
  - 2.2.3 Perturbation theory treatment of degenerate energy levels — 66
  - 2.2.4 Spin components of wavefunctions and the Slater determinants — 68
  - 2.2.5 Variation modification of the Slater determinants — 69
- 2.3 The Hartree-Fock self-consistent field method — 70
  - 2.3.1 The Hartree self-consistent field method — 70
  - 2.3.2 The Hartree-Fock SCF method for molecules — 73
  - 2.3.3 The matrix elements of the Fock operator and calculation of physically meaningful quantities — 78
- 2.4 Configuration interactions — 82
- 2.5 The Møller-Plesset (MP) perturbation theory — 89
- 2.6 The coupled-cluster approximation — 94
- 2.7 Basis function sets — 98
- 2.8 Ab initio software packages and their use — 101
- 2.9 The virtual synthesis method — 103
- References — 107

Part Two Applications: Electronic Structure of Small Systems at Equilibrium — 111

3 Quantum Dots of Traditional III–V Semiconductor Compounds — 113

- 3.1 Introduction — 113
- 3.2 Virtual synthesis setup — 115
- 3.3 The smallest 3D molecule of In and As atoms — 117
- 3.4 Pre-designed and vacuum $In_{10}As_4$ molecules — 121
- 3.5 "Artificial" molecules $[In_{10}As_4]_{Ga}$ — 125
- 3.6 $Ga_{10}As_4$ molecules — 129
- 3.7 Spin density distributions of the studied molecules — 133
- 3.8 Electron charge delocalization and bonding in the studied molecules — 135
- 3.9 Conclusions — 139
- References — 141

4 Quantum Dots of Gallium and Indium Arsenide Phosphides: Opto-electronic Properties, Spin Polarization and a Composition Effect of Quantum Confinement — 147

- 4.1 Introduction — 147
- 4.2 Virtual synthesis procedure — 151
- 4.3 Ga-As molecules with one and two phosphorus atoms — 153

	4.4	In – As molecules with one and two atoms of phosphorus	162
	4.5	More about composition effects of quantum confinement: small molecules of In-As –based phosphides "imbedded" into a model Ga-As confinement	173
	4.6	Conclusions	184
		References	186

5 Quantum Dots of Diluted Magnetic Semiconductor Compounds — 191

	5.1	Introduction	191
	5.2	Virtual synthesis of small quantum dots of diluted magnetic semiconductor compounds	195
	5.3	Pre-designed and vacuum $In_{10}As_3Mn$ molecules	197
	5.4	Pre-designed and vacuum $In_{10}As_3V$ molecules	206
	5.5	$Ga_{10}As_3V$ molecules with one vanadium atom	214
	5.6	InAs - and GaAs - based molecules with two vanadium atoms	222
	5.7	Conclusions	228
		References	232

6 Quantum Dots of Indium Nitrides — 239

	6.1	Introduction	239
	6.2	Virtual synthesis of small indium nitride QDs	241
	6.3	Pyramidal InAs-based molecules with one nitrogen atom	243
	6.4	Pyramidal InAs-based molecules with two nitrogen atoms	252
	6.5	Pyramidal molecules $In_{10}N_4$	260
	6.6	Hexagonal molecules In_6N_6	265
	6.7	Conclusions	272
		References	277

7 Nickel Oxide Quantum Dots and Nanopolymer Quantum Wires — 283

	7.1	Introduction	283
	7.2	Molecules derived from Ni_2O cluster	285
	7.3	Molecules derived from Ni_2O_2 cluster	292
	7.4	Quantum dots derived from larger Ni-O clusters	296
	7.5	Ni-O nanopolymer quantum wires	306
	7.6	Discussion and conclusions	310
		References	313

8 Quantum Dots of Indium Nitrides with Special Magneto-Optic Properties — 317

	8.1	Introduction	317
	8.2	Virtual synthesis procedure for small indium nitride QDs doped with Ni or Co atoms	318
	8.3	Ni-doped molecules derived from unconstrained $In_{10}As_2N_2$ molecule	320

8.4	Ni-doped molecules derived from the pre-designed $In_{10}N_4$ molecule	330
8.5	Co-doped In-As-N and In-N molecules	337
8.6	Conclusions	344
	References	348

Appendix   Examples of Virtual Templates of Small Quantum Dots and Wires of Semiconductor Compound Elements   **351**

Index   **355**

Preface

A fundamental role of quantum many-body theory, quantum statistical mechanics and quantum chemistry methods in advanced nanomaterials and media design, and in particular, in the development of novel approaches to electronic structure engineering, makes it highly desirable for graduate students and researchers working in physics, chemistry, various fields of engineering, applied mathematics and computer science to understand major ideas and practical applications of such methods. This book has been written with a goal to help readers familiar with foundations of quantum theory at least at a graduate level link their theoretical background to the state of the art in the field of virtual (that is, quantum many-body theory-based, computational) synthesis of materials and media by design, its numerical methods and its software, to be able to apply their theoretical knowledge to solve practical problems.

Material included in this book is not intended as a review in the fields of science and engineering concerning synthesis of materials and media by design. Instead, the book provides only necessary conceptual, methodological, and software information to enable a reader to use the virtual synthesis method in practice. Part One of the book is devoted to theoretical foundations and consists of two chapters. Rigorous quantum statistical mechanical methods allow calculation of measurable properties of quantum systems using information on their structure, composition and conditions of their synthesis (such as the presence of quantum confinement). Descriptions of such methods and their results can be found in extensive literature accumulated since the late 19th century. Unfortunately, the majority of these methods has been derived to address spatially uniform systems, and thus is not applicable without the use of additional *ad hoc* assumptions to the case of small and/or strongly inhomogeneous systems, such as small quantum dots or wires, where a characteristic size of the system is of the same order as that of its constitutive units. This makes results obtained by applications of the standard quantum statistical methods to spatially non-uniform systems ambiguous and inconclusive. The only existing family of quantum statistical mechanical methods capable of providing reliable description of strongly spatially non-uniform systems is the first-principle projection operator methods. These methods and their results are discussed in Chapter 1. The chapter is unique in that that it links two of the first-principle projection operator methods and discusses their results from a standpoint of their use in quantum system design.

Notably, applications of all quantum statistical mechanical methods (projection operator-based, or otherwise) for purposes of (nano)materials synthesis by design rely on quantum many-body theoretical methods that provide information on the structure factors (such as electronic energies, correlation functions, Green's functions, etc.) of studied systems. The latter are used in quantum statistical mechanical formulae (derived using projection operator or other methods) to calculate measurable properties of many-body systems. Quantum chemistry codes realize some of such quantum many-body theoretical methods and implement them in the form of so-called first-principle quantum chemistry software. Chapter 2 of this book contains an overview of quantum many-body theoretical methods used in quantum chemistry, and their corresponding implementation. It also briefly outlines the existing first-principle quantum chemistry software. Quantum many-body theoretical methods of Chapter 2 can also be used on their own, if only the basic electronic and structural information on a system of interest is needed.

Part Two of this book is devoted to applications of the first-principle methods discussed in Part One to design virtually about 40 small systems composed of semiconductor compound and transition metal atoms, and Ni- and Co- oxides. The equilibrium structure, electronic energy level structure (ELS), molecular orbits, the charge and spin density distributions (CDDs and SDDs) of the nanostructures [or molecules, from a chemical point of view] of about 1 to 6 nm in linear dimensions virtually synthesized in model quantum confinement or in "vacuum" (*i.e.*, in the absence of any external fields or foreign atoms) are discussed in detail. Each of Chapters 3 to 8 included into Part Two can be worked with independently of the rest of the book content. However, an inexperienced reader is strongly advised to familiarize him/herself at least with Chapter 2 of Part One, to understand physical and chemical meaning of the results discussed in Chapters 3 to 8. Readers who are well familiar with quantum statistical mechanical, many-body theoretical and computational quantum chemistry methods will benefit from a unique description of projection operator-based quantum statistical methods outlined in Chapter 1, overview of computational methods in Chapter 2, and their applications described in Part Two of the book.

The book website booksite.elsevier.com/9780123969842/ and the corresponding webpage of the website www.PermaNature.com contain an additional Chapter 9 available only electronically for a free download. This chapter describes results of virtual synthesis by design of small cobalt oxide quantum dots. Chapter 9 also is independent of other chapters included in Part Two. The website www.PermaNature.com has a forum page and contact pages that can be used to communicate with the author. Moreover, both websites include free downloads containing 14 files (a brief description of these files can be found in Appendix "Examples of Virtual Templates of Small Quantum Dots and Wires of Semiconductor Compound Elements" at the end of the book) that can be used to configure input files for GAMESS software modules. Readers are encouraged to use such input files to (1) repeat some results of the author, (2) attempt higher-order approximations as applied to the systems investigated by the author, and (3) modify author's setups and attempt virtual design of other systems of interest.

My research discussed in this book significantly benefitted from advice and encouragement of outstanding physicists and chemists whom I had a privilege to collaborate with. In particular, the content of Chapter 1 has been refined in enlightening discussions with my teacher Yu. A. Tserkovnikov (Steklov Institute of Mathematics, Russian Academy of Sciences, Moscow, Russia) who was one of the brightest students of Nikolai N. Bogoliubov, and one of the greatest scientists ever worked in the fields of quantum statistical mechanics and its mathematical foundations. My interest in applications of the quantum many-body theoretical methods to semiconductor compound systems has been stimulated by F. Szmulowicz (University of Dayton Research Institute, Dayton, OH) and W. C. Mitchel (Air Force Research Laboratory, Dayton, OH). Enthusiasm and support of A. T. Yeates and D. Dudis of the Air Force Research Laboratory in Dayton, OH, encouraged my involvement with quantum chemistry, and my appreciation of its virtues and exquisite charms. The virtual synthesis method has been developed and realized in close collaboration with these esteemed colleagues. I had many insightful discussions of magnetic properties of nanosystems with Y. Qiang, and mathematical methods of many-body quantum theory with R. Machleidt and F. Sammarruca-Machleidt during my years with the University of Idaho in Moscow, ID. Several of my students, notably C. Mavromichalis (BoiseLAN, Boise, ID), contributed to research described in Chapters 7 and 9. Encouragement and understanding of my husband G. M. Tsoy (University of Alabama at Birmingham and IPG Photonics, Birmingham, AL) have made my work on this book, and our joint life in physics in general, an exciting and unforgettable experience. My heartfelt gratitude goes to these great scientists, colleagues and friends.

Liudmila A. Pozhar
ChiefScientist@PermaNature.com
LPozhar@yahoo.com
PermaNature, Sterrett, AL

Part One

Quantum Statistical Mechanics Fundamentals

Transport Properties of Spatially Inhomogeneous Quantum Systems From the First Principles

1.1 Introduction

Since Gibbs and Boltzmann, statistical mechanics has been focused on the first-principle prediction of thermodynamic and transport properties of many particle systems. The majority of models and mathematical methods of statistical mechanics are designed to work in so-called thermodynamic limit where the number of particles (N) and the system volume (V) simultaneously tend to infinity, while their ratio remains finite. Another important concept concerns the initial state of a many-particle system that is assumed to be the thermodynamic equilibrium corresponding to the *minimum minimorum* of the total energy of the system. These two concepts validate rigorously the use of theory of stochastic processes and mathematical statistics methods to reduce a system of 6N coupled equations of motion (in the simplest case) for the system particles to one equation of motion of the entire system formulated with respect to the N-particle distribution function of the system (classical Liouville equation) or the N-particle density matrix (the von Neumann, or quantum Liouville equation), respectively. At the next step, perturbation theory, DFT or projection operator methods are used to reduce the Liouville, or von Neumann (in the quantum case) equations to the so-called master equation for collective dynamical variables or observables, respectively, that can be further reduced to a manageable system of coupled equations for correlation functions or Green's functions (GFs). Further on, the equations for conserved collective dynamical variables (observables) are derived, and the thermodynamic and/or transport properties are identified in terms of the correlation functions and/or GFs. Thus, once the correlation functions or GFs are determined, the thermodynamic and transport properties of the N-particle system can be calculated directly.

With advent of novel technologies of materials and media synthesis there is a growing demand for updating the fundamental basis of statistical mechanics to account for small and/or strongly spatially inhomogeneous systems. Such first-principle statistical mechanical foundation is especially important to design novel materials for quantum electronics, spinstronics, quantum computing, communication, information processing and storage technologies. In particular, fundamental understanding of coherent, polarized, and entangled charge and spin states of quantum particles, their dynamics, and their contributions to quantum spin/charge transport properties at realistic materials synthesis conditions, such as quantum confinement, is paramount [1–5] to establish novel electronic materials technologies. In other words, relations between the structure, and thermodynamic and transport properties of materials and media, must be established using first principle quantum statistical mechanical methods.

At present, solid state electronic structure theory [6–10] largely employs somewhat modified statistical mechanical foundation specific to bulk solid lattices and various half-heuristic methods [11] to identify systems exhibiting new electronic properties, such as quantum wells, large quantum wires [12] and quantum dots (QDs) [13]. In the case of small structures, such as small QDs, where statistical mechanical approaches so modified do not work, computational methods, such as DFT- and Hartree-Fock-based (HF) methods, self-consistent field (SCF) approximations, configuration interaction (CI) methods, complete active space SCF (CASSCF), multi configuration SCF (MCSCF), Møller-Plesset - and coupled-clusters approximations are used to calculate the electronic energy level structure directly solving the Schrödinger equation numerically. [These methods will be briefly discussed in Chapter 2.] At present, the corresponding software packages, such as GAMESS, NWChem [14], GAUSSIAN or Molpro, allow the electronic structure calculations for systems in equilibrium at zero temperature. In addition, equilibrium and non-equilibrium molecular dynamics (MD) simulations are widely used to study structure-property relations at non-zero temperature. In particular, the spin/charge transport processes are simulated using MD or Monte-Carlo means in the Born-Oppenheimer approximation. In the case of numerical calculations and MD simulations, correlations between the electronic structure and the transport properties are introduced heuristically, adjusting the statistical mechanical and semi-phenomenological approaches developed for large systems. Such computations, on their own, do not permit first-principle predictions of the spin/charge transport properties of small QDs and molecules. Yet accurate first-principle predictions are crucial to manipulate with electron spins and quantum states of energy and information carriers in small QD/QW systems.

In their turn, existing *semi-heuristic* modifications [15–23] of various theoretical models developed originally for much larger systems at low temperature conditions and applied to characterize charge and spin transport in small systems often lead to physically incorrect predictions even for mesoscopic tunneling junctions [24]. Even the best of such models do not include adequate description of system-to-confinement coupling, such as quantum confinement effects. At the same time, such coupling is one of the major sources of both decoherence and coherence [25,26] of states of quantum particles, such as the electron charge and spin states [27–29]. Thus, the nature of such models does not allow, in principle, first-principle predictions of electronic and spin/charge transport properties of small and strongly spatially inhomogeneous systems.

As already mentioned in the beginning of this section, first-principle predictions of electronic transport properties imply the use of specifically tailored quantum statistical mechanical methods to derive self-consistently the spin/charge transport theory from the quantum Liouville (von Neumann) equation. Thus far, this formidable task has been properly addressed only for mesoscale systems where non-equilibrium GF (NGF)-based methods [such as Keldysh's two-time NGF [30–32] and more recent DFT-NGF approaches [33]] are among the most adequate statistical mechanical techniques used. Unfortunately, these methods have several major disadvantages. In particular, the field-theoretical NGFs used in the majority of these approaches are not directly related to the transport coefficients. Thus, despite the availability of a system of coupled equations for NGFs that can be reduced by *controlled approximations* to a manageable master equation (such as the Kadanoff-Dyson equation, etc.), these methods require further approximations to link NGFs to the transport coefficients. This is done using the major *uncontrolled approximation* - introducing the (Wigner's) distribution functions (DFs) [34,35], and deriving and solving the corresponding kinetic equations for DFs with (semi-heuristic) boundary and initial conditions. Using these solutions, one can establish semi-heuristic structure-property relations for the transport coefficients in terms of DFs that are related to NGFs.

Some of the NGF-based approaches correctly use thermodynamic NGFs that can be related to the transport properties directly [36] in the case of large systems, but still have to make use of semi-heuristic considerations when applied to small or spatially inhomogeneous systems. Even more important disadvantage of both first-principle and semi-heuristic NGF-based approaches is that NGFs have to be calculated for the same system every time when any process parameter changes, which is completely impractical for applications. To date, NGF-based methods have been successfully used to justify a semi-phenomenological theory of conductance due to Landauer, Imry, Buttiker *et al.* [35,37–40] for mesoscale systems. However, applications of these methods to electron transport in small QDs and molecules requires even more uncontrolled approximations (such as formulations of the boundary conditions for the distribution functions in atomic-scale gates, etc.). This results in a large number of case-specific, semi-phenomenological descriptions of transport in mesoscopic systems, and leads to progressively less reliable evaluations for the electron transport properties of such systems.

Many other existing approaches to charge and spin transport in mesoscale systems originate both from rigorous methods, such as R-matrix theory and supersymmetry-based techniques [41,42], semi-heuristic and heuristic master [43–45] and/or kinetic equation methods [46–53]. Yet all of these approaches rely on semi-phenomenological considerations to relate the structure, electronic structure or statistical properties to transport properties they are expected to predict.

Note again, that GF-based quantum-statistical mechanical methods [36,54–56] provide the only reliable *ab initio* tools that are smoothly integrated into all first-principle statistical mechanical approaches to transport properties of quantum systems. In particular, such integration has already been instrumental in derivation of both existing rigorous linear response theories of the charge and spin transport by Baranger and Stone [57], and Pozhar [58–61], applicable to small and spatially inhomogeneous quantum systems. The latter approach is notable for using two-time temperature GFs defined with regard to equilibrium or steady states (so-called equilibrium GFs, or EGFs) that can be calculated to any desirable degree of accuracy using Zubarev-Tserkovnikov's (ZT) projection operator method [36,62,63]. This method can also be used to develop an *ab initio* quantum transport theory *free of uncontrolled approximations* (i.e., without introduction of the distribution functions). The latter can be done if the ZT system of projection operators is defined in terms of observables, rather than EGFs, in a way similar to formulation of the Mori projection operator method. As applied to observables, Zubarev-Tserkovnikov projection operator method is, in fact, the Mori projection operator method [64] rigorously derived and generalized to be applicable to spatially homogeneous quantum and classical systems in a systematic fashion. Another rigorous generalization of Mori's method to include strongly inhomogeneous classical systems was introduced by Pozhar in the end of 1980's [65–69]. This method has been successfully applied to predict transport properties of tens fluidic systems in atomically-narrow pores of porous solids with atomically structured pore walls.

The major advantages of the ZT projection operator method are that it is (i) the only first-principle method that can be applied, upon some generalization, to strongly inhomogeneous and small finite systems, (ii) can be formulated in terms of NGFs and EGFs, and also in terms of collective dynamical variables or observables, (iii) permits to avoid uncontrolled and many controlled approximations, thus (iv) sparing significant efforts required to develop a quantum kinetic theory. Both NGFs and EGFs have a transparent physical meaning, as they are directly related to non-equilibrium and equilibrium susceptibilities, respectively, that can be measured experimentally. At the same time, ZT transport coefficients can be explicitly related to EGFs (for systems in their equilibrium or steady states, respectively), or NGF (for systems in their higher excited states). The corresponding GFs are calculated solving the ZT system of coupled equations for GFs and using controlled approximations. An added bonus of this approach is that GFs used in the ZT expressions for the transport coefficients can also be calculated from experimental data or by applying molecular dynamic simulations methods, which is very convenient in practice. Fortunately, for the majority of applications, including spin-charge transport in low-dimensional quantum systems, one is primarily interested in close to equilibrium system states, and thus concerned only with calculation of two-time temperature EGFs. These EGFs need be calculated only once for a given system [while NGFs have to be calculated anew for the same system once process conditions or parameters change].

The major disadvantage of the ZT projection operator method is that the closure of the method (that is, truncation of the infinite system of coupled equations for GFs or dynamical variables/observables) is case-specific and based upon physical and chemical nature of a system under consideration. While several methods have been proposed [70,71] to streamline the closure procedure, those methods are not immediately derivable from the first principles in a general case, and have to be justified using, for example, perturbation theory considerations, in each particular case.

In this chapter the first-principle statistical mechanical methods available at present are demonstrated in the case of the electron transport properties of strongly spatially inhomogeneous quantum systems near equilibrium or steady states. Thus, the only existing first-principle quantum charge transport theory [58–60] derived in the linear approximation with regard to external electromagnetic fields and without any assumptions concerning the number of particles, charge density and other observables, is outlined in section 1.2. Conservation equations for the charge density, momentum and energy observables so derived feature explicit expressions for the tensorial conductivity, dielectric permittivity, magnetic permeability and generalized susceptibility expressed in terms of two-time temperature EGFs. The explicit expression for the tensor of refractive indices is derived in section 1.3. Section 1.4 outlines calculation of the charge density – charge density EGFs using the ZT projection operator method, and section 1.5 describes foundations and major results of this method in conjunction with calculation of EGFs of higher orders specific to finite, spatially inhomogeneous quantum systems. Other existing descriptions of charge and spin transport in spatially inhomogeneous quantum systems including semi-phenomenological NGF- and quantum field-theoretical GF-based methods, are widely discussed in literature and can be found elsewhere.

1.2 Charge and spin transport in spatially inhomogeneous quantum systems

A system of N particles possessing charges and spins positioned in confinement or at interface can be considered as a system in an external electromagnetic field created by atoms of the confinement or interface and characterized by the vector intensities $\mathbf{E}(\mathbf{r},t)$ and $\mathbf{H}(\mathbf{r},t)$, respectively, dependent of the position $\mathbf{r}$ in the system and time t. In condensed matter physics the majority of considered systems are non-relativistic, so these fields are treated classically and defined in terms of the vector and scalar potentials $\mathbf{A}(\mathbf{r},t)$ and $\varphi(\mathbf{r},t)$, respectively, through the local Maxwell's equations:

$$\mathbf{E}(\mathbf{r},t) = -\frac{1}{c}\frac{\partial \mathbf{A}(\mathbf{r},t)}{\partial t} - \nabla_\mathbf{r} \varphi(\mathbf{r},t),$$
$$\mathbf{H}(\mathbf{r},t) = \mathbf{curl}\, \mathbf{A}(\mathbf{r},t),$$
(1)

The subscript $\mathbf{r}$ at the gradient sign $\nabla_\mathbf{r}$ in Eq. (1) indicates that the gradient operator is applied to the vector $\mathbf{r}$, and c is the speed of light. By their nature, the external electromagnetic fields created by confinement atoms are weak, and at this point, are defined with regard to an arbitrary gauge. Magnetization of the system is considered small and therefore, a contribution to the total energy of the system due to particle spins' interactions with the external fields can be neglected. All particles carry the same spins σ and charges e. With these assumptions the Hamiltonian $\mathcal{H}$ can be represented as a sum of 3 terms,

$$\mathcal{H} = \frac{1}{2m}\sum_{i=1}^{N}\left(\mathbf{p}_i - \frac{e}{c}\mathbf{A}(\mathbf{r}_i,t)\right)^2 + H_{\text{int}} + e\sum_{i=1}^{N}\varphi(\mathbf{r}_i,t),$$
(2)

where the first term is the "kinetic energy" operator defined by contributions due to the momentum operators $\mathbf{p}_k = (\hbar/i)\nabla_{\mathbf{r}_k} - (e/c)\mathbf{A}(\mathbf{r}_k,t)$ of each particle (the index k runs through all particles) affected by the vector-potential of the external field, the last term is the "potential energy" operator due to interaction of the charges on the particles with the external scalar potential, and H_{int} is the energy operator that includes interactions of the system particles between themselves and with the particles of the environment. For further technical considerations it is convenient to introduce quantum field operators $\psi^+(\mathbf{r})$ and $\psi(\mathbf{r})$ defined in terms of the creation and annihilation operators of the charge/spin carriers, $a^+_{k\sigma}$ and $a_{k\sigma}$ (where **k** is the wave vector), respectively,

$$\psi^+(\mathbf{r}) = \frac{1}{\sqrt{V}} \sum_{k,\sigma} e^{-i\mathbf{k}\mathbf{r}} \delta_{\sigma\sigma_z} a^+_{k\sigma},$$
$$\psi(\mathbf{r}) = \frac{1}{\sqrt{V}} \sum_{k,\sigma} e^{i\mathbf{k}\mathbf{r}} \delta_{\sigma\sigma_z} a_{k\sigma}, \quad (3)$$

where σ_z denotes the z-component of the spin, and $\delta_{\sigma\sigma_z}$ is the Kronecker delta. Using the field operators one can write the total Hamiltonian of Eq. (2) in the second-quantized representation,

$$\mathcal{H} = \frac{1}{2m} \int d\mathbf{r} \, \psi^+(\mathbf{r}) \left(\frac{\hbar}{i} \nabla_\mathbf{r} - \frac{e}{c} \mathbf{A}(\mathbf{r},t) \right)^2 \psi(\mathbf{r}) + H_{int} + e \int d\mathbf{r} \, \psi^+(\mathbf{r}) \psi(\mathbf{r}) \varphi(\mathbf{r},t). \quad (4)$$

Using a definition of the current density operator as the Fréchet derivative of the total Hamiltonian with respect to the vector-potential, $\mathbf{j}(\mathbf{r},t) = -e \frac{\delta \mathcal{H}}{\delta \mathbf{A}(\mathbf{r},t)}$, one can find from Eq. (4) $\mathbf{j}(\mathbf{r},t)$ explicitly:

$$\mathbf{j}(\mathbf{r},t) = \frac{e\hbar}{2mi} \{ \psi^+(\mathbf{r}) \nabla_\mathbf{r} \psi(\mathbf{r}) - [\nabla_\mathbf{r} \psi^+(\mathbf{r})] \psi(\mathbf{r}) \} - \frac{e^2}{mc} \mathbf{A}(\mathbf{r},t) \psi^+(\mathbf{r}) \psi(\mathbf{r}). \quad (5)$$

Introducing the number density operator $n(\mathbf{r})$ of the charge carriers, the current density operator $\mathbf{j}_0(\mathbf{r})$ in the absence of the field,

$$\mathbf{j}_0(\mathbf{r}) = \frac{e\hbar}{2mi} \{ \psi^+(\mathbf{r}) \nabla_\mathbf{r} \psi(\mathbf{r}) - [\nabla_\mathbf{r} \psi^+(\mathbf{r})] \psi(\mathbf{r}) \}, \quad (6)$$

and the charge density operator,

$$\rho(\mathbf{r}) = e\psi^+(\mathbf{r}) \psi(\mathbf{r}) = en(\mathbf{r}), \quad (7)$$

in the absence of the external field, one can represent the Hamiltonian [4] as

$$\mathcal{H} = H_0 + H_1 + H_2. \quad (8)$$

The first term in this expression,

$$H_0 = \frac{1}{2m} \int d\mathbf{r} \psi^+(\mathbf{r}) \left[\frac{\hbar}{i} \nabla_\mathbf{r} \right]^2 \psi(\mathbf{r}) + H_{int}, \quad (9)$$

is the Hamiltonian in the absence of the external electromagnetic field, and the second term,

$$H_1 = -\frac{1}{c}\int d\mathbf{r}\, \mathbf{j}_0(\mathbf{r})\cdot \mathbf{A}(\mathbf{r},t) + \int d\mathbf{r}\rho(\mathbf{r})\varphi(\mathbf{r},t), \tag{10}$$

is the perturbation Hamiltonian, which is linear in the field potentials. [Here and below the dot "·" denotes the inner product.] The last term,

$$H_2 = \frac{e}{2mc^2}\int d\mathbf{r}\rho(\mathbf{r})\mathbf{A}^2(\mathbf{r},t), \tag{11}$$

is the second-order perturbation with respect to the vector-potential, and can be neglected because the external field is assumed to be weak. As a result, the total Hamiltonian linearized with respect to the external field potentials is

$$\mathcal{H} = H_0 + H_1. \tag{12}$$

1.2.1 Expectation values of the charge and current densities

Expectation values $\langle \mathbf{O}(\mathbf{r},t)\rangle$ of observables $\mathbf{O}$ of a system characterized by a Hamiltonian linear in an external field potential can be suitably linearized using Kubo's procedure generalized to include spatially inhomogeneous systems. The latter is justified due to a quasi-local nature [72] of the field operators [3]. Thus, introducing the equilibrium or steady state retarded two-time temperature Bogoliubov-Tiablikov's GFs (EGFs, or TTGFs),

$$\langle\langle \mathbf{O}(\mathbf{r},t)\,|\,\mathbf{K}(\mathbf{r}',t')\rangle\rangle_0 \equiv \vartheta(t-t')\frac{1}{i\hbar}\langle[\mathbf{O}(\mathbf{r},t),\mathbf{K}(\mathbf{r}',t')]\rangle_0, \tag{13}$$

one can write the expectation values in the form

$$\langle \mathbf{O}(\mathbf{r},t)\rangle = \langle \mathbf{O}(\mathbf{r})\rangle_0 + \int_{-\infty}^{\infty} dt'\,\langle\langle \mathbf{O}(\mathbf{r},t)\,|\,H_1(t')\rangle\rangle_0. \tag{14}$$

In Eqs. (13) and (14) $\vartheta(t-t')$ denotes the step function, $[\ldots,\ldots]$ is the commutator, and the expectation values are defined as $\langle \mathbf{O}(\mathbf{r},t)\rangle = Tr(\wp \mathbf{O})$, with Tr being the trace of an operator, and the subscript 0 means the expectation value calculated with regard to the equilibrium or steady state of the system (that is, with respect to the equilibrium or steady state statistical operator $\wp_0$). The statistical operator $\wp$ of the system is the solution of the quantum Liouville (von Neumann's) equation with the Hamiltonian of Eq. (12) and the initial condition $\wp(t=-\infty)\equiv \wp_0 = e^{-\beta H_0}/Tre^{-\beta H_0}$, where $\beta=1/k_BT$ and k_B is the Boltzmann constant.

Using Eq. (14) for the charge density observable, $\rho(\mathbf{r},t) = e^{iH_0t/\hbar}\rho(\mathbf{r})e^{-iH_0t/\hbar}$, and Eq. (10), one can immediately derive the local charge conservation equation for time t:

$$\langle \rho(\mathbf{r},t) \rangle = \rho_0(\mathbf{r}) - \frac{1}{c}\int d\mathbf{r}' \int_{-\infty}^{\infty} dt' \langle\langle \rho(\mathbf{r},t)_0 | \mathbf{j}_0(\mathbf{r}') \rangle\rangle_0 \cdot \mathbf{A}(\mathbf{r}',t')$$
$$+ \int d\mathbf{r}' \int_{-\infty}^{\infty} dt' \langle\langle \rho(\mathbf{r},t) | \rho(\mathbf{r}') \rangle\rangle_0 \varphi(\mathbf{r}',t'), \qquad (15)$$

where $\rho_0(\mathbf{r})$ denotes the charge density at the equilibrium or steady state.

Applying Eq. (13) to the current density $\mathbf{j}(\mathbf{r},t)$ of Eq. (5), and using Eq. (10) again, one can derive the local current density conservation equation,

$$\langle \mathbf{j}(\mathbf{r},t) \rangle = -\frac{1}{c}\int d\mathbf{r}' \int_{-\infty}^{\infty} dt' \langle\langle \mathbf{j}(\mathbf{r},t) | \mathbf{j}_0(\mathbf{r}') \rangle\rangle_0 \cdot \mathbf{A}(\mathbf{r}',t')$$
$$+ \int d\mathbf{r}' \langle\langle \mathbf{j}(\mathbf{r},t) | \rho(\mathbf{r}') \rangle\rangle_0 \varphi(\mathbf{r}',t') - \frac{e}{mc}\rho_0(\mathbf{r})\mathbf{A}(\mathbf{r},t), \qquad (16)$$

where it was taken into account that the expectation value of the unperturbed current of Eq. (6) at equilibrium is zero, $\langle \mathbf{j}_0(\mathbf{r}) \rangle_0 = 0$. Using Eq. (16) one can prove, that for the current density of Eq. (5) the microscopic continuity equation still holds:

$$\frac{\partial \rho(\mathbf{r})}{\partial t} = -div\ \mathbf{j}(\mathbf{r},t).$$

1.2.2 Space-time Fourier transforms of the expectation values of the charge and current densities

Space-time Fourier transforms (STFTs) $\rho(\mathbf{k},\omega)$ and $\mathbf{j}(\mathbf{k},\omega)$ of the expectation values of the charge and current density observables, respectively, often occur more convenient for applications than those of Eqs. (15) and (16). Explicit expressions for these STFTs are:

$$\rho(\mathbf{k},\omega) = en_0(\mathbf{k})\delta(\omega) - \frac{1}{c}\sum_{\mathbf{l}}\langle\langle \rho_\mathbf{k} | \mathbf{j}_{-\mathbf{l}} \rangle\rangle_{0,\omega} \cdot \mathbf{A}(\mathbf{l},\omega)$$
$$+ \sum_{\mathbf{l}}\langle\langle \rho_\mathbf{k} | \rho_{-\mathbf{l}} \rangle\rangle_{0,\omega} \varphi(\mathbf{l},\omega), \qquad (17)$$

$$\mathbf{j}(\mathbf{k},\omega) = -\frac{e^2}{mc}\sum_{\mathbf{l}}n_0(\mathbf{k}-\mathbf{l})\mathbf{A}(\mathbf{l},\omega) - \frac{1}{c}\sum_{\mathbf{l}}\langle\langle \mathbf{j}_\mathbf{k} | \mathbf{j}_{-\mathbf{l}} \rangle\rangle_{0,\omega} \cdot \mathbf{A}(\mathbf{l},\omega)$$
$$+ \sum_{\mathbf{l}}\langle\langle \mathbf{j}_\mathbf{k} | \rho_{-\mathbf{l}} \rangle\rangle_{0,\omega} \varphi(\mathbf{l},\omega), \qquad (18)$$

In these equations brackets $\langle\langle ...|... \rangle\rangle_{0,\omega}$ denote STFTs of the corresponding EGFs, ω is the frequency, $\mathbf{k}$ and $\mathbf{l}$ are wave vectors (the summation runs over all wave vectors from $-\infty$ to $+\infty$), $\delta(\omega)$ denotes Dirac's delta function, and $n_0(\mathbf{k}) = (1/V)\left\langle \sum_{\mathbf{q},\sigma} a^+_{(\mathbf{q}-\mathbf{k}),\sigma} a_{\mathbf{q},\sigma} \right\rangle_0$

is STFT of the equilibrium charge number density that does not depend on ω. The quantities

$$\rho_1 = \frac{e}{\sqrt{V}} \sum_{q,\sigma} a^+_{q-l,\sigma} a_{q,\sigma} \tag{19}$$

and

$$\mathbf{j}_1 = \frac{e\hbar}{m\sqrt{V}} \sum_{q,\sigma} (\mathbf{q} - \frac{\mathbf{l}}{2}) a^+_{q-l,\sigma} a_{q,\sigma} \tag{20}$$

are space-time Fourier transforms of the microscopic charge and current density operators, respectively, where σ denotes summation over the z-component of the particle spins. Note, that for spatially homogeneous systems only terms with $\mathbf{l} = \mathbf{k}$ are non-zero, and the equilibrium charge number density is a constant. In this case Eqs. (17) and (18) reduce to those derived by Zubarev [73] in 1960 for large, spatially homogeneous systems [see also Ref. 54].

1.2.3 Space-time Fourier transforms of the generalized susceptibility and microcurrent-microcurrent EGFs

The first two terms in the right hand side of Eq. (18) help identify STFT of the second rank Cartesian tensor χ with components $\chi_{\alpha\beta}(\mathbf{k},\mathbf{l},\omega)$ (where α and β indices run from 1 to 3) expressed in terms of the $\alpha\beta$-components of STFT of the microcurrent-microcurrent EGFs, $\langle\langle \mathbf{j}^\alpha_\mathbf{k} | \mathbf{j}^\beta_{-l} \rangle\rangle_{0,\omega}$:

$$\chi_{\alpha\beta}(\mathbf{k},\mathbf{l},\omega) = \frac{e^2}{mc} n(\mathbf{k}-\mathbf{l})\delta_{\alpha\beta} + \langle\langle \mathbf{j}^\alpha_\mathbf{k} | \mathbf{j}^\beta_{-l} \rangle\rangle_{0,\omega}, \tag{21}$$

where $\delta_{\alpha\beta}$ is Kronecker's symbol. The Fourier transforms of the EGFs $\langle\langle \mathbf{j}^\alpha_\mathbf{k} | \mathbf{j}^\beta_{-l} \rangle\rangle_{0,\omega}$ are bilinear forms of the vectors $\mathbf{k}$ and $\mathbf{l}$, and in the most general case can be represented in the form

$$\langle\langle \mathbf{j}^\alpha_\mathbf{k} | \mathbf{j}^\beta_{-l} \rangle\rangle_{0,\omega} = \chi^{lon}(\mathbf{k},\mathbf{l},\omega) \frac{k_\alpha l_\beta}{kl} + \left[C(\mathbf{k},\mathbf{l},\omega)\delta_{\alpha\beta} - \frac{k_\alpha l_\beta}{kl} \right] \chi^{tr}(\mathbf{k},\mathbf{l},\omega), \tag{22}$$

where

$$\chi^{lon}(\mathbf{k},\mathbf{l},\omega) = \frac{\langle\langle \mathbf{k}\cdot\mathbf{j}_\mathbf{k} | \mathbf{l}\cdot\mathbf{j}_{-l} \rangle\rangle_{0,\omega}}{kl} \tag{23}$$

is STFT of the generalized scalar longitudinal susceptibility, $\mathbf{I}$ denotes the unit matrix, and k and l are the absolute values of the vectors $\mathbf{k}$ and $\mathbf{l}$, respectively. From Eq. (22)

one can derive the identity relating STFT of the second rank Cartesian tensor of the transversal susceptibility $\chi^{tr}(\mathbf{k},\mathbf{l},\omega)$ and the (arbitrary) scalar function $C(\mathbf{k},\mathbf{l},\omega)$:

$$\sum_{\alpha,\beta}\langle\langle j_{\mathbf{k}}^{\alpha} | j_{-\mathbf{l}}^{\beta}\rangle\rangle_{0,\omega} \left[C(\mathbf{k},\mathbf{l},\omega)\delta_{\alpha\beta} - \frac{k_\beta l_\alpha}{kl} \right] = \sum_{\alpha,\beta} \chi^{lon}(\mathbf{k},\mathbf{l},\omega) \frac{k_\alpha l_\beta}{kl} \left[C(\mathbf{k},\mathbf{l},\omega)\delta_{\alpha\beta} - \frac{k_\beta l_\alpha}{kl} \right]$$
$$+ \left[3C^2(\mathbf{k},\mathbf{l},\omega) - 2\frac{\mathbf{k}\cdot\mathbf{l}}{kl}C(\mathbf{k},\mathbf{l},\omega) + \frac{(\mathbf{k}\cdot\mathbf{l})^2}{k^2 l^2} \right] \chi^{tr}(\mathbf{k},\mathbf{l},\omega). \quad (24)$$

Using the expression of Eq. (23) for $\chi^{lon}(\mathbf{k},\mathbf{l},\omega)$, it is easy to transform the first term in the right hand side (r.h.s.) of Eq. (24) that contains a summation over the indices α and β to the form

$$C(\mathbf{k},\mathbf{l},\omega)\frac{\mathbf{k}\cdot\mathbf{l}}{k^2 l^2}\langle\langle \mathbf{k}\cdot\mathbf{j_k} | \mathbf{l}\cdot\mathbf{j_{-l}}\rangle\rangle_{0,\omega} - \frac{(\mathbf{k}\cdot\mathbf{l})^2}{k^3 l^3}\langle\langle \mathbf{k}\cdot\mathbf{j_k} | \mathbf{l}\cdot\mathbf{j_{-l}}\rangle\rangle_{0,\omega}. \quad (25)$$

The arbitrary scalar function $C(\mathbf{k},\mathbf{l},\omega)$ can be chosen in the form

$$C(\mathbf{k},\mathbf{l},\omega) = \frac{\mathbf{k}\cdot\mathbf{l}}{kl}, \quad (26)$$

that does not affect generality of the above consideration, but permits setting the first term in the r.h.s. of Eq. (24) to zero. Thus, the r.h.s. of the identity (24) reduces to

$$2\frac{(\mathbf{k}\cdot\mathbf{l})^2}{k^2 l^2}\chi^{tr}(\mathbf{k},\mathbf{l},\omega). \quad (27)$$

Further on, one can use the identity $[\mathbf{k}\times\mathbf{j_k}]\cdot[\mathbf{l}\times\mathbf{j_{-l}}] = (\mathbf{k}\cdot\mathbf{l})(\mathbf{j_k}\cdot\mathbf{j_{-l}}) - (\mathbf{k}\cdot\mathbf{j_{-l}})(\mathbf{l}\cdot\mathbf{j_k})$, where the square brackets $[\ldots\times\ldots]$ denote the vector cross-product, to transform the left hand side (l.h.s.) of Eq. (24). Defining EGF of the observable $[\mathbf{k}\times\mathbf{j_k}]\cdot[\mathbf{l}\times\mathbf{j_{-l}}] = (\mathbf{k}\cdot\mathbf{l})(\mathbf{j_k}\cdot\mathbf{j_{-l}}) - (\mathbf{k}\cdot\mathbf{j_{-l}})(\mathbf{l}\cdot\mathbf{j_k})$, one can prove that l.h.s. of Eq. (24) reduces to $(1/kl)\langle\langle[\mathbf{k}\times\mathbf{j_k}]|[\mathbf{l}\times\mathbf{j_{-l}}]\rangle\rangle_{0,\omega}$. Therefore, using Eq. (27) in the r.h.s. of Eq. (24), one can reduce Eq. (24) to the form:

$$\frac{1}{kl}\langle\langle[\mathbf{k}\times\mathbf{j_k}]|[\mathbf{l}\times\mathbf{j_{-l}}]\rangle\rangle_{0,\omega} = 2\frac{(\mathbf{k}\cdot\mathbf{l})^2}{k^2 l^2}\chi^{tr}(\mathbf{k},\mathbf{l},\omega) \quad (28)$$

and determine the scalar transversal susceptibility $\chi^{tr}(\mathbf{k},\mathbf{l},\omega)$,

$$\chi^{tr}(\mathbf{k},\mathbf{l},\omega) = \frac{kl}{2(\mathbf{k}\cdot\mathbf{l})^2}\langle\langle[\mathbf{k}\times\mathbf{j_k}]|[\mathbf{l}\times\mathbf{j_{-l}}]\rangle\rangle_{0,\omega}. \quad (29)$$

Finally, one can use Eq. (26) for the scalar function $C(\mathbf{k},\mathbf{l},\omega)$ to simplify STFT $\langle\langle \mathbf{j}_\mathbf{k}^\alpha | \mathbf{j}_{-\mathbf{l}}^\beta \rangle\rangle_{0,\omega}$ of the $\alpha\beta$-component of the microcurrent-microcurrent EGF of Eq. (22),

$$\langle\langle \mathbf{j}_\mathbf{k}^\alpha | \mathbf{j}_{-\mathbf{l}}^\beta \rangle\rangle_{0,\omega} = \chi^{lon}(\mathbf{k},\mathbf{l},\omega)\frac{k_\alpha l_\beta}{kl} + \chi^{tr}(\mathbf{k},\mathbf{l},\omega)\left[\frac{(\mathbf{k}\cdot\mathbf{l})}{kl}\delta_{\alpha\beta} - \frac{k_\alpha l_\beta}{kl}\right]. \quad (30)$$

Note, that both scalar susceptibilities $\chi^{lon}(\mathbf{k},\mathbf{l},\omega)$ and $\chi^{tr}(\mathbf{k},\mathbf{l},\omega)$ are explicitly defined by Eqs. (23) and (29), respectively. The equations (23), (29) and (30) generalize Zubarev's formulae for the STFTs of the generalized scalar susceptibilities and the microcurrent-microcurrent EGF [54,73], respectively, to include the case of spatially inhomogeneous systems. They reduce to the corresponding Zubarev's expressions in the case of spatially homogeneous systems.

1.2.4 Generalized longitudinal sum rule

Linearity of STFTs of EGFs of various observables with regard to wave vectors was thoroughly used in the above section 1.2.3 to derive explicit expressions for those quantities, in correspondence with the approximation of Eq. (12) for the total Hamiltonian of a spatially inhomogeneous system in a weak external electromagnetic field. Similar considerations can be applied to split STFTs of the following EGFs into scalar and vector factors:

$$\langle\langle \mathbf{j}_\mathbf{k}^\alpha | \rho_{-\mathbf{l}} \rangle\rangle_{0,\omega} = \langle\langle (\mathbf{k}\cdot\mathbf{j}_\mathbf{k}) | \rho_{-\mathbf{l}} \rangle\rangle_{0,\omega}\frac{k_\alpha}{k^2}, \quad (31)$$

$$\langle\langle \rho_\mathbf{k} | \mathbf{j}_{-\mathbf{l}}^\beta \rangle\rangle_{0,\omega} = \langle\langle \rho_{-\mathbf{k}} | (\mathbf{l}\cdot\mathbf{j}_{-\mathbf{l}}) \rangle\rangle_{0,\omega}\frac{l_\beta}{l^2}. \quad (32)$$

The STFTs $\langle\langle (\mathbf{k}\cdot\mathbf{j}_\mathbf{k}) | \rho_{-\mathbf{l}} \rangle\rangle_{0,\omega}$ and $\langle\langle \rho_{-\mathbf{k}} | (\mathbf{l}\cdot\mathbf{j}_{-\mathbf{l}}) \rangle\rangle_{0,\omega}$ are scalar functions of the vectors $\mathbf{k}$ and $\mathbf{l}$, and the subscripts α and β denote $\alpha-$ and $\beta-$ components of the corresponding Cartesian vectors. Equations (31) and (32) help transform Eqs. (17) and (18) for the expectation values of the charge and microcurrent density observables to the form:

$$\rho(\mathbf{k},\omega) = e\sum_\mathbf{l} n_0(\mathbf{k})\delta(\omega)\delta_{\mathbf{k}\mathbf{l}} - \frac{1}{c}\sum_\mathbf{l}\langle\langle \rho_\mathbf{k} | (\mathbf{l}\cdot\mathbf{j}_{-\mathbf{l}}) \rangle\rangle_{0,\omega}\frac{1}{l^2}\mathbf{l}\cdot\mathbf{A}(\mathbf{l},\omega)$$
$$+\sum_\mathbf{l}\langle\langle \rho_\mathbf{k} | \rho_{-\mathbf{l}} \rangle\rangle_{0,\omega}\varphi(\mathbf{l},\omega), \quad (33)$$

$$\mathbf{j}(\mathbf{k},\omega) = -\frac{e^2}{mc}\sum_{\mathbf{l}} n_0(\mathbf{k}-\mathbf{l})\mathbf{A}(\mathbf{l},\omega)$$

$$-\frac{1}{c}\sum_{\mathbf{l}}\left\{x^{lon}(\mathbf{k},\mathbf{l},\omega)\frac{\mathbf{kl}}{kl} + x^{tr}(\mathbf{k},\mathbf{l},\omega)\left[\frac{(\mathbf{k}\cdot\mathbf{l})}{kl}\mathbf{I} - \frac{\mathbf{k}\cdot\mathbf{l}}{kl}\right]\right\}\cdot\mathbf{A}(\mathbf{l},\omega) \qquad (34)$$

$$+\sum_{\mathbf{l}}\langle\langle(\mathbf{k}\cdot\mathbf{j}_{\mathbf{k}})|\rho_{-\mathbf{l}}\rangle\rangle_{0,\omega}\frac{\mathbf{k}}{k^2}\varphi(\mathbf{l},\omega),$$

respectively. For the case of pairwise additive interparticle interactions, it can be proven that $\rho_{\mathbf{k}}$ and the interaction Hamiltonian H_{int} commute, and that the commutator of $\rho_{\mathbf{k}}$ and the kinetic part of the total Hamiltonian (2) is equal to $\hbar(\mathbf{k}\cdot\mathbf{j})$, so that the equation of motion for the charge density operator $\rho_{\mathbf{k}}$ can be written in the form that includes only the Hamiltonian H_0 of Eq. (9):

$$\dot{\rho}_{\mathbf{k}} = \frac{1}{i\hbar}[\rho_{\mathbf{k}},H_0] = -i(\mathbf{k}\cdot\mathbf{j}_{\mathbf{k}}), \qquad (35)$$

where $\dot{\rho}_{\mathbf{k}}$ is the STFT of the time derivative of $\rho_{\mathbf{k}}$. This equation can be used to calculate the commutator $(1/i\hbar)[\rho_{\mathbf{k}},\dot{\rho}_{-\mathbf{l}}]$, thus establishing *the generalized longitudinal sum rule* that governs evolution of the charge density operator specific to spatially inhomogeneous systems in weak external electromagnetic fields:

$$\frac{1}{i\hbar}[\rho_{\mathbf{k}},\dot{\rho}_{-\mathbf{l}}] = \frac{e}{m}(\mathbf{l}\cdot\mathbf{k})\rho_{\mathbf{k}-\mathbf{l}} \qquad (36)$$

where $\rho_{\mathbf{k}-\mathbf{l}} = (e/V)\sum_{\mathbf{q},\sigma} a^+_{\mathbf{q}-\mathbf{k}+\mathbf{l},\sigma}a_{\mathbf{q},\sigma}$. For spatially homogeneous systems, only terms with $\mathbf{l} = \mathbf{k}$ survive, so in this case that Eq. (36) reduces to that derived by Zubarev [54]:

$$\frac{1}{i\hbar}[\rho_{\mathbf{k}},\dot{\rho}_{-\mathbf{k}}] = \frac{Ne}{mV}k^2, \qquad (37)$$

The notation N stands here for the charge carrier number operator $N = \sum_{\mathbf{q},\sigma}a^+_{\mathbf{q},\sigma}a_{\mathbf{q},\sigma}$, where the summations is over all wave vectors $\mathbf{q}$ and z-components of the spin, σ.

1.2.5 Dielectric permittivity of a spatially inhomogeneous quantum system in a weak external electromagnetic field

To derive explicit expressions for the dielectric permittivity tensor, one has to relate explicitly the expectation value of the charge density operator, Eq. (33), and the induced field intensity. For this purpose, Eq. (33) is transformed to the more convenient form using Eqs. (35), (31) and (32),

$$\rho(\mathbf{k},\omega) = e\sum_{\mathbf{l}} n_0(\mathbf{k})\delta(\omega)\delta_{kl} + \frac{i}{c}\sum_{\mathbf{l}}\frac{1}{l^2}\langle\langle\rho_\mathbf{K}|\dot\rho_{-\mathbf{l}}\rangle\rangle_{0,\omega}(\mathbf{l}\cdot\mathbf{A}(\mathbf{l},\omega)) \\ + \sum_{\mathbf{l}}\langle\langle\rho_\mathbf{k}|\rho_{-\mathbf{l}}\rangle\rangle_{0,\omega}\varphi(\mathbf{l},\omega).$$ (38)

Noticing that $\langle\langle\rho_\mathbf{k}|\dot\rho_{-\mathbf{l}}\rangle\rangle_{0,\omega} = \int_{-\infty}^{\infty} dt\, e^{i\omega t}\langle\langle\rho_\mathbf{k}|\dot\rho_{-\mathbf{l}}\rangle\rangle_0 = i\omega\langle\langle\rho_\mathbf{k}|\rho_{-\mathbf{l}}\rangle\rangle_{0,\omega}$ and $(1/i\hbar)<[\rho_\mathbf{k},\rho_{-\mathbf{l}}]>_0 = 0$, one can further transform Eq. (38) to read:

$$\rho(\mathbf{k},\omega) = e\sum_{\mathbf{l}} n_0(\mathbf{k})\delta(\omega)\delta_{kl} \\ + \sum_{\mathbf{l}}\frac{1}{l^2}\langle\langle\rho_\mathbf{k}|\rho_{-\mathbf{l}}\rangle\rangle_{0,\omega}\left\{-\frac{\omega}{c}(\mathbf{l}\cdot\mathbf{A}(\mathbf{l},\omega))+l^2\varphi(\mathbf{l},\omega)\right\}.$$ (39)

Because induced charges screen only the longitudinal component of the external electric field [54,75–77], one can write a simple relation $\mathbf{l}\cdot\mathbf{D}(\mathbf{l},\omega)=\mathbf{l}\cdot\mathbf{E}(\mathbf{l},\omega)$ between the space-time Fourier transforms of the induced and external fields, $\mathbf{D}(\mathbf{l},\omega)$ and $\mathbf{E}(\mathbf{l},\omega)$, respectively. The inner product $\mathbf{l}\cdot\mathbf{E}(\mathbf{l},\omega)$ can be found from the space-time Fourier transformation applied to the Maxwell equation (1) for the electric field intensity, and subsequently replaced by $\mathbf{l}\cdot\mathbf{D}(\mathbf{l},\omega)$, leading to the expression

$$i\mathbf{l}\cdot\mathbf{D}(\mathbf{l},\omega) = -\frac{\omega}{c}(\mathbf{l}\cdot\mathbf{A}(\mathbf{l},\omega))+l^2\varphi(\mathbf{l},\omega).$$ (40)

The r.h.s. of this expression is exactly the term in curly brackets in the r.h.s. of Eq. (39), so Eq. (39) now contains explicitly STFT of the induced field intensity,

$$\rho(\mathbf{k},\omega) = e\sum_{\mathbf{l}} n_0(\mathbf{k})\delta(\omega)\delta_{kl} \\ + \sum_{\mathbf{l}}\frac{i}{l^2}\langle\langle\rho_\mathbf{k}|\rho_{-\mathbf{l}}\rangle\rangle_{0,\omega}\mathbf{l}\cdot\mathbf{D}(\mathbf{l},\omega).$$ (41)

In a linear approximation with regard to the field $\mathbf{E}(\mathbf{r},t)$, and in view of the causality condition, the induced field $\mathbf{D}(\mathbf{r},t)$ can be written in terms of the second rank tensor $\mathbf{F}(\mathbf{r}',t')$ (unknown, at this point) in a general form as

$$\mathbf{D}(\mathbf{r},t) = \int_0^\infty dt'\int d\mathbf{r}'\left[\delta(\mathbf{r}-\mathbf{r}')\delta(t-t')\mathbf{I}+\mathbf{F}(\mathbf{r}',t')\right]\cdot\mathbf{E}(\mathbf{r}-\mathbf{r}',t-t'),$$ (42)

where $\varepsilon(\mathbf{r}',t') \equiv [\delta(\mathbf{r}-\mathbf{r}')\delta(t-t')\mathbf{I}+\mathbf{F}(\mathbf{r}',t')]$ is a formal representation for the second rank Cartesian tensor of dielectric permittivity. The space-time Fourier transformation of Eq. (42) results in a simple correlation between the external and induced fields,

$$\mathbf{D}(\mathbf{l},\omega) = \varepsilon(\mathbf{l},\omega) \cdot \mathbf{E}(\mathbf{l},\omega), \tag{43}$$

with $\varepsilon(\mathbf{l},\omega)$ being STFT of the dielectric permittivity tensor. The difference between the induced and external fields, $\mathbf{D}(\mathbf{r},t)$ and $\mathbf{E}(\mathbf{r},t)$, respectively, defines the electric polarization vector $\mathbf{P}(\mathbf{r},t)$ at a position $\mathbf{r}$ and time t,

$$\mathbf{P}(\mathbf{r},t) = \frac{1}{4\pi}[\mathbf{D}(\mathbf{r},t) - \mathbf{E}(\mathbf{r},t)]. \tag{44}$$

This polarization is caused by the induced charge density $\rho_{ind}(\mathbf{r},t) = -\nabla_\mathbf{r} \mathbf{P}(\mathbf{r},t)$. The STFT $\rho_{ind}(\mathbf{k},\omega)$ of this density can be found from Eq. (44):

$$\rho_{ind}(\mathbf{k},\omega) = -\frac{i}{4\pi}\mathbf{k} \cdot [\mathbf{D}(\mathbf{k},\omega) - \mathbf{E}(\mathbf{k},\omega)]. \tag{45}$$

Alternatively, the induced charge density can be determined from Eq. (41):

$$\begin{aligned}\rho_{ind}(\mathbf{k},\omega) &= \rho(\mathbf{k},\omega) - e\sum_\mathbf{l} n_0(\mathbf{K})\delta(\omega)\delta_{\mathbf{k}\mathbf{l}} \\ &= \sum_\mathbf{l} \frac{i}{l^2} \ll \rho_\mathbf{K} | \rho_{-\mathbf{l}} \gg_{0,\omega} \mathbf{D}(\mathbf{l},\omega) \cdot \mathbf{l}.\end{aligned} \tag{46}$$

Using Eqs. (45), (46), (44) and linearity considerations, similar to those resulted in Eqs. (23) and (29) for STFTs of the generalized scalar susceptibilities, one can relate STFT of the dielectric permittivity tensor to STFT $\langle\langle \rho_\mathbf{k} | \rho_{-\mathbf{l}}\rangle\rangle_{0,\omega}$ of the charge density-charge density EGF. First, Eq. (43) has to be transformed to take the form linear in the wave vectors k and l. The most general form of STFT of the dielectric permittivity tensor $\varepsilon(\mathbf{l},\omega)$ linear in the wave vectors $\mathbf{k}$ and $\mathbf{l}$ is:

$$\varepsilon(\mathbf{k},\omega) = \sum_\mathbf{l}\left\{\varepsilon^{lon}(\mathbf{k},\mathbf{l},\omega)\frac{\mathbf{k}\mathbf{l}}{kl} + \varepsilon^{tr}(\mathbf{k},\mathbf{l},\omega)\left[\frac{(\mathbf{k}\cdot\mathbf{l})}{kl}\mathbf{I} - \frac{\mathbf{k}\mathbf{l}}{kl}\right]\right\}, \tag{47}$$

with $\varepsilon^{lon}(\mathbf{k},\mathbf{l},\omega)$ and $\varepsilon^{tr}(\mathbf{k},\mathbf{l},\omega)$ being (unknown) scalar functions of the absolute values k, l and the inner product $(\mathbf{k}\cdot\mathbf{l})$. Correspondingly, the field $\mathbf{E}(\mathbf{k},\omega)$ can be represented as a vector sum of two contributions that are parallel and orthogonal to the wave vector $\mathbf{k}$, respectively:

$$\mathbf{E}(\mathbf{k},\omega) = \frac{\mathbf{k}}{k^2}(\mathbf{k}\cdot\mathbf{E}(\mathbf{k},\omega)) + \frac{1}{k^2}[\mathbf{k}\times\mathbf{E}(\mathbf{k},\omega)\times\mathbf{k}]. \tag{48}$$

Changing the notation **k** to **l** in Eq. (47) and using Eqs. (47) and (48), one can obtain from Eq. (43) the following expression:

$$\mathbf{k} \cdot \mathbf{D}(\mathbf{k},\omega) = \sum_{\mathbf{l}} \left[\varepsilon^{lon}(\mathbf{k},\mathbf{l},\omega) - \varepsilon^{tr}(\mathbf{k},\mathbf{l},\omega) \right] \frac{k}{l} \times$$
$$\left\{ \frac{(\mathbf{l} \cdot \mathbf{k})}{k^2}(\mathbf{k} \cdot \mathbf{E}(\mathbf{k},\omega)) + \frac{\mathbf{l} \cdot [\mathbf{k} \times \mathbf{E}(\mathbf{k},\omega) \times \mathbf{k}]}{k^2} \right\} \quad (49)$$
$$+ \sum_{\mathbf{l}} \varepsilon^{tr}(\mathbf{k},\mathbf{l},\omega) \frac{\mathbf{l} \cdot \mathbf{k}}{kl}(\mathbf{k} \cdot \mathbf{E}(\mathbf{k},\omega)).$$

This expression can be simplified further. Thus, application of the space-time Fourier transformation procedure to the Maxwell equation $\nabla_r \times \mathbf{E}(\mathbf{r},t) = -(1/c)[\partial \mathbf{B}(\mathbf{r},t)/\partial t]$ that relates the induced magnetic field $\mathbf{B}(\mathbf{r},t)$ to the external electric field produces the following correlation:

$$[i\mathbf{k} \times \mathbf{E}(\mathbf{k},\omega)] = \frac{i\omega}{c} \mathbf{B}(\mathbf{k},\omega). \quad (50)$$

This result leads to the relation $\mathbf{l} \cdot [\mathbf{k} \times \mathbf{E}(\mathbf{k},\omega) \times \mathbf{k}] = (\omega/c)\mathbf{l} \cdot [\mathbf{B}(\mathbf{k},\omega) \times \mathbf{k}]$. Noticing once again, that only the longitudinal component of the electrical field defines the induced charge [75–77], one concludes that the term $\{\mathbf{l} \cdot [\mathbf{k} \times \mathbf{E}(\mathbf{k},\omega) \times \mathbf{k}]\}/k^2$ in the r.h.s. of Eq. (49) can be neglected in a linear approximation, because it is equal to the small term $(\omega/ck^2)\mathbf{l} \cdot [\mathbf{B}(\mathbf{k},\omega) \times \mathbf{k}]$. Thus, Eq. (49) reduces to become

$$\mathbf{k} \cdot \mathbf{D}(\mathbf{k},\omega) = \varepsilon(\mathbf{k},\omega)(\mathbf{k} \cdot \mathbf{E}(\mathbf{k},\omega)), \quad (51)$$

where $\varepsilon(\mathbf{k},\omega) \equiv \sum_{\mathbf{l}} \varepsilon^{lon}(\mathbf{k},\mathbf{l},\omega)[(\mathbf{l} \cdot \mathbf{k})/kl]$ denotes the scalar dielectric permittivity that indeed contains only the longitudinal part of the dielectric permittivity tensor and is a scalar function of the wave vectors. Notably, this reduced form of Eq. (49) holds only in the case of a system in a weak external electromagnetic field.

From Eqs. (45) and (51) one can derive the following expression for the induced charge density:

$$\rho_{ind}(\mathbf{k},\omega) = \frac{1}{4\pi} \left[\frac{1}{\varepsilon(\mathbf{k},\omega)} - 1 \right] i\mathbf{k} \cdot \mathbf{D}(\mathbf{k},\omega), \quad (52)$$

to be used further in Eq. (46). This allows identification of STFT of the scalar dielectric permittivity in terms of STFT of the charge density - charge density EGFs:

$$\varepsilon^{-1}(\mathbf{k},\omega) = 1 + 4\pi \sum_{\mathbf{l}} \frac{1}{l^2} \langle\langle \rho_{\mathbf{k}} | \rho_{-\mathbf{l}} \rangle\rangle_{0,\omega}. \quad (53)$$

For spatially homogeneous systems $\langle\langle\rho_\mathbf{k}|\rho_{-\mathbf{l}}\rangle\rangle_{0,\omega} = \langle\langle\rho_\mathbf{k}|\rho_{-\mathbf{k}}\rangle\rangle_{0,\omega}\delta_{\mathbf{k}\mathbf{l}}$, so that Eq. (53) reduces to the well-known form of Zubarev's Eq. (18.21) of Ref. 54,

$$\varepsilon^{-1}(\mathbf{k},\omega) = 1 + \frac{4\pi}{l^2}\langle\langle\rho_\mathbf{k}|\rho_{-\mathbf{k}}\rangle\rangle_{0,\omega}. \tag{54}$$

Notably, equation (53) [obtained by neglecting the term proportional to $(\omega/c)\mathbf{l}\cdot[\mathbf{B}(\mathbf{k},\omega)\times\mathbf{k}]$ in Eq. (49)] is an approximate equation even in the framework of the linear response theory of spatially inhomogeneous systems developed here, while Zubarev's Eq. (54) is an exact equation of the linear response theory of spatially homogeneous systems. More accurate expressions for the dielectric permittivity tensor can be derived from Eqs. (43), (45) and (46) using Eq. (49) where the term proportional to $\mathbf{l}\cdot[\mathbf{B}(\mathbf{k},\omega)\times\mathbf{k}]$ is appropriately simplified, rather than neglected.

The Eq. (53) for STFT of the dielectric permittivity can be used in Eq. (41) to explicitly relate STFT of the charge density $\rho(\mathbf{k},\omega)$ and the electric field intensity $\mathbf{E}(\mathbf{k},\omega)$:

$$\rho(\mathbf{k},\omega) = e\sum_\mathbf{l} n_0(\mathbf{k})\delta(\omega)\delta_{\mathbf{k}\mathbf{l}} + \sum_\mathbf{l} \frac{i}{l^2} \frac{\langle\langle\rho_\mathbf{k}|\rho_{-\mathbf{l}}\rangle\rangle_{0,\omega}}{\left\{1 + 4\pi\sum_\mathbf{s}\frac{1}{s^2}\langle\langle\rho_\mathbf{l}|\rho_{-\mathbf{s}}\rangle\rangle_{0,\omega}\right\}}\mathbf{l}\cdot\mathbf{E}(\mathbf{l},\omega). \tag{55}$$

This expression is approximate even in the framework of the linear response theory derived here, because the approximate Eq. (53) for the dielectric permittivity was used to derive it. In the case of spatially homogeneous systems $\langle\langle\rho_\mathbf{k}|\rho_{-\mathbf{l}}\rangle\rangle_{0,\omega} = \langle\langle\rho_\mathbf{k}|\rho_{-\mathbf{k}}\rangle\rangle_{0,\omega}\delta_{\mathbf{k}\mathbf{l}}$, so that Eq. (55) reduces to the corresponding exact equation derived in Ref. 54.

1.2.6 Generalized susceptibility of a spatially inhomogeneous quantum system in a weak external electromagnetic field

To derive an explicit expression for the longitudinal generalized susceptibility, Eq. (35) is used to obtain STFT of microcurrent-microcurrent EGFs featuring in Eqs. (23) and (34):

$$\langle\langle\mathbf{k}\cdot\mathbf{j}_\mathbf{k}|\mathbf{l}\cdot\mathbf{j}_{-\mathbf{l}}\rangle\rangle_{0,\omega} = \langle\langle\dot\rho_\mathbf{k}|\dot\rho_{-\mathbf{l}}\rangle\rangle_{0,\omega}, \tag{56}$$

$$\langle\langle\mathbf{k}\cdot\mathbf{j}_\mathbf{k}|\rho_{-\mathbf{l}}\rangle\rangle_{0,\omega} = i\langle\langle\dot\rho_\mathbf{k}|\rho_{-\mathbf{l}}\rangle\rangle_{0,\omega}, \tag{57}$$

and then Eq. (56) is used in the r.h.s. of Eq. (23) to derive the relation

$$\chi^{lon}(\mathbf{k},\mathbf{l},\omega) = \frac{\langle\langle\dot{\rho}_\mathbf{k}|\dot{\rho}_{-\mathbf{l}}\rangle\rangle_{0,\omega}}{kl}. \tag{58}$$

EGFs appearing in Eqs. (56) and (58) can be significantly simplified. First, using integration by parts similar to that used to obtain the relation

$\langle\langle\rho_\mathbf{k}|\dot{\rho}_{-\mathbf{l}}\rangle\rangle_{0,\omega} = \int_{-\infty}^{\infty} dt\, e^{i\omega t} \langle\langle\rho_\mathbf{k}|\dot{\rho}_{-\mathbf{l}}(t)\rangle\rangle_0 = i\omega\langle\langle\rho_\mathbf{k}|\rho_{-\mathbf{l}}\rangle\rangle_{0,\omega}$, one can derive the following expression:

$$\langle\langle\dot{\rho}_\mathbf{k}|\dot{\rho}_{-\mathbf{l}}\rangle\rangle_{0,\omega} = \omega^2 \langle\langle\rho_\mathbf{k}|\rho_{-\mathbf{l}}\rangle\rangle_{0,\omega} - \frac{1}{i\hbar}\langle[\rho_\mathbf{k},\dot{\rho}_{-\mathbf{l}}]\rangle_0. \tag{59}$$

Then, using the generalized longitudinal sum rule of Eq. (36), one can obtain from this expression a simple result for EGF of the time derivatives of the charge density:

$$\langle\langle\dot{\rho}_\mathbf{k}|\dot{\rho}_{-\mathbf{l}}\rangle\rangle_{0,\omega} = \omega^2 \langle\langle\rho_\mathbf{k}|\rho_{-\mathbf{l}}\rangle\rangle_{0,\omega} - \frac{e^2}{m}(\mathbf{k}\cdot\mathbf{l})n_0(\mathbf{k}-\mathbf{l}). \tag{60}$$

With the use of this result the expression of Eq. (58) for the longitudinal susceptibility reduces to the form:

$$\chi^{lon}(\mathbf{k},\mathbf{l},\omega) = \frac{1}{kl}\left\{\omega^2\langle\langle\rho_\mathbf{k}|\rho_{-\mathbf{l}}\rangle\rangle_{0,\omega} - \frac{e^2}{m}n_0(\mathbf{k}-\mathbf{l})(\mathbf{k}\cdot\mathbf{l})\right\}. \tag{61}$$

From the expression $\langle\langle\rho_\mathbf{k}|\dot{\rho}_{-\mathbf{l}}\rangle\rangle_{0,\omega} = i\omega\langle\langle\rho_\mathbf{k}|\rho_{-\mathbf{l}}\rangle\rangle_{0,\omega}$ one can immediately obtain

$$\langle\langle\dot{\rho}_\mathbf{k}|\rho_{-\mathbf{l}}\rangle\rangle_{0,\omega} = -i\omega\langle\langle\rho_\mathbf{k}|\rho_{-\mathbf{l}}\rangle\rangle_0, \tag{62}$$

and use Eqs. (62), (57), and (61), to derive from Eq. (34) the following simplified expression for STFT of the local current density:

$$\begin{aligned}\mathbf{j}(\mathbf{k},\omega) = &-\frac{e^2}{mc}\sum_\mathbf{l} n_0(\mathbf{k}-\mathbf{l})\mathbf{A}(\mathbf{l},\omega) \\ &- \frac{1}{c}\sum_\mathbf{l}\frac{1}{k^2 l^2}\left\{\omega^2\langle\langle\rho_\mathbf{k}|\rho_{-\mathbf{l}}\rangle\rangle_{0,\omega} - \frac{e^2}{m}n_0(\mathbf{k}-\mathbf{l})(\mathbf{k}\cdot\mathbf{l})\right\}(\mathbf{l}\cdot\mathbf{A}(\mathbf{l},\omega))\mathbf{k} \\ &+ \sum_\mathbf{l}\frac{\omega}{k^2}\langle\langle\rho_\mathbf{k}|\rho_{-\mathbf{l}}\rangle\rangle_{0,\omega}\varphi(\mathbf{l},\omega)\mathbf{k} \\ &- \frac{1}{c}\sum_\mathbf{l}\chi^{tr}(\mathbf{k},\mathbf{l},\omega)\left\{\frac{(\mathbf{k}\cdot\mathbf{l})}{kl}\mathbf{A}(\mathbf{l},\omega) - \frac{1}{kl}(\mathbf{l}\cdot\mathbf{A}(\mathbf{l},\omega))\mathbf{k}\right\}.\end{aligned} \tag{63}$$

This equation can be significantly reduced if one relates the divergence of the current density to the charge density using the continuity equation as follows. First, from Eqs. (35) and (19) one can obtain STFT of the longitudinal component of the current density in the form:

$$\mathbf{k} \cdot \mathbf{j}(\mathbf{k},\omega) = -\frac{e^2}{mc}\sum_{\mathbf{l}} n_0(\mathbf{k}-\mathbf{l})\, \mathbf{k} \cdot \mathbf{A}(\mathbf{l},\omega) - \frac{i}{c}\sum_{\mathbf{l}}\langle\langle \dot{\rho}_{\mathbf{k}} | \mathbf{j}_{-\mathbf{l}}\rangle\rangle_{0,\omega} \cdot \mathbf{A}(\mathbf{l},\omega) \\ + i\sum_{\mathbf{l}}\langle\langle \dot{\rho}_{\mathbf{k}} | \rho_{-\mathbf{l}}\rangle\rangle_{0,\omega}\, \varphi(\mathbf{l},\omega). \quad (64)$$

Then, integrating by parts in the r.h.s. of $\langle\langle \dot{\rho}_{\mathbf{k}} | \mathbf{j}_{-\mathbf{l}}\rangle\rangle_{0,\omega} = \int_{-\infty}^{\infty} dt\, e^{i\omega t}\langle\langle \dot{\rho}_{\mathbf{k}}(t) | \mathbf{j}_{-\mathbf{l}}\rangle\rangle_{0}$ one can establish that

$$\langle\langle \dot{\rho}_{\mathbf{k}} | \mathbf{j}_{-\mathbf{l}}\rangle\rangle_{0,\omega} = -\frac{1}{i\hbar}\langle[\rho_{\mathbf{k}}, \mathbf{j}_{-\mathbf{l}}]\rangle_{0} - i\omega\langle\langle \rho_{\mathbf{k}} | \mathbf{j}_{-\mathbf{l}}\rangle\rangle_{0}. \quad (65)$$

Using Eqs. (65), (62) and (17), and accounting for the fact that in the absence of the external field the system is electroneutral, one can rewrite Eq. (64) in the form:

$$\mathbf{k} \cdot \mathbf{j}(\mathbf{k},\omega) = -\frac{e^2}{mc}\sum_{\mathbf{l}} n_0(\mathbf{k}-\mathbf{l})\, \mathbf{k} \cdot \mathbf{A}(\mathbf{l},\omega) + \frac{1}{c\hbar}\sum_{\mathbf{l}}\langle[\rho_{\mathbf{k}}, \mathbf{j}_{-\mathbf{l}}]\rangle_{0} \cdot \mathbf{A}(\mathbf{l},\omega) + \omega\rho(\mathbf{k},\omega). \quad (66)$$

The commutator in the r.h.s. of this equation can be reduced using Eqs. (19) and (20) to the following expression:

$$\frac{1}{\hbar}[\rho_{\mathbf{k}}, \mathbf{j}_{-\mathbf{l}}] = \frac{e}{mV}\mathbf{k}\rho_{\mathbf{k}-\mathbf{l}}. \quad (67)$$

Realizing that $<\rho_{\mathbf{k}-\mathbf{l}}>_0 = en_0(\mathbf{k}-\mathbf{l})$ and using Eq. (66), one can prove that the sum of the first two terms in the r.h.s. of Eq. (63) is equal to zero. Thus, one recovers STFT of the continuity equation in the simple form:

$$\mathbf{k} \cdot \mathbf{j}(\mathbf{k},\omega) = \omega\rho(\mathbf{k},\omega). \quad (68)$$

This result signifies that in addition to the continuity equation for the charge density operator, a similar continuity equation (68) holds for STFT of the expectation value of this operator. Thus, from Eq. (68) one concludes that the induced charge density is defined by the induced current density. *Notably, the converse statement is not correct*, because only the longitudinal component of the induced current density is determined by the induced charge density.

From Eqs. (68), (39) and (63), one can derive the equality:

$$\frac{1}{c}\chi^{tr}(\mathbf{k},\mathbf{l},\omega)\left\{\frac{(\mathbf{k}\cdot\mathbf{l})}{kl}+\frac{e^2}{m}n_0(\mathbf{k}-\mathbf{l})\right\}(\mathbf{k}\cdot\mathbf{A}(\mathbf{l},\omega))$$
$$=\frac{1}{c}\left\{\chi^{tr}(\mathbf{k},\mathbf{l},\omega)+\frac{e^2}{m}\frac{(\mathbf{k}\cdot\mathbf{l})}{kl}n_0(\mathbf{k}-\mathbf{l})\right\}\frac{k^2}{kl}(\mathbf{l}\cdot\mathbf{A}(\mathbf{l},\omega)).$$

Using this equality and expressing the inner product $(\mathbf{l}\cdot\mathbf{A}(\mathbf{l},\omega))$ in terms of $(\mathbf{l}\cdot\mathbf{D}(\mathbf{l},\omega))$ from Eq. (39), the local current density conservation equation (63) reduces to the form:

$$\mathbf{j}(\mathbf{k},\omega)=\sum_{\mathbf{l}}\frac{i\omega}{k^2 l^2}\langle\langle\rho_\mathbf{k}|\rho_{-\mathbf{l}}\rangle\rangle_{0,\omega}(\mathbf{l}\cdot\mathbf{D}(\mathbf{l},\omega))\mathbf{k}$$
$$-\frac{1}{c}\sum_{\mathbf{l}}\left\{\left[\chi^{tr}(\mathbf{k},\mathbf{l},\omega)\frac{(\mathbf{k}\cdot\mathbf{l})}{kl}+\frac{e^2}{m}n_0(\mathbf{k}-\mathbf{l})\right]\right. \qquad (69)$$
$$\left.\times\left[\mathbf{A}(\mathbf{l},\omega)-\frac{1}{k^2}(\mathbf{k}\cdot\mathbf{A}(\mathbf{l},\omega))\mathbf{k}\right]\right\}$$

The transversal component of the vector-potential $\mathbf{A}(\mathbf{r},t)$ defines the induced magnetic field $\mathbf{B}(\mathbf{r},t)$ via the second equation of the Maxwell's system of equations (1), and the induced magnetic field defines the transversal component of the current density. Correspondingly, for STFT $\mathbf{B}(\mathbf{k},\omega)$ of $\mathbf{B}(\mathbf{r},t)$ one has:

$$\mathbf{B}(\mathbf{k},\omega)=[i\mathbf{k}\times\mathbf{A}(\mathbf{k},\omega)]. \qquad (70)$$

The STFTs of $\mathbf{B}(\mathbf{k},\omega)$ and the curl of the electric field, $[\mathbf{k}\times\mathbf{E}(\mathbf{k},\omega)]$, are related by the Maxwell's equation (50). The term

$$\mathbf{B}_\mathbf{k}(\mathbf{l},\omega)=\frac{1}{k^2}[i\mathbf{k}\times i\mathbf{k}\times\mathbf{A}(\mathbf{l},\omega)]=\mathbf{A}(\mathbf{l},\omega)-\frac{[\mathbf{k}\cdot\mathbf{A}(\mathbf{l},\omega)]}{k^2}\mathbf{k} \qquad (71)$$

in the r.h.s. of Eq. (69) is the component of STFT of the vector-potential orthogonal to the wave vector $\mathbf{k}$. Using Eq. (70) this vector can be transformed to the form: $(1/k^2)[i\mathbf{l}\times\mathbf{B}(\mathbf{l},\omega)]+\{[\mathbf{l}\cdot\mathbf{A}(\mathbf{l},\omega)]/l^2\}\mathbf{l}-\{[\mathbf{k}\cdot\mathbf{A}(\mathbf{l},\omega)]/k^2\}\mathbf{k}$.
Because the inner product of this vector with the wave vector $\mathbf{k}$ is zero, one can prove that the following relation holds:

$$\mathbf{k}\cdot\mathbf{A}(\mathbf{l},\omega)=\frac{1}{l^2}\mathbf{k}\cdot[i\mathbf{l}\times\mathbf{B}(\mathbf{l},\omega)]+\frac{(\mathbf{l}\cdot\mathbf{k})}{l^2}[\mathbf{l}\cdot\mathbf{A}(\mathbf{l},\omega)]. \qquad (72)$$

Using this equality and Maxwell's Eq. (50), one can finally derive from Eq. (69) a simple expression for STFT of the local current density:

$$\mathbf{j}(\mathbf{k},\omega) = i\omega \sum_{\mathbf{l}} \frac{\langle\langle \rho_{\mathbf{k}} | \rho_{-\mathbf{l}} \rangle\rangle_{0,\omega}}{k^2 l^2} (\mathbf{l} \cdot \mathbf{D}(\mathbf{l},\omega)) \mathbf{k}$$
$$- \frac{1}{c} \sum_{\mathbf{l}} \frac{1}{l^2} \left[\chi^{tr}(\mathbf{k},\mathbf{l},\omega) \frac{(\mathbf{k} \cdot \mathbf{l})}{kl} + \frac{e^2}{m} n_0(\mathbf{k}-\mathbf{l}) \right]$$
$$\times \left[\mathbf{I} - \frac{\mathbf{k}\mathbf{k}}{k^2} \right] \cdot \{ [i\mathbf{l} \times \mathbf{B}(\mathbf{l},\omega)] - \mathbf{l}(\mathbf{l} \cdot \mathbf{A}(\mathbf{l},\omega)) \} \cdot \tag{73}$$

This expression reduces to Zubarev's Eq. (18.24) of Ref. 54 in the case of homogeneous systems, because in this case the sums over $\mathbf{l}$ in the r.h.s. of Eq. (73) reduce to one term with $\mathbf{l} = \mathbf{k}$ each.

In the case of homogeneous systems the appropriately simplified Eq. (73) can be further transformed to the form that does not include explicitly STFT of the vector-potential $\mathbf{A}(\mathbf{k},\omega)$. Unfortunately, in the case of inhomogeneous systems, the summation over $\mathbf{l}$ in the second term in the r.h.s. of Eq. (73) does not allow the corresponding transformation without an additional condition. The most convenient (but not unique) choice of such a condition is the Lorentz *anzatz* that is universally used in electrodynamics to ensure that Maxwell's equations in vacuum are gauge-invariant:

$$i\mathbf{l} \cdot \mathbf{A}(\mathbf{l},\omega) = \frac{i\omega}{c} \varphi(\mathbf{l},\omega) \cdot \tag{74}$$

Using Eq. (74) and Eq. (40) one can relate the longitudinal component of STFT of the vector-potential and that of the field intensity $\mathbf{D}(\mathbf{l},\omega)$:

$$i\mathbf{l} \cdot \mathbf{D}(\mathbf{l},\omega) = \frac{i\omega}{c} \left\{ 1 - \frac{c^2 l^2}{\omega^2} \right\} i\mathbf{l} \cdot \mathbf{A}(\mathbf{l},\omega) \cdot \tag{75}$$

Determining $\mathbf{l} \cdot \mathbf{A}(\mathbf{l},\omega)$ from this equation and substituting it in Eq. (73), one can transform Eq. (73) to the form that does not include explicitly the external electromagnetic field potentials:

$$\mathbf{j}(\mathbf{k},\omega) = \sum_{\mathbf{l}} \frac{i\omega}{k^2 l^2} \langle\langle \rho_{\mathbf{k}} | \rho_{-\mathbf{l}} \rangle\rangle 0, \omega (\mathbf{l} \cdot \mathbf{D}(\mathbf{l},\omega)) \mathbf{k}$$
$$- \sum_{\mathbf{l}} \frac{1}{l^2} \left[\chi^{tr}(\mathbf{k},\mathbf{l},\omega) \frac{(\mathbf{k} \cdot \mathbf{l})}{kl} + \frac{e^2}{m} n_0(\mathbf{k}-\mathbf{l}) \right]$$
$$\times \left[\mathbf{I} - \frac{\mathbf{k}\mathbf{k}}{k^2} \right] \cdot \left\{ \frac{1}{c} [i\mathbf{l} \times \mathbf{B}(\mathbf{l},\omega)] + \frac{i\mathbf{l}}{\omega \left(1 - \frac{c^2 l^2}{\omega^2}\right)} (\mathbf{l} \cdot \mathbf{D}(\mathbf{l},\omega)) \right\} \cdot \tag{76}$$

1.2.7 Longitudinal quantum conductivity of a spatially inhomogeneous system in a weak external electromagnetic field

Using Maxwell's equation (50) and Eqs. (51) and (53), STFT of the current density in the form of Eq. (76) can be now expressed explicitly in terms of STFT of the electric field intensity $\mathbf{E}(\mathbf{l},\omega)$ only:

$$\mathbf{j}(\mathbf{k},\omega) = \frac{i\omega}{4\pi}[1-\varepsilon(\mathbf{k},\omega)]\frac{\mathbf{kk}}{k^2} \cdot \mathbf{E}(\mathbf{l},\omega)$$
$$-\sum_{\mathbf{l}} \frac{i}{\omega\, l^2}\left[\chi^{tr}(\mathbf{k},\mathbf{l},\omega)\frac{(\mathbf{k}\cdot\mathbf{l})}{kl} + \frac{e^2}{m}n_0(\mathbf{k}-\mathbf{l})\right]$$
$$\times\left[\mathbf{I} - \frac{\mathbf{kk}}{k^2}\right]\cdot\left\{\mathbf{l}\times\mathbf{l}\times\mathbf{E}(\mathbf{l},\omega) + \frac{\varepsilon(\mathbf{l},\omega)}{\left(1-\frac{c^2 l^2}{\omega^2}\right)}\mathbf{l}(\mathbf{l}\cdot\mathbf{E}(\mathbf{l},\omega))\right\} \quad (77)$$

The first term in this expression permits to identify explicitly STFT of the (diagonal) second rank tensor of longitudinal conductivity in terms of the dielectric permittivity,

$$\sigma^{lon}(\mathbf{k},\omega)\frac{\mathbf{kk}}{k^2} = \frac{i\omega}{4\pi}\{1-\varepsilon(\mathbf{k},\omega)\}\frac{\mathbf{kk}}{k^2}. \quad (78)$$

Using Eq. (53) one can rewrite Eq. (78) in terms of STFTs of the charge density-charge density EGFs:

$$\sigma^{lon}(\mathbf{k},\omega) = \frac{i\omega\sum_{\mathbf{l}}\frac{1}{l^2}\langle\langle\rho_{\mathbf{k}}|\rho_{-\mathbf{l}}\rangle\rangle_{0,\omega}}{1+4\pi\sum_{\mathbf{l}}\frac{1}{l^2}\langle\langle\rho_{\mathbf{k}}|\rho_{-\mathbf{l}}\rangle\rangle_{0,\omega}}. \quad (79)$$

The obtained Eq. (79) reduces to the corresponding expression derived by Zubarev in Ref. 54 for the case of spatially homogeneous systems [the first equation in the system of Eqs. (18.29a)], because in that case only the terms with $\mathbf{l} = \mathbf{k}$ in the sums are non-zero. However, in contrast to Zubarev's result, Eq. (79) is approximate, because it has been derived using the approximate explicit expression, Eq. (53), for the dielectric permittivity that holds only in the case of weak external electromagnetic fields. Simplicity of the form and diagonality of the tensor of longitudinal conductivity is due to the used approximation for the dielectric permittivity.

The approximation of Eq. (53) for the dielectric permittivity and linearity of the theoretical approach developed here with regard to the magnitude of the external electromagnetic field [i.e., the use of the linearized Hamiltonian of Eq. (12)] lead to a simple Eq. (79) for STFT of the longitudinal conductivity that occurs completely defined by STFTs of the microscopic charge density - charge density EGFs. Other approximations for the dielectric permittivity tensor (still in the framework of the linear response theory with regard to the external fields) may be developed using Eqs. (43), (45) and (46), and will undoubtedly lead to appearance of non-diagonal components of the longitudinal conductivity tensor. However, values of such components will be of the next order in magnitude compared to those of the diagonal components. The development of such approximations is postponed to the near future.

Despite simplicity of the derived approximate expressions, the tensors of dielectric permittivity and longitudinal conductivity of Eqs. (53) and (79), respectively, still depend on the entire set of STFTs of the microscopic charge density – charge density EGFs, which is due to spatial inhomogeneity of the system.

1.2.8 Transversal conductivity of a spatially inhomogeneous quantum system in a weak external electromagnetic field

1.2.8.1 Induced magnetic moment and the magnetic permeability

The STFT of the generalized susceptibility $\chi^{tr}(\mathbf{k},\omega)$ of spatially homogeneous systems depends only on the wave vector $\mathbf{k}$ and is negative for the majority of such systems. Correspondingly, Lev Landau proved that diamagnetic effect described by $\{\chi^{tr}(\mathbf{k},\omega)+(e^2/m)n_0\}$ is small, and increases only in the case of superconducting systems due to the existence of the gap in the spectrum of the elementary excitations. This consideration leads to a conclusion that for inhomogeneous systems with small magnetization in small external magnetic fields the quantity $\{\chi^{tr}(\mathbf{k},\mathbf{l},\omega)[(\mathbf{k}\cdot\mathbf{l})/kl]+(e^2/m)n_0(\mathbf{k}-\mathbf{l})\}$ in Eq. (77) should also be small, albeit non-negligible. Thus, considering the linear response with regard to the external fields, spatial inhomogeneity effects manifest themselves only through a dependence of the quantity $\{\chi^{tr}(\mathbf{k},\mathbf{l},\omega)[(\mathbf{k}\cdot\mathbf{l})/kl]+(e^2/m)n_0(\mathbf{k}-\mathbf{l})\}$ on both $\mathbf{k}$ and $\mathbf{l}$ wave vectors.

Magnetic properties of a system in a magnetic field are described by the induced magnetic moment $\mathbf{M}(\mathbf{r},t)$ that is proportional to the difference between the induced and external magnetic fields. Correspondingly, STFT $\mathbf{M}(\mathbf{k},\omega)$ of the induced magnetic moment is:

$$\mathbf{M}(\mathbf{k},\omega) = \frac{1}{4\pi}[\mathbf{B}(\mathbf{k},\omega) - \mathbf{H}(\mathbf{k},\omega)]. \tag{80}$$

For systems without significant magnetization, STFT of the local current density $\mathbf{j}^{tr}(\mathbf{k},\omega)$ described by the second term in the r.h.s. of Eq. (77) is defined by STFT of the induced magnetic moment,

$$\mathbf{j}^{tr}(\mathbf{k},\omega) = c[i\mathbf{k}\times\mathbf{M}(\mathbf{k},\omega)] \tag{81}$$

Following a line of reasoning similar to that resulted in Eq. (43), and using the causality condition, one can show in the linear approximation with regard to the external field that the STFTs of the induced and applied fields are related by the expression

$$\mathbf{B}(\mathbf{k},\omega) = \mu(\mathbf{k},\omega) \cdot \mathbf{H}(\mathbf{k},\omega), \qquad (82)$$

where $\mu(\mathbf{k},\omega)$ is the second-rank tensor of magnetic permeability. Arguments similar to those outlined in section 1.2.3 bring about the following expression for STFT of the magnetic permeability tensor in the linear approximation with regard to the applied field:

$$\mu(\mathbf{k},\omega) = \sum_{\mathbf{l}} \left\{ \mu^{lon}(\mathbf{k},\mathbf{l},\omega) \frac{\mathbf{kl}}{kl} + \mu^{tr}(\mathbf{k},\mathbf{l},\omega) \left[\frac{(\mathbf{k}\cdot\mathbf{l})}{kl}\mathbf{1} - \frac{\mathbf{kl}}{kl} \right] \right\}. \qquad (83)$$

In this expression STFTs of the scalar longitudinal and transversal magnetic permeabilities, $\mu^{lon}(\mathbf{k},\mathbf{l},\omega)$ and $\mu^{tr}(\mathbf{k},\mathbf{l},\omega)$, respectively, are (unknown) functions of the absolute values k, l, and the inner product $(\mathbf{k}\cdot\mathbf{l})$ only. Combining Eqs. (83) and (82), one can relate the induced and external magnetic fields as follows:

$$\begin{aligned}\mathbf{B}(\mathbf{k},\omega) &= \sum_{\mathbf{l}} \left\{ \mu^{lon}(\mathbf{k},\mathbf{l},\omega) - \mu^{tr}(\mathbf{k},\mathbf{l},\omega) \right\} \frac{\mathbf{k}}{kl}(\mathbf{l}\cdot\mathbf{H}(\mathbf{k},\omega)) \\ &+ \sum_{\mathbf{l}} \mu^{tr}(\mathbf{k},\mathbf{l},\omega) \frac{\mathbf{k}\cdot\mathbf{l}}{kl} \mathbf{H}(\mathbf{k},\omega).\end{aligned} \qquad (84)$$

Using this equation, one can determine the cross product $[i\mathbf{k}\times\mathbf{B}(\mathbf{k},\omega)]$,

$$[i\mathbf{k}\times\mathbf{H}(\mathbf{k},\omega)] = \mu^{-1}(\mathbf{k},\omega)[i\mathbf{k}\times\mathbf{B}(\mathbf{k},\omega)], \qquad (85)$$

and identify STFT of the scalar magnetic permeability, $\mu(\mathbf{k},\omega)$:

$$\mu(\mathbf{k},\omega) = \sum_{\mathbf{l}} \mu^{tr}(\mathbf{k},\mathbf{l},\omega) \frac{\mathbf{k}\cdot\mathbf{l}}{kl}. \qquad (86)$$

To derive an explicit expression for the magnetic permeability and its relation to the conductivity, one can use Eq. (69) for the induced current density. In particular, using Eqs. (70), (81), (85) and (86), the transversal component of the current density, $\mathbf{j}^{tr}(\mathbf{k},\omega)$ [which is the second term in the r.h.s. of Eq. (69)], can be related explicitly to the vector-potential $\mathbf{A}(\mathbf{k},\omega)$ producing the first line in the r.h.s. of the following Eq. (87):

$$\begin{aligned}\mathbf{j}^{tr}(\mathbf{k},\omega) &= \sum_{\mathbf{l}} \frac{i}{ck^2}\left[\chi^{tr}(\mathbf{k},\mathbf{l},\omega)\frac{\mathbf{k}\cdot\mathbf{l}}{kl} + \frac{e^2}{m}n_0(\mathbf{k}-\mathbf{l}) \right][\mathbf{k}\times\mathbf{k}\times\mathbf{A}(\mathbf{l},\omega)] \\ &= \frac{c}{4\pi}\left[1 - \mu^{-1}(\mathbf{k},\omega) \right][i\mathbf{k}\times\mathbf{B}(\mathbf{l},\omega)].\end{aligned} \qquad (87)$$

The second line in the r.h.s. of Eq. (87) is obtained from the first line using Eqs. (70) and (85). An explicit expression for the scalar magnetic permeability $\mu(\mathbf{k},\omega)$ follows from comparison of the two lines in the r.h.s., and the use of Eq. (70) again:

$$\mu^{-1}(\mathbf{k},\omega) = 1 + \frac{4\pi}{c^2 k^2} \sum_{\mathbf{l}} \left[\chi^{tr}(\mathbf{k},\mathbf{l},\omega) \frac{\mathbf{k} \cdot \mathbf{l}}{kl} + \frac{e^2}{m} n_0(\mathbf{k}-\mathbf{l}) \right]. \tag{88}$$

1.2.8.2 Transversal conductivity of a spatially inhomogeneous quantum system in a weak external electromagnetic field

The second term in the r.h.s. of Eq. (77) can be transformed using the second line in the r.h.s. of Eq. (87) and Eq. (50) to be explicitly expressed in terms of the external electric field:

$$\mathbf{j}(\mathbf{k},\omega) = \frac{i\omega}{4\pi}\left[1-\varepsilon(\mathbf{k},\omega)\right]\frac{\mathbf{k}}{k^2}\mathbf{k}\cdot\mathbf{E}(\mathbf{k},\omega) - \frac{c^2 k^2}{4\pi i\omega}\left[1-\mu^{-1}(\mathbf{k},\omega)\right]\frac{1}{k^2}\left[\mathbf{k}\times\mathbf{k}\times\mathbf{E}(\mathbf{k},\omega)\right]. \tag{89}$$

The obtained expression permits to identify STFT of the scalar transversal conductivity $\sigma^{tr}(\mathbf{k},\omega)$ as the coefficient with the term $(1/k^2)\left[\mathbf{k}\times\mathbf{k}\times\mathbf{E}(\mathbf{k},\omega)\right]$:

$$\sigma^{tr}(\mathbf{k},\omega) = \frac{c^2 k^2}{4\pi i\omega}\left\{1-\mu^{-1}(\mathbf{k},\omega)\right\}. \tag{90}$$

It is expressed in terms of the scalar magnetic permeability, and thus is uniquely defined by STFTs of the charge-charge and the microcurrent-microcurrent density EGFs via Eq. (88) for the magnetic permeability $\mu(\mathbf{k},\omega)$. Using Eq. (90) one can transform Eq. (77) for STFT of the induced current density to the *form* it takes in a linear approximation for the homogeneous system case,

$$\begin{aligned}\mathbf{j}(\mathbf{k},\omega) &= \sigma^{lon}(\mathbf{k},\omega)\frac{\mathbf{k}\mathbf{k}}{k^2}\cdot\mathbf{E}(\mathbf{k},\omega) \\ &- \sigma^{tr}(\mathbf{k},\omega)\frac{1}{k^2}[\mathbf{k}\times\mathbf{k}\times\mathbf{E}(\mathbf{k},\omega)].\end{aligned} \tag{91}$$

Notably, the explicit equations (78) and (90) for STFTs of the scalar longitudinal and transversal conductivities, respectively, appearing in this equation differ markedly from those specific to the homogeneous system case. In particular, in the case of spatially inhomogeneous systems, STFTs of the quantum conductivities depend on the infinite set of STFTs of the microscopic charge and microcurrent density EGFs via Eqs. (78), (53) and (29), respectively, rather than only on the two EGFs corresponding to the wave vectors $\mathbf{k}$ and $-\mathbf{k}$.

In the homogeneous system case the sums over **l** in the r.h.s. of Eq. (79) for STFT of the longitudinal quantum conductivity reduce to one term each with **l** = **k**, and thus Eq. (79) reduces to its homogeneous case counterpart. Similarly, from Eqs. (90), (88), (86) and (27) one can recover the corresponding expression for STFT of the transversal quantum conductivity for spatially homogeneous systems. The reduced expression so obtained coincides with Zubarev's Eq. (18.29a) of Ref. 54.

1.3 Optical properties: the tensor of refractive indices

In section 1.2 quantum transport coefficients of spatially inhomogeneous systems have been derived in a linear approximation with regard to an external electromagnetic field. Due to this approximation, all such tensors are diagonal. Using the fact that the dielectric permittivity and magnetic permeability tensors are diagonal, one can define STFT $N(\mathbf{k},\omega)$ of the tensor of refractive indices (TRI) $\mathbf{N}(\mathbf{r},t)$ as a diagonal tensor in the linear approximation with regard to an external field:

$$N(\mathbf{k},\omega)\mathbf{I} = \pm\sqrt{\varepsilon(\mathbf{k},\omega)\mu(\mathbf{k},\omega)}\,\mathbf{I}, \tag{92}$$

The corresponding explicit expression for this tensor in terms of STFTs of the charge and microcurrent density EGFs have been derived in Ref. 77 using Eqs. (54), (88) and (29):

$$N^{-2}(\mathbf{k},\omega) = 1 + 2\pi\sum_{\mathbf{l}}\left\{\frac{2}{l^2}\langle\langle\rho_{\mathbf{k}}|\rho_{-\mathbf{l}}\rangle\rangle_{0,\omega} + \frac{1}{c^2k^2(\mathbf{k}\cdot\mathbf{l})}\langle\langle[\mathbf{k}\times\mathbf{j}_{\mathbf{k}}]\cdot[\mathbf{l}\times\mathbf{j}_{-\mathbf{l}}]\rangle\rangle_{0,\omega} + \frac{2e^2}{mc^2k^2}n_0(\mathbf{k}-\mathbf{l})\right\}$$
$$+\frac{8\pi^2}{c^2k^2}\sum_{\mathbf{l},\mathbf{p}}\frac{1}{l^2}\langle\langle\rho_{\mathbf{k}}|\rho_{-\mathbf{l}}\rangle\rangle_{0,\omega}\left\{\frac{1}{(\mathbf{k}\cdot\mathbf{p})}\langle\langle[\mathbf{k}\times\mathbf{j}_{\mathbf{k}}]\cdot[\mathbf{p}\times\mathbf{j}_{-\mathbf{p}}]\rangle\rangle_{0,\omega} + \frac{2e^2}{m}n_0(\mathbf{k}-\mathbf{p})\right\}. \tag{93}$$

The STFTs of the charge density-charge density EGFs in the r.h.s. of Eq. (93) can be evaluated [see Ref. 79 and section 1.4 for more details] using a generalized continuous fraction method [80,82] to take the form:

$$\langle\langle\rho_{\mathbf{k}}|\rho_{-\mathbf{l}}\rangle\rangle_{0,\omega} = \frac{(e\hbar^2/m)(\mathbf{k}\cdot\mathbf{l})\langle\rho_{\mathbf{k}-\mathbf{l}}\rangle_0}{(\hbar\omega)^2 - (e\hbar^2/m)(\mathbf{k}\cdot\mathbf{l})((R_{\mathbf{k}},R_{-\mathbf{l}}))_{2,\omega} - E^2_{\mathbf{k},-\mathbf{l}}}. \tag{94}$$

In this expression ω is the frequency, $\langle\rho_{\mathbf{k}-\mathbf{l}}\rangle_0$ is STFT of the charge density, and the quantities $((R_{\mathbf{k}},R_{-\mathbf{l}}))_{2,\omega}$ and $E^2_{\mathbf{k},-\mathbf{l}}$ are related to charge density fluctuations of higher orders. Evaluation of STFT of the microcurrent – microcurrent EGFs $\langle\langle[\mathbf{k}\times\mathbf{j}_{\mathbf{k}}]\cdot[\mathbf{l}\times\mathbf{j}_{-\mathbf{l}}]\rangle\rangle_{0,\omega}$ can be done using the Zubarev-Tserkovnikov projection operator method [36] generalized appropriately to include into consideration spatially inhomogeneous quantum systems. This method and the major ideas concerning its generalization are discussed in section 1.5.

It's important to point out here that STFT of the charge-charge and microcurrent – microcurrent density EGFs can also be evaluated using quantum statistical mechanical simulations and/or experimental data. These routes of EGF evaluation are especially valuable in the case of composite materials and complex media for which analytical calculations of GFs are very difficult, if at all possible at present.

Analysis of explicit expressions for the quantum transport property tensors in Eq. (93) shows [58,77,78] that the third and fifth terms in Eq. (93) containing $\langle\langle[\mathbf{k}\times\mathbf{j_k}]\cdot[\mathbf{p}\times\mathbf{j_{-p}}]\rangle\rangle_{0,\omega} / (\mathbf{k}\cdot\mathbf{p})$ provide for the major contributions to the local TRI when the wave vectors $\mathbf{l}$ or $\mathbf{p}$ are close to $\mathbf{k}$. Notably, for spatially homogeneous systems only the terms with $\mathbf{l} = \mathbf{p} = \mathbf{k}$ are non-zero in Eq. (93), and contributions with $\mathbf{l}\neq\mathbf{k}$, $\mathbf{p}\neq\mathbf{k}$ are absent.

Another type of important contributions to the local TRI are provided by terms with small inner products $\mathbf{k}\cdot\mathbf{l}$ and $\mathbf{k}\cdot\mathbf{p}$ (i.e., the vectors $\mathbf{l}$ and $\mathbf{p}$ are almost orthogonal to $\mathbf{k}$), when the microcurrent-microcurrent density EGFs are finite. In this case the products $\langle\langle[\mathbf{k}\times\mathbf{j_k}]\cdot[\mathbf{p}\times\mathbf{j_{-p}}]\rangle\rangle_{0,\omega} / (\mathbf{k}\cdot\mathbf{p})$ are large. If, at the same time, charge density-related contributions to the local TRI are negligibly small, STFTs of the magnetic permeability and the local values of TRI are defined by STFTs of the microcurrent-microcurrent density EGFs only, that is, by the microcurrent density and its fluctuations.

From this analysis one can conclude that there exist two refraction regimes of significant importance for applications.

(1) Both $(\mathbf{k}\cdot\mathbf{l})$ and $\langle\langle[\mathbf{k}\times\mathbf{j_k}]\cdot[\mathbf{p}\times\mathbf{j_{-1}}]\rangle\rangle_{0,\omega}$ are small. In this case the factor $\langle\langle[\mathbf{k}\times\mathbf{j_k}]\cdot[\mathbf{p}\times\mathbf{j_{-1}}]\rangle\rangle_{0,\omega} / (\mathbf{k}\cdot\mathbf{l})$ can be close to 1, and STFT of TRI from Eq. (93) is defined by terms containing STFTs of the charge density-charge density EGFs $(2/l^2)\langle\langle\rho_k|\rho_{-1}\rangle\rangle_{0,\omega}$ only. For large wave vectors $\mathbf{l}$ such terms are small, and local values of TRI approach the unity, so refraction of the waves with large wave vectors $\mathbf{l}$ is negligibly small. This regime may be engineered in spatially inhomogeneous systems where the characteristic size of inhomogeneity is about $\lambda=2\pi/k$ in the directions orthogonal to that of the incoming waves of the wave vector $\mathbf{l}$.

(2) The product $\langle\langle[\mathbf{k}\times\mathbf{j_k}]\cdot[\mathbf{p}\times\mathbf{j_{-1}}]\rangle\rangle_{0,\omega} / (\mathbf{k}\cdot\mathbf{l})$ is negative, and its absolute value is reasonably large (but smaller then 1). In this regime a spatially inhomogeneous system should exhibit a sharp peak of TRI for the incoming waves with the wave vectors $\mathbf{l}$ almost orthogonal to the direction of the inhomogeneity ($\mathbf{k}$).

It follows from Eq. (93) that STFTs of the local TRI are non-linear in the wave vectors $\mathbf{k}$ and $\mathbf{l}$. This non-linearity concerns both the incoming wave vectors and characteristic dimensions of inhomogeneity, and persists even in the case of small external electromagnetic fields. Non-linearity of TRI with respect to the incoming wave vectors exists for any spatially inhomogeneous system and is due to TRI's dependence on the local charge density - charge density correlations. Non-linearity of TRI with respect to the characteristic dimensions of inhomogeneity is specific to spatially inhomogeneous magnetic systems, and most likely, can be negligibly small in other systems.

The linear response Eq. (93) can be used to evaluate the local TRI values for quantum systems of any level of spatial inhomogeneity in small external electromagnetic fields. Such systems may include any number of constitutive particles and range from molecules to small quantum dots, magnetic nanoparticles or structural units of metamaterials, and to their assemblies, composite materials and metamaterials. In all cases, the formula [93] should be applied after appropriate adjustments of the physical meaning and values of the local charge and microcurrent densities used to calculate STFTs of the corresponding EGFs.

1.4 Calculation of equilibrium Green's functions

As follows from Eqs. (90), (79), (88), (61), (53), (29) and (93), transport properties of quantum systems in a linear approximation with regard to external fields are expressed in terms of charge – charge density and microcurrent – microcurrent density EGFs. More accurate approximations include higher order EGFs of these quantities and those of their fluctuations. The majority of methods used to calculate such EGFs are not universal in a sense that they rely on model Hamiltonians describing particular systems under consideration and the use of the corresponding diagrammatic techniques in some approximations (see, for example, Ref. 79 for introduction to the most common diagrammatic methods). Generally, this approach to EGF calculation does not permit to relate in a systematic fashion the higher and lower order GFs to parameters describing the system structure and composition, so the validity of any approximation used is judged only on the basis of comparison of the theoretical results and experimental data for the transport coefficients.

At present, the only general method that provides a systematic tool to relate higher order and lower order GFs without the use of model Hamiltonians is the projection operator method [36,71,73,80] due to Zubarev and Tserkovnikov (ZT). In its present form, this method is applicable to spatially homogeneous quantum systems, and thus needs some modification and generalization before it can be applied to spatially inhomogeneous ones. In this section, the ZT method is demonstrated in the case of calculation of STFTs of the charge density – charge density Kubo's relaxation functions (KRF) $((\rho_k, \rho_{-l}))$ introduced in Ref. 81. The STFTs of these functions are directly related to "shifted" STFTs of the charge density – charge density EGFs of Eq. (13) by the expression

$$((\rho_k, \rho_{-l})) = \frac{1}{\hbar\omega}\left\{\langle\langle\rho_k | \rho_{-l}\rangle\rangle_{0,\omega} - \langle\langle\rho_k | \rho_{-l}\rangle\rangle_{0,0}\right\}, \tag{95}$$

where $\langle\langle\rho_k | \rho_{-l}\rangle\rangle_{0,0}$ is the value of $\langle\langle\rho_k | \rho_{-l}\rangle\rangle_{0,\omega}$ calculated at $\omega = 0$.

The ZT method introduces an orthonormal basis in a Hilbert space of observables chosen from physical considerations and used for determining irreducible projections of the EGFs. To calculate STFTs $((\rho_k, \rho_{-l}))$ in the simplest approximation, the basis set composed of STFTs of the charge density operators, ρ_k, and their time derivatives,

$\dot{\rho}_{-\mathbf{k}}$, is sufficient. Using this set, one can obtain from Eqs. (2.15a) and (2.15b) of Ref. 36 the correlations

$$\left(\left(\rho_{\mathbf{k}},\rho_{-1}\right)\right)_{0,\omega} = \frac{\left(\rho_{\mathbf{k}},\rho_{-1}\right)_0}{\hbar\omega - \left\{\left(i\hbar\dot{\rho}_{\mathbf{k}},\rho_{-1}\right)_0 - \left(\left(i\hbar\dot{\rho}_{\mathbf{k}},-i\hbar\dot{\rho}_{-1}\right)\right)_{1,\omega}\right\}\left(i\hbar\rho_{\mathbf{k}},\rho_{-1}\right)_0^{-1}} \qquad (96)$$

and

$$\left(\left(i\hbar\dot{\rho}_{\mathbf{k}},i\hbar\dot{\rho}_{-1}\right)\right)_{1,\omega} = \frac{\left(i\hbar\dot{\rho}_{\mathbf{k}},-i\hbar\dot{\rho}_{-1}\right)_1}{\hbar\omega - \frac{\left((i\hbar)^2\ddot{\rho}_{\mathbf{k}},-i\hbar\dot{\rho}_{-1}\right)_1}{\left(i\hbar\dot{\rho}_{\mathbf{k}},-i\hbar\dot{\rho}_{-1}\right)_1} + \frac{\left(i\hbar\dot{\rho}_{\mathbf{k}},\rho_{-1}\right)_0}{\left(\rho_{\mathbf{k}},\rho_{-1}\right)_0} - \frac{\left(\left((i\hbar)^2\ddot{\rho}_{\mathbf{k}},(i\hbar)^2\ddot{\rho}_{-1}\right)\right)_{2,\omega}}{\left(i\hbar\dot{\rho}_{\mathbf{k}},-i\hbar\dot{\rho}_{-1}\right)_1}} \qquad (97)$$

between STFTs of KRFs of the charge density, $\left(\left(\rho_{\mathbf{k}},\rho_{-1}\right)\right)_{0,\omega}$, and STFTs of the higher order Kubo charge density relaxation functions, $\left(\left(i\hbar\dot{\rho}_{\mathbf{k}},i\hbar\dot{\rho}_{-1}\right)\right)_{1,\omega}$ and $\left(\left((i\hbar)^2\ddot{\rho}_{\mathbf{k}},(i\hbar)^2\ddot{\rho}_{-1}\right)\right)_{2,\omega}$. Subscripts 1 and 2 in these expressions denote irreducible parts of STFTs of these relaxation functions in the sense of the ZT method. The round parentheses with the subscript 0, such as $\left(\rho_{\mathbf{k}},\rho_{-1}\right)_0$, denote the initial values of STFTs of the corresponding KRFs that coincide with the initial values of STFTs of EGFs:

$$\left(\rho_{\mathbf{k}},\rho_{-1}\right)_0 \equiv \langle\langle\rho_{\mathbf{k}}|\rho_{-1}\rangle\rangle_0. \qquad (98)$$

Straightforward calculations show that $\left(i\hbar\dot{\rho}_{\mathbf{k}},\rho_{-1}\right)_0 = 0$ and $\left((i\hbar)^2\ddot{\rho}_{\mathbf{k}},-i\hbar\dot{\rho}_{-1}\right)_1 = 0$, where Eq. (2.9) of Ref. 36 was used to obtain the latter STFT of KRF of the charge density fluctuations. Using these results, and Eqs. (98) and (95), one can reduce Eqs. (96) and (97) to the form:

$$\langle\langle\rho_{\mathbf{k}}|\rho_{-1}\rangle\rangle_{0,\omega} = \frac{\left(i\hbar\dot{\rho}_{\mathbf{k}},i\hbar\dot{\rho}_{-1}\right)_{1,\omega}}{\hbar\omega - \left(\left(i\hbar\dot{\rho}_{\mathbf{k}},-i\hbar\dot{\rho}_{-1}\right)\right)_{1,\omega}\left(\rho_{\mathbf{k}},\rho_{-1}\right)_0^{-1}}, \qquad (99)$$

$$\left(\left(i\hbar\dot{\rho}_{\mathbf{k}},i\hbar\dot{\rho}_{-1}\right)\right)_{1,\omega} = \frac{\left(i\hbar\dot{\rho}_{\mathbf{k}},-i\hbar\dot{\rho}_{-1}\right)_1}{\hbar\omega - \frac{\left(\left((i\hbar)^2\ddot{\rho}_{\mathbf{k}},(i\hbar)^2\ddot{\rho}_{-1}\right)\right)_{2,\omega}}{\left(i\hbar\dot{\rho}_{\mathbf{k}},-i\hbar\dot{\rho}_{-1}\right)_1}}. \qquad (100)$$

Considering electrons as the only charge carriers in the system and using the longitudinal sum rule of Eq. (36), one can prove that $\left(i\hbar\dot{\rho}_{\mathbf{k}},i\hbar\dot{\rho}_{-1}\right)_1 = (e\hbar^2/m\sqrt{V})(\mathbf{l}\cdot\mathbf{k})\langle\rho_{\mathbf{k}-1}\rangle_0$, where and e and m are the electron charge and mass, respectively. The STFT of the KRF $\left(\left((i\hbar)^2\ddot{\rho}_{\mathbf{k}},(i\hbar)^2\ddot{\rho}_{-1}\right)\right)_{2,\omega}$ in Eq. (100) can be evaluated using Eq. (2.14b) of Ref. 36:

$$\lim_{\hbar\omega\to\infty} \hbar\omega\left(\left((i\hbar)^2 \ddot{\rho}_\mathbf{k}, (i\hbar)^2 \ddot{\rho}_{-1}\right)\right)_{2,\omega} = \left((i\hbar)^2 \ddot{\rho}_\mathbf{k}, (i\hbar)^2 \ddot{\rho}_{-1}\right)_2. \tag{101}$$

Further calculation of the irreducible component of the initial value in the numerator of the r.h.s. of Eq. (100) requires an explicit expression for the Hamiltonian of the considered system. At the same time, $\left((i\hbar)^2 \ddot{\rho}_\mathbf{k}, (i\hbar)^2 \ddot{\rho}_{-1}\right)_{2,\omega}$ can be related to the initial value to $(\rho_\mathbf{k}, \rho_{-1})_0$ using Eq. (2.12) of Ref. 36. Thus,

$$\left((i\hbar)^2 \ddot{\rho}_\mathbf{k}, (i\hbar)^2 \ddot{\rho}_{-1}\right)_{2,\omega} = \left\langle (i\hbar)^2 \ddot{\rho}_\mathbf{k}, -i\hbar\dot{\rho}_{-1}\right\rangle_0 - \left\langle [i\hbar\dot{\rho}_\mathbf{k}, \rho_{-1}]\right\rangle_0 (\rho_\mathbf{k}, \rho_{-1})_0^{-1} \left\langle [\rho_\mathbf{k}, -i\hbar\dot{\rho}_{-1}]\right\rangle_0. \tag{102}$$

Calculating $\left\langle [\rho_\mathbf{k}, -i\hbar\dot{\rho}_{-1}]\right\rangle_0 = \left\langle [i\hbar\dot{\rho}_\mathbf{k}, \rho_{-1}]\right\rangle_0 = (e\hbar^2/m\sqrt{V})(\mathbf{1}\cdot\mathbf{k})\left\langle \rho_{\mathbf{k}-1}\right\rangle_0$ straightforwardly, and using Eqs. (101) and (102), one can determine $(\rho_\mathbf{k}, \rho_{-1})_0$. This means, that in the ZT method the initial values $(\rho_\mathbf{k}, \rho_{-1})_0$ are calculated self-consistently, rather than introduced as *ad hoc* adjustable parameters known from experiment or other considerations, which is typical for other methods of KRF/EGF calculations. Thus, in contrast to other approaches, the ZT method permits to develop increasingly accurate approximations for STFTs of KRFs and EGFs using a self-consistent approximation procedure.

Note again, that the initial value $(\rho_\mathbf{k}, \rho_{-1})_0$ can be obtained directly using quantum chemistry software packages, such as GAMESS, NWChem, GAISSIAN or Molpro, to compute $\rho_\mathbf{k}$ and ρ_{-1} for a model of a studied system. [Such computations, however, require an addition of a simple module (that is not available at present) for calculation of EGFs from the computational data that the software already provide.] Other initial values, such as $\left\langle (i\hbar)^2 \ddot{\rho}_\mathbf{k}, -i\hbar\dot{\rho}_{-1}\right\rangle_0$ featuring in Eq. (102), can be obtained from computational data in a similar fashion.

Alternatively, the initial value $\left\langle [(i\hbar)^2 \ddot{\rho}_\mathbf{k}, -i\hbar\dot{\rho}_{-1}]\right\rangle_0$ can be evaluated analytically in terms of the inner products of STFTs of the initial values of the charge density fluctuations, $(\delta\rho_\mathbf{k}, \delta\rho_{-1})_0$, if the Hamiltonian of the studied system is explicitly specified. To do so, Eq. (102) should be re-written in the form:

$$(\rho_\mathbf{k}, \rho_{-1})_0^{-1} = \frac{1}{\dfrac{e\hbar^2}{m\sqrt{V}}(\mathbf{1}\cdot\mathbf{k})\langle \rho_{\mathbf{k}-1}\rangle_0} E_{\mathbf{k},-1}^2 - (R_\mathbf{k}, R_{-1})_2, \tag{103}$$

where notations $E_{\mathbf{k},-1}^2 \equiv \{\langle [(i\hbar)^2 \ddot{\rho}_\mathbf{k}, -i\hbar\dot{\rho}_{-1}]\rangle_0\}/\{(e\hbar^2/m\sqrt{V})(\mathbf{1}\cdot\mathbf{k})\langle \rho_{\mathbf{k}-1}\rangle_0\}$ and $i\hbar\ddot{\rho}_\mathbf{k} \equiv (e\hbar^2/m\sqrt{V})(\mathbf{1}\cdot\mathbf{k})\langle \rho_{\mathbf{k}-1}\rangle_0 R_\mathbf{k}$ are introduced, with the quantities $R_\mathbf{k}$ being related to the charge density fluctuations $\delta\rho_\mathbf{k}$. Using Eqs. (103) and (102), one can derive from Eq. (99) an expression that approximates STFTs of the charge density-charge density EGFs, $\langle\langle \rho_\mathbf{k}|\rho_{-1}\rangle\rangle_{0,\omega}$, to the second order in $\delta\rho_\mathbf{k}$:

$$\langle\langle \rho_\mathbf{k}|\rho_{-1}\rangle\rangle_{0,\omega} = \frac{(e\hbar^2/m\sqrt{V})(\mathbf{1}\cdot\mathbf{k})\langle \rho_{\mathbf{k}-1}\rangle_0}{(\hbar\omega)^2 - (e\hbar^2/m\sqrt{V})(\mathbf{1}\cdot\mathbf{k})((R_\mathbf{k}, R_{-1}))_{2,\omega} - E_{\mathbf{k},-1}^2}. \tag{104}$$

The term in the numerator in the r.h.s. of this equation contains STFT of the local charge density. This term defines the major contribution to the STFT $\langle\langle \rho_k | \rho_{-1} \rangle\rangle_{0,\omega}$ of the charge density-charge density EGF, and therefore, to the generalized longitudinal susceptibility of Eq. (61), the scalar dielectric permittivity of Eq. (53), the longitudinal quantum conductivity of Eq. (79), and TRI of Eq. (93). However, the initial values of EGFs of the higher order charge density fluctuations included in the denominator of Eq. (104) via $E_{k,-1}^2$ and R_k may dramatically affect $\langle\langle \rho_k | \rho_{-1} \rangle\rangle_{0,\omega}$ values specific to strongly spatially inhomogeneous systems, such as small quantum dots. This fact makes accurate calculation of EGFs for strongly spatially inhomogeneous systems very important, and thus necessitates the corresponding upgrade of the ZT method to account fully for charge density fluctuation EGFs of higher orders in such systems.

Similarly, STFTs of the microcurrent density-microcurrent density EGFs define transversal components of the quantum transport coefficients of Eqs. (29), (88), (90) and several crucial contributions to TRI of Eq. (93). They include EGFs of the microcurrent density fluctuations that may significantly impact local values of the transport coefficients and TRI specific to strongly inhomogeneous systems, and have to be adequately accounted for.

Thus, first-principle predictions of the transport coefficients and TRI of spatially inhomogeneous quantum systems of any nature can be reduced to a tractable computational problem using a properly generalized transport theory (such as the linear response theory of section 1.2), a generalized version of the analytical method (such as the ZT-method) to calculate EGFs and the correspondingly upgraded quantum chemistry software. The explicit expressions for the quantum transport coefficients and TRI derived in section 1.2 of this chapter in the linear approximation with regard to the external electromagnetic field are applicable to inhomogeneous systems of any nature and degree of inhomogeneity, such as molecules, atomic clusters, quantum dots, etc. At present, no other method, not even in a linear approximation, allows derivation of such explicit expressions from the first principles without any *ad hoc* assumptions. Simplicity of the derived explicit expressions for STFTs of the quantum transport coefficients and TRI is entirely due to the linear approximation with regard to the external fields used. Such linearization may not be adequate for many strongly inhomogeneous systems (such as "magnetic" quantum dots and molecules with large magnetic moment, or any charged systems) subject to moderate and stronger external electromagnetic fields. At the same time, some of the derived equations, such as Eq. (34) for STFT of the quantum current density, do not use any assumptions or conditions concerning correlations between the scalar and vector field potentials and therefore, are the most general expressions applicable to any spatially inhomogeneous system with negligibly small non-linear effects. In particular, Eq. (34) provides a quite general starting point for non-linear theory developments, including those that may require the use of specific gauges for the field potentials. Finally, when inhomogeneity of a system is negligibly small, all results derived in the above linear response theory of spatially inhomogeneous quantum systems reduce to the corresponding Zubarev's expressions derived in the framework of the linear response theory of homogeneous quantum systems.

1.5 Zubarev-Tserkovnikov's pojection operator method

In sections 1.2 and 1.3 of this chapter it has been shown that similar to the homogeneous system case, the linear response transport coefficients of spatially inhomogeneous quantum systems are expressed in terms of the charge density - charge density and microcurrent density-microcurrent density two-time temperature EGFs. Calculations of these EGFs and description of non-linear phenomena in such systems involves two-time temperature EGFs of higher order, including those of charge density and microcurrent fluctuations, and EGFs of other observables. Because of importance of GFs of various types for applications in condensed matter physics and quantum field theory, there exists extensive literature devoted to GF calculations and ranging from advanced monographs containing chapters on GFs [82–90] to introductory textbooks for wide circles of readers, such as Refs. 55 and 79. Among those, there are significantly less publications concerning specifically two-time temperature EGFs introduced originally by Bogoliubov and Tyablikov [91]. While these particular GFs are much more convenient for applications in statistical physics and condensed matter theory, the absence of diagrammatic methods to calculate such GFs have slowed down their use. At the same time, two-time temperature GFs have been intensively studied in numerous works by Zubarev (see, for example Refs. 36, 73 and 54, and references therein), and especially Tserkovnikov [36,63,71,80,92–96] who established a general projection operator method (below called the ZT method) to calculate in a self-consistent fashion two-time temperature GFs for spatially homogeneous quantum and classical systems with a desirable accuracy. It has become possible to reduce the problem of the GF calculation to finding a solution of a system of coupled *algebraic* equations [36] for irreducible two-time temperature GFs that can be decoupled by the use of the generalized continued fraction method. This decoupling method permits to express irreducible two-time temperature GFs in terms of the charge/spin carrier energies specific to the equilibrium or steady state of the studied system, thus linking the GFs to the energy spectrum of the charge/spin carriers, and correlators of the charge and spin density fluctuations of higher orders. Thus, the ZT method makes explicit calculations of two-time temperature GFs feasible not only analytically, but also computationally. In fact, given that a simple software interface developed to calculate such charge and spin correlators from data provided by the existing software (GAMESS, Gaussian, NWChem, Molpro), numerical calculation of two-time temperature GFs becomes preferable in the case of complex systems. In this section the structure and results of the ZT method are discussed having in mind a prospect of generalization of this method to include spatially inhomogeneous quantum systems. The major notations of Ref. 36 are retained for convenience of comparison.

1.5.1 Definitions and the major properties of two-time temperature GFs used in statistical physics

Following Ref. 36, let's introduce definitions of two-time temperature GFs most widely used in statistical physics. First, they include retarded, advanced and causal GFs: $\langle\langle A(t), B^+(t') \rangle\rangle_r$, $\langle\langle A(t), B^+(t') \rangle\rangle_a$ and $\langle\langle A(t), B^+(t') \rangle\rangle_c$, respectively:

$$\langle\langle A(t), B^+(t')\rangle\rangle_r \equiv \theta(t-t')\frac{1}{i\hbar}\langle[A(t), B^+(t')]_\eta\rangle \qquad (105)$$

$$\langle\langle A(t), B^+(t')\rangle\rangle_a \equiv -\theta(t'-t)\frac{1}{i\hbar}\langle[A(t), B^+(t')]_\eta\rangle, \qquad (106)$$

and $\langle\langle A(t), B^+(t')\rangle\rangle_c \equiv \frac{1}{i\hbar}\{\theta(t-t')\langle A(t)B^+(t')\rangle + \eta\theta(t'-t)\langle B^+(t')A(t)\rangle\},$ (107)

where t and t' are time variables, $\hbar$ is the Planck constant, $\theta(t) = \begin{cases} 0, t < 0 \\ 1, t > 0 \end{cases}$ is the Heaviside theta-function, and the square brackets $[A(t), B^+(t')]_\eta = A(t)B^+(t') - \eta B^+(t')A(t)$ denote either the commutator ($\eta = 1$) or anticommutator ($\eta = -1$) of the operators $A(t)$ and $B^+(t')$. The operators $A(t)$ and $B^+(t')$ are in the Heisenberg representation, $A(t) = \exp[(i/\hbar)\mathcal{H}t]A\exp[-(i/\hbar)\mathcal{H}t]$. The average $\langle...\rangle \equiv Q^{-1}\mathrm{Tr}\{\exp(-\beta\mathcal{H})...\}$, with $Q \equiv \mathrm{Tr}\{\exp(-\beta\mathcal{H})\}$ being the partition function, is defined with regard to the equilibrium grand canonical ensemble; Tr denotes the trace of a matrix, $\beta \equiv 1/kT$, and T is the absolute temperature. The grand canonical Hamiltonian is $\mathcal{H} = H - \mu N$, where μ and N are the chemical potential and the particle number operator, respectively.

Another two bilinear combinations of operators $A(t)$ and $B^+(t')$, called Kubo's relaxation function [81] and the isothermal susceptibility, can also be used in the same capacity as two-time temperature GFs:

$$\langle\langle A(t), B^+(t')\rangle\rangle^K \equiv \theta(t-t')\frac{1}{\hbar^2}\int_t^\infty dt_1 \langle[A(t_1), B^+(t')]_-\rangle \qquad (108)$$

and $\langle\langle A(t), B^+(t')\rangle\rangle^T \equiv \theta(t-t')\frac{1}{i\hbar}\int_0^\beta d\tau \langle A(t), B^+(t'+i\hbar\tau)\rangle,$ (109)

respectively. The square brackets $[A(t), B^+(t')]_- = A(t)B^+(t') - B^+(t')A(t)$ denote the commutator of the operators $A(t)$ and $B^+(t')$ ($\eta = 1$), and the integration in the r.h.s. of Eq. (109) is over purely imaginary time τ. These functions are convenient to analyze thermodynamic properties of ergodic systems in thermodynamic limit.

The EGFs of Eqs. (105) – (107) satisfy the equation of motion:

$$i\hbar\frac{d}{dt}\langle\langle A(t), B^+(t')\rangle\rangle \equiv \delta(t-t')\frac{1}{i\hbar}\langle[A(t), B^+(t')]_\eta\rangle + \langle\langle[A(t), \mathcal{H}]_-, B^+(t')\rangle\rangle$$
$$= \delta(t-t')\frac{1}{i\hbar}\langle[A, B^+]_\eta\rangle + \langle\langle[A(t), \mathcal{H}]_-, B^+(t')\rangle\rangle, \qquad (110)$$

where the second line takes into account that the averaging is over the equilibrium distribution, and thus depends only on the time difference $t - t'$, and $A \equiv A(0)$ and $B^+ \equiv B^+(0)$ denote the initial values of the operators A and B. Note, that the second line of Eq. (110) also holds for averaging over a steady state distribution where the Hamiltonian does not depend on time explicitly. If in Eq. (110) one replaces $\left\langle \left[A, B^+ \right]_\eta \right\rangle$ in the first term in r.h.s. of the second line by the initial values

$$(A, B^+)^K \equiv (A(0), B^+(0))^K = \frac{i}{\hbar} \int_0^\infty dt_1 \left\langle \left[A(t_1), B^+(0) \right]_- \right\rangle \equiv \chi_{AB^+}. \tag{111}$$

$$\text{or } (A, B^+)^T \equiv (A(0), B^+(0))^T = \int_0^\beta d\tau \left\langle A(0), B^+(i\hbar\tau) \right\rangle \equiv \chi_{AB^+}^T, \tag{112}$$

one would obtain the equation of motion describing the EGFs of Eqs. (108) and (109), respectively. Notably, the time correlation functions, $\left\langle A(t), B^+(t') \right\rangle$, satisfy the same Eq. (110), but without the first term on the right, $\delta(t - t')(1/i\hbar)\left\langle \left[A, B^+ \right]_\eta \right\rangle$. Equations of motion similar to Eq. (110) can also be derived for the corresponding higher order EGFs. In a general case, such equations for the lower order EGFs contain the corresponding higher order EGFs, much the same way as equations of motion for the lower order correlation functions contain the corresponding higher order correlation functions in perturbation theory, but without small parameters that simplify solving such systems of equations in the case of perturbation theory. On the other hand, the very absence of small parameters, or any other assumptions that restrict validity of the systems of coupled equations for EGFs to some classes of quantum many-body systems, is an advantage that permits to apply those systems of equations to finite and/or spatially inhomogeneous systems. In all cases, one has to solve an infinite system of coupled equations for EGFs using a self-consistent method of approximation that would reduce in a controlled fashion the infinite chain of equations to a manageable closed system of the first few of these equations. The ZT method considered later in this section is such a self-consistent method.

In the case of spatially homogeneous quantum systems all EGFs of Eqs. (105) – (109) do not depend on positions in a system. This is not true in the case of spatially inhomogeneous systems. Even at equilibrium ensemble averages of dynamical variables (observables) in such systems depend on a position in a system. [In fact, any spatially inhomogeneous system can be described as a system in an external field, with its Hamiltonian containing the corresponding contributions due to interactions of the systems' particles with the external fields produced by constitutive particles of the confinement or boundaries.] Thus, for spatially inhomogeneous systems all EGFs of Eqs. (105) – (108), the corresponding higher order EGFs, and their equations of motions, have to include explicit dependence on coordinates, for example, $\left\langle \left\langle A(\mathbf{r},t), B^+(\mathbf{r}',t') \right\rangle \right\rangle_a \equiv -\theta(t'-t)(1/i\hbar)\left\langle \left[A(\mathbf{r},t), B^+(\mathbf{r}',t') \right]_\eta \right\rangle$, etc. While such dependence significantly complicates a procedure of solving the systems of equations for EGFs, it also indicates, at the same time, that spatial properties of EGFs should be investigated and used to develop effective solution algorithms. In particular, a generalization of the ZT projection operator method to include spatially inhomogeneous systems must use this opportunity to its advantage.

One of the most useful properties of the retarded and advanced EGFs is revealed when one calculates the equilibrium averages in the right hand sides of Eqs. (105) and (106) explicitly. Thus, choosing complete infinite sets $\{|n\rangle\}$ and $\{|m\rangle\}$ of eigenvectors of the Hamiltonian H and using the definition of the ensemble averaging, one can obtain [see Refs. 54 (§16, Ch. III) and 97] the following results for the time correlation functions (TCFs) $\langle B^+(t'), A(t)\rangle$, $\langle A(t), B^+(t')\rangle$ and EGFs:

$$\langle B^+(t'), A(t)\rangle \equiv \frac{1}{Q}\sum_{n,m}^{\infty}\langle n|B^+|m\rangle\langle m|A|n\rangle \exp(-E_n\beta)\exp\left\{-\frac{i}{\hbar}(E_n-E_m)(t-t')\right\}, \quad (113)$$

$$\langle A(t), B^+(t')\rangle \equiv \frac{1}{Q}\sum_{n,m}^{\infty}\langle n|B^+|m\rangle\langle m|A|n\rangle \exp(-E_m\beta)\exp\left\{-\frac{i}{\hbar}(E_n-E_m)(t-t')\right\}, \quad (114)$$

$$\langle [A(t), B^+(t')]_-\rangle \equiv \frac{1}{Q}\sum_{\substack{n,m;\\E_n\neq E_m}}\langle n|B^+|m\rangle\langle m|A|n\rangle \{\exp(-E_m\beta)-\exp(-E_n\beta)\}$$
$$\times \exp\left\{-\frac{i}{\hbar}(E_n-E_m)(t-t')\right\}, \quad (115)$$

$$\langle [A(t), B^+(t')]_+\rangle \equiv \frac{1}{Q}\sum_{n,m}^{\infty}\langle n|B^+|m\rangle\langle m|A|n\rangle \{\exp(-E_m\beta)+\exp(-E_n\beta)\}$$
$$\times \exp\left\{-\frac{i}{\hbar}(E_n-E_m)(t-t')\right\}, \quad (116)$$

where the subscripts "−" and " + " at the square brackets in the left hand sides (l.h.s) of Eqs. (115) and (116) indicate the commutator ($\eta = 1$) and anticommutator ($\eta = -1$), respectively, and E_n and E_m are the eigenvalues of the Hamiltonian corresponding to the eigenvectors $|n\rangle$ and $|m\rangle$, respectively.

From Eqs. (105), (106) and (115) it follows that the commutatorial EGFs that are proportional to $\langle [A(t), B^+(t')]_-\rangle$ do not contain diagonal elements of the operators A and B^+ ($E_n \neq E_m$), because the term in the sum in r.h.s. of Eq. (115) corresponding to $E_n = E_m$ is zero. At the same time, the anticommutatorial EGFs of Eqs. (105) and (106) contain such diagonal elements, because the corresponding term in the sum in r.h.s. of Eq. (116) is non-zero. Considering that commutatorial EGFs describe the generalized susceptibility [81], which signifies a response of a system to external perturbations, one concludes that such a response involves only non-diagonal matrix elements of observables. This indicates that the linear response theory should be modified in the cases when matrices of observables have non-zero diagonal elements.

The time Fourier transforms (TFTs) $\langle\langle A | B^+ \rangle\rangle^{\eta}_{(r,a)_E}$ of the retarded and advanced two-time temperature EGFs of Eqs. (105) and (106) are:

$$\langle\langle A | B^+ \rangle\rangle^{\eta}_{(r,a)_E} = \frac{1}{2\pi}\int_{-\infty}^{\infty}\langle\langle A(t-t'), B^+ \rangle\rangle^{\eta}_{(r,a)} \exp\left[\frac{i}{\hbar}E(t-t')\right]d(t-t'). \quad (117)$$

Because this result is the same for the retarded and advanced EGFs, one can define a complex-valued function $\langle\langle A | B^+ \rangle\rangle^{\eta}_{E}$ of the *complex energy* E that coincides with $\langle\langle A | B^+ \rangle\rangle^{\eta}_{(r)E}$ in the upper half of the complex E-plane, and with $\langle\langle A | B^+ \rangle\rangle^{\eta}_{(a)E}$ in the lower half of the complex E-plane for each $\eta = \pm 1$ (that is, for each pair of commutatorial or anticommutatorial EGFs). It must be noted, that in order to introduce TFTs of Eq. (117), the integral in r.h.s. of this equation must exist. As discussed below, the existence of such integrals is ensured in the thermodynamic limit for homogeneous quantum systems, but is not at all obvious for finite and/or spatially homogeneous ones.

Substituting Eqs. (115) or (116) into Eq. (117), one can derive the following results:

$$\langle\langle A | B^+ \rangle\rangle^{(-)}_{E} = \frac{1}{2\pi Q}\sum_{\substack{n,m;\\E_n \neq E_m}}^{\infty}\frac{\langle n | B^+ | m\rangle\langle m | A | n\rangle}{E+E_m-E_n}\{\exp(-E_m\beta)-\exp(-E_n\beta)\}, \quad (118)$$

and $\langle\langle A | B^+ \rangle\rangle^{(+)}_{E} = \frac{1}{2\pi Q}\sum_{n,m}^{\infty}\frac{\langle n | B^+ | m\rangle\langle m | A | n\rangle}{E+E_m-E_n}\{\exp(-E_m\beta)+\exp(-E_n\beta)\}. \quad (119)$

The functions $\langle\langle A | B^+ \rangle\rangle^{\eta}_{E}$ of Eqs. (118), (119) are analytical for Im$E \neq 0$, and have singularities on the real axis. For Re$E \neq 0$ such singularities describe excitations [98]. At $E \to 0$ the limit

$$\lim_{E\to 0}\langle\langle A | B^+ \rangle\rangle^{(-)}_{E} = \frac{1}{2\pi Q}\sum_{\substack{n,m;\\E_n \neq E_m}}^{\infty}\frac{\langle n | B^+ | m\rangle\langle m | A | n\rangle}{E_m-E_n}\{\exp(-E_m\beta)-\exp(-E_n\beta)\} \quad (120)$$

is finite and defines the value of the commutatorial EGF at the origin. Therefore, the commutatorial EGF is regular at the origin,

$$\lim_{E\to 0}\left\{E\langle\langle A | B^+ \rangle\rangle^{(-)}_{E}\right\} = 0. \quad (121)$$

Note here, that when the TFTs of Eq. (117) exist in the case of finite and/or spatially inhomogeneous quantum systems, they carry on their dependence on coordinates. Thus, the limit of Eq. (121) in this case should be written in the form

$$\lim_{E \to 0} \left\{ E \langle\langle A(\mathbf{r}) | B^+(\mathbf{r'}) \rangle\rangle_E^{(-)} \right\} = 0, \quad (122)$$

and in fact, includes 3 conditions concerning each of the dependences of the commutatorial EGFs on x, y and z coordinates of *any* pair of vectors $\mathbf{r}$ and $\mathbf{r'}$. These conditions serve as boundary conditions to reduce a set of possible solutions of a system of coupled equations for commutatorial EGFs in the case of finite and/or strongly spatially inhomogeneous quantum systems to a few manageable options.

In the case of the anticommutatorial EGFs the corresponding limit [97] is:

$$2C_{B^+A} \equiv \lim_{E \to 0} \left\{ E \langle\langle A | B^+ \rangle\rangle_E^{(+)} \right\} = \frac{1}{\pi Q} \sum_{\substack{n,m; \\ E_n = E_m}} \langle n | B^+ | m \rangle \langle m | A | n \rangle \exp(-E_m \beta). \quad (123)$$

Once again, in the case of finite and/or spatially inhomogeneous quantum systems this limit depends on coordinates, and thus may differ for different positions in the system. From Eq. (119) it follows that the anticommutatorial EGF has a pole at $E = 0$ with the residue $2C_{B+A}$.

Similar to Eq. (117), one can define TFTs $J_{B^+A}(\omega)$, or the spectral densities, of the time correlations functions (TCFs) $\langle B^+(t'), A(t) \rangle$:

$$\langle B^+(t'), A(t) \rangle = \int_{-\infty}^{\infty} J_{B^+A}(\omega) e^{-i\omega(t'-t)} d\omega, \quad (124)$$

with ω being a complex variable. To enforce convergence of the integral in Eq. (124), TCFs are assumed to tend to zero as $|t - t'|$ tends to infinity. This condition is satisfied in the thermodynamic limit for homogeneous systems, but may not be satisfied in the case of finite or spatially inhomogeneous systems, where time correlation functions may oscillate [99–101] as $|t - t'| \to \infty$. Quantum dynamics of finite and/or spatially inhomogeneous systems, and in particular strongly correlated systems, is non-ergodic [81,102–104], so one has to formulate some less restrictive condition to ensure that the integral in the r.h.s. of Eq. (124) reproduces correct functional properties of TCFs in the limit $|t - t'| \to \infty$. Moreover, for such systems the spectral densities of Eq. (124) carry on their dependence on coordinates inherent from that of their TCF originals. The functional form of coordinate dependence of TCFs and their TFTs is defined by the structure of each particular system and the nature of the external fields.

Comparing Eqs. (113) and (124) one can derive the following explicit expression for the spectral intensity $J_{B^+A}(\omega)$:

$$J_{B^+A}(\omega) = \frac{2\pi}{Q} \sum_{n,m} \langle n | B^+ | m \rangle \langle m | A | n \rangle \exp(-E_n \beta) \, \delta\!\left(\omega - \frac{E_n}{\hbar} + \frac{E_m}{\hbar}\right), \quad (125)$$

where $\delta[\omega-(E_n/\hbar)+(E_m/\hbar)]$ is the Dirac δ-function, and establish a useful symmetry property, $J_{AB^+}(-\omega) = J_{B^+A}(\omega) e^{\hbar\omega\beta}$. Using Eqs. (125) and (114), one can write the TCF $\langle A(t), B^+(t') \rangle$ in the form:

$$\langle A(t), B^+(t') \rangle = \int_{-\infty}^{\infty} J_{B^+A}(\omega) e^{\hbar\omega\beta} e^{-i\omega(t'-t)} d\omega, \qquad (126)$$

Substituting Eqs. (124) and (126) in Eq. (117), one can obtain the spectral representation (TFT) $\langle\langle A | B^+ \rangle\rangle_E$ of the retarded (Im $E > 0$) and advanced (Im $E < 0$) EGFs [36,54]:

$$\langle\langle A | B^+ \rangle\rangle_E \equiv \frac{1}{i\hbar} \int_0^\infty dt\, e^{\frac{i}{\hbar}Et} \langle [A(t), B^+]_\eta \rangle$$
$$= \int_{-\infty}^\infty d\omega' \frac{(e^{\hbar\beta\omega'} - \eta)}{E - \hbar\omega'} J_{B^+A}(\omega'). \qquad (127)$$

At a given η=±1, Eq. (127) holds for both retarded and advanced two-time temperature EGFs. With a cut along the real axis, $\langle\langle A | B^+ \rangle\rangle_E$ can be considered as an analytical function of the complex variable E having two branches. One branch is defined for ImE >0 and coincides with the spectral density of the retarded EGF, and the other branch is defined for ImE <0 and coincides with the spectral density of the advanced EGF. Notably, the spectral densities of the retarded and advanced EGFs are not analytical continuations of each other, as $\langle\langle A | B^+ \rangle\rangle_E$ exhibits a jump on the real axis. In particular, using simple analytical properties of this function, from Eq. (127) one can derive the following explicit correlation [54], sometimes called *the spectral theorem*, between the jump of $\langle\langle A | B^+ \rangle\rangle_E$ on the real E-axis and the spectral intensity $J_{B^+A}(\omega)$:

$$\lim_{\varepsilon \to 0}\left\{\langle\langle A | B^+ \rangle\rangle_{\omega+i\varepsilon} - \langle\langle A | B^+ \rangle\rangle_{\omega-i\varepsilon}\right\} = \frac{2\pi}{i\hbar}(e^{\hbar\beta\omega} - \eta) J_{B^+A}(\omega),\ 0 < \varepsilon \ll 1. \qquad (128)$$

In this expression ω is real, and $J_{B^+A}(\omega)$ is considered to be real.

From Eqs. (127) and (128), using one of representations of the Dirac δ-function, $\delta(\omega-\omega') = \lim_{\varepsilon\to 0}(1/2\pi i)\{[1/(\omega-\omega'-i\varepsilon)] - [1/(\omega-\omega'+i\varepsilon)]\}$, and the Sokhotski-Plemeli identity [105], $[1/(\omega-\omega' \pm i\varepsilon)] = P[1/(\omega-\omega')] \mp i\pi\delta(\omega-\omega')$, where $P(...)$ denotes the Cauchy principal value, and ω and ω´ are both real, one can derive several useful relations as follows:

$$\frac{2\pi}{i\hbar}(e^{\hbar\beta\omega} - \eta)J_{B^+A}(\omega) = 2i\,\text{Im}\langle\langle A | B^+ \rangle\rangle_{(r),\omega}^{\eta} = -2i\,\text{Im}\langle\langle A | B^+ \rangle\rangle_{(a),\omega}^{\eta}, \qquad (129)$$

$$\operatorname{Re}\langle\langle A|B^+\rangle\rangle_{(r),\omega} = \frac{1}{\pi} P \int_{-\infty}^{\infty} d\omega' \frac{\operatorname{Im}\langle\langle A|B^+\rangle\rangle_{(r),\omega}}{\omega' - \omega}, \tag{130}$$

$$\operatorname{Re}\langle\langle A|B^+\rangle\rangle_{(a),\omega} = -\frac{1}{\pi} P \int_{-\infty}^{\infty} d\omega' \frac{\operatorname{Im}\langle\langle A|B^+\rangle\rangle_{(a),\omega}}{\omega' - \omega}, \tag{131}$$

where in Eq. (129) ω is real, and in *the dispersion relations* [Eqs. (130) and (131)] ω and ω' may be real or complex. Finally, from Eqs. (105), (106) and (117) one can derive sum rules for the retarded and advanced two-time temperature EGFs:

$$\int_{-\infty}^{\infty} d\omega \langle\langle A|B^+\rangle\rangle_{(r),\omega} = \frac{\pi}{i\hbar} \langle [A(0), B^+(0)] \rangle, \tag{132}$$

$$\int_{-\infty}^{\infty} d\omega \langle\langle A|B^+\rangle\rangle_{(a),\omega} = -\frac{\pi}{i\hbar} \langle [A(0), B^+(0)] \rangle, \tag{133}$$

where $A(0)$ and $B(0)$ are initial values of the operators A and B at $t = 0$.

Returning to the spectral theorem of Eq. (128) and solving it for the spectral intensity $J_{B^+A}(\omega)$, one observes that in the case of the commutatorial EGFs ($\eta=1$) such a solution is not unique,

$$\begin{aligned} J_{B^+A}(\omega) &= \frac{i\hbar}{2\pi} \frac{\lim_{\varepsilon \to 0}\left\{\langle\langle A|B^+\rangle\rangle_{\omega+i\varepsilon} - \langle\langle A|B^+\rangle\rangle_{\omega-i\varepsilon}\right\}}{\left(e^{\hbar\beta\omega} - \eta\right)} + C_{B^+A}\delta(\omega) \\ &\equiv J_{B^+A}^{(-)}(\omega) + C_{B+A}\delta(\omega),\ 0 < \varepsilon \ll 1, \end{aligned} \tag{134}$$

where C_{B^+A} is a constant. In a general case, Ramos and Gomes [97] proved that $C_{B^+A} = (1/2)\lim_{E \to 0}\left\{E\langle\langle A|B^+\rangle\rangle_E^{(+)}\right\}$, and its explicit value is defined by Eq. (123). If the anticommutatorial EGF $\langle\langle A|B^+\rangle\rangle_E^{(+)}$ admits a pole at $E = 0$, then the value of the constant C_{B^+A} is equal to the half of the residue of $\langle\langle A|B^+\rangle\rangle_E^{(+)}$ at $E = 0$. This constant can also be directly related to the difference between the isothermal and isolated susceptibilities [97,106], $\chi_{AB^+}^T$ and χ_{AB^+}, respectively:

$$C_{B^+A} = \frac{1}{\beta}\left[\chi_{AB^+}^T - \chi_{AB^+}(0)\right] = \lim_{T \to \infty} \frac{1}{T}\int_0^T \langle \Delta A, \Delta B^+(t)\rangle \, dt, \tag{135}$$

where T and t are time variables, $\Delta A \equiv A(0) - \langle A(0)\rangle$, $\Delta B^+ \equiv B^+(0) - \langle B^+(0)\rangle$, the isothermal susceptibility

$$\chi_{AB^+}^T(\omega) = \int_0^\beta \langle \Delta A, \Delta B^+(i\hbar\tau)\rangle d\tau \tag{136}$$

can be related to the thermodynamic two-time temperature EGF of Eq. (109) using purely imaginary time τ, and the isolated susceptibility is

$$\chi_{AB^+}(\omega) = -\langle\langle A | B^+\rangle\rangle_{\hbar\omega}^{(-)}. \tag{137}$$

In the case of ergodic systems Mazur [107] proved that $C_{B^+A} = 0$.

Using Eq. (134), one can establish an explicit correlation between the average value of a product of two operators and the spectral intensity $J_{B^+A}(\omega)$ as follows:

$$\langle B^+A\rangle = \int_{-\infty}^\infty d\omega\, J_{AB^+}^{(-)}(\omega) + C_{B^+A}. \tag{138}$$

Abrikosov *et al.* [86] and Tserkovnikov [108] proved that the contour of integration over complex ω's can be shifted, and the integral in the r.h.s. of Eq. (138) reduced to an infinite sum of values of the spectral densities of EGFs calculated at points $i\omega_{2n}$ =$2\pi i\, n\theta$ located along the imaginary axis of the complex ω-plane:

$$\langle B^+A\rangle = -\theta\langle\langle A | B^+\rangle\rangle_{E=0}^{(-)} - \theta\sum_{\substack{n=-\infty \\ n\neq 0}}^\infty \langle\langle A | B^+\rangle\rangle_{E=i\hbar\omega_{2n}} - \frac{1}{2}\langle[A, B^+]\rangle + C_{B^+A}, \tag{139}$$

thus providing a closed formal solution to the problem of calculation of TFTs of the average values of operators in terms of the spectral densities of two-time temperature EGFs.

Foreseeable problems with applications of the above method to finite and spatially inhomogeneous systems are related to the facts that (1) all ensemble averages in such systems depend on coordinates, and (2) finite systems and many spatially inhomogeneous ones are non-ergodic [99–102]. Thus, for such systems all quantities defined by Eq. (105) and on, including the TFTs (the spectral intensity and spectral densities of EGFs) of Eqs. (117), (124), (126), (127), and limiting procedures of Eqs. (120), (121), (122), (123), (128), (134) and (135), are functions of coordinates, whose existence for any pairs of (**r**,**r**′) coordinates in a system must be ensured. Generally, time Fourier transformations are defined for large sets of functions, so that the spectral intensity and spectral densities are likely to be well-defined for finite and/or spatially inhomogeneous systems. Non-ergodicity of such systems, however, implies that in Eq. (134) C_{B^+A} is non-zero, apart from a few incidental zeros it may adopt. Therefore, C_{B^+A} and the isothermal susceptibility (now functions of $\mathbf{r}$ and $\mathbf{r'}$) may not be well-defined and unique functions of $(\mathbf{r},\mathbf{r'})$ for the entire set of the spectral intensities $J_{B^+A}(\omega;\mathbf{r},\mathbf{r'})$. This problem poses a formidable difficulty for the derivation and use of an explicit correlation similar to Eq. (134) between a jump of the spectral density of EGFs on the real axis and the spectral intensity. Thus, the existence and uniqueness of the formal solution for the ensemble averages in terms of the spectral densities of EGFs similar to that of Eq. (139) have to be proven. At this time, there is no remedy to manage this problem in general. A possible route to overcome above difficulties of two-time temperature EGF-based methods in the case of their applications to finite and/or spatially inhomogeneous systems is to work directly with the EGFs in their "coordinate representation", rather than with their time Fourier transforms. However, in this approach, many advantages of working with TFTs will be lost, and it remains to be seen whether or not technical difficulties acquired by such an approach are manageable.

1.5.2 ZT projection operator method: energy-dependent representation

In sections 1.2 to 1.4 it was demonstrated that quantum transport coefficients and ensemble averages of operators at equilibrium can be represented in terms of two-time temperature EGFs. Thus, the problem of prediction of thermodynamic and transport properties, and correlation between those and the structure and composition of a system, is reduced, at least formally, to calculation of the EGFs. For realistic systems, especially finite and/or spatially inhomogeneous ones, such calculations are a formidable problem that requires specific methods to be solved. One of the universal methods to calculate Green's functions (and generally, correlation functions of any nature) is to derive a chain of coupled equations where contributions to GF's of interest are expressed in terms decreasing in value and containing GFs of progressively higher order. Then a physically reasonable condition would help truncate and decouple that chain, and solve the remaining system of few equations. For example, chains of equations for slow changing averages of observables, such as particle density, momentum or energy density, can be written as a hierarchy of coupled equations in such a way, that progressively smaller contributions from faster processes described by many-body correlation functions (that involve progressively more particles) are included into the corresponding equations for such higher order correlation functions. The major requirement in such approaches is to provide a systematic procedure that allows unambiguous evaluation of the higher order contributions. One of such methods is that of Feynman diagrams that is used to calculate field-theoretical GFs at zero absolute temperature [79,87–89], and has also been adjusted to calculate Matsubara's temperature Green's functions at zero temperature [86].

However, there is no diagrammatic technique for two-time temperature EGFs calculations. In this case, the ZT method provides a powerful, systematic and universal tool for such calculations [36]. This method originates from Mori's projection operator approach [64] widely applied to calculate collective dynamic variables of classical fluids [109–113], and also two-time temperature GFs [114,115]. The original Mori's approach does not provide a systematic procedure to calculate higher order contributions to collective dynamic variables (ensemble averages) due to projection operators of evolution, and thus needs additional assumptions. In 1990's Pozhar and Gubbins [65–69,116] generalized Mori's approach to include such contributions explicitly, thus providing a necessary tool to evaluate those contributions without the use of *ad hoc* assumptions. Using their generalized functional perturbation theory, they developed a very successful transport theory of classical fluids that is applicable to a wide range of classical systems, including strongly inhomogeneous fluids and finite classical systems [117,118].

The ZT-method [36,63,92–96,108] is another generalization of Mori's approach applied to calculation of EGFs. It provides a procedure for systematic calculations of two-time temperature GFs to a desirable degree of accuracy for infinite, spatially homogeneous quantum and classical systems. Making use of differentiation with respect to two time variables, in this method a chain of equations for two-time temperature EGFs is derived and re-arranged to be written in terms of contributions from so-called irreducible EGFs. The irreducible EGFs are projections of the original ones on mutually orthogonal subspaces of the Hilbert space of complex-valued holomorphic functions defined on the complex energy plane. Calculations are greatly simplified due to the fact that the inner product of two such irreducible EGFs that do not belong to the same subspace of the Hilbert space is zero. Often such functions are meromorphic (that is, have only pole-like singularities), which further simplifies the calculations.

Initially, one has to identify a sequence of the column operators $\{A_n\}_{n=1}^{\infty}$ chosen for each particular system from physical considerations, to be used as a basis, and a sequence of the row operators $\{B_n^+\}_{n=1}^{\infty}$, also chosen for the considered system, to be a Hermitian conjugated basis (the operators B_n for which B_n^+ are Hermitially conjugate, may not necessarily coincide with operators A_n). A system of equation of motion for all basis operators A_n and B_n^+ is written in a general form that holds for all observables defined for Hamiltonian systems:

$$i\hbar \dot{A}_n = [A_n, \mathcal{H}]_- = \omega(n,n) A_n + U(n,n+1) A_{n+1}, \qquad (140a)$$

$$-i\hbar \dot{B}_n^+ = [\mathcal{H}, B_n^+]_- = B_n^+ v^+(n,n) + B_{n+1}^+ V^+(n+1,n), \qquad (140b)$$

where the subscript n runs over all operators of the bases (that may be infinite). The matrices $\omega(n,n)$ and $v^+(n,n)$ describe a part of motion of the physical system that does not include interparticle interaction, that is, they describe those contributions to the full derivatives of the operators A_n and B_n^+ that depend only on the operators A_n and B_n^+, respectively. The matrices $U(n,n+1)$ and $V^+(n+1,n)$ describe the contributions to the time derivatives of the operators A_n and B_n^+, respectively, due to interparticle interactions. In this system of equations the usual "column by row" matrix multiplication also includes convolutions with regard to variables describing the state of the system (such as coordinates and spins of the particles), similar to those that take place in equations for correlation functions in perturbation theory. The dimensions of columns of A_n and rows of B_n^+ may not coincide, so that Green's functions $\langle\langle A_n(t), B_n^+(t')\rangle\rangle$ and their time Fourier transforms (TFTs), that is, spectral densities $\langle\langle A_n | B_n^+\rangle\rangle_E$, can be rectangular matrices.

In what follows, the brackets $\langle\langle A(t), B^+(t')\rangle\rangle$ denote any of the two-time temperature Green's functions defined by Eqs. (105) through (109) of section 1.4, and the notation $\langle A | B^+\rangle$ signifies the corresponding initial condition: (i) $(1/i\hbar)\langle[A, B^+]_\eta\rangle$, in the case of the retarded, advanced and causal EGFs of Eqs. (105) – (107), (ii) $(A, B^+) = \chi_{AB^+}$ for the Kubo's relaxation function of Eq. (108), and (iii) $(A, B^+)^T = \chi_{AB^+}^T$ for the isothermal susceptibility of Eq. (109).

Differentiating the EGF $\langle\langle A(t), B^+(t')\rangle\rangle$ with regard to each of the time variables t and t' in accordance with Eqs. (140) one can prove the following correlations:

$$i\hbar \frac{d}{dt}\langle\langle A(t), B^+(t')\rangle\rangle = -i\hbar \frac{d}{dt'}\langle\langle A(t), B^+(t')\rangle\rangle, \tag{141}$$

$$i\hbar \frac{d}{dt}\langle\langle A(t), B^+(t')\rangle\rangle = \delta(t-t')\langle A | B^+\rangle + \langle\langle i\hbar \dot{A}(t), B^+(t')\rangle\rangle,$$
$$= \delta(t-t')\langle A | B^+\rangle + \langle\langle A(t), -i\hbar \dot{B}^+(t')\rangle\rangle. \tag{142}$$

Taking the integral on the r.h.s. of Eq. (117) by parts, one can obtain the time Fourier transforms of the EGFs:

$$E\langle\langle A | B^+\rangle\rangle_E = \langle A | B^+\rangle + \langle\langle i\hbar \dot{A} | B^+\rangle\rangle_E,$$
$$= \langle A | B^+\rangle + \langle\langle A | -i\hbar \dot{B}^+\rangle\rangle_E, \tag{143}$$

where E denotes the energy variable. Below the subscript E is omitted, unless it is necessary to clarify notations. Equation (143) hold for any observables A and B^+. Choosing $A = A_1$ and $B = B_1^+$, one can obtain from Eq. (143) the following correlations:

$$\left\{E - \langle\langle i\hbar \dot{A}_1 | B_1^+\rangle\rangle\langle\langle A_1 | B_1^+\rangle\rangle^{-1}\right\}\langle\langle A_1 | B_1^+\rangle\rangle = \langle A_1 | B_1^+\rangle, \tag{144a}$$

$$\langle\langle A_1 | B_1^+\rangle\rangle\left\{E - \langle\langle A_1 | B_1^+\rangle\rangle^{-1}\langle\langle A_1 | -i\hbar \dot{B}_1^+\rangle\rangle\right\} = \langle A_1 | B_1^+\rangle, \tag{144b}$$

where the inverse matrices are assumed to exist and to be defined as follows.

$$\langle\langle A_1 | B_1^+ \rangle\rangle \langle\langle A_1 | B_1^+ \rangle\rangle^{-1} = \delta_A(11),$$ where $\delta_A(11)$ is the "right" Kronecker delta symbol:

$$\langle\langle A_1 | B_1^+ \rangle\rangle^{-1} \delta_A(11) = \langle\langle A_1 | B_1^+ \rangle\rangle^{-1};$$
$$\delta_A(11)\langle\langle A_1 | B_1^+ \rangle\rangle = \langle\langle A_1 | B_1^+ \rangle\rangle; \delta_A(11)\delta_A(11) = \delta_A(11);$$
(145a)

$$\langle\langle A_1 | B_1^+ \rangle\rangle^{-1} \langle\langle A_1 | B_1^+ \rangle\rangle = \delta_B(11),$$ where $\delta_B(11)$ is the "left" Kronecker delta symbol:

$$\langle\langle A_1 | B_1^+ \rangle\rangle \delta_B(11) = \langle\langle A_1 | B_1^+ \rangle\rangle;$$
$$\delta_B(11)\langle\langle A_1 | B_1^+ \rangle\rangle^{-1} = \langle\langle A_1 | B_1^+ \rangle\rangle^{-1}; \delta_B(11)\delta_B(11) = \delta_B(11).$$
(145b)

Once again, matrix multiplication in Eqs. (145a) and (145b) is defined as convolution over the matrix indices and over all other variables describing quantum states. The two types of the generalized Kronecker's symbols $\delta_A(11)$ and $\delta_B(11)$ are introduced to account for the fact that dimensions of A_1 and B_1^+ matrices may differ. These symbols are chosen in conjunction with the operators A_1 and B_1^+ and depend on the nature of these operators. If $B_1^+ = A_1^+$, these symbols coincide and are equal to the regular Kronecker delta, $\delta(11)$.

Using Eqs. (145b), (144b), and the last line in the r.h.s. of Eq. (143), where A was replaced by $i\hbar \dot{A}_1$ and B^+ by B_1^+, one can prove that:

$$\langle\langle i\hbar \dot{A}_1 | B_1^+ \rangle\rangle \langle\langle A_1 | B_1^+ \rangle\rangle^{-1} = \left\{ \langle\langle i\hbar \dot{A}_1 | B_1^+ \rangle + \langle\langle i\hbar \dot{A}_1 | -i\hbar \dot{B}_1^+ \rangle\rangle \right\}$$
$$- \langle\langle i\hbar \dot{A}_1 | B_1^+ \rangle\rangle \langle\langle A_1 | B_1^+ \rangle\rangle^{-1} \langle\langle A_1 | -i\hbar \dot{B}_1^+ \rangle\rangle \langle\langle A_1 | B_1^+ \rangle\rangle^{-1}.$$
(146)

Using this equation, one can exclude the product $\langle\langle i\hbar \dot{A}_1 | B_1^+ \rangle\rangle \langle\langle A_1 | B_1^+ \rangle\rangle^{-1}$ in l.h.s. of Eq. (144a) to obtain the Dyson equation for $\langle\langle A_1 | B_1^+ \rangle\rangle$:

$$\left\{ E - \langle i\hbar \dot{A}_1 | B_1^+ \rangle \langle A_1 | B_1^+ \rangle^{-1} - \langle\langle i\hbar \dot{A}_1 | -i\hbar \dot{B}_1^+ \rangle\rangle_1 \langle A_1 | B_1^+ \rangle^{-1} \right\} \langle\langle A_1 | B_1^+ \rangle\rangle = \langle A_1 | B_1^+ \rangle, \quad (147)$$

where $\langle\langle A | B^+ \rangle\rangle_1 \equiv \langle\langle A | B^+ \rangle\rangle - \langle\langle A | B_1^+ \rangle\rangle \langle\langle A_1 | B_1^+ \rangle\rangle^{-1} \langle\langle A_1 | B^+ \rangle\rangle$ (148)

(with $A = i\hbar \dot{A}_1$ and $B^+ = -i\hbar \dot{B}_1^+$) is a so-called irreducible EGF, that is, a combination of EGFs that does not contain any contributions linear in A_1 and B_1. Thus, the irreducible EGF $\left\langle\!\left\langle i\hbar \dot{A}_1 \mid -i\hbar \dot{B}_1^+ \right\rangle\!\right\rangle_1$ does not contain any contributions linear in A_1 and B_1. For the first time, this equation was derived by Tserkovnikov in Ref. 119. The equation conjugate to Eq. (147) can be derived from Eqs. (145a), (144a) and the first line in the r.h.s. of Eq. (143) in a similar fashion to read:

$$\left\langle\!\left\langle A_1 \mid B_1^+ \right\rangle\!\right\rangle \left\{ E - \left\langle A_1 \mid B_1^+ \right\rangle^{-1} \left\langle A_1 \mid -i\hbar \dot{B}_1^+ \right\rangle - \left\langle A_1 \mid B_1^+ \right\rangle^{-1} \left\langle\!\left\langle i\hbar \dot{A}_1 \mid -i\hbar \dot{B}_1^+ \right\rangle\!\right\rangle_1 \right\} = \left\langle A_1 \mid B_1^+ \right\rangle. \quad (149)$$

The mass operators in Eqs. (147) and (149) are now expressed in terms that contain only irreducible parts of the EGF $\left\langle\!\left\langle A \mid B^+ \right\rangle\!\right\rangle$.

Applying Eqs. (140a) and (140b) to $A = A_1$ and $B = B_1^+$, respectively, and substituting the derivatives $i\hbar \dot{A}_1$ and $-i\hbar \dot{B}_1^+$ from these equations into Eq. (147), one can re-write Eq. (147) in the form:

$$\left\{ E - \omega(1,1) - U(1,2) \left\langle A_2 \mid B_1^+ \right\rangle \left\langle A_1 \mid B_1^+ \right\rangle^{-1} - U(1,2) \left\langle\!\left\langle A_2 \mid B_2^+ \right\rangle\!\right\rangle_1 V^+(2,1) \left\langle A_1 \mid B_1^+ \right\rangle^{-1} \right\} \left\langle\!\left\langle A_1 \mid B_1^+ \right\rangle\!\right\rangle$$
$$= \left\langle A_1 \mid B_1^+ \right\rangle, \quad (150)$$

where $\left\langle\!\left\langle A_2 \mid B_2^+ \right\rangle\!\right\rangle_1$ is the irreducible part of the EGF $\left\langle\!\left\langle A_2 \mid B_2^+ \right\rangle\!\right\rangle$ defined by Eq. (148) with $A = A_2$ and $B = B_2^+$,

$$\left\langle\!\left\langle A_2 \mid B_2^+ \right\rangle\!\right\rangle_1 \equiv \left\langle\!\left\langle A_2 \mid B_2^+ \right\rangle\!\right\rangle - \left\langle\!\left\langle A_2 \mid B_1^+ \right\rangle\!\right\rangle \left\langle\!\left\langle A_1 \mid B_1^+ \right\rangle\!\right\rangle^{-1} \left\langle\!\left\langle A_1 \mid B_2^+ \right\rangle\!\right\rangle.$$

Thus, only terms that do not include contributions to $\left\langle\!\left\langle A_2 \mid B_2^+ \right\rangle\!\right\rangle$ linear in A_1 and B_1^+ feature in the mass operator of Eq. (150). Considering the basis sequences of the operators $\{A_n\}_{n=1}^{\infty}$ and $\{B_n^+\}_{n=1}^{\infty}$, for future use it's convenient to define a chain of such irreducible contributions to EGFs $\left\langle\!\left\langle A \mid B^+ \right\rangle\!\right\rangle$ and the corresponding initial conditions $\left\langle A \mid B^+ \right\rangle$ as follows:

$$\left\langle\!\left\langle A \mid B^+ \right\rangle\!\right\rangle_n \equiv \left\langle\!\left\langle A \mid B^+ \right\rangle\!\right\rangle_{n-1} - \left\langle\!\left\langle A \mid B_n^+ \right\rangle\!\right\rangle_{n-1} \left\langle\!\left\langle A_n \mid B_n^+ \right\rangle\!\right\rangle_{n-1}^{-1} \left\langle\!\left\langle A_n \mid B^+ \right\rangle\!\right\rangle_{n-1}, \quad (151)$$

and $\left\langle A \mid B^+ \right\rangle_n \equiv \left\langle A \mid B^+ \right\rangle_{n-1} - \left\langle A \mid B_n^+ \right\rangle_{n-1} \left\langle A_n \mid B_n^+ \right\rangle_{n-1}^{-1} \left\langle A_n \mid B^+ \right\rangle_{n-1}, \quad (152)$

respectively (no summation over repeated indices is assumed). In Eqs. (151) and (152) $n = 0, 1, 2,\ldots$, the inverse matrices $\left\langle\!\left\langle A_n \mid B_n^+ \right\rangle\!\right\rangle_{n-1}^{-1}$ and $\left\langle A_n \mid B_n^+ \right\rangle_{n-1}^{-1}$ are assumed to exist and satisfy the conditions of Eqs. (145), notations $\left\langle\!\left\langle A \mid B^+ \right\rangle\!\right\rangle_0 = \left\langle\!\left\langle A \mid B^+ \right\rangle\!\right\rangle$ and

$\langle A | B^+ \rangle_0 = \langle A | B^+ \rangle$ are introduced, and $\langle\langle A | B^+ \rangle\rangle_{-1} = \langle A | B^+ \rangle_{-1} \equiv 0$ for any A's and B's. The irreducible EGFs similar to those of Eq. (151) for $n = 1$ were introduced for the first time by Kalashnikov in Ref. 120. One can easily prove that the basis operators possess orthogonality properties in the sense of Eqs. (151) and (152) listed below:

$$\langle\langle A_k | B^+ \rangle\rangle_n = 0, \langle\langle A | B_k^+ \rangle\rangle_n = 0, 1 \leq k \leq n, \tag{153}$$

$$\langle A_k | B^+ \rangle_n = 0, \langle A | B_k^+ \rangle_n = 0, 1 \leq k \leq n. \tag{154}$$

Ones Eq. (143) (for $n = 0$) was derived, Tserkovnikov [63] proved by induction the following relations for the irreducible contributions to EGFs, Eq. (151), and initial values, Eq. (152), for any operators A and B, and any $n \geq 1$:

$$E\langle\langle A | B^+ \rangle\rangle_n = \langle A | B^+ \rangle_n - \langle A | B_n^+ \rangle_{n-1} \langle A_n | B_n^+ \rangle_{n-1}^{-1} \langle\langle i\hbar \dot{A}_n | B^+ \rangle\rangle_n + \langle\langle i\hbar \dot{A} | B^+ \rangle\rangle_n, \tag{155}$$

$$E\langle\langle A | B^+ \rangle\rangle_n = \langle A | B^+ \rangle_n - \langle\langle A | -i\hbar \dot{B}_n^+ \rangle\rangle_n \langle A_n | B_n^+ \rangle_{n-1}^{-1} \langle A_n | B^+ \rangle_{n-1} + \langle\langle A | -i\hbar \dot{B}^+ \rangle\rangle_n. \tag{156}$$

These equations reduce to Eq. (143) for $n = 0$. A re-arrangement procedure similar to that used to derive Eqs. (147) and (149) from Eq. (143), permits to re-write Eq. (155) and its conjugate Eq. (156) for $(n-1)$-th irreducible contribution to EGFs of the basis operators:

$$E\langle\langle A_n | B_n^+ \rangle\rangle_{n-1} = \langle A_n | B_n^+ \rangle_{n-1} + \langle i\hbar \dot{A}_n | B_n^+ \rangle_{n-1} \langle A_n | B_n^+ \rangle_{n-1}^{-1} \langle\langle A_n | B_n^+ \rangle\rangle_{n-1}$$
$$- \langle A_n | B_{n-1}^+ \rangle_{n-2} \langle A_{n-1} | B_{n-1}^+ \rangle_{n-2}^{-1} \langle\langle i\hbar \dot{A}_{n-1} | B_n^+ \rangle\rangle_{n-1}$$
$$+ \langle\langle i\hbar \dot{A}_n | -i\hbar \dot{B}_n^+ \rangle\rangle \langle A_n | B_n^+ \rangle_{n-1}^{-1} \langle\langle A_n | B_n^+ \rangle\rangle_{n-1}, \tag{157}$$

$$E\langle\langle A_n | B_n^+ \rangle\rangle_{n-1} = \langle A_n | B_n^+ \rangle_{n-1} + \langle\langle A_n | B_n^+ \rangle\rangle_{n-1} \langle A_n | B_n^+ \rangle_{n-1}^{-1} \langle A_n | -i\hbar \dot{B}_n^+ \rangle_{n-1}$$
$$- \langle\langle A_{n-1} | -i\hbar \dot{B}_n^+ \rangle\rangle_{n-1} \langle A_{n-1} | B_{n-1}^+ \rangle_{n-2}^{-1} \langle A_{n-1} | B_n^+ \rangle_{n-2}$$
$$+ \langle\langle A_n | B_n^+ \rangle\rangle_{n-1} \langle A_n | B_n^+ \rangle_{n-1}^{-1} \langle\langle i\hbar \dot{A}_n | -i\hbar \dot{B}_n^+ \rangle\rangle_n, \tag{158}$$

respectively. These equations reduce to their respective counterparts, Eqs. (147) and (149), for $n = 1$. Using the equations of motion, Eqs. (140), and the orthogonality conditions, Eqs. (153) and (154), to exclude the EGFs of the time derivative operators, one can further simplify Eq. (157) to the form:

$$E\langle\langle A_n | B_n^+\rangle\rangle_{n-1} = \langle A_n | B_n^+\rangle_{n-1} + \{\omega(n,n) + U(n,n-1)\}\langle A_{n+1} | B_n^+\rangle_{n-1}\langle A_n | B_n^+\rangle_{n-1}^{-1}$$
$$-\langle A_n | B_{n-1}^+\rangle_{n-2}\langle A_n | B_n^+\rangle_{n-2}^{-1} U(n-1,n) + U(n,n+1)\langle\langle A_{n+1} | B_{n+1}^+\rangle\rangle_n \times$$
$$V^+(n+1,n)\langle A_n | B_n^+\rangle^{-1}\langle\langle A_n | B_n^+\rangle\rangle_{n-1}. \qquad (159)$$

Only the last term on the right of this equation includes EGF of the next order. A similar equation for $n = 2$ was derived by Mazenko in Ref. 121 for classical correlation functions. One can further clarify the physical meaning of this equation introducing the interaction operator,

$$Y(n,n) \equiv -\langle A_n | B_{n-1}^+\rangle_{n-2}\langle A_{n-1} | B_{n-1}^+\rangle_{n-2}^{-1} U(n-1,n)$$
$$+ U(n,n+1)\langle A_{n+1} | B_n^+\rangle_{n-1}\langle A_n | B_n^+\rangle_{n-1}^{-1}, \qquad (160)$$

the null-EGF, $\langle\langle A_n | B_n^+\rangle\rangle_{n-1}^0$, that satisfies the equation:

$$E\langle\langle A_n | B_n^+\rangle\rangle_{n-1}^0 \equiv \langle A_n | B_n^+\rangle_{n-1} + \{\omega(n,n) + Y(n,n)\}\langle\langle A_n | B_n^+\rangle\rangle_{n-1}^0, \qquad (161)$$

and the mass operator,

$$M_E(n,n) \equiv \langle A_n | B_n^+\rangle_{n-1}^{-1} U(n,n+1)\langle\langle A_{n+1} | B_{n+1}^+\rangle\rangle_n V^+(n+1,n)\langle A_n | B_n^+\rangle_{n-1}^{-1}. \qquad (162)$$

Using these notations, one can rearrange Eq. (159), so that it takes the form

$$\langle\langle A_n | B_n^+\rangle\rangle_{n-1} = \langle\langle A_n | B_n^+\rangle\rangle_{n-1}^0 + \langle\langle A_n | B_n^+\rangle\rangle_{n-1}^0 M_E(n,n)\langle\langle A_n | B_n^+\rangle\rangle_{n-1}, \qquad (163)$$

where the first term in the r.h.s. contains a contribution to the $(n-1)$-th irreducible EGF due to the operator $Y(n,n)$ which is of the first order in interactions [see Eq. (160)], and the second term includes the mass operator $M_E(n,n)$ of the second order in interactions. The Eq. (163) is a short-hand for Eq. (159), where the meaning of each term as to its order with regard to interactions is revealed. Notably, the differentiation of EGFs with respect to each of two times enabled the use of the two equations of motion, Eqs. (140), simultaneously and finally, helped exclusion of any EGFs containing time derivatives in the final equations (159) or (163). Any of these equations generates an infinite chain of coupled equations for irreducible EGFs each of which includes an irreducible EGF of the next order through the mass operator.

One can truncate the chain of Eq. (163) using some approximation for $\langle\langle A_n | B_n^+ \rangle\rangle_{n-1}$, such as $\langle\langle A_n | B_n^+ \rangle\rangle_{n-1} = \langle\langle A_n | B_n^+ \rangle\rangle_{n-1}^0$, and substitute it in Eq. (163) to solve for $\langle\langle A_{n+1} | B_{n+1}^+ \rangle\rangle_n$ included in the mass operator [see Eq. (162)]. Continuing this procedure step by step, one can derive an equation for $\langle\langle A_1 | B_1^+ \rangle\rangle$ in the form of a generalized continued fraction. In the case of the standard equations of motion (140), where $\omega(n,n) = v^+(n,n) = 0$ and $U(n,n+1) = V^+(n+1,n) = 1$, for a non-matrix EGF $\langle\langle A_1 | B_1^+ \rangle\rangle_E$ such a generalized continued fraction chain reduces to a simple continued fraction:

$$\langle\langle A_1 | B_1^+ \rangle\rangle_E = \cfrac{\Delta_1}{E - \Omega_1 - \cfrac{\Delta_2}{E - \Omega_2 - \cfrac{\Delta_3}{E - \Omega_3 - \ldots}}} . \qquad (164)$$

The quantities Δ's and Ω's denote combinations of the initial values as follows:

$$\Delta_1 = \langle A_1 | B_1^+ \rangle, \quad \Delta_2 = \langle A_2 | B_2^+ \rangle_1 \langle A_1 | B_1^+ \rangle_1^{-1},$$
$$\Delta_3 = \langle A_3 | B_3^+ \rangle_2 \langle A_2 | B_2^+ \rangle_1^{-1}, \quad \Delta_4 = \langle A_4 | B_4^+ \rangle_3 \langle A_3 | B_3^+ \rangle_2^{-1}, \ldots$$

$$\Omega_1 = \langle A_2 | B_1^+ \rangle \langle A_1 | B_1^+ \rangle^{-1}, \quad \Omega_2 = \langle A_3 | B_2^+ \rangle_1 \langle A_2 | B_2^+ \rangle_1^{-1} - \langle A_2 | B_1^+ \rangle \langle A_1 | B_1^+ \rangle^{-1},$$
$$\Omega_3 = \langle A_4 | B_3^+ \rangle_2 \langle A_3 | B_3^+ \rangle_2^{-1} - \langle A_3 | B_2^+ \rangle_1 \langle A_2 | B_2^+ \rangle_1^{-1} + \langle A_2 | B_1^+ \rangle \langle A_1 | B_1^+ \rangle^{-1}, \ldots$$

The continued fraction Eq. (164) was obtained originally by Saraswati [115] and also Kruglov [122]. One can see that for large energies E the r.h.s. of Eq. (164) can be expanded into a series with respect to $1/E$ to provide an alternative route to calculate Δ's and Ω's by equating the coefficients factoring the same powers of $1/E$. With this simplification, Eq. (164) reduces to the form:

$$\langle\langle A_1 | B_1^+ \rangle\rangle_E = \frac{\langle A_1 | B_1^+ \rangle}{E} + \frac{\langle A_2 | B_1^+ \rangle}{E^2} + \cdots + \frac{\langle A_m | B_1^+ \rangle}{E^m} + \cdots, \qquad (165)$$

where it was taken into account that at equilibrium $\langle A_3 | B_1^+ \rangle = \langle A_2 | B_2^+ \rangle, \langle A_4 | B_1^+ \rangle = \langle A_3 | B_2^+ \rangle, \ldots$.

Due to simplifications introduced to cut-off the higher order contributions in r.h.s. of the continued fraction Eq. (164), this representation of EGF $\langle\langle A_1 | B_1^+ \rangle\rangle_E$ contains a large number of false poles. A representation of EGFs as generalized continued fractions of Eq. (159) does not contain false poles and correctly reproduces the analytical properties of the EGF of Eq. (127) in the complex E-plain for any order of an approximation used. Thus, the ZT method not only provides a self-consistent calculation tool, Eq. (159), to obtain EGFs to any desirable order of accuracy, but also ensures that the approximate EGFs so calculated have only those poles in the complex E-plane that are specific to the "true" EGFs.

1.5.3 ZT projection operator method: time-dependent representation

It's often convenient to re-formulate the ZT method in such a way, that the energies of elementary excitations and excitation damping are explicitly expressed in terms of correlation functions. For this and other reasons (such as the existence of very fast and robust processes in the system of interest), Tserkovnikov [92–94] suggested a version of the ZT method where projections depend on time, rather than on energy [that is, the time Fourier transformations are not used to derive a system of coupled equations for irreducible EGFs]. A similar approach was also developed by Tokuyama and Mori [123]. In the foundation of this version of the ZT method is the following identity [31–33] for the time derivative of the time correlation function (TCF) $\langle A_1(t) | B_1^+ \rangle$:

$$i\hbar \frac{d}{dt}\langle A_1(t) | B_1^+ \rangle = \langle i\hbar \dot{A}_1(t) | B_1^+ \rangle \langle A_1(t) | B_1^+ \rangle^{-1} \langle A_1(t) | B_1^+ \rangle$$

$$= \left\{ \langle i\hbar \dot{A}_1 | B_1^+ \rangle \langle A_1 | B_1^+ \rangle^{-1} - \frac{i}{\hbar} \int_0^t i\hbar \frac{d}{dt'} \left[\langle i\hbar \dot{A}_1 | B_1^+(-t') \rangle \langle A_1 | B_1^+(-t') \rangle^{-1} \right] dt' \right\} \quad (166)$$

$$\times \langle A_1(t) | B_1^+ \rangle.$$

Using the expression $\frac{d}{dt'}\langle A_1 | B_1^+(-t') \rangle^{-1} = -\langle A_1 | B_1^+(-t') \rangle^{-1} \frac{d}{dt'}\langle A_1 | B_1^+(-t') \rangle \times \langle A_1 | B_1^+(-t') \rangle^{-1}$ for the derivative of the inverse function, one can re-arrange Eq. (166) to the form:

$$i\hbar \frac{d}{dt}\langle A_1(t) | B_1^+ \rangle = \left\{ \langle i\hbar \dot{A}_1 | B_1^+ \rangle \langle A_1 | B_1^+ \rangle^{-1} - \frac{i}{\hbar} \int_0^t dt' \langle i\hbar \dot{A}_1(t') | -i\hbar \dot{B}_1^+ \rangle_1 \langle A_1(t') | B_1^+ \rangle^{-1} \right\}$$

$$\times \langle A_1(t) | B_1^+ \rangle, \quad (167)$$

where in similarity to Eq. (152) the irreducible TCF $\langle A(t) | B^+ \rangle_1$ of the two operators A and B is defined by the relation:

$$\langle A(t) | B^+ \rangle_1 = \langle A(t) | B^+ \rangle - \langle A(t) | B_1^+ \rangle \langle A_1(t) | B_1^+ \rangle^{-1} \langle A_1(t) | B^+ \rangle. \quad (168)$$

Notably, $\langle A_1(t) | B^+ \rangle_1 = \langle A(t) | B_1^+ \rangle_1 = 0$. Now one can use the equations of motion, Eqs. (140), to exclude TCF of the time derivatives in the r.h.s. of Eq. (167), thus obtaining the following equation:

$$i\hbar\frac{d}{dt}\langle A_1(t)|B_1^+\rangle = \left\{\omega(1,1) + U(1,2)\langle A_2|B_1^+\rangle\langle A_1|B_1^+\rangle^{-1}\right.$$
$$\left. -\frac{i}{\hbar}\int_0^t dt' U(1,2)\langle A_2(t')|B_2^+\rangle_1 V^+(2,1)\langle A_1(t')|B_1^+\rangle^{-1}\right\}\langle A_1(t)|B_1^+\rangle. \tag{169}$$

In a similar fashion, one can derive the corresponding equation for the irreducible TCF $\langle A_2(t)|B_2^+\rangle_1$, and so on. It's convenient to introduce the time-dependent projection operators $\mathbf{P}(t) = \mathbf{P}^2(t)$ and $\mathbf{Q}(t) = \mathbf{Q}^2(t)$ defined by the equations

$$\mathbf{P}(t) = |B^+\rangle\langle A_1(t)|B_1^+\rangle^{-1}\langle A_1(t)| \tag{170}$$

and $\mathbf{Q}(t) = \mathbf{1} - \mathbf{P}(t) = \mathbf{1} - |B^+\rangle\langle A_1(t)|B_1^+\rangle^{-1}\langle A(t)|,$ (171)

respectively. The operator $\mathbf{P}(t)$ realizes the time-dependent projection of Eq. (168) of the TCF $\langle A(t)|B^+\rangle$ on the subspace spanned by the operator $|B^+\rangle\langle A(t)|$ of the Hilbert space of observables.

It can be shown [92–94] that in the limit $t\to\infty$ Eq. (167) reduces to the form:

$$\lim_{t\to\infty}\left\{i\hbar\left[\frac{d}{dt}\langle A_1(t)|B_1^+\rangle\right]\langle A_1(t)|B_1^+\rangle^{-1}\right\} \equiv E_1 - i\Gamma_1$$
$$= \langle i\hbar\dot{A}_1|B_1^+\rangle\langle A_1|B_1^+\rangle^{-1} - \frac{i}{\hbar}\int_0^\infty dt'\langle i\hbar\dot{A}_1(t')|-i\hbar\dot{B}_1^+\rangle_1\langle A_1(t')|B_1^+\rangle^{-1}, \tag{172}$$

where E_1 and Γ_1 are directly related to the energy of elementary excitations and their damping.

1.5.4 Advantages and shortcomings of the ZT projection operator method

The major advantage of the ZT method is that it provides a self-consistent algorithm [Eqs. (159) and (163) or (167) and (168)] to calculate EGFs with any desirable accuracy, while retaining all analytical properties of the "true" EGFs (for example, only true poles in the complex E-plain are included in the EGFs so approximated).

Tserkovnikov's papers, Ref. 36 and other publications provide a number of examples of applications of both complex energy (E-based) and time (T-based) versions of the ZT method. These examples prove that ZT-based analytical methods of the generalized continuous fraction type can be developed in each particular case to approximate EGFs self-consistently, thus providing a convenient replacement of the diagrammatic technique used to calculate other types of GFs. In addition, the ZT method offers some advantages over the diagrammatic technique. Thus, in the case of the diagrammatic technique GFs or their mass operators are derived as an infinite series in the interaction parameter(s) with its terms being put in the form of a diagram. Selection of the major contributions (that is, a sub-series of diagrams) to those quantities, and summation of the selected sub-series of diagrams, is the major problem of the diagrammatic approach, and the major reason for the loss of its self-consistency and accuracy. The matter is that a truncation of GFs or their mass operators to the corresponding subseries of diagrams comes with a price of loss of solutions that depend non-analytically on interaction. In contrast, both realizations of the ZT method are free from such a shortcoming.

In similarity with the diagrammatic methods that suffer from inadequacy of the partial diagram summation, the ZT method suffers from the absence of small (or large) interaction parameters to justify truncation of the infinite systems of coupled equations for EGFs [Eqs. (159) and (163) or (167) and (168)] and/or the generalized continuous fraction, that is necessary to provide for EGF solutions in the closed form. This "closure" problem has been addressed in a number of publications since the original papers on the ZT method, and is typical for all projection operator methods [see a brief review by Zubarev and Rudoĭ in Ref. 124].

The root of this problem lies in the absence of any general procedure to identify the basis operators $\{A_n\}_{n=1}^{\infty}$ and $\{B_n^+\}_{n=1}^{\infty}$ with the necessary properties [see Eqs. (153) and (154)], and to decide how many of these operators have to be kept in calculations to ensure a desirable accuracy of EGF truncation. Obviously, the type of the basis operators depends on the nature of a physical problem and has to be selected in each particular case. The task of identifying the size of the basic operator sets is usually reduced to the task of truncation of the system of coupled equations for irreducible EGFs using some approximation. Obviously, the major contribution to the mass operator in the first equations of the chains of Eq. (159) or Eq. (163) is linear in the operators A_1 and B_1^+, if the equations of motion, Eqs. (140), are linear in A_1 and B_1^+, respectively. This class of approximations [125–135] that differ by coefficients in the equations of motion is called a generalized Hartree-Fock approximation (GHFA). Using any of several possible GHFA approximations, one can linearize the mass operator in the first equation ($n = 1$) of Eq. (159) to obtain the reduced first equation of the form:

$$\langle\langle A_1(t) | B_1^+ \rangle\rangle_E = \langle A_1(t) | B_1^+ \rangle \left\{ E - \left[\omega(1,1) + U(1,2)\langle A_2(t) | B_1^+ \rangle \langle A_1(t) | B_1^+ \rangle^{-1} \right] \right\}^{-1} \quad (173)$$

Any attempt to construct an approximation beyond GHFA using the linearized equations of motion for the operators A_1 and B_1^+ produces zero result. One can also suggest non-trivial approximations for EGF $\langle\langle A_2(t) | B_2^+ \rangle\rangle_E$ that is of higher order than $\langle\langle A_1(t) | B_1^+ \rangle\rangle_E$. Of course, the most general procedure of calculation of EGFs is that Eq. (159) of the ZT method should be appended with a closure (that is, an approximation for the highest order EGF kept in calculations) formulated on the basis of the properties of a physical system under consideration. As already mentioned, there is no systematic procedure to formulate such closures, and this is a drawback of the ZT method, that is otherwise almost perfect for the purpose of calculation of EGFs.

1.5.5 Prospects of applications of the ZT projection operator method to include finite and/or spatially inhomogeneous quantum systems

A generalization of the ZT method to include finite and/or spatially inhomogeneous systems faces many challenges steaming both from Green's function-based nature of the method and details of the existing versions of the method. The former type of challenges has been outlined in the course of a discussion of analytical properties of Green's functions, their spectral densities and the spectral intensity in section 1.5.1. Some of these challenges, such as those related to analytical properties of the spectral densities and spectral intensity, can be avoided using the T-representation of the method. Other EGF-related problems, such as the existence of inverse matrices and EGF derivatives with regard to time variables at any point in a system, remain even for T-representation of the method.

The other type of challenges, namely, the challenges related to mathematical nature of the ZT method method itself persist for both representations of the method. Thus, dependence of EGFs, correlation functions and their respective initial values on a position in a system creates a spectrum of problems. In particular, even in the case of classical systems, inverse matrices may contain all zero elements for some positions in a system (for example, in those regions of the system, where there are very few particles). In the quantum case, more trouble is added due to interference properties of the wave packages. Thus, one has to reformulate or renormalize the ZT method in a way that such situations could not cause any technical concern.

Further on, the ZT method introduces the matrix multiplication operation that is appended by convolution over coordinates in the case of inhomogeneous systems. Such convolutions may produce zero results, or be undefined for such systems. Moreover, dependence of EGFs and correlation functions on coordinates brings about a dependence of the transport properties on coordinates. Thus, it is highly undesirable to have coordinates and spins integrated over, because this would result in the transport properties averaged over the system, as opposed to quasi-local values of such properties that are of the major interest for quantum materials design.

Yet another point of concern is the existence of time derivatives of TCFs and EGFs at any point in a system and convergence of integrals with regard to the time variable. Given nano- and atomic scale of finite quantum systems of interest for applications, the uncertainty principle may interfere with formulation of the initial and boundary conditions for EGFs and correlation functions.

Despite of the above concerns, at present there is no other method that provides mathematical means to derive theoretically structure-property relations for finite and/or spatially inhomogeneous systems. The problem is that whatever physical nature of such a system, there are no parameters that would be universally small or large throughout the system, so perturbation theoretical methods cannot be used. Despite of numerous challenges, the only first-principle methods that may be developed to predict properties of such systems must be of projection operator type, of which the ZT method is the most advanced and developed.

It has to be specifically noted, that many problems concerning analytical properties of EGFs and time correlation functions, boundary and initial value problems, and convergence of integrals of these functions with regard to time may be avoided if such functions are computed using already existing quantum chemistry and/or molecular simulation software. A relatively small library of codes to compute such time correlation functions using the charge and spin density distributions provided by such software packages is necessary to realize this option. The TCFs and initial values so computed can be immediately used in ZT system of coupled equations for irreducible EGFs to obtain such EGFs numerically, and then to calculate the transport properties from theoretical correlations, such as those discussed in section 1.2 of this chapter. The existing supercomputers are powerful enough to realize real-time full-quantum computations for systems of about a hundred many-electron atoms, as shown in Chapters 3 to 8 of this book. In Chapter 2 available quantum chemistry software that can be used to realize such quantum computations is briefly discussed.

References

[1] Chen, G., Diao, Z. J., Kim, J. U., Neogi, A., Urtekin, K., and Zhang, Z. G. (2006). Quantum dot computing gates. *Int. J. Quant. Inform.* **4**, 233–296.
[2] Bandyopadhyay, S. (2005). Computing with spins: from classical to quantum computing. *Superlattices and Microstructures* **37**, 77–86.
[3] Radtke, T., and Fritzsche, S. (2005). Simulation of n-qubit quantum systems. I. Quantum registers and quantum gates. *Comp. Phys. Commun.* **173**, 91–113.
[4] Mani, R. G., Johnson, W. B., Narayanamutri, V., Privman, V., and Zhang, Y.-H. (2002). Nuclear spin memory and logic in quantum Hall semiconductor nanostructures for quantum computing applications. *Physica E* **12**, 152–156.
[5] Burkard, G., and Loss, D. (1998). Coupled quantum dots as quantum gates. *Phys. Rev. B* **59**, 2070–2078; DiVincenzo, D. P. (1994). Two-bit gates are universal for quantum computation. *Phys. Rev. A* **51**, 1015–1022.
[6] See, for example, Chelikowsky, J. R. (2000). The pseudopotential-density functional method applied to nanostructures. *J. Phys. D* **33**, R33–R50.
[7] Alhassid, Y. (2000).The statistical theory of quantum dots. *Rev. Mod. Phys.* **72**, 895–968.
[8] Smith, D. L., and Mailhiot, C. (1990). Theory of semiconductor superlattice electronic structure. *Rev. Mod. Phys.* **62**, 173–234.
[9] Landsberg, E. (Ed.) (1996). "Quantum Theory of Real Materials", Kluwer, Boston, MA.
[10] Wu, R. Q., and Freeman, A. J. (1995). Metal-ceramic interfaces: overlap-induced reconstruction and magnetism of 4D transition-metal monolayers. *Phys. Rev. B* **51**, 5408–5412.
[11] Ashkroft, N. M., and Mermin, N. D. (1976). "Solid State Physics". Holt, Rinehart&Winston, New York, NY.
[12] Ivchenko, E. L., and Pikus, G. E. (1997). "Superlattices and Other Heterostructures", 2nd ed. Springer-Verlag, Berlin.
[13] See, for example, Tang, J. M., and Flatte, M. E. (2004). Multiband tight-binding model of local magnetism in $Ga_{1-x}Mn_xAs$. *Phys. Rev. Lett.* **92**, 047201.
[14] http://www.msg.ameslab.gov/GAMESS;
http://www.emsl.pnl.gov/docs/nwchem/nwchem.html.
[15] DiVincenzo, D. P., and Loss, D. (1999). Quantum computers and quantum coherence. *J. Magnet. Magnet. Mater.* **200**, 202–218.

[16] Hu, X., and Das Sarma, S. (2000). Hilbert-space structure of a solid-state quantum computer: two-electron states of a double–quantum-dot artificial molecule. *Phys. Rev. A* **61**, 062301.
[17] Recher, P., Sukhorukov, E. V., and Loss, D. (2000). Quantum dot as spin filter and spin memory. *Phys. Rev. Lett.* **85**, 1962–1965.
[18] Meier, F., Levy, J., and Loss, D. (2003). Quantum computing with antiferromagnetic spin clusters. *Phys. Rev. B* **68**, 134417.
[19] Scrola, V. W., and Das Sarma, S. (2005). Exchange gate in solid-state quantum computation: the applicability of the Heisenberg model. *Phys. Rev. A* **71**, 032340.
[20] Mizel, A., and Lidar, D. A. (2004). Exchange interaction between three and four coupled quantum dots: theory and applications to quantum computing. *Phys. Rev. B* **70**, 115310.
[21] Meier, F., Cerletti, V., Gywat, O., Loss, D., and Awschalom, D. D. (2004). Molecular spintronics: coherent spin transfer in coupled quantum dots. *Phys. Rev. B* **69**, 195315.
[22] Lehmann, J., and Loss, D. (2006). Cotunneling current through quantum dots with phonon-assisted spin-flip processes. *Phys. Rev. B* **73**, 045328.
[23] Mahan, G. D. (1993). "Many Particles Physics", 2nd ed. Plenum Press, New York.
[24] Miller, C. W., Li, Z.-P., Schuller, I. K., Dave, R. W., Slaughter, J. M., and Akerman, J. (2006). Origin of the breakdown of Wentzel-Kramets-Brillouin-based tunneling models. *Phys. Rev. B* **74**, 212404.
[25] Nogues, J., Sort, J., Langlais, V., Skumryev, V., Sarinach, S., Munoz, J. S., and Baro, M. D. (2005). Exchange bias in nanostructures. *Phys. Reports* **422**, 65–117.
[26] Cullity, B.D. (1972). "Introduction to Magnetic Materials." Addison-Wesley, Reading, MA.
[27] Koppens, F. H. L., Buizert, C., Tielrooij, K. J., Vink, I. T., Nowack, K. C., Meunier, T., Kouwenhoven, L. P., and Vandersypen, L. M. K. (2006). Driven coherent oscillations of a single electron spin in a quantum dot. *Nature* **442**, 766–771.
[28] Kitchens, D., Richardella, A., Tang, J.-M., Flatte, M. E., and Yazdani, A. (2006). Atom-by-atom substitution of Mn in GaAs and visualization of their hole-mediated interactions. *Nature* **442**, 436–439.
[29] Dykman, M. I., Santos, L. F., and Shapiro, M. (2005). Many-particle confinement by constructed disorder and quantum computing. *J. Optics B* **7**, S363–S370.
[30] Keldysh, L. V. (1965). Diagram technique for nonequilibrium processes. *Sov. Phys. JETP* **20**, 1018–1026.
[31] Averin, D. V., and Likharev, K. K. (1991). Theory of single electron charging of quantum wells and dots. *Phys. Rev. B* **44**, 6199–6211.
[32] Altshuler, B. L., Lee, P. A., and Webb, R .A. (Eds.) (1991) "Mesoscopic Phenomena in Solids". North-Holland, Amsterdam; Timp, G. (Ed.) (1998). "Nanotechnology". AIP, New York.
[33] Xue, Y., Datta, S., and Ratner, M. A. (2002). First-principle based matrix Green's function approach to molecular electronic devices: general formalism. *Chem. Phys.* **251**, 151–170.
[34] See, for example, Rammer, J. (1991). Quantum transport theory of electronic solids: a single particle approach. *Rev. Mod. Phys.* **63**, 781–817, and references therein.
[35] Fujita, S. (1986). "Introduction to Non-Equilibrium Quantum Statistical Mechanics." W.B. Saunders, Philadelphia.
[36] See Zubarev, D. N., and Tserkovnikov, Yu. A. (1986). The method of two-time temperature Green functions in equilibrium and nonequilibrium statistical mechanics. *Proc. Steklov Inst. Math.* **175**, 139–185, and references therein.
[37] Imry, Y., and Landauer, R. (1999). Conductance viewed as transmission. *Rev. Mod. Phys.* **71**, S306–S312.
[38] Beenakker, C. W. J. (1991). Theory of Coulomb-blockade oscillations in the conductance of a quantum dot. *Phys. Rev. B* **44**, 1646–1656.
[39] Imry, Y. (1986). Physics of mesoscopic systems. *In* "Directions in Condensed Matter Physics" (G. Grinstein and G. Mazenko, Eds.), Vol. 1, pp. 101–163. World Scientific, Singapore.

[40] Reimann, S. M., and Manninen, M. (2002). Electronic structure of quantum dots. *Rev. Mod. Phys.* **74**, 1283–1327.
[41] Bohigas, O., and Giannoni, M.-J. (1984). Chaotic motion and random matrix theories. In "Mathematical and Computational methods in Nuclear Physics. Lecture Notes in Physics" (J. S. Dehesa, J. M. G. Gomez and A. Polls, Eds.), Vol. 209, pp. 1–99. Springer, Berlin.
[42] Altshuler, B. L., Khmel'nitzkii, D., Larkin, A. I., and Lee, P. A. (1980). Magnetoresistance and Hall effect in a disordered two-dimensional electron gas. *Phys. Rev. B* **22**, 5142–5153.
[43] Sukhorukov, E. V., and Loss, D. (1999). Noise in multiterminal diffusive conductors: Universality, nonlocality and exchange effects. *Phys. Rev. B* **59**, 13054–13065.
[44] Li, X.-Q., Zhang, W.-K., Cui, P., Shao, J., Ma, Z., and Yan, Y.-J. (2004). Quantum measurement of solid-state qubit: a unified quantum master equation approach. *Phys. Rev. B* **69**, 085315.
[45] Moya-Cessa, H. (2006). Decoherence in atom-field interactions: A treatment using superoperator techniques. *Phys. Reports* **432**, 1–41.
[46] Qiao, B., Xing, X. S., and Ruda, H. E. (2005). Kinetic equations for quantum information. *Physica A* **355**, 319–332.
[47] Qiao, B., and Ruda, H. E. (2004). Nonlinear Liouville equation, project Green function and stabilizing quantum computing. *Physica A* **333**, 197–224.
[48] Todorov, T. N. (2000). Non-linear conductance of disordered quantum wires. *J. Phys: Condens. Matter* **12**, 8995–9006.
[49] Buttiker, M. (2002). Charge densities and noise in mesoscopic conductors. *Pramana J. Phys.* **58**, 241–257.
[50] Brandbyge, M., Moroz, J.-L., Ordejon, P., Taylor, J., and Stokbro, K. (2002). Density functional methods for non-equilibrium electron transport. *Phys. Rev. B* **65**, 165401.
[51] Buttiker, M. (1986). Role of quantum coherence in series resistors. *Phys. Rev. B* **33**, 3020–3026.
[52] D'Amato, J. L., and Pastawski, H. (1990). Conductance of a disordered linear chain including inelastic scattering events. *Phys. Rev. B* **41**, 7411–7420.
[53] van Hauten, H., Beenakker, C. W. J., and Starring, A. A. M. (1991). Coulomb-blocade oscillations in semiconductor nanostructures. In "Single Electron Tunneling" (H. Grabert, and M. H. Devoret, Eds.), pp. 167–216. Plenum Press, New York.
[54] Zubarev, D. N. (1974). "Nonequilibrium Statistical Thermodynamics". Consultants Bureau, New York; Zubarev, D.N. (1974). "Nonequilibrium Statistical Thermodynamics". Plenum Press, NY.
[55] Lindenberg, J., and Öhrn, Y. (2004). "Propagators in Quantum Chemistry". Wiley-Interscience, Hoboken, New Jersey.
[56] Blaizot, J.-P., and Ripka, G. (1986). "Quantum Theory of Finite Systems". MIT Press, Cambridge, Massachusetts.
[57] Baranger, H. U., and Stone, A. D. (1989). Electrical linear-response theory in an arbitrary magnetic field: a new Fermi-surface formation. *Phys. Rev. B* **40**, 8169–8193.
[58] Pozhar, L. A., and Mitchel, W. C. (2009). Virtual synthesis of electronic nanomaterials: fundamentals and prospects. In "Toward Functional Nanomaterials. Lecture Notes in Nanoscale Science and Technology" (Z. Wang, Ed.), Vol. 5, pp. 423–474. Springer, NY; see also arXiv:cond-mat/0502476) [cond-mat.mes-hall]; ISBN: 978-0-387-77717-7.
[59] Pozhar, L. A. (2005). Charge transport in inhomogeneous quantum systems in weak electro-magnetic fields: two-time Green's function approach. http://www.arXiv.org/cond-mat/0502476, 60 p.
[60] Pozhar, L. A. (2004). Quantum conductivity of spatially Inhomogeneous systems. *Mat. Res. Soc. Proc.* **789**, 49–54.
[61] Pozhar, L. A. (2006). Linearized quantum conductivity of atomic clusters and artificial molecules. *Mater. Res. Soc. Proc.* **900**E, O03-02.1(6).

[62] Zubarev, D. N. (1980). Modern methods of theory of non-equilibrium processes. *In* "Modern Problems of Mathematics", Vol. 15, pp. 131–226. Moscow, VINITI (in Russian).
[63] Tserkovnikov, Yu. A. (1981). On the method of solution of infinite systems of equations for two-time temperature Green's functions. *Teoreticheskaya i Matematicheskaya Fizika* **49**, 219–233. [English translation in *Theor. Math. Phys.* **49** (1981)].
[64] Mori, H. (1965). Transport, Collective motion and Brownian motion. *Progr. Theor. Phys.* **33**, 423–455.
[65] Pozhar, L. A. (1994). "Transport Theory of Inhomogeneous Fluids." World Scientific, New Jersey, 1994. ISBN: 9810217501.
[66] Pozhar, L. A., and Gubbins, K. E. (1993). Transport theory of dense inhomogeneous fluids. *J. Chem. Phys.* **99**, 8970–8996.
[67] Pozhar, L. A., and Gubbins, K. E. (1991). Dense inhomogeneous fluids: Functional perturbation theory, the generalized Langevin equation, and kinetic theory. *J. Chem. Phys.* **94**, 1367–1384.
[68] Pozhar, L. A. (1989). A master equation for dynamical systems with thermal disturbances. *Ukrainian Phys. J.* **34**, 779–788 (in Russian).
[69] Pozhar, L. A. (1988). "Generalization of the Mori Projection Operator Method for Dynamical Systems with Thermal Disturbances." Preprint # 12-88, Institute for Low Temperature Physics and Engineering, Ukrainian Academy of Sciences, Kharkov, USSR.
[70] See a discussion and references in: Zubarev, D. N., and Rudoǐ, Yu. G. (1993). On the evolution of the correlation functions in quantum statistical physics. *Physics-Uspekhi* **36**, 188–191.
[71] Tserkovnikov, Yu. A. (2006). Optimal method for truncation of a chain of equation for two-time temperature Green's functions. *Teoreticheskaya i Matematicheskaya Fizika* **147**, 868–875 [English translation in *Theor. Math. Phys.* **147** (2006)].
[72] Akhieser, A. I., and Peletminskii, S. V. (1977). "Methods of Statistical Physics." Nauka, Moscow (in Russian).
[73] Zubarev, D. N. (1960/61). Double-time Green functions in statistical physics. *Soviet Phys. Uspekhi* **3**, 320–357.
[74] Ewald, P. P. (1917). Berechnung optischer und electrostatischer Gitterpotentiale. *Ann. der Physik* **54**, 519–597.
[75] Agranovich, V. M., and Ginzburg, V. L. (1965). "Optical Properties of Crystalline Solids, Space Dispersion and the Theory of Excitons." Glavnaya Redaktsiya Fiz-Mat. Literatury, Moscow (in Russian).
[76] Klimontovich, Yu. L. (1964). "Statistical Theory of Non-Equilibrium Processes in Plasma." Glavnaya Redaktsiya Fiz-Mat. Literatury, Moscow (in Russian).
[77] Pozhar, L. A. (2010). Refraction indices of spatially non-uniform magnetic systems and nanostructures from the first principles. *J. Appl. Phys.* **107**, 09E151.
[78] Pozhar, L. A. (2010). Refraction indices of non-uniform systems from the first principles. *Mater. Res. Soc. Proc.* **1223**, E03.12 (6).
[79] Economou, E. N. (1983). "Green's Functions in Quantum Physics." Springer, Berlin; Mattuck, R. D. (1967). "A Guide to Feynman Diagrams in the Many-Body Problem." McGraw-Hill, London.
[80] Tserkovnikov, Yu. A. (2001). Two-time temperature Green's functions in kinetic theory and molecular hydrodynamics: III. Taking the interaction of hydrodynamic fluctuations into account. *Teoreticheskaya i Matematicheskaya Fizika* **129**, 1669–1693. [English translation in *Theor. Math. Phys.* **129** (2001)].
[81] Kubo, R. (1957). Statistical-mechanical theory of irreversible processes. I. *J. Phys. Soc. Japan* **12**, 570–586.
[82] Bogolyubov, N. N., Tolmachev, V. V., and Shirkov, D. V. (1959). "New Methods in the Theory of Superconductivity." Chapman & Hall, London.

[83] Bonch-Bruevich, V. L., and Tyablikov, S. V. (1962). "The Green Function Method in Statistical mechanics." Interscience, New York.
[84] Kirzhnitz, D. A. (1963). "Field-Theoretical Methods in Many-Body Theory." GosAtomIzdat, Moscow (in Russian).
[85] Tyablikov, S. V. (1967). "Methods in the Quantum Theory of Magnetism." Plenum Press, NY.
[86] Abricosov, A. A., Gorkov, L. P., and Dzyaloshinski, I. E. (1968). "Methods of Quantum Field Theory in Statistical Physics." Dover, New York.
[87] Bogolyubov, N. N. (1982). "Introduction to Quantum Statistical Mechanics." World Scientific, NJ.
[88] Bogolyubov, N. N., Logunov, A. A., Oksak, A. I., and Todorov, I. T. (1987). "General Principles of Quantum Field Theory." Nauka, Moscow (in Russian).
[89] Bogolubov, N. N., Jr., and Kurbatov, A. M. (Eds.) (1991). "N.N. Bogolubov: Selected Works." Gordon and Breach, NY.
[90] Rickayzen, G. (1980). "Green's Functions and Condensed Matter." Academic Press, London.
[91] Bogoliubov, N. N., and Tyablikov, S. V. (1959). Retarded and advanced Green functions in statistical physics. *Dokl. Acad. Nauk SSSR* **126**, 53–56 (in Russian). [English translation in *Soviet Phys. Dokl.* **4**, 1959/60.]
[92] Tserkovnikov, Yu. A. (1971). Decoupling of chains of equations for two-time Green functions. *Teoreticheskaya i Matematicheskaya Fizika,* **7**, 250–261 [English translation in *Theor. Math. Phys.* **7** (1971)].
[93] Tserkovnikov, Yu. A. (1975). Calculation of correlation functions in the case of degenerate statistical equilibrium state. *Teoreticheskaya i Matematicheskaya Fizika* **23**, 221–237 [English translation in *Theor. Math. Phys.* **23** (1975)].
[94] Tserkovnikov, Yu. A. (1976). Calculation of correlation functions of the non-ideal Bose gas in the framework of the collective variables method. *Teoreticheskaya i Matematicheskaya Fizika* **26**, 77–95 [English translation in *Theor. Math. Phys.* **26** (1976)].
[95] Tserkovnikov, Yu. A. (1982). Two-time temperature Green functions for commuting dynamic variables. *Teoreticheskaya i Matematicheskaya Fizika* **50**, 261–271 [English translation in *Theor. Math. Phys.* **50** (1982)].
[96] Tserkovnikov, Yu. A. (1985). Two-time Green function method in molecular hydrodynamics. *Teoreticheskaya i Matematicheskaya Fizika* **63**, 440–456 [English translation in *Theor. Math. Phys.* **63** (1985)].
[97] Ramos, J. G., and Gomes, A. A. (1971). Remarks on the retarded, advanced and thermodynamic Green's functions. *Nuovo Cimento* **3A**, 441–455.
[98] Pike, E. R. (1964). A new identity in the Green's function approach to the many-body problem. *Proc. Phys. Soc.* **84**, 83–88.
[99] Bocciery, P., and Loinger, A. (1957). Quantum recurrence theorem. *Phys. Rev.* **107**, 337–338.
[100] Percival, I. C. (1961). Almost periodicity and the quantal H theorem. *J. Math. Phys.* **2**, 235–239.
[101] Hogg, T., and Huberman, B. A. (1982). Recurrence phenomena in quantum dynamics. *Phys. Rev. Lett.* **48**, 711–714.
[102] Khintchin, A. I. (1949). "Mathematical Foundation of Statistical Mechanics." Dover, New York.
[103] Stevens, K. W. H., and Toombs, G. A. (1965). Green functions in solid state physics. *Proc. Phys. Soc.* **85**, 1307–1312.
[104] Avella, A., and Mancini, F. (2012). The composite operator method. *In* "Strongly Correlated Systems: Theoretical Methods. Springer Series in Solid State Sciences" (A. Avella and F. Mancini, Eds.), Vol. **171**, pp. 103–139. Springer, Berlin. ISBN 978-3-642-21830-9.
[105] Kanwal, R. P. (1983). "Generalized Functions. Theory and Technique." Academic Press, New York.

[106] Suzuki, M. (1971). Ergodicity, constants of motion and bounds for susceptibilities. *Physica* **51**, 277–291.
[107] Mazur, P. (1960). On the statistical mechanical theory of irreversible behavior. In "Termodinamicadei Processi Irreversibili. Rend. Scuola Internaz. Fis. Enrico Fermi. Corso X, Varenna, 1959" (S. R. de Groot, Ed.), pp. 282–293. Zanichelli, Bologna.
[108] Tserkovnikov, Yu. A. (1979). Construction of the correlation functions of the Heisenberg ferromagnet in the ferromagnetic and paramagnetic regions *Teoreticheskaya i Matematicheskaya Fizika* **40**, 251–268. [English translation in *Theor. Math. Phys.* **40** (1979)].
[109] Götze, W., and Lücke, M. (1975). Dynamical current correlation functions of simple classical liquids for intermediate wave numbers. *Phys. Rev. A* **11**, 2173–2190.
[110] Bosse, J., Götze, W., and Lücke, M. (1978). Mode-coupling theory of simple classical liquids. *Phys. Rev. A* **17**, 434–446.
[111] Bosse, J., Götze, W., and Zippelius, A. (1978). Velocity autocorrelation spectrum of simple classical liquids. *Phys. Rev. A* **18**, 1214–1221.
[112] Sung, W., and Dahler, J. S. (1983). Two-fluid theory of coupled kinetic equations for a test particle and a background fluid. *J. Chem. Phys.* **78**, 6264–6279.
[113] Sung, W., and Dahler, J. S. (1984). Theory of transport process in dense fluids. *J. Chem. Phys.* **80**, 3025–3037.
[114] Ichiyanagi, M. (1972). Projection operators in the theory of two-time Green functions. *J. Phys. Soc. Japan* **32**, 604–609.
[115] Saraswati, D. K. (1976). A continued fraction representation of the mass operator. *Canad. J. Phys.* **54**, 26–33.
[116] Pozhar, L. A., and Gubbins, K. E. (1997). Quasihydrodynamics of nanofluid mixtures. *Phys. Rev. E* **56**, 5367–5396.
[117] Pozhar, L. A. (2000). Structure and dynamics of nanofluids: theory and simulations to calculate viscosity. *Phys. Rev. E* **61**, 1432–1446.
[118] MacElroy, J. M. D., Pozhar, L. A., and Suh, S.-H. (2001). Self-diffusion in a fluid confined within a model nanopore structure. *Colloids and Surfaces* **A187-A188**, 493–507.
[119] Tserkovnikov, Yu. A. (1962). Theory of a perfect Bose gas at a temperature different from zero. *Dokl. Akadem. Nauk SSSR* **143**, 832–835. [English translation in *Soviet Phys. Dokl.* **7** (1962)].
[120] Kalashnikov, V. P. (1978). Linear relaxation equations in the nonequilibrium statistical operator method. *Theoreticheskaya i Matematicheskaya Fizika* **34**, 412–425. [English translation in *Theor. Math. Phys.* **34** (1978)].
[121] Mazenko, G. F. Fully renormalized kinetic theory. I, II, III. *Phys. Rev. A* **7**, 209–222 (1973); *Ibid.* **7**, 222–233 (1973); *Ibid.* **9**, 360–387 (1974).
[122] Kruglov, V. I. (1979). A complete system of equations for the mass functions in the two-time formalism. *Vesti Akad. Navuk BSSR, Ser. Fiz.-Mat. Navuk* **2**, 55–62 (in Russian).
[123] Tokuyama, M., and Mori, H. (1976). Statistical mechanical theory of random frequency modulations and general Brownian motions. *Progr. Theor. Phys.* **55**, 411–429.
[124] Zubarev, D. N., and Rudoĭ, Yu. G. (1992). On the evaluation of the correlation functions in quantum statistical physics. *Physics-Uspekhi* **36**, 188–193.
[125] Ross, L. M. (1968). New method for linearizing many-body equations of motion in statistical mechanics. *Phys. Rev. Lett.* **20**, 1431–1434.
[126] Plakida, N. M. (1970). Decoupling of two-time Green functions and diagram technique. *Teoreticheskaya i Matematicheskaya Fizika* **5**, 147–153. [English translation in *Theor. Math. Phys.* **5**, 1047–1952 (1970)]; (1972). Free energy of an anharmonic crystal. *Ibid.* **12**, 135–146. [English translation in *Theor. Math. Phys.* **12**, 712–719 (1972)].

[127] Kalashnikov, O. K., and Fradkin, E. S. (1968). Theory of the Ising model. *Zhurnal Eksperimental'noj i Teoreticheskoj Fiziki* **55**, 1845–1851. [English translation in *Sov. Phys. JETP* **28**, 976–983 (1969)].
[128] Tahir-Kheli, R. A., and Jarrett, H. S. (1969). Moment conserving decoupling procedure for many-body problems. *Phys. Rev.* **180**, 544–552; Tahir-Kheli, R. A. (1970). Self-consistent moment-conserving decoupling scheme and its applications to the Heisenberg ferromagnet. *Phys. Rev. B* **1**, 3163–3171.
[129] Wallace, D. C. (1966). Renormalization and statistical mechanics in many-particle systems. I. Hamiltonian perturbation method. *Phys. Rev.* **152**, 247–256.
[130] Sawada, K. (1970). Many-body variation theory. II. Equation-of-motion method, an extension of the Hartree-Fock method. *Progr. Theor. Phys.* **43**, 1199–1203.
[131] Young, R. A. (1969). Comments on Roth's method of linearizing many-body equations of motion. *Phys. Rev.* **184**, 601–603.
[132] Oguchi, A. (1970). Variational theory of the Heisenberg ferromagnet. *Progr. Theor. Phys.* **44**, 1548–1556; (1971). Phase transition of the Heisenberg ferromagnet. *Ibid.* **46**, 63–76.
[133] Mubayi, V., and Lange, R. V. (1969). Phase transition in the two-dimensional Heisenberg ferromagnet. *Phys. Rev.* **178**, 882–894.
[134] Kenan, R. P. (1970). Comment on "Phase transition in the two-dimensional ferromagnet". *Phys. Rev. B* **1**, 3205–3211.
[135] Sarry, M. F. (1991). Analytical methods of calculations of correlation functions in quantum statistical physics. *Uspekhi Fizicheskih Nauk* **161**, 47–92. [English translation in *Soviet Phys. Uspekhi* **34**, 958–1003 (1991)].

Quantum Properties of Small Systems at Equilibrium: First Principle Calculations

2.1 Introduction

A possibility to express quantum transport properties of strongly spatially inhomogeneous systems in terms of equilibrium structure factors [that is, structure parameters, correlation functions (ECF), and/or EGFs (Chapter 1)] offers several options for applications. Thus, ECFs and EGFs can be calculated analytically to a desirable degree of accuracy (using, for example, the ZT-projection operator method), numerically, and determined from molecular simulations or experimental data. Notably, one still has to apply case-specific statistical mechanical considerations to extract ECFs and EGFs from simulations or experiment. At the same time, theoretical accuracy of analytical ECFs and EGFs are difficult to predict and control for realistic spatially inhomogeneous systems, as it directly depends on ability of particular theoretical models to account for all atomistic structural details specific to such systems. The most practical way to obtain ECFs and EGFs is to calculate these quantities numerically using *ab initio* (or first-principle) algorithms for detailed characterization of the ground state of a system of interest (often called *ab initio* quantum chemistry methods).

In this chapter only the first-principle, many-body quantum theoretical calculations of the ground state that use the extended variation method based on the variation theorem (see Chapter 8 of Ref. 1) are discussed. Such calculations use exact Hamiltonians and do not use any fitting parameters or experimental data, save for the fundamental physics constants. The ground state wavefunction of a system is the wavefunction that corresponds to the minimum eigenvalue of the Hamiltonian of the system. It is usually represented in the form of a Slater determinant or a composition of such determinants each of which is built of compositions of one-electron atomic orbits or spin-orbits. [The coupled-cluster (CC) calculations seek the true ground state wavefunction in the form of $e^{\hat{T}} \Psi_{HF}$, where Ψ_{HF} is the normalized ground state Hartree-Fock wavefunction, $e^{\hat{T}}$ is the exponential operator, and $\hat{T}$ is the so-called cluster operator.] Once the ground state wavefunction is computed, the physical properties are found as the ground state averages of the desirable observables. Below the first-principle many-body quantum mechanical methods of the ground state wavefunction calculation are considered in detail necessary to understand and use the existing *ab initio* quantum chemistry software packages. Section 2.8 of this chapter contains a brief overview of such software packages and some recommendations concerning their efficient use. The last section of this chapter is dedicated to the virtual synthesis method that uses basic symmetry properties of atomic systems to ascertain the most favorable input configuration of the system atoms to facilitate more efficient *ab initio* computations.

There exists a variety of non-variational methods to characterize the ground state of a system, with the most prominent of them being the density functional theory (DFT). DFT-based computational methods do not compute the ground state wavefunction. Rather, they calculate the electron probability density and energy of a studied system. DFT-based methods are widely discussed in literature and are not addressed in this chapter.

Similarly, semi-empirical quantum-mechanical methods that use simplified model Hamiltonians and a range of fitting parameters extracted from experimental data or molecular simulations, and molecular-mechanics computation methods (that are not quantum mechanical methods) are beyond of the scope of this chapter. Finally, quantum molecular simulations methods constitute a separate field of modeling research that is not a subject of this book.

2.2 Variational methods

2.2.1 The variation theorem and extended variation method

So-called variation methods of numerical solution of the time-independent Schrödinger equation are based on the variation theorem. This theorem states that for any normalized, square-integrable, complex function Ψ dependent of coordinates of system particles and satisfying boundary conditions applied to the system the following variation integral provides the upper bound for the system's ground state energy:

$$\int d\vec{r}\, \Psi^*(\vec{r})\hat{H}\Psi(\vec{r}) \geq E_{GS}, \quad \int d\vec{r}\, |\Psi(\vec{r})|^2 = 1, \tag{1}$$

where $\hat{H}$ is the Hamiltonian of the system, $\vec{r} = \vec{r}_1, ..., \vec{r}_N$ denotes coordinates of the centers of mass of all N particles of the systems, E_{GS} is the system's ground state energy, and integration is over the entire coordinate space $\mathbf{R}^N$. Reasoning serving as a proof of this theorem in the case when the trial wavefunctions depend only on coordinates of quantum particles can be found in every textbook on quantum mechanics or quantum chemistry (see, for example, Chapter 8 of Ref. 1). It involves an expansion of $\Psi(\vec{r})$ in terms of a complete orthonormal set of stationary-state eigenfunctions $\{\varphi_k\}_1^\infty$ of the Hamiltonian $\hat{H}$, and this expansion is required to satisfy the same boundary conditions as those specified for the functions $\{\varphi_k\}_1^\infty$. Rigorous mathematical proofs that include spin component of the trial functions $\Psi(\vec{r})$ and eigenfunctions of the Hamiltonian are derivable in the framework of calculus of variations [2] using relationships between global geometry on the one hand, and the action principle and universal conservation laws, on the other [modern generalizations [3–9] of Noether's theorems [10–12]; see also a historic sketch [13] containing a complete list of references on this and related subjects].

One can select numerous trial variation functions $\Psi(\vec{r})$ and use them to calculate the variation integral of Eq. (1). A particular $\Psi_{GS}(\vec{r})$ that provides for the lowest value of the variation integral is the best approximation for the ground state wavefunction.

To test various trial functions, one can chose some known function as a zero-order approximation for a trial function and develop an algorithm of improvement on that approximation requesting that at any step of the algorithm an improved trial function $[\Psi_{GS}(\vec{r})]_n$ produces a lower value of the variation integral than its predecessor $[\Psi_{GS}(\vec{r})]_{n-1}$. Once a desirable accuracy of calculations is achieved [that is, the improved values $(E_{GS})_n$ fall into the error brackets near some minimum value E_{GS}] the process ends, and the ground state energy is considered to be E_{GS}. To test the ground state wavefunctions so obtained the developed algorithms are applied to problems for which eigenfunctions of the Hamiltonian are known. This allows investigation of convergence of the trial function sequences to the known solution. In reality, however, the sequence of $(E_{GS})_n$ converges to the known ground state value E_{GS} much faster than the sequence of the improved wavefunctions $\{[\Psi_{GS}(\vec{r})]_n\}$ to the ground state wavefunction $\Psi_{GS}(\vec{r})$. The reason for that is integration over the coordinate domains in Eq. (1) that cancels out small deviations while calculating $(E_{GS})_n$ for subsequent steps of the algorithm. Thus, although the zero-order trial function may be poorly chosen and the sequence $\{[\Psi_{GS}(\vec{r})]_n\}$ slow converging to $\Psi_{GS}(\vec{r})$, the sequence of $\{(E_{GS})_n\}$ usually converges to E_{GS} fast enough to make an algorithm practical. To accelerate this convergence even further, modern algorithms of building the sequence $\{[\Psi_{GS}(\vec{r})]_n\}$ include specific accelerators (such as the fastest descent method) that manipulate a search for a suitable improvement in the n-th order trial function $[\Psi_{GS}(\vec{r})]_n$ using a feedback from evaluations for the expected fastest route of convergence of the $\{(E_{GS})_n\}$ sequence. All modern software packages include such accelerators.

The described variational approach can be extended to include excited states of a system. Once the ground state wavefunction $\Psi_{GS}(\vec{r})$ is obtained, the mutual orthogonality of the eigenfunctions of the Hamiltonian is used to establish the minimum value E_2 of the variation integral of Eq. (1) calculated on trial functions orthogonal to $\Psi_{GS}(\vec{r})$. Such a minimum value is the upper bound for the lowest excited state, and the corresponding wavefunction(s) is an approximation for the wavefunction $\Phi_2(\vec{r})$ of that state. Similarly, at the k-th step of this search, choosing trial functions orthogonal to all calculated approximations for the first $k-1$ eigenfunctions of the Hamiltonian one can establish the upper bound E_k for the energy of the k-th excited state, and the corresponding approximation $\Phi_k(\vec{r})$ for its wavefunction:

$$\int d\vec{r}\ \Phi_k^*(\vec{r})\hat{H}\Phi_k(\vec{r}) \geq E_k, \quad (2)$$
$$\int d\vec{r}|\Phi_k(\vec{r})|^2 = 1, \int d\vec{r}\ \Psi_{GS}^*(\vec{r})\Phi_k(\vec{r}) = \int d\vec{r}\ \Phi_i^*(\vec{r})\Phi_k(\vec{r}) = 0, i = 2,...,k-1.$$

Despite seeming simplicity of this extension of the variation method, it is rather cumbersome even in the case when wavefunctions depend on coordinates of the system particles only, because for many practical systems the wavefunctions $\Phi_i(\vec{r})$ of the excited states have nods. Thus, for even functions of coordinates $\Phi_i^*(\vec{r})$ the integrals $\int d\vec{r}\Phi_i^*(\vec{r})\Phi_k(\vec{r}) = 0, i = 2,...,k-1$, are zero for any odd trial function of coordinates $\Phi_k(\vec{r})$. A similar situation also arises for degenerate ground states of many systems and is further aggravated for all systems that include particles with spin. Thus, to calculate E_k's and $\Phi_k(\vec{r})$'s much more sophisticated methods based on perturbation theory for quantum systems (see, for example, Chapter 9 of Ref. 1 and numerous other textbooks) are used.

2.2.2 Non-degenerate perturbation theory and the variation-perturbation method

For realistic systems Hamiltonians have a complex structure, their eigenfunctions are complicated complex functions of coordinates and spins of system particles, and their eigenvalues may coincide (for degenerate states) or lie closely to each other. In such cases the direct variation method of the previous sub-section generates a lot of problems, and thus needs be replaced by much more reliable approaches. Perturbation theory-based methods are such well-developed and reliable tools that permit calculations of the electronic structure of realistic model systems.

Perturbation-theoretical treatment of non-degenerate states of a system are relatively simple and relies on representation of the Hamiltonian $\hat{H}$ of a model system as a sum of the non-perturbed Hamiltonian $\hat{H}_0$ responsible for the major contribution to the total system energy and a (supposedly small) perturbation described by the Hamiltonian $\lambda \hat{H}_1$, where the parameter λ permits to manipulate with strength of the perturbation ($\lambda=0$ corresponds to the unperturbed system). The values of the eigenfunctions $\{\psi_n\}_1^\infty$ and the corresponding eigenvalues $\{E_n\}_1^\infty$ of the Hamiltonian $\hat{H}$ are considered to be perturbed near the values $\{\varphi_n^0\}_1^\infty$ and $\{E_n^0\}_1^\infty$ of the eigenfunctions and eigenvalues, respectively, of the unperturbed Hamiltonian $\hat{H}_0$, and thus can be expanded in the corresponding Taylor series in the parameter λ near their unperturbed values:

$$\psi_n(\lambda, \vec{r}) = \psi_n^{(0)}(\lambda, \vec{r})\bigg|_{\lambda=0} + \frac{\partial \psi_n(\lambda, \vec{r})}{\partial \lambda}\bigg|_{\lambda=0} \lambda + \frac{1}{2!} \frac{\partial^2 \psi_n(\lambda, \vec{r})}{\partial \lambda^2}\bigg|_{\lambda=0} \lambda^2 + \ldots + \frac{1}{k!} \frac{\partial^k \psi_n(\lambda, \vec{r})}{\partial \lambda^k}\bigg|_{\lambda=0} \lambda^k + \ldots, \quad (3)$$

$$E_n = E_n(\lambda, \vec{r})\bigg|_{\lambda=0} + \frac{\partial E_n}{\partial \lambda}\bigg|_{\lambda=0} \lambda + \frac{1}{2!} \frac{\partial^2 E_n}{\partial \lambda^2}\bigg|_{\lambda=0} \lambda^2 + \ldots + \frac{1}{k!} \frac{\partial^k E_n}{\partial \lambda^k}\bigg|_{\lambda=0} \lambda^k + \ldots .$$

Here all ψ-functions contain only coordinate-dependent parts of the eigenfunctions, for simplicity of the discussion, so the notation $\vec{r}$ only includes all coordinates of the system particles. Substituting these expansions into the Schrödinger equation,

$$\hat{H}\psi_n(\lambda;\vec{r}) \equiv (\hat{H}_0 + \lambda \hat{H}_1)\psi_n(\lambda;\vec{r}) = E_n \psi_n(\lambda;\vec{r}), \quad (4)$$

and collecting terms with the like powers of λ, one can obtain systems of equations for corrections $\psi_n^{(k)}(\lambda, \vec{r}) \equiv \dfrac{1}{k!} \dfrac{\partial^k \psi_n(\lambda, \vec{r})}{\partial \lambda^k}\bigg|_{\lambda=0}$ and $E_n^{(k)} \equiv \dfrac{1}{k!} \dfrac{\partial^k E_n}{\partial \lambda^k}\bigg|_{\lambda=0}$, $k = 0, 1, \ldots$, to the unperturbed eigenfunctions and eigenvalues, respectively, and solve those systems of equations. The final results for the perturbed eigenfunctions $\{\psi_n\}_1^\infty$ and eigenvalues $\{E_n\}_1^\infty$ to the k-th order in λ are:

$$\psi_n = \psi_n^{(0)} + \lambda \psi_n^{(1)} + \lambda^2 \psi_n^{(2)} + \ldots + \lambda^k \psi_n^{(k)} + \ldots, \tag{5}$$

$$E_n = E_n^{(0)} + \lambda E_n^{(1)} + \lambda^2 E_n^{(2)} + \ldots + \lambda^k E_n^{(k)} + \ldots, \tag{6}$$

where

$$\psi_n^{(0)} = \varphi_n^0 \text{ and } E_n^{(0)} = E_n^0 \text{ are the unperturbed values,} \tag{7}$$

and $\psi_n^{(1)} = \sum_{m \neq n} \dfrac{\langle \psi_m^{(0)} | \hat{H}_1 | \psi_n^{(0)} \rangle}{E_n^{(0)} - E_m^{(0)}} \psi_m^{(0)}$ and

$E_n^{(1)} = \langle \psi_n^{(0)} | \hat{H}_1 | \psi_n^{(0)} \rangle$ are the first-order corrections. $\tag{8}$

Explicit expressions for the higher order corrections can be found in Ref. 14 and elsewhere in literature. In Eqs. (5) to (8) the arguments of the functions ψ_n, $\psi_n^{(1)}$, *etc.*, were dropped for convenience, $\langle \psi_m^{(0)} | \hat{H}_1 | \psi_n^{(0)} \rangle \equiv \int d\vec{r}\, \psi_m^{(0)*} \hat{H}_1 \psi_n^{(0)}$, and the integration is over the entire $3N$-dimensional coordinate space of the system of N quantum particles. To represent the first-order corrections in the above form, the functions $\psi_n^{(0)}$ are chosen to be normalized, $\langle \psi_n^{(0)} | \psi_n^{(0)} \rangle \equiv \int d\vec{r}\, \psi_n^{(0)*} \psi_n^{(0)} = 1$, and ψ_n not normalized and satisfying the condition $\langle \psi_n^{(0)} | \psi_n \rangle = 1$. With this choice all corrections $\psi_n^{(k)}$, $k = 1, 2, \ldots$, are orthogonal to the corresponding unperturbed functions $\psi_n^{(0)} = \varphi_n^0$, and all resulting formulae for such corrections become significantly simplified. This approach is called Rayleigh-Schödinger perturbation theory, and is one of the most convenient among many other representations of perturbation theory.

The corresponding variation-perturbation method based on the Rayleigh-Schödinger perturbation theory deals with accurate evaluation of the second order and higher-order corrections to the ground state energy E_{GS} and excited state energies without computing the sums similar to that in Eq. (8). Instead, the method evaluates such correction using inequalities [15] similar to the one below written for the ground state energy correction of the second order:

$$\langle \psi_{GS}^{(1)} | \hat{H}_0 - E_{GS}^{(0)} | \psi_{GS}^{(1)} \rangle + \langle \psi_{GS}^{(1)} | \hat{H}_1 - E_{GS}^{(1)} | \Psi_{GS}^{(0)} \rangle + \langle \Psi_{GS}^{(0)} | \hat{H}_1 - E_{GS}^{(1)} | \psi_{GS}^{(1)} \rangle \geq E_{GS}^{(2)}, \tag{9}$$

where the function $\psi_{GS}^{(1)}$ is any square-integrable function satisfying the same boundary conditions as those of the ground state wavefunction Ψ_{GS}. [It can be shown that such a function is the first-order correction to Ψ_{GS}.] Once the correction $E_{GS}^{(2)}$ to the ground state energy has been determined, the function $\psi_{GS}^{(1)}$ is further used to calculate $E_{GS}^{(3)}$, and so on.

2.2.3 Perturbation theory treatment of degenerate energy levels

In the case of an unperturbed, m-fold degenerate energy level of the energy $E_D^{(0)}$ there exist m eigenfunctions $\varphi_{1D}^{(0)}, \varphi_{2D}^{(0)}, \ldots, \varphi_{mD}^{(0)}$ of the unperturbed Hamiltonian satisfying the unperturbed Schrödinger equation $\hat{H}_0 \psi_D^{(0)} = E_D^{(0)} \psi_D^{(0)}$ with the same eigenvalue $E_D^{(0)} = E_{1D}^{(0)} = E_{2D}^{(0)} = \ldots = E_{mD}^{(0)}$. The corresponding perturbed Schrödinger equation

$$\hat{H}\psi_D \equiv (\hat{H}_0 + \lambda\hat{H}_1)\psi_D = E_D \psi_D \tag{10}$$

must reduce to the unperturbed one when $\lambda \to 0$, so that $\lim_{\lambda \to 0}(E_D) = E_D^{(0)}$, but ψ_D may or may not approach one of the unperturbed eigenfunctions $\varphi_{1D}^{(0)}, \varphi_{2D}^{(0)}, \ldots, \varphi_{mD}^{(0)}$. Instead, it may approach any of infinite number of linearly independent combinations of m normalized functions $\varphi_{1D}^{(0)}, \varphi_{2D}^{(0)}, \ldots, \varphi_{mD}^{(0)}$,

$$\lim_{\lambda \to 0} \psi_D \equiv \sum_1^m c_k \varphi_{kD}^{(0)}, \ 1 \le k \le m. \tag{11}$$

The wavefunction $\psi_D^{(0)}$ satisfying Eq. (11) is called the zero-order wave function for the perturbation $\hat{H}_1$. Notably, perturbation may or may not lift degeneracy, or lift it only partially. The rest of the perturbation-theoretical treatment of a degenerate energy level follows a route similar to that of non-degenerate one with necessary modifications of Eqs. (5) and (7). Thus, for the m-fold degenerate energy level number n these equations take the form

$$\psi_{nD} = \psi_{nD}^{(0)} + \lambda \psi_{nD}^{(1)} + \lambda^2 \psi_{nD}^{(2)} + \ldots + \lambda^k \psi_{nD}^{(k)} + \ldots, \tag{12}$$

and $E_n = E_D^{(0)} + \lambda E_n^{(1)} + \lambda^2 E_n^{(2)} + \ldots + \lambda^k E_n^{(k)} + \ldots,$ (13)

where $\psi_{nD}^{(0)}$ satisfies Eq. (11). Similar to the non-degenerate case, these expressions have to be substituted into the perturbed Schrödinger equation (10) (where ψ_{nD} replaces ψ_D), and the coefficients at the like powers of λ equated. Choosing the zero-th order wavefunctions $\varphi_{1n}^{(0)}, \varphi_{2n}^{(0)}, \ldots, \varphi_{mn}^{(0)}$ orthonormal, one can derive a system of linear homogeneous algebraic equations for the first-order correction $E_n^{(1)}$ that include unknown coefficients c_k entering due to Eq. (11) for $\psi_{nD}^{(0)}$. This system of equations possesses a non-trivial solution if and only the determinant of its coefficients is equal to zero,

$$\begin{vmatrix} H_1^{11} - E_n^{(1)} & H_1^{12} & \cdots & H_1^{1m} \\ H_1^{21} & H_1^{22} - E_n^{(1)} & \cdots & H_1^{2m} \\ \cdots & \cdots & \cdots & \cdots \\ H_1^{m1} & H_1^{m2} & \cdots & H_1^{mm} - E_n^{(1)} \end{vmatrix} = 0, \tag{14}$$

where $H_1^{ik} \equiv \int d\vec{r}\, \varphi_{iD}^{(0)*} \hat{H}_1 \varphi_{kD}^{(0)}, i,k = 1, \ldots, m$ are matrix elements of the perturbation Hamiltonian calculated with regard to the eigenfunctions of the unperturbed Hamiltonian corresponding to the degenerate energy level $E_D^{(0)}$. The secular equation, Eq. (14), is an algebraic equation of the order m in $E_n^{(1)}$, and thus has m roots $E_{1n}^{(1)}, E_{2n}^{(1)}, \ldots, E_{mn}^{(1)}$. If all these roots are different, then the degeneracy has been removed in the first order, and the m-fold degenerate energy level split into m perturbed energy levels. If several (or all) roots are equal, then the degeneracy is not removed in the first order, and one has to perform a procedure similar to the above one to remove the degeneracy in the second order, and so on. For simplicity, in what follows below in this sub-section, the degeneracy is considered removed in the first order, so that m roots $E_{1n}^{(1)}, E_{2n}^{(1)}, \ldots, E_{mn}^{(1)}$ are all different. These roots are further substituted into the original system of algebraic equations [that produced the secular Eq. (14)] to determine the coefficients c_k in Eq. (11) written for $\psi_n^{(0)}$. Once the first-order corrections $E_{1n}^{(1)}, E_{2n}^{(1)}, \ldots, E_{mn}^{(1)}$ and $\psi_{kD}^{(0)}$, $k = 1, \ldots, m$, are found for each of the first-order perturbed energy levels $E_D^{(0)} + E_{1n}^{(1)}, E_D^{(0)} + E_{2n}^{(1)}, \ldots, E_D^{(0)} + E_{mn}^{(1)}$, one can determine the first-order corrections to the corresponding eigenfinctions, and the second-order correction to the energy of each of the m energy levels using the procedure outlined in the above sub-section 2.2.2 for a non-degenerate energy level and described in detail in Refs. 14 (volume 1) and 15.

The secular Eq. (14) can be significantly simplified if some or all of the off-diagonal elements in the determinant on the left are zero. In the latter case this equation takes the form

$$\left(H_1^{11} - E_n^{(1)}\right)\left(H_1^{22} - E_n^{(1)}\right)\cdots\left(H_1^{mm} - E_n^{(1)}\right) = 0, \tag{15}$$

so that its solutions are $E_{1n}^{(1)} = H_1^{11}, E_{2n}^{(1)} = H_1^{22}, \ldots, E_{mn}^{(1)} = H_1^{mm}$. \tag{16}

In this case the system of equations for the coefficients c_k in the linear combination in the r.h.s. of Eq. (11) has the trivial solution:

$$c_1 = 1, c_k = 0, k = 2, 3, \ldots, m, \tag{17}$$

where the coefficient $c_1 = 1$ was found from the normalization condition applied to the function $\psi_{1D}^{(0)}$. Therefore, the zero-order function for the perturbation H_1^{11} is $\psi_{1D}^{(0)} = \varphi_{1D}^{(0)}$. A similar consideration can be applied to the rest of the linear combinations

ψ_{kD}, $k = 2, \ldots, m$, to find that all zero-order functions $\psi_{kD}^{(0)}$'s corresponding to the perturbations H_1^{kk}'s, respectively, are equal to the corresponding functions $\varphi_{kD}^{(0)}$: $\psi_{kD}^{(0)} = \varphi_{kD}^{(0)}$, $k = 2, \ldots, m$. That is, the diagonality of the secular determinant means that the "guess" functions $\varphi_{kD}^{(0)}$ in the r.h.s. of Eq. (11) are correct zero-order wavefunctions for the perturbation $\hat{H}_1$. The converse statement is also correct.

Certainly, for a realistic system it is unlikely, that all of the off-diagonal terms in the secular Eq. (14) are zero, and thus the zero-order perturbed wavefunctions of degenerate energy levels are not so directly related to the unperturbed wavefunctions. Much more often the perturbation can be written in such a form that the secular determinant of Eq. (14) becomes block-diagonal, and the secular equation splits in a set of several secular equations that have to be satisfied simultaneously. In this case it is easy to prove that the zero-order wavefunctions $\psi_{kD}^{(0)}$ corresponding to a degenerate energy level of the unperturbed Hamiltonian can be expressed as relatively simple linear combinations of the wavefunctions $\varphi_{kD}^{(0)}$ corresponding to that energy level. The above described treatment of a degenerate energy level of the unperturbed Hamiltonian can be immediately extended to include excited states of a system. Once again, perturbation may or may not fully lift the degeneracy. In the letter case, one has to precede with determination of the next order corrections to the energies and wavefunctions to make sure that the nature of the studied energy level is properly understood. As demonstrated in Chapters 3 to 8 of this book, many molecules possess degenerate ground and excited states. In fact, for some nanosystems possessing higher spatial symmetry a degenerate ground state is much more common than a non-degenerate one.

2.2.4 Spin components of wavefunctions and the Slater determinants

First-principle-based computational design of novel materials usually deals with strongly correlated electron systems, such as those in molecules, quantum dots and wires, and other condensed matter systems. Such systems are described by wavefunctions that fully specify the state of electrons, and therefore depend not only on coordinates of electrons, but also on electron spins. At the same time, the Hamiltonian of a single electron in the absence of magnetic fields can be well-approximated by operators that act only on coordinates and do not act on spin variables of the electron. This means that the eigenfunctions of the single-electron Hamiltonian can be well-approximated by a product of two functions, one of which, $\psi(\vec{r})$, depends on coordinates and is called a one-electron spatial orbital (or an electron orbit), and the other, $\chi(m_s)$, on the spin of the electron, and is called a one-electron spinor: $\Phi(\vec{r}, m_s) = \psi(\vec{r})\chi(m_s)$, with m_s denoting the z-component of the electron spin equal to $1/2$ or $-1/2$. Thus, for a single electron each electron energy state E_n is doubly degenerate. The wavefunction $\Phi_n(\vec{r}, m_s) = \left[\psi(\vec{r})\chi(m_s)\right]_n$, where n is an integer denoting the energy state, is called a spin-orbital or spin-orbit. Because m_s takes only two values, the function $\chi(m_s)$ is usually denoted with α for $m_s = 1/2$ (or spin up), and β for $m_s = -1/2$ (or spin down). To shorten notations, the spin-orbit

$\Phi_n(\vec{r},m_s)$ is often written as $\Phi_n(\vec{r},m_s) = n\psi(\vec{r})$ when $\chi(m_s) = \alpha$ (i.e., $m_s = 1/2$, or the spin is up) and $\Phi_n(\vec{r},m_s) = n\overline{\psi}(\vec{r})$ (with the bar over the spatial function) when $\chi(m_s) = \beta$ (i.e., $m_s = -1/2$, or the spin is down). [Notably, n here is a notation, not a factor.] To skip the coordinate notation $\vec{r}$ entirely, one simply places the particle number (k, in the example below) as the argument of both the space function and spinor in the spin-orbit: $\Phi_n(k) = n\psi(k)\chi(k)$.

Taking into account the above representation of a single-electron wavefunction, one can expect that to a good approximation a wavefunction of a system of electrons would be a product of two functions one of which depends on coordinates of electrons, and the other on their spins. Due to fermionic nature of electrons manifested by the Pauli antisymmetry requirement, a wavefunction of a system of electrons must also be antisymmetric with regard to interchange of any two electrons. This latter requirement can be satisfied in a general case if a wavefunction of a system of N electrons in the energy states 1 to N is written in the form of the Slater determinant:

$$\Psi(\vec{r}_1,\ldots,\vec{r}_N, m_s^1,\ldots,m_s^N) = \frac{1}{\sqrt{N!}} \begin{vmatrix} \Phi_1(1) & \Phi_2(1) & \cdots & \Phi_N(1) \\ \Phi_1(2) & \Phi_2(2) & \cdots & \Phi_N(2) \\ \cdots & \cdots & \cdots & \cdots \\ \Phi_1(N) & \Phi_2(N) & \cdots & \Phi_N(N) \end{vmatrix}, \quad (18)$$

where $\Phi_i(K)$ denotes the K-th electron ($K = 1, \ldots, N$) in the spin-orbit corresponding to the i-th energy state ($i = 1, \ldots, N$), and where the factor $1/\sqrt{N!}$ is due to normalization of the function $\Psi(\vec{r}_1,\ldots,\vec{r}_N, m_s^1,\ldots,m_s^N)$. Several examples of Slater determinants, and further details regarding perturbation and variation treatments of the ground state of simple atoms can be found in Chapter 10 of Ref. 1.

2.2.5 Variation modification of the Slater determinants

The zeroth-order wavefunctions usually are Slater determinants of the type reflected by Eq. (18) where hydrogen-like wavefunctions with fitting parameters are used as the one-electron spatial orbits $\psi(k)$. For example, for the ground state of lithium hydrogen-like electron orbits for 1s and 2s states take the form

$$\psi_{1s}(r) = \frac{1}{\sqrt{\pi}} \left(\frac{b_{1s}}{a_0}\right)^{3/2} e^{-b_{1s} r / a_0}, \quad (19)$$

$$\psi_{2s}(r) = \frac{1}{4\sqrt{2\pi}} \left(\frac{b_{2s}}{a_0}\right)^{3/2} \left(2 - \frac{b_{2s} r}{a_0}\right) e^{-b_{2s} r / 2 a_0}, \quad (20)$$

where a_0 is the Bohr radius, $a_0 = \dfrac{\hbar^2}{m_e e^2}$, with m_e and e being the electron mass and charge, respectively, and b_{1s} and b_{2s} are the variational parameters. Then the zeroth-order trial wavefunction for this state is approximated as

$$\Psi(r_1, r_2, r_3) = \frac{1}{\sqrt{6!}} \begin{vmatrix} \psi_{1s}(r_1)\alpha(1) & \psi_{1s}(r_1)\beta(1) & \psi_{2s}(r_1)\alpha(1) \\ \psi_{1s}(r_2)\alpha(2) & \psi_{1s}(r_2)\beta(2) & \psi_{2s}(r_2)\alpha(2) \\ \psi_{1s}(r_3)\alpha(3) & \psi_{1s}(r_3)\beta(3) & \psi_{2s}(r_3)\alpha(3) \end{vmatrix}. \qquad (21)$$

Naturally, this function is not normalized. The best values of the variation parameters b_{1s} and b_{2s} are those that minimize the variation integral of Eq. (1):

$$\frac{\partial \left(\int d\vec{r}\, \Psi^*(r_1, r_2, r_3) \hat{H} \Psi(r_1, r_2, r_3) \right)}{\partial b_{1s}} = 0,$$

$$\frac{\partial \left(\int d\vec{r}\, \Psi^*(r_1, r_2, r_3) \hat{H} \Psi(r_1, r_2, r_3) \right)}{\partial b_{2s}} = 0,$$

and they happened to be $b_{1s} = 2.686$ and $b_{2s} = 1.776$ [16]. Notably, the use of the variation theorem with the trial function of Eq. (21) based on one-electron spin-orbits allows finding only the upper bound for the energy of the ground state. The true value of the ground state energy can be calculated more accurately using further modifications of the trial wavefunction. In particular, linear combinations of several Slater determinants are used to account more precisely for configuration interactions of electrons, thus obtaining the ground state energy value approximations very close to its true value.

2.3 The Hartree-Fock self-consistent field method

2.3.1 The Hartree self-consistent field method

The simplest Hamiltonian of an n-electron atom accounts for inter-electronic repulsion and has the form

$$\hat{H} = -\frac{\hbar^2}{2m_e} \sum_{i=1}^{n} \nabla_i^2 - \sum_{i=1}^{n} \frac{Ze^2}{r_i} + \sum_{i,j=1; i \neq j}^{n} \frac{e^2}{r_{ij}}, \qquad (22)$$

where ∇_i is the gradient operator acting on coordinates of the i-th electron whose distance from the nucleus is r_i, the point-wise nucleus has the positive charge Ze, and the separation of i-th and j-th electrons is r_{ij}. Spin-orbit, spin-spin and other possible interactions are weaker than those included in the Hamiltonian of Eq. (22), and thus omitted in the above approximation. Due to the last term in Eq. (22) the wavefunction of an atom is not separated in a product of one-electron hydrogen-like orbits. However, such a product can still be used as the zeroth-order approximation for the spatial part of the wavefunction:

$$\psi^{(0)}(\vec{r}_1, \vec{r}_2, \ldots, \vec{r}_n) = \psi_1(\vec{r}_1)\psi_2(\vec{r}_2)\cdots\psi_n(\vec{r}_n), \tag{23}$$

where similar to hydrogen-like orbits, each of one-electron wavefunctions $\psi_i(\vec{r}_i)$, $i = 1,\ldots, n$, is a product of a radial factor $R_{kl}(r_i)$ and a spherical harmonic $Y_{l_i}^m(\theta_i, \varphi_i)$, and where r_i and θ_i, φ_i are the radial and angular coordinates of the i-th electron, respectively, with the origin centered at the nucleus,

$$\psi_i(\vec{r}_i) = R_{kl}(r_i) Y_{l_i}^m(\theta_i, \varphi_i) \tag{24}$$

The explicit expression for the radial factor is

$$R_{kl}(r) = r^l e^{-Zr/ka} \sum_{j=0}^{k-l-1} b_j r^j, \quad b_{j+1} = \frac{2Z}{ka}\frac{j+l+1-k}{(j+1)(j+2l+2)}, \quad a \equiv \frac{\hbar^2}{\mu e^2}, \quad \mu \equiv \frac{m_e M_N}{m_e + M_N}, \tag{25}$$

with M_N being the nucleus mass, and the spherical harmonics are

$$Y_l^m(\theta, \varphi) = \left[\frac{2l+1}{4\pi}\frac{(l-|m|)!}{(l+|m|)!}\right]^{1/2} P_l^{|m|}(\cos\theta)\, e^{im\varphi}, \tag{26}$$

where $P_l^{|m|}(\cos\theta)$ is the normalized associated Legendre polynomials. According to the Pauli exclusion principle, in the ground state each of the lowest orbits is occupied by two electrons with opposite spins. The trial function of Eq. (23) is qualitatively correct, but it does not include screening of an electron by the presence of other electrons, so it fails quantitatively. Thus, it has to be replaced by a function

$$\psi(\vec{r}_1, \vec{r}_2, \ldots, \vec{r}_n) = \Phi_1(\vec{r}_1)\Phi_2(\vec{r}_2)\cdots\Phi_n(\vec{r}_n), \tag{27}$$

where each of the functions $\Phi_i(\vec{r}_i)$, $i = 1, \ldots, n$, has the angular dependence described by an appropriate spherical harmonic of Eq. (26), but the radial dependence modified by (unknown at this step) function $h_i(r_i)$,

$$\Phi_i(\vec{r}_i) = h_i(r_i) Y_l^m(\theta_i, \varphi_i). \tag{28}$$

Notably, the trial function of Eq. (28), as well as that of Eq. (27), is not normalized.

In 1928 Hartree introduced a self-consistent field (SCF) method that allows calculation of the functions $h_i(r_i)$ based on the variation theorem of Eq. (1) with two major simplifying assumptions. Firstly, the electron is considered as moving in a charge cloud composed of other electrons. Secondly, *the central-field assumption* replaces the actual potential of the field created by the nucleus and all other electrons in the atom in which a given electron moves by an effective potential that depends only on the radial coordinate of the electron. With these two assumptions the variation theorem of Eq. (1), written for the trial functions of Eq. (28), reduces to the following expression for the total energy E of the atom:

$$E = \sum_{i=1}^{n} \varepsilon_i - \sum_{i=1}^{n-1} \sum_{j=i+1}^{n} J_{ij}, \quad J_{ij} \equiv \int\int \frac{e^2}{r_{ij}} |\Phi_i(\vec{r}_i)|^2 |\Phi_j(\vec{r}_j)|^2 d\vec{r}_i d\vec{r}_j, \qquad (29)$$

where ε_i is the value of the energy of the i-th electron in the central-field potential

$$V_i(r_i) = \frac{1}{4\pi} \int_0^\pi d\theta_i \int_0^{2\pi} d\varphi_i V_i(r_i,\theta_i,\varphi_i), \quad V_i(r_i,\theta_i,\varphi_i) \equiv \sum_{j\neq i} \int\int \frac{e^2}{r_{ij}} |\Phi_j(\vec{r}_j)|^2 d\vec{r}_i d\vec{r}_j - \frac{Ze^2}{r_i}, \qquad (30)$$

corresponding to the radial orbits $\xi_i(r_i)$ found from a one-electron Schrödinger equation

$$\left[-\frac{\hbar^2}{2m_e} \nabla_i^2 + V_i(r_i) \right] \xi_i(\vec{r}_i) = \varepsilon_i \xi_i(\vec{r}_i). \qquad (31)$$

The one-electron orbits $\xi_i(r_i)$ are considered in the Hartree procedure as improved electron orbits at the next stage of the approximation. Due to spherical symmetry of the potential $V_i(r_i)$ these orbits split into products of the radial and angular parts, where the radial factors $R(\vec{r}_i)$ are the solutions of the one-dimensional Schrödinger equation with the center-symmetric potential, and the angular parts are spherical harmonics of Eq. (26) involving quantum numbers l and m. Finding the solutions of the above Schrödinger equation in the boundaries $r = 0$ and $r \to \infty$, one obtains a set of the radial factors $R(\vec{r}_i)$'s with the integer number of nodes k inside the boundaries that begin from zero for the lowest energy and increases by 1 thereafter. This allows introduction of the total energy quantum number $n = l + 1 + k$, with $k = 0, 1, 2, \ldots$, and thus orbits 1s, 2s, 2p, etc., similar to those of the hydrogen-like atom, with the same number $n - l - 1$ of radial nodes. At the same time, the set of the radial factors $R(\vec{r}_i)$'s differ from the hydrogen-like functions, because the potential $V_i(r_i)$ is not the Coulomb potential.

Using the Hartree orbits of Eq. (28) and the spherical harmonic addition theorem, one can calculate the Hartree probability densities of electrons in the filled (n,l)-th subshell that provides the major equation to ensure self-consistency of calculations of the Hartee's method:

$$2 \sum_{m=-l}^{l} |h_{n,l}(r)|^2 |Y_l^m(\theta,\varphi)|^2 = 2 |h_{n,l}(r)|^2 \frac{(2l+1)}{4\pi}, \qquad (32)$$

where $h_{n,l}(r)$ denotes the Hartree radial function $h_i(r)$ of an electron in the (n,l)-th subshell. Note, that for a half-filled subshell the factor 2 in Eq. (32) must be omitted. The Hartree orbits of Eq. (28) account for the fact that there are no more than two electrons in every spatial orbit via the radial functions $h_{n,l}(r)$ that must satisfy the self-consistency condition of Eq. (32).

The Hartree wavefunction of Eq. (27) possesses a significant shortcoming, because it does not contain any explicit dependence on the electron spins. In 1930 Fock suggested to use antisymmetrized spin-orbits in the form of the Slatter determinants instead of the Hartree wavefunctions of Eq. (27) to alleviate for this shortcoming. Thus, one starts from the trial Slater determinant of Eq. (18) built of one-electron spin orbits $\Phi_i(k) = i\psi(k)\chi(k)$, where $\psi(k)$ is a spatial orbit and $\chi(k)$ is the corresponding spinor specific to the k-th electron in the i-th energy state, and then uses the Hartree's SCF procedure to obtain the ground state energy from a generalization of Eq. (29), where the one-electron energies ε_i's and spin-orbits now satisfy the Hartree-Fock one-electron equation that replaces Eq. (31). The major steps of the Hartree-Fock (HF) method are outlined as applied to molecules (a reduction of this procedure for atoms is straightforward) in the following subsection 2.3.2.

2.3.2 The Hartree-Fock SCF method for molecules

Neglecting small spin-orbit and relativistic interactions, the Hamiltonian specific to a molecule that consists of Γ multielectron nuclei and has totally N electrons can be written in the following form:

$$\hat{H} = -\frac{\hbar^2}{2}\sum_{\gamma=1}^{\Gamma}\frac{1}{m_\gamma}\nabla_\gamma^2 - \frac{\hbar^2}{2m_e}\sum_{i=1}^{N}\nabla_i^2 + \sum_{\gamma=1}^{\Gamma-1}\sum_{\beta>\gamma}^{\Gamma}\frac{Z_\gamma Z_\beta e^2}{r_{\gamma\beta}} - \sum_{\gamma}^{\Gamma}\sum_{i=1}^{N}\frac{Z_\gamma e^2}{r_{i\gamma}} + \sum_{j=1}^{N-1}\sum_{i>j}\frac{e^2}{r_{ij}}, \quad (33)$$

where Latin subscripts refer to electrons and Greek subscripts to nuclei, m_γ and m_e denote the masses of the γ-th nuclear and the i-th electron, respectively, ∇_γ and ∇_i are the gradient operators acting on the coordinates of the γ-th nuclear and the i-th electron, respectively, $r_{\gamma\beta}$ is the distance between the nuclei with the atomic numbers Z_γ and Z_β, $r_{i\gamma}$ is the distance between the i-th electron and the γ-th nuclear, and r_{ij} is the distance between the i-th and j-th electrons. Because the nuclei are several orders of magnitude more massive than electrons and therefore, their motion is much slower than that of electrons, the term accounting for the kinetic energy of the nuclei can be neglected in the Hamiltonian of Eq. (33). Further on, separating the so-called purely electronic part $\hat{H}_e$ of the Hamiltonian in Eq. (33),

$$\hat{H}_e \equiv -\frac{\hbar^2}{2m_e}\sum_{i=1}^{N}\nabla_i^2 - \sum_{\gamma}^{\Gamma}\sum_{i=1}^{N}\frac{Z_\gamma e^2}{r_{i\gamma}} + \sum_{j=1}^{N-1}\sum_{i>j}\frac{e^2}{r_{ij}}, \quad (34)$$

and the nuclear repulsion term V_{nuc},

$$V_{nuc} = \sum_{\gamma}^{\Gamma-1} \sum_{\beta>\gamma} \frac{Z_\gamma Z_\beta e^2}{r_{\gamma\beta}}, \tag{35}$$

one can write the Schrödinger equation for electrons in a molecule in the form

$$\left(\hat{H}_e + V_{nuc}\right)\psi_e(\vec{r},\vec{R}) = W\psi_e(\vec{r},\vec{R}). \tag{36}$$

The electronic wavefunctions $\psi_{e,n}(\vec{r},\vec{R})$ and energies $W_n(\vec{R})$ satisfying Eq. (36) correspond to the set of the electronic quantum numbers (symbolized by the subscript n) and depend parametrically on the nuclear configuration (symbolized by the notation $\vec{R}$). The wavefunctions $\psi_{e,n}(\vec{r},\vec{R})$ also are functions of coordinates of all electrons (symbolized by the notation $\vec{r}$) that are variables in Eq. (36). The eigenvalues $W_n(\vec{R})$ are electronic energies that include internuclear repulsion and do not depend on the electron coordinates. The nuclear repulsion term V_{nuc} of Eq. (35) is constant for any given nuclear configuration, and thus does not affect the eigenfunctions $\psi_{e,n}(\vec{r},\vec{R})$ of the Schrödinger equation (36), while decreasing the eigenvalues by the value of V_{nuc}. Thus, the term V_{nuc} in Eq. (33) can be omitted to reduce Eq. (36) to the form

$$\hat{H}_e \psi_e(\vec{r},\vec{R}) = E_e \psi_e(\vec{r},\vec{R}), \tag{37}$$

where $E_e = W - V_{nuc}$. \hfill (38)

The electronic energies E_e do not include the internuclear repulsion energy, but depend parametrically on the nuclear configuration $\vec{R}$ and the set of the electron quantum numbers n, $E_e \equiv E_{e,n}(\vec{R})$.

In the case of diatomic molecules where the ratio of the electron to nuclear mass satisfies the inequality $(m_e/m_\gamma)^{1/4} \ll 1$ the electronic and nuclear motion can be separated. In this case the true wavefunctions $\psi(\vec{r},\vec{R})$ of a molecule in its ground state can be approximated with a good accuracy as a product of the electronic wavefunction $\psi_{e,n}(\vec{r},\vec{R})$ and the nuclear one, $\psi_{nuc,n}(\vec{R})$, where the latter depends only on the nuclear configuration:

$$\psi(\vec{r},\vec{R}) = \psi_{e,n}(\vec{r},\vec{R})\psi_{nuc,n}(\vec{R}). \tag{39}$$

This approximation for the wavefucntion is called the *Born-Oppenheimer approximation* [17]. While this approximation is quite adequate for the great majority of diatomic molecules in their ground states, it is less accurate for their excited states. In the case of many-atomic molecules this approximation is even less accurate, but errors introduced by this approximation still are much smaller than those due to other approximations used to solve the electronic Schrödinger equation (36). Thus, the majority of quantum chemistry calculations use the Born-Oppenheimer approximation for molecular wavefunctions.

In the Hartree-Fock approximation the value of the molecular electronic energy E_{HF} is found from the variation theorem used with the Slater determinant electron wavefunctions of Eq. (39) and the Hamiltonian defined by Eqs. (34) to (36):

$$E_{HF} = \int d\vec{r} \int d\vec{R}\, \psi^*_{e,n}(\vec{r},\vec{R})\left(\hat{H}_e + V_{nuc}\right)\psi_{e,n}(\vec{r},\vec{R}) \equiv \langle \psi_{e,n} | \hat{H}_e + V_{nuc} | \psi_{e,n} \rangle. \quad (40)$$

In the last expression in the r.h.s. of Eq. (40) Dirac's brackets are used to denote the inner product. For convenience of further calculations, the Slater determinant wavefunction $\psi_{e,n}(\vec{r},\vec{R})$ is normalized before it is used in Eq. (40). The wavefunction $\psi_{e,n}(\vec{r},\vec{R})$ depends on the nuclear coordinates only parametrically (that is, inexplicitly), so in what follows the notation $\vec{R}$ is dropped from the list of arguments of this Slater determinant. Thus, in Eq. (40) $\langle \psi_{e,n} | V_{nuc} | \psi_{e,n} \rangle = V_{nuc} \langle \psi_{e,n} | \psi_{e,n} \rangle = V_{nuc}$. The electronic Hamiltonian $\hat{H}_e$ is a sum of one-electron operator

$$\hat{H}^{core}(1) = -\frac{\hbar^2}{2m_e}\nabla_i^2 - \sum_\gamma \frac{Z_\gamma e^2}{r_{i\gamma}} \quad (41)$$

and the two-electron operator $\hat{H}(12) = \dfrac{e^2}{r_{ij}}$, respectively, where the subscript i and j refer to the i-th and j-th electrons. Therefore, Eq. (40) can be written in the form:

$$E_{HF} = \left\langle \psi_{e,n} \left| 2\sum_{i=1}^{N/2}\hat{H}^{core}(1) + \sum_{i=1}^{N-1}\sum_{j>i}\hat{H}(12) \right| \psi_{e,n} \right\rangle + V_{nuc}. \quad (42)$$

Introducing notations

$$H^{core}_{ii} \equiv \langle \psi_i(1) | \hat{H}^{core}(1) | \psi_i(1) \rangle = \left\langle \psi_i(1) \left| -\frac{\hbar^2}{2m_e}\nabla_i^2 - \sum_\gamma \frac{Z_\gamma e^2}{r_{i\gamma}} \right| \psi_i(1) \right\rangle, \quad (43)$$

$$J_{ij} \equiv \left\langle \psi_i(1)\psi_j(2) \left| \frac{e^2}{r_{12}} \right| \psi_i(1)\psi_j(2) \right\rangle, \quad (44)$$

and $K_{ij} \equiv \left\langle \psi_i(1)\psi_j(2) \left| \dfrac{e^2}{r_{12}} \right| \psi_j(1)\psi_i(2) \right\rangle, \quad (45)$

where $\psi_i(k)$ is the one-electron spatial component of a *molecular orbit* (MO) specific to the k-th electron in the i-th quantum state, one can re-write Eq. (42) for polyatomic molecules with only closed shells in the short form:

$$E_{HF} = 2\sum_{i=1}^{N/2} H_{ii}^{core} + \sum_{i=1}^{N/2}\sum_{j=1}^{N/2}\left(2J_{ij} - K_{ij}\right) + V_{nuc}. \tag{46}$$

The set of MOs $\psi_i(k)$ are taken to be orthonormalized: $\langle\psi_i(k)|\psi_j(k)\rangle = \delta_{ij}$, where δ_{ij} is the Kronecker delta. The core Hamiltonian $\hat{H}^{core}(1)$ of Eq. (41) describes the total energy of a single electron in the electric field created by all nuclei, ignoring the presence of all other electrons. The *Coulomb integrals* J_{ij} and *exchange integrals* K_{ij} account for inter-electron repulsion. In particular, the exchange integrals account for exchange effects that are specific to fermions only.

The task of the HF method is to determine such orthonormal MOs $\psi_i(k)$ that minimize the variation integral E_{HF} of Eq. (46). This rather complicated procedure is described in detail in many textbooks, such as Refs. 18–20, and boils down to the fact that closed-shell orthonormal MOs $\psi_i(k)$ satisfy the following equation:

$$\hat{F}(1)\psi_i(1) = \varepsilon_i\psi_i(1), \tag{47}$$

where ε_i is the orbital energy (called an *electronic energy level*) of an electron in the MO $\psi_i(k)$, and the Hartree-Fock operator $\hat{F}(1)$ is defined as follows:

$$\hat{F}(1) = \hat{H}^{core}(1) + \sum_{j=1}^{N/2}\left[2\hat{J}_j(1) - \hat{K}_j(1)\right], \tag{48}$$

with the core Hamiltonian being defined by Eq. (41), where the subscript i is omitted. The *Coulomb operator* $\hat{J}_j(1)$ and *exchange operator* $\hat{K}_j(1)$ satisfy the following equations:

$$\hat{J}_j(1)f(1) = f(1)\int d\vec{r}_2 \frac{e^2}{r_{12}}|\psi_j(2)|^2, \tag{49}$$

$$\hat{K}_j(1)f(1) = \psi_j(1)\int d\vec{r}_2 \frac{e^2}{r_{12}}\psi_j(2)f(2). \tag{50}$$

The function $f(1)$ is an arbitrary function of the coordinates of an electron, and Eqs. (49) and (50) must be satisfied simultaneously for the same functions $f(1)$ and ψ_j, where ψ_j are eigenfunctions of the same HF operator $\hat{F}(1)$. The integration over $\vec{r}_2$ includes the entire coordinate space. The factor 2 at the Coulomb operator in Eq. (48) accounts for 2 electrons in each of the MOs (all electron shells are considered closed). The Hartree-Fock Eq. (47) contains the exchange operator $\hat{K}_j(1)$ that is absent in Hartree's Eq. (29). This operator ensures that only antisymmetric MOs with regard to electron exchange are considered. The HF operator $\hat{F}(1)$ is defined through its own eigenfinctions ψ_j that are included in the integrands in Eqs. (49) and (50). This exceptional property of the HF operator steams from a complex approximation procedure that leads from Eq. (46) featuring two-electron Coulomb and exchange integrals to one-electron Coulomb and exchange operators included in Eqs. (47) and (48). Because the eigenfunctions $\psi_j(1)$ are unknown, the HF equations (47), (49) and (50) must be solved using an iteration process.

The sum of the electron orbital energies ε_i can be obtained by multiplying Eq. (47) by $\psi_j^*(1)$, integrating over the coordinates of the electron 1, and summation over all of the occupied orbitals. The resulting equation is:

$$\sum_{i=1}^{N/2} \varepsilon_i = \sum_{i=1}^{N/2} H_{ii}^{core} + \sum_{i=1}^{N/2}\sum_{j=1}^{N/2} \left(2J_{ij} - K_{ij}\right). \tag{51}$$

Solving this equation for $\sum_{i=1}^{N/2} H_{ii}^{core}$ and substituting the result into Eq. (46) one can express the HF energy E_{HF} in terms of electron energies ε_i :

$$E_{HF} = 2\sum_{i=1}^{N/2} \varepsilon_i - \sum_{i=1}^{N/2}\sum_{j=1}^{N/2} \left(2J_{ij} - K_{ij}\right) + V_{nuc}. \tag{52}$$

To minimize the r.h.s. of Eq. (52) it is convenient to express molecular spatial orbits $\psi_i(1)$ as linear combinations of one-electron basis functions ζ_α (numbered for convenience with Greek letters) that form a complete set:

$$\psi_i(1) = \sum_{\alpha=1}^{\infty} c_{\alpha i} \zeta_\alpha. \tag{53}$$

The infinite sum in Eq. (53) has to be replaced with a finite sum over M basis functions in such a way that MOs $\psi_i(1)$ are represented by such a finite sum with a very small error. Substituting the MOs in the form of Eq. (53) into the HF equation (47), multiplying Eq. (47) by ζ_β^* and integrating over the coordinates of the electron leads to the system of M linear homogeneous equations (known as *the Hartree-Fock-Roothaan equations*) for M unknown coefficients $c_{\alpha i}$:

$$\sum_{\alpha=1}^{M} c_{\alpha i}\left(F_{\beta\alpha} - \varepsilon_i S_{\beta\alpha}\right) = 0, \quad \beta = 1, 2, \dots, M, \tag{54}$$

where $F_{\beta\alpha} \equiv \left\langle \zeta_\beta \left| \hat{F} \right| \zeta_\alpha \right\rangle$ \hfill (55a) are Fock matrix elements, and

$$S_{\beta\alpha} \equiv \langle \zeta_\beta | \zeta_\alpha \rangle \tag{55b}$$

are the *overlap integrals*. The system of Eq. (54) has a non-trivial solution (that is, not all $c_{\alpha i}$ of the set M of the coefficients $c_{\alpha i}$ are equal to zero), if and only if

$$\det(F_{\beta\alpha} - \varepsilon_i S_{\beta\alpha}) = 0. \tag{56}$$

Similar to the HF equation (47), the Hartree-Fock-Roothaan Eq. (54) must be solved via an iteration process, because the HF operator $\hat{F}$ included into $F_{\beta\alpha}$ of Eq. (55a) depends upon $\psi_i(1)$'s that in their turn depend on the coefficients $c_{\alpha i}$'s. In practice, one choses one-electron Slater determinants representing spatial parts of atomic orbits (AOs) as a set of the basis functions ζ_α. This initial guess is used to calculate the matrix elements of Eq. (55a), and to solve Eq. (56) for ε_i's. The one-electron orbital energies ε_i then are used in the HF-Roothaan Eq. (54) to find an improved set of the coefficients $c_{\alpha i}$'s, and the latter are used to improve the linear combinations for the MOs. These improved set of MOs is then used to calculate the HF operator and its matrix elements, and to proceed to the next step of calculations of improved ε_i's. The process ends when the coefficients $c_{\alpha i}$ and the energies ε_i at some step of calculation cycle lie within an acceptable error brackets from their respective values found in the previous cycle of calculations.

Despite a formal requirement that necessitates an infinite number of AOs in a basis set to represent the HF MOs [see Eq. (53)], in practice a good approximation of the MOs can be achieved using a rather small number of appropriate Slater AOs. Notably,

---

a minimal basis set of AOs necessary for the above molecular **self-consistent field (SCF)** calculations is composed of a single basis function for each inner shell AO and each valence-shell AO of each atom in the molecule.

To perform more accurate calculations, extended basis sets much larger than the corresponding minimal basis sets are usually used. The SCF wave functions calculated using the above iteration process constitute approximations to the (unique) Hartree-Fock wavefunction that is written as a Slater determinant of electronic AOs.

2.3.3 The matrix elements of the Fock operator and calculation of physically meaningful quantities

The matrix elements $F_{\beta\alpha} \equiv \langle \zeta_\beta | \hat{F} | \zeta_\alpha \rangle$ of the Fock operator in the basis set $\{\zeta_\gamma\}_{\gamma=1}^{M}$ can be written as follows [1]:

$$\begin{aligned}
F_{\beta\alpha} &= H_{\beta\alpha}^{\text{core}} + \sum_{\gamma=1}^{M}\sum_{\delta=1}^{M}\sum_{j=1}^{N/2} c_{\gamma j}^* c_{\delta j} \left[2\langle \beta\alpha | \gamma\delta \rangle - \langle \beta\delta | \gamma\alpha \rangle \right] \\
&= H_{\beta\alpha}^{\text{core}} + \sum_{\gamma=1}^{M}\sum_{\delta=1}^{M} P_{\gamma\delta} \left[2\langle \beta\alpha | \gamma\delta \rangle - \frac{1}{2}\langle \beta\delta | \gamma\alpha \rangle \right],
\end{aligned} \tag{57}$$

where $\langle \beta\alpha | \gamma\delta \rangle$ are *the two-electron repulsion integrals*,

$$\langle \beta\alpha | \gamma\delta \rangle \equiv \int d\vec{r}_1 \int d\vec{r}_2 \frac{\zeta_\beta^*(1)\zeta_\alpha(1)\zeta_\gamma^*(2)\zeta_\delta(2)}{r_{12}}, \quad (58)$$

and $P_{\gamma\delta} \equiv 2\sum_{j=1}^{N/2} c_{\gamma j}^* c_{\delta j}, \quad \gamma,\delta = 1, 2, \ldots, M$ (59)

are the elements of *the density matrix* of a closed-shell molecule. The electron probability density ρ is directly related to $P_{\gamma\delta}$,

$$\rho \equiv 2\sum_{j=1}^{N/2} \psi_j^* \psi_j = 2\sum_{\alpha=1}^{M}\sum_{\beta=1}^{M}\sum_{j=1}^{N/2} c_{\beta j}^* c_{\alpha j} \zeta_\beta^* \zeta_\alpha = \sum_{\alpha=1}^{M}\sum_{\beta=1}^{M} P_{\beta\alpha} \zeta_\beta^* \zeta_\alpha. \quad (60)$$

Using explicit expressions for the matrix elements of the Fock operator, Eq. (57), and the core Hamiltonian H_{ii}^{core},

$$H_{ii}^{core} \equiv \langle \psi_i(1) | \hat{H}^{core}(1) | \psi_i(1) \rangle = \sum_{\beta=1}^{M}\sum_{\alpha=1}^{M} c_{\beta i}^* c_{\alpha i} \langle \zeta_\beta | \hat{H}^{core}(1) | \zeta_\alpha \rangle \equiv \sum_{\beta=1}^{M}\sum_{\alpha=1}^{M} c_{\beta i}^* c_{\alpha i} H_{\beta\alpha}^{core}, \quad (61)$$

one can derive from the variation theorem of Eq. (52) the following expression for the HF energy E_{HF} [1]:

$$E_{HF} = \sum_{i=1}^{N/2} \varepsilon_i + \frac{1}{2}\sum_{\beta=1}^{M}\sum_{\alpha=1}^{M} P_{\beta\alpha} H_{\beta\alpha}^{core} + V_{nuc}. \quad (62)$$

Expressing the one-electron energies ε_i in terms of the matrix elements of the Fock operator, Eq. (57), one can derive from Eq. (47):

$$\varepsilon_i = \sum_{\beta=1}^{M}\sum_{\alpha=1}^{M} c_{\beta i}^* c_{\alpha i} \langle \zeta_\beta | \hat{F} | \zeta_\alpha \rangle = \sum_{\beta=1}^{M}\sum_{\alpha=1}^{M} c_{\beta i}^* c_{\alpha i} F_{\beta\alpha}, \quad (63)$$

and rewrite Eq. (62) in the form

$$E_{HF} = \frac{1}{2}\sum_{\beta=1}^{M}\sum_{\alpha=1}^{M} P_{\beta\alpha}\left(F_{\beta\alpha} + H_{\beta\alpha}^{core}\right) + V_{nuc}. \quad (64)$$

This expression permits to relate directly the HF energy of a closed-shell molecule to the density and matrix elements of the Fock operator and core Hamiltonian.

For the purposes of numerical calculations, the HF-Roothaan equations can be solved most effectively using methods of matrix algebra. To do so, it's convenient to introduce the $M \times M$ matrix $\mathbf{C}$ composed of the coefficients $c_{\beta i}$, and the (diagonal) matrix $\boldsymbol{\varepsilon}$ of the one-electron orbital energies with the matrix elements $\varepsilon_{\beta i} = \delta_{\beta i}\varepsilon_i$. Using the introduced matrices, the HF-Roothaan equations can be written in the matrix form:

$$\sum_{\alpha=1}^{M} F_{\beta\alpha} c_{\alpha i} = \sum_{\alpha=1}^{M} S_{\beta\alpha} (\mathbf{C}\boldsymbol{\varepsilon})_{\alpha i}, \tag{65}$$

where the matrix elements $F_{\beta\alpha}$ and $S_{\beta\alpha}$ are defined by Eqs. (55a) and (55b). From Eq. (65) it follows that the matrix form of the HF-Roothaan equations is

$$\mathbf{FC} = \mathbf{SC}\boldsymbol{\varepsilon}, \tag{66}$$

where the $M \times M$ matrices $\mathbf{F}$ and $\mathbf{S}$ are composed of the corresponding matrix elements defined by Eqs. (55a) and (55b). A procedure of solving the HF-Roothaan equations is simplified further by using Schmidt's method to orthonormalize the basis set $\{\zeta_\beta\}_{\beta=1}^{M}$. The matrix elements of the matrix $\mathbf{S}$ of Eq. (55b) calculated in the new orthonormal basis $\{\varphi_\beta\}_{\beta=1}^{M}$, $\varphi_\beta = \sum_{\gamma=1}^{M} a_{\beta\alpha} \zeta_\alpha$, form the unit matrix of the order M: $S_{\beta\alpha}(\varphi) = \langle \varphi_\beta | \varphi_\alpha \rangle = \delta_{\beta\alpha}$. Denoting $\mathbf{A}$ the $M \times M$ matrix composed of the coefficients $a_{\beta\alpha}$, one can find the matrix elements of the matrices $\mathbf{F}$ and $\mathbf{C}$ in the new orthonormal basis and compose the corresponding matrices $\mathbf{F}(\varphi)$ and $\mathbf{C}(\varphi)$:

$$\mathbf{F}(\varphi) = \mathbf{A}^+ \mathbf{F} \mathbf{A} \text{ and } \mathbf{C} = \mathbf{A}\mathbf{C}(\varphi), \tag{67}$$

respectively, where $\mathbf{A}^+$ denotes the matrix Hermitian conjugate to the matrix $\mathbf{A}$.

The Fock operator is Hermitian and therefore, both $\mathbf{F}$ and $\mathbf{F}(\varphi)$ matrices are Hermitian. Correspondingly, the eigenvector matrix $\mathbf{C}(\varphi)$ can be chosen to be unitary: $\mathbf{C}^{-1}(\varphi) = \mathbf{C}^+(\varphi)$. With this choice of $\mathbf{C}(\varphi)$, the matrix HF-Roothaan equations take the form:

$$\mathbf{C}^+(\varphi)\mathbf{F}(\varphi)\mathbf{C}(\varphi) = \boldsymbol{\varepsilon}, \tag{68}$$

where the matrix $\boldsymbol{\varepsilon}$ of the one-electron orbital energies remains the same, as it does not depend on the basis functions. All calculation algorithms realized as computer codes use this form of the HF-Roothaan equations and apply matrix algebra to solve it efficiently.

Using Eq. (59) one can prove that the density matrix can be written in the form

$$\mathbf{P} = 2\mathbf{C}\mathbf{C}(\varphi), \tag{69}$$

that is used in numerical calculations.

An approximate calculation of the HF wavefunction, HF energy and one-electron HF orbital energies begins with guessing the initial eigenvectors of the Fock operator of Eq. (53), that is, choosing appropriate basis functions and coefficients $c_{\beta i}$, and using them to calculate the matrix element of the "guess" density matrix $\mathbf{P}_0$ from Eq. (59), those of the core Hamiltonian from Eq. (61), the two-electron integrals from Eq. (58), and finally, the matrix elements of the Fock matrix from Eq. (57). Then the initial basis functions are orthonormalized and the matrix $\mathbf{A}$ obtained. This matrix is further used to calculate the matrix elements of the Fock matrix $\mathbf{F}(\varphi)$ and those of the eigenvector matrix $\mathbf{C}(\varphi)$ in the new basis, and to calculate the one-electron energies using Eq. (68). Once this is done, an improved eigenvector matrix C_{impr} is calculated with regard to the initial basis functions using the equation $C_{impr} = \mathbf{A}\mathbf{C}(\varphi)$. Then this improved matrix C_{impr} is used to calculate the corresponding improved density matrix $\mathbf{P}_{impr}$ using Eq. (69). Once calculated, the matrix elements of the improved density matrix $\mathbf{P}_{impr}$ are compared with those of the initially computed "guess" density matrix $\mathbf{P}_0$. If the values of the matrix elements of $\mathbf{P}_{impr}$ differ from those of the corresponding matrix elements of $\mathbf{P}_0$ less than initially introduced small quantities (or the error brackets), then the procedure is called converged and calculations stopped. In practice this never occurs after just one cycle of the calculations. That is, after the initial cycle of calculations the matrix elements of $\mathbf{P}_{impr}$ are vastly different from those of $\mathbf{P}_0$, so the procedure is not converged. To improve the results further, the matrix elements of $\mathbf{P}_{impr}$ are used instead of those of $\mathbf{P}_0$, and the calculation cycle is repeated again and again, until the procedure converges. For clarity, the HF SCF calculations procedure is outlined below as a series of steps.

1. Choose the molecule and its geometry (that is, chose the origin and assign positions to the centers of atoms in the molecule).
2. Choose the initial basis set $\{\zeta_\beta\}_{\beta=1}^M$.
3. Calculate the matrix elements $H_{\beta\alpha}^{core}$ from Eq. (61), overlap integrals $S_{\beta\alpha}$ from Eq. (55b), and two-electron repulsion integrals $\langle\beta\alpha|\gamma\delta\rangle$ from Eq. (58).
4. Using Schmidt's procedure, orthogonalize the basis set $\{\zeta_\beta\}_{\beta=1}^M$, thus establishing the basis $\{\varphi_\beta\}_{\beta=1}^M$ and the matrix $\mathbf{A}$.
5. Introduce initial guess coefficients $c_{\alpha i}$, compose the corresponding initial guess MOs $\psi_i(1) = \sum_{\alpha=1}^\infty c_{\alpha i}\zeta_\alpha$ and calculate matrix elements of the density matrix $\mathbf{P}$ from Eq. (59). [This is the "initial" density matrix denoted above as $\mathbf{P}_0$].
6. Calculate the corresponding matrix elements $F_{\beta\alpha}$ of the Fock matrix $\mathbf{F}$ from Eq. (57).
7. Using the matrix $\mathbf{A}$, calculate the Fock matrix $\mathbf{F}(\varphi)$ from the first equation in Eq. (67).
8. Diagonalize the matrix $\mathbf{F}(\varphi)$ and find its eigenvalues (the matrix $\boldsymbol{\varepsilon}$), and calculate the eigenvectors [the matrix $\mathbf{C}(\varphi)$].
9. Calculate the matrix $\mathbf{C}$ using the equation $\mathbf{C} = \mathbf{A}\mathbf{C}(\varphi)$.
10. Calculate the density matrix $\mathbf{P}$ using Eq. (69). [This is the improved density matrix denoted as $\mathbf{P}_{impr}$ above.]
11. Compare the matrix elements of the improved density matrix (calculated in the step 10) with those of the "initial" density matrix elements (calculated in the step 5). If the matrix elements differ by more than allowed quantities [error brackets], go to the step 5 and use the improved density matrix of the step 10 instead of the initial density matrix (of the step 5).

Follow the rest of the steps 6 to 10. Repeat the cycle: step 10 → steps 5 to 9 → step 10 until the difference between the matrix elements of the improved matrix of the step 10 and that of the step 5 falls in the allowed error brackets. At this point the procedure is considered converged, and the density matrix final.

12. Use the final density matrix to calculate physical properties.

It's important to realize that to increase efficiency of HF SCF calculations by orders of magnitude one has to begin the calculations with determination of the equilibrium geometry of a molecule. This can be done by guessing the initial positions of the atomic centers and calculating the density matrix and the Hessian matrix following the above described calculation cycles. The Hessian matrix is composed of the characteristic frequencies specific to a molecule. The absence of any frequencies with imaginary values indicates that the geometry of the molecule is equilibrium (see literature for further details on the Hessian matrix and its properties).

There are many ways to help initial guessing, and thus to speed up SCF computations. In particular, one can begin from semi-phenomenological calculations of the initial density matrix. More often, the initial density matrix of a molecule is constructed using the density matrices of the constitutive atoms. Modern first-principle quantum chemistry software packages already incorporate a great number of choices of the initial basis functions and algorithms of the initial density matrix calculations.

2.4 Configuration interactions

The HF SCF calculations provide a good (although overestimated) evaluation of the total energy of a molecule, but usually grossly overestimated energies of electrons in the occupied MOs, and a rather inaccurate Slater determinant HF wavefunction of even small molecule (sometimes it is represented as several Slater determinants). The HF energies corresponding to unoccupied MOs are calculated with large errors (sometimes up to an order of magnitude in value), and usually are quite unrealistic. In the order of their importance, the major fundamental sources of the HF computational errors include inadequate treatment of electro-electron and electron-nuclei correlations in the framework of the HF method, incompleteness of the basis set of spin-orbits (caused by the basis set truncation) used to calculate HF MOs, neglect or inadequate treatment of relativistic effects, and roughness of the Bohr-Oppenheimer approximation. In addition, the molecular geometry optimization algorithms realized in modern software packages do not ensure that the calculations converge to atomic coordinates corresponding to the true equilibrium configuration of the atoms in the molecule, rather than to a steady state one. Such algorithms provide for calculations of the initial total HF energy E_{HF} of a molecule, and then change slightly one of the coordinates of an atom in the molecule and calculate a new value of E_{HF} corresponding to the new configuration of the atoms. [This procedure is repeated multiple times, and the obtained values of E_{HF} are compared at each step to identify the smallest of them as the HF ground state energy.] There are several reasons for such results of the geometry optimization. In the first place, (1) the "guessed" initial positions of the atoms in the molecule are likely to be far from those corresponding to the equilibrium atomic configuration. Furthermore, (2) in the majority of studied cases to-date, the total energy surfaces of molecules are rather flat in the vicinity of the equilibrium and steady state atomic configurations, so the steepest descent-based search has to be abbreviated to complete the calculations within a realistic computational time allowance. While the virtual design method (discussed later in this chapter) addresses effectively the problem of the optimal initial configuration, none of the existing steepest descent algorithms ensures that the resulting atomic configuration is the equilibrium configuration, rather than a steady state one.

Shortcomings of the HF wavefunction and energy calculations have been addressed since introduction of the HF method at both fundamental and algorithmic levels in the order of their importance. As the major source of such errors, inadequate accounting for electron-electron correlations has been of prime concern. The easiest way to improve the HF approach to the HF wavefunction calculations is to expend the number and type of the basis functions $\{\zeta_\beta\}_{\beta=1}^{M}$ used in calculations of the one-electron MOs $\psi_i(k)$ of Eq. (53). Notably, a specific choice of the dimension M of the basis and the types of functions included in the basis set limit the number of the resulting HF MOs (to M) and their type. The Slater determinant molecular wavefunction is built of such MOs, and therefore is directly affected by errors of MO calculations. To appreciate the influence of the type of chosen basis functions ζ_β, note that if these functions are of s type, one can only obtain σ-type of MOs, while the true MOs can be of π-, δ- and other types. This consideration makes it clear that a choice of the basis set to a large degree defines accuracy of calculations of the values of electron energy levels (translated into the electronic level structure, or ELS), the Slater determinant wavefunction, and the total energy of the molecule.

One can improve the HF approximation immediately introducing configuration interactions (CI) via formation of the Slater determinant wavefunction of a molecule from only those MOs $\psi_i(k)$ of Eq. (53) that possess the same symmetry properties as those of the molecular state under consideration [21]. This approach is based on a basic theorem of operator algebra stating that an eigenfunction Φ of a Hermitian operator $\mathbf{O}$ corresponding to its eigenvalue E can be represented as a linear combination of only those eigenfunctions of $\mathbf{O}$ that correspond to the same eigenvalue E. It can be shown that in the case of a system of N electrons and M basis functions the total number of configuration wavefunctions that satisfy the above symmetry requirements is proportional to M^N. For realistic molecules this is a huge number of MOs, so even for small molecules *full* CI calculations that include into $\Psi_{CI} = \sum_{i=1}^{\approx M^N} a_i \Xi_i$ all possible *configuration state functions* (CSF) Ξ_i (that are Slater determinants) possessing the Ψ-state-matching symmetry cannot be realized within a realistic duration of the computations. Therefore, one has to choose only those CSFs that lead to the largest contributions to the molecular Slater determinant Ψ. The unexcited configuration function (*the HF SCF wavefunction* Ψ_{HF}) is assumed to make the largest contribution to Ψ_{CI}. Contributions from *singly, doubly, triply, etc.* excited configuration functions have to be estimated on the basis of effects produced by instantaneous electron correlations perturbing Ψ_{CI}. [Excites states of a molecule are classified according to the number of electrons excited from occupied to unoccupied, or virtual, spin-orbits.]

As discussed further in this section, only doubly-excited configuration functions make the first-order contributions to Ψ_{CI} of a closed-shell configuration state. At the same time, singly excited configurations contribute significantly to one-electron properties, such as electron charge density or the dipole moment. Thus, being much smaller than the contributions from doubly excited configuration functions, singly excited configuration functions still are included into the CI molecular wavefunction. The second-order contributions to Ψ_{CI} are due to singly, doubly, triply and quadruple-excited configuration functions.

In a general case [22–27], the CI molecular wavefunction Ψ_{CI} is a linear combination of CSFs, each of which is a linear combination of several Slater determinants composed of spin-orbits of single electrons. Each of the CSFs is an eigenfunction of the spin operator $\hat{S}$ and its $\hat{S}_z$ component, and satisfies all symmetry conditions applied to the molecule. For a molecule possessing N electrons in a ground state with the spin quantum number $S = 0$ and all symmetry requirements ignored, and M basis functions used, the number N_{CI} of CSFs necessary for the full CI calculations is proportional to [28]

$$N_{CI} = \frac{M!(M+1)!}{\left(\frac{1}{2}N\right)!\left(\frac{1}{2}N+1\right)!\left(M-\frac{1}{2}N\right)!\left(M-\frac{1}{2}N+1\right)!}. \tag{70}$$

Thus, using a basis of 50 spin-orbits for calculation of molecular wavefunctions of small molecules with the number of electrons below 20 and no symmetry restrictions, one has to calculate about 10^{18} CSFs. While this is feasible at present only due to a boost in supercomputer efficacy achieved in recent years, such computations are impractical, and become even more demanding for larger molecules and larger basis sets. At the same time, the use of large basis sets significantly improves accuracy of calculations of the electron correlation energy. There exists a number of basis sets that have been used for decades, from those that include tens spin-orbits, such as STO-3G basis set for small molecules, to those including hundreds and thousands spin-orbits, such as SBKJC basis set [29], that have to be used in the case of molecules containing atoms up to the fifth row (see Chapters 3 to 8 of this book, for numerous examples).

Full CI (FCI) calculations for large molecules also require large basis sets, and become impractical for molecules containing about 100 many-electron atoms even with the use of advanced supercomputers. Thus, various simplifications of the CI procedure have been developed and used over the years. In many cases, contributions to Ψ_{CI} can be adequately approximated by using only configuration state functions describing excited states of valence shell electrons and those of a few topmost inner shell electrons. This consideration is widely used to develop simplified, or *limited*, CI calculation methods.

The simplest types of CI computations include only singly and doubly excited configuration functions (SDCI, CISD or CI-SD) and are size-dependent. Often such limited CI calculations use an additional frozen-core (FC) approximation, where excitations of electrons from the inner-shell MOs are not accounted for. While such excitations provide for a noticeable contribution to the molecular energy, they do not change much with environment, and thus may be neglected in many cases. For the majority of calculations, such as molecular excitation, dissociation, bond formation, etc., the inner-shell electrons are affected only slightly, so their correlation energies do not change much, and the use of smaller basis sets are justifiable. However, one of significant drawbacks of the use of smaller basis sets is that such CI calculations are not size-consistent, that is, the error in energy calculations for such calculations is not correlated with the size of a molecule. Among other disadvantages, this affects size consistency for infinitely separated systems, and thus makes impossible a comparison of calculations applied to molecules of significantly different sizes. [Notably, both HF SCF and full CI calculations are size consistent.]

Calculations of SCF MOs from the basis functions and their use to construct CSFs are remarkable for their slow convergence rate, because a great number of CSFs must be incorporated to obtain accurate results. For example, for large molecules from 10^8 to 10^{10} CSFs must be calculated for the CI results to be reasonably good, and the larger the system the greater the number of CSFs should be included. The major reason for slow convergence of the CI procedure is that the electron charge probability density described by excited spin-orbits is significant in the space regions far from the nuclei, while the ground state HF SCF wavefunction (which is usually used as the zero-order approximation) describes the electron charge density distributed much closer to the nuclei. In CI calculations it is not necessary to use HF SCF wavefunctions as the zero-order approximation, so that sometimes well-chosen non-SCF CSFs provide much faster convergence to the "true" molecular wavefunction.

There exist two other procedures that permit to account more accurately for configuration interactions called multiconfiguration SCF (or MCSCF) method and the method of natural orbitals. Both procedures allow more flexible approach to molecular wavefunction calculations than that of the "direct" CI method. Similar to the CI procedure, in the framework of the MCSCF method [22–27] a molecular wavefunction is represented as a linear combination $\Psi_{MCSCF} = \sum_{i=1}^{\approx M^N} a_i \Xi_i$ of CSFs Ξ_i. However, in contrast to the CI molecular wavefunction, in the molecular MCSCF wavefunction Ψ_{MCSCF} both the coefficients a_i and the shape of the MOs ψ_i [that is, a representation of ψ_i in terms of the basis functions, Eq. (53)] in the Slater determinants Ξ_i are varied. The procedure of finding optimum MCSCF spin-orbits realizes an iterative algorithm similar, to a degree, to that of HF SCF method. The number of varied CSFs in the MCSCF procedure can be larger than that of (non-varied) CSFs in the CI procedure, and the convergence of the MCSCF procedure is usually slow. At the same time, inner flexibility of this method, the existence of advanced MCSCF calculation algorithms [22–27], and efficient algorithms and codes utilizing these methods made MCSCF calculations and related methods a preferred pathway to obtain accurate molecular wavefunctions both for the ground and excited states.

The complete active space (CAS) SCF (CASSCF) method [30] is the most widely used MCSCF method. In similarity to the SCF method, CASSCF MOs are written as linear combinations of the basis functions, $\psi_i = \sum_{\alpha=1}^{M} c_{\alpha i} \zeta_\alpha$ and are used to write CSFs, again, as linear combinations of the MOs. The MOs in the CSFs are divided into *active* and *inactive* spin-orbits. The inactive spin-orbits are kept doubly occupied in all CSFs. Electrons in active spin-orbits are called active electrons. The molecular wavefunction Ψ_{CASSCF} is constructed as a linear combination $\Psi_{CASSCF} = \sum_{i=1}^{K} a_i \Xi_i$ of all CSFs Ξ_i each of which is formed by distributing all active electrons over all active spin-orbits in all possible ways, and have the same spin and symmetry eigenvalues as those of the investigated state Ψ_{CASSCF}. The coefficients $c_{\alpha i}$ [see Eq. (53)] and a_i are varied so that Ψ_{CASSCF} would minimize the variation integral. Active spin-orbits are usually selected among valence spin-orbits of atoms that form the molecule.

One of widely used methods to improve accuracy of CI calculations is the *multireference* CI (MRCI) method that incorporates important quadruple electron excitations, and thus reduces the size-inconsistency of the limited CI method, and also speeds up the convergence. MRCI calculations are popular even at present time for preliminary investigations of energy surfaces, while SDCI calculations are almost extinct. The MRCI method combines advantages of the MCSCF method and those of the conventional CI ones. In particular, the CI calculations begin with the HF SCF wavefunction Ξ_1 (called the *reference function*), and then electrons are moved from occupied Ξ_1 one-electron spin-orbits to unoccupied one-electron spin-orbits. These produces a number of CSFs $\Xi_2, \Xi_3, ..., \Xi_K$, and the molecular wavefunction is written as a linear combination of these CSFs. The MRCI molecular wavefunction is a linear combination of a MCSCF molecular wavefunction and various CSFs Ξ_i obtained by excitation of electrons from occupied spin-orbits of the MCSCF molecular wavefunction into virtual spin-orbits, $\Psi_{MRCI} = \Psi_{MCSCF} + \sum_{i=1}^{K} a_i \Xi_i$. The coefficients a_i are varied to minimize the variation integral. In many modern software packages CASSCF wavefunctions are used as the reference function in MRCI calculations.

Similar to MCSCF, *the natural orbitals method* is used to begin CI calculations from a starting point that is not SCF wavefunctions. Before going into the CI calculation process, one uses components ψ_i of the Slater determinant CSFs to write the electron charge probability density, $\rho(\vec{r}) = \sum_i \sum_j a_{ij} \psi_i^*(\vec{r}) \psi_j(\vec{r})$. Then this probability density is diagonalized to take the form

$$\rho(\vec{r}) = \sum_i \lambda_i |\chi_i(\vec{r})|^2. \tag{71}$$

The MOs $\chi_i(\vec{r})$ so obtained are called *natural orbitals*. The eigenvalues λ_i are taking values between 0 and 2, and are called *occupation numbers*. These new MOs $\chi_i(\vec{r})$, that are linear combinations of the former MOs $\psi_i(\vec{r})$, are then used to write the Slater determinant CSFs. Experience proves that CI calculations need significantly fewer natural spin-orbital-based CSFs and converge much faster than CI calculations based on CSFs formed of SCF MOs. A disadvantage of the natural orbital method is that the natural orbitals can be calculated only after CI calculations based on SCF MOs are finished. The *iterative natural-orbital* (INO) *method* and other algorithms were developed over the years to overcome this shortcoming. In the INO approach the CI wavefunction Ψ_{CI} is calculated using a fewer number of CSFs, and a set of approximate natural orbitals is calculated from this CI wavefunction. Then again, an improved version of Ψ_{CI} is calculated using these natural orbitals, and a new, improved set of the natural orbitals is calculated again. The process is repeated until a desired accuracy of Ψ_{CI} is reached, and Ψ_{CI} is written as a linear combination of Slater determinant CSFs composed of natural spin-orbitals.

CI calculations include many intermediate tasks that grow into serious problems with an increase in the number of atoms in a molecule. One of such problems is time-consuming calculations of Coulomb energy integrals $\langle \psi_i(1)\psi_j(2) | e^2 / r_{12} | \psi_i(1)\psi_j(2) \rangle$ over SCF MOs, Eqs. (44) and (45), from already computed integrals over the basis functions, Eq. (58). For the number M of the basis functions there are M SCF MOs ψ_i, and over M^7 computation steps [1] necessary to calculate each of the Coulomb energy integrals. This huge number of computations is reduced by the use of skillfully developed algorithms [31] that allow reduction of the number of computation steps by several orders of magnitude.

Yet another computational problem arises in the process of accurate CI computations in conjunction with solving the secular equation

$$\det(H_{ij} - ES_{ij}) = 0 \qquad (72)$$

for eigenvalues and eigenfunctions of the interaction part of the Hamiltonian $\hat{H}$, where $H_{ij} = \langle \psi_i | \hat{H} | \psi_j \rangle$ and $S_{ij} = \langle \psi_i | \psi_j \rangle$, and the related set of the coefficients $c_{i\alpha}$, Eq. (53). With the basis set of M functions one has to solve Eq. (72) for the eigenfunctions and eigenvectors of the $M \times M$ matrix. In SCF calculations this problem is solved easily by the standard matrix algebra methods even for a relatively large number of basis functions. In a typical CI calculation run for a small molecule about 10^6 CSFs are used to mimic the molecular wavefunction, the order of the matrix in the secular equations is at least 10^6, and the standard matrix algebra methods to solve the secular equation become too cumbersome and time consuming. Thus, a number of more sophisticated methods to solve the secular problem for very large matrices has been suggested over the years [32], with the most popular being the Davison method [33].

In 1972 Roos [24] developed an ingenuous technique called *the direct configuration interaction* (DCI) *method* that takes care of calculations of the Coulomb and exchange integral matrix elements $H_{ij} = \langle \Xi_i | \hat{H} | \Xi_j \rangle$ of the secular Eq. (72) written with respect to CSFs. For the number of CSFs being about 10^6 in the case of small molecules, the number of such matrix elements is about 10^{12}, so for larger molecules calculations of such matrix elements take the largest portion of computation time. The DCI method allows computations of the matrix elements $H_{ij} = \langle \Xi_i | \hat{H} | \Xi_j \rangle$, and the eigenvalues and eigenvectors in Eq. (72), without explicit calculations of the matrix elements and without explicit solving of the secular equation. Rather than providing for such calculations, the DCI method computes the CI expansion coefficients a_i for expansions of the molecular wavefunction $\Psi_{MCSCF} = \sum_{i=1}^{\approx M^N} a_i \Xi_i$ over CSFs Ξ_i, and then computes one- and two-electron integrals $\langle \beta\alpha | \gamma\delta \rangle$ of Eq. (58) expanded over the basis functions ζ_α [34]. The DCI method extended CI calculations capabilities to include over 10^8 CSFs, and MCSCF calculations [26,35] to involve about 10^6 CSFs. Among other popular methods to speed up CI calculations is so-called *the graphical unitary group approach* (GUGA) [36]. This method is integrated into the direct CI method and is included in the vast majority of the first-principle quantum chemistry software packages [37].

At present, SCF algorithm is a routine method that does not require a special care to use. At the same time, CI calculations, with many speed up procedures integrated into the CI computation process, often require special care to perform in the case of large molecules. It is advisable to consult numerous available literature sources to use this method efficiently and to understand physical and chemical meaning of the obtained results.

Treatment of excited states of molecules characterized by the low charge probability density values in many cases needs a thoughtful approach. One of the simplest techniques that can be used for the purpose is the CI-singles (CIS) method [38]. In this method one choses a fixed molecular geometry known either from reliable preliminary calculations of the ground state of the molecule, or from experimental data. For the molecular geometry so chosen the single Slater determinant molecular wavefunction Ψ_0 formed of SCF MOs is calculated. The SCF calculations also provide a set of virtual spin-orbits. Knowing the virtual spin-orbits, one can replace in Ψ_0 one of such orbits (for example, i) with another orbit (α) to form a new Slater determinant Ψ_i^α. Going through all occupied and all virtual orbits, one can establish a range of such Slater determinant Ψ_i^α's and build a linear combination of the Ψ_i^α's, which is the CIS linear variation function $\Psi_{CIS} = \sum_\alpha \sum_i b_{i\alpha} \Psi_i^\alpha$. The coefficients $b_{i\alpha}$ are variation coefficients. This function is then used to minimize the variation integral (where the electronic Hamiltonian corresponds to the chosen molecular geometry), thus deriving the corresponding secular equation, and solving it for the several lowest energy roots and the corresponding coefficients $b_{i\alpha}$.

All of the energies so calculated are approximations to the energy of the excited state of the molecule obtained while its geometry was fixed. Given that the mass of an electron is orders of magnitude smaller than that of a nucleus, an excited state of a molecule obtained in a geometry that is close to the equilibrium configuration of electrons in the ground state of the molecule has larger probability than that of excited states produced in other electronic geometries. Because the equilibrium ground state geometry is used to model the excited state, it is not a real excited state. The difference in the energies of the excited and the ground states of the molecule so calculated is called the vertical excitation energy. The frequencies calculated from the CIS vertical excitation energies are rather rough evaluations of the maximum-intensity frequencies of the electronic absorption spectra of molecules.

Once the variation coefficients $b_{i\alpha}$ are found, one can recover the CIS molecular wavefunction. This function and its analytical gradients are used to optimize geometry of each excited state, and the corresponding excitation energies and vibration frequencies. The newly obtained vibration frequencies should be more accurate that the vertical excitation frequencies, because they are obtained for excited state geometries.

The CIS reference function is always the SCF ground state wavefunction, but it does not contribute to the CIS excited state wavefunctions, making the CIS excited state wavefunctions orthogonal to the CIS reference function. [In contrast, the reference function of the standard CI method is the SCF wavefunction corresponding to the studied state, and it provides the largest contribution to the CI molecular wavefunction.] This useful property attracted further attention to the CIS method, so its generalizations (XCIS method) that use reference functions other than the SCF ground state one, and selected double MO substitutions, were proposed [34]. In recent years XCIS method was applied to study excited states of moderate and large molecular systems [39].

2.5 The Møller-Plesset (MP) perturbation theory

Quantum field theory and quantum statistical mechanics offer a range of mathematical methods to calculate properties of systems composed of many interacting particles. The most mathematically accurate of them is *many-body perturbation theory* (MBPT) that offers a general first-principle algorithm to calculate the major contributions to such properties and corrections to the major contributions that allows improvement of the property calculations. Each of such corrections is proportional to a power of an initially chosen small parameter (usually identified on the basis of known properties of the studied system, such as a value of the averaged density or interaction energy). The use of the small parameters permits to systematize corrections to the major contribution to a property in the order of their diminishing values, and thus to calculate from the first principles any properties of interest to any desirable degree of accuracy. In the 1930's Møller and Plesset adjusted MBPT methods to calculate electronic properties of atoms and molecules and developed the MBPT-based first-principle approach called the Møller-Plesset (MP) perturbation theory. To be useful in practical computations, the approach needs developed algorithms, software and hardware capabilities. Such conditions have been only created in the 1970's, when many effective algorithms were implemented into efficient computer codes (by Pople, Bartlett, and their co-workers) designed for relatively powerful supercomputers that were already used for quantum chemical computing.

In MP theory the major contribution to the molecular wavefunction is delivered by the HF SCF wavefunction Ψ_0 (see section 2.3.2), which is formed of one or several Slater determinants. The SCF calculations also provide the HF energy, E_{HF}, and a set of virtual spin-orbits. However, the variation functional is now minimized using spin-orbits, rather than spatial orbits. Thus, Eqs. (43) to (50) are now modified to account for the presence of spin-components in the spin-orbits. Correspondingly, the Fock operator of Eq. (47) for the s-th electron [denoted as $\hat{\mathbf{f}}(s)$, for further convenience] now takes the form [40]:

$$\hat{\mathbf{f}}(s) = \hat{\mathrm{H}}^{core}(s) + \sum_{j=1}^{N}\left[\hat{\mathbf{j}}_j(s) - \hat{\mathbf{k}}_j(s)\right], \tag{73}$$

where N denotes, as always, the number of electrons in the molecule, $\hat{\mathrm{H}}^{core}(s)$ is the core Hamiltonian of Eq. (41) written for the s-th electron,

$\hat{H}^{core}(s) = -(\hbar^2/2m_e)\nabla_s^2 - \sum_\gamma (Z_\gamma e^2/r_{s\gamma})$, and the operators $\hat{j}_j(s)$ and $\hat{k}_j(s)$ are the Coulomb and exchange operators satisfying the equations

$$\hat{j}_j(s)v(s) = v(s)\sum_{\sigma_2} \int d\vec{r}_2 \frac{e^2}{r_{s2}} |u_j(2)|^2 \tag{74}$$

and

$$\hat{k}_j(s)v(s) = u_j(s)\sum_{\sigma_2} \int d\vec{r}_2 \frac{e^2}{r_{s2}} u_j^*(2)v(s), \tag{75}$$

respectively, that are somewhat different from Eqs. (49) and (50). In particular, in Eqs. (74) and (75) spin-orbits u_j replace spatial orbits ψ_j of Eqs. (49) and (50), summation over σ_2 (the spin of the electron 2) is added, and the functions $v(s)$ are arbitrary spin-orbits, rather than arbitrary spatial orbits of Eqs. (49) and (50). The Hartree-Fock equations for the Fock operator $\hat{f}(s)$ of Eq. (73) are:

$$\hat{f}(s)u_i(s) = \varepsilon_i u_i(s). \tag{76}$$

The unperturbed Hamiltonian of the N-electron system in MP theory is a sum of the one-electron Fock operators,

$$\hat{H}_0 = \sum_{s=1}^N \hat{f}(s), \tag{77}$$

and the Slater determinant HF molecular wavefunction Ψ_0, that is an eigenfunction of the unperturbed Hamiltonian $\hat{H}_0$, is a linear combination of $N!$ terms $|u_1 u_2 ... u_N|$, with the spin-orbits u_i being eigenfunctions of the Fock operator $\hat{f}(s)$. Each of the $N!$ terms $|u_1 u_2 ... u_N|$ is an eigenfunction of $\hat{H}_0$ corresponding to the same eigenvalue $\sum_{s=1}^N \varepsilon_s$:

$$\hat{H}_0 \Psi_0 = \left(\sum_{s=1}^N \varepsilon_s\right) \Psi_0. \tag{78}$$

The ground state HF wavefunction Ψ_0 is the zero-th order eigenfunction of $\hat{H}_0$, and the reference function of the MP theory. Because the Fock operator is Hermitian, it has the complete set of eigenfunctions, and therefore, there exists an infinite set of spin-orbits that are eigenfunctions of this operator. While the occupied (N) spin-orbits form $N!$ products $|u_1 u_2 ... u_N|$ whose linear combination constitutes the ground-state HF wavefunction Ψ_0, the rest of the zero-th order eigenfunctions of $\hat{H}_0$ are all possible Slater determinants composed of products of any N unoccupied spin-orbits (this number corresponds to the total number of electrons, of course) each of which is an eigenfunction of the Fock operator $\hat{f}(s)$.

The difference between the actual electronic Hamiltonian $\hat{H}$ of a molecule and its HF Hamiltonian $\hat{H}_0$ is called the perturbation Hamiltonian $\hat{H}'$,

$$\hat{H}' \equiv \hat{H} - \hat{H}_0 = \sum_k \sum_{s>k} \frac{e^2}{r_{ks}} - \sum_{s=1}^{N} \sum_{j=1}^{N} \left[\hat{j}_j(s) - \hat{k}_j(s) \right], \tag{79}$$

where the expression in the r.h.s. was obtained using Eqs. (34), (76), (77). [This expression is only valid for the non-relativistic case, which is considered here for simplicity.] The first term in this expression is an interelectronic repulsion, and the second term the average HF interelectronic potential.

The energy of a state described by the eigenfunction of the HF Hamiltonian, $\varphi_0^{(0)}$, of the zero-th order is: $E_0^{(0)} + E_0^{(1)} = \langle \varphi_0^{(0)} | \hat{H}_0 + \hat{H}' | \varphi_0^{(0)} \rangle$, where $E_0^{(1)}$ is the first-order correction to the zero-th order energy specific to the zero-th order eigenfunction $\varphi_0^{(0)}$ of $\hat{H}_0$. Choosing $\varphi_0^{(0)}$ to be the ground state wavefunction Ψ_0, one obtains the following equation for the first order MP correction to the HF SCF energy of the ground state,

$$E_0^{(0)} + E_0^{(1)} = \langle \varphi_0^{(0)} | \hat{H}_0 + \hat{H}' | \varphi_0^{(0)} \rangle = \langle \Psi_0 | \hat{H}_0 + \hat{H}' | \Psi_0 \rangle = \langle \Psi_0 | \hat{H} | \Psi_0 \rangle = E_{HF},$$

where the last term on the right follows from the fact that $\langle \Psi_0 | \hat{H} | \Psi_0 \rangle$ is the variational integral calculated for the HF wavefunction Ψ_0 which minimizes this integral giving it the value of E_{HF}. Thus,

$$E_0^{(0)} + E_0^{(1)} = E_{HF}. \tag{80}$$

At the same time, from Eq. (78) it follows that $E_0^{(0)} \equiv \langle \Psi_0 | \hat{H}_0 | \Psi_0 \rangle = \sum_{s=1}^{N} \varepsilon_s$, and therefore, the first-order MP correction to the ground state energy is:

$$E_0^{(1)} = E_{HF} - \sum_{s=1}^{N} \varepsilon_s. \tag{81}$$

The second-order MP correction to the ground state energy can be found using the corresponding formula of the perturbation theory:

$$E_0^{(2)} = \sum_{m \neq 0} \frac{\left| \langle \varphi_m^{(0)} | \hat{H}' | \Psi_0 \rangle \right|^2}{E_0^{(0)} - E_m^{(0)}}, \tag{82}$$

where the summation is over all possible unperturbed Slater determinant wavefunctions $\varphi_m^{(0)}$ (i.e., the eigenfinctions of the unperturbed Hamiltonian $\hat{H}_0$) composed of products of N different spin-orbits each of which is an eigenfunction of the Fock operator. The excited Slater determinants $\varphi_m^{(0)}$ are usually systematized according to the number of occupied spin-orbits (denoted with the Latin indices $i, j, k, \ldots$) that are replaced with the virtual spin-orbits (denoted with the Latin indices $a, b, c, \ldots$): $\varphi_i^{(0)a}$, $\varphi_{ij}^{(0)ab}$, $\ldots$. The Slater determinant $\varphi_i^{(0)a}$ is constructed from the closed-shell (the ground state) Ψ_0 by replacement of the i-th occupied spin-orbit by the a-th virtual one, the Slater determinant $\varphi_{ij}^{(0)ab}$ is obtained from Ψ_0 replacing i-th and j-th occupied spin-orbits with a-th and b-th virtual ones, respectively, *etc.* Noticing that in the sum on the right of Eq. (82) the matrix elements $\langle \varphi_m^{(0)} | \hat{H}' | \Psi_0 \rangle$ with $\varphi_m^{(0)}$'s containing one, three and more virtual spin-orbits vanish due to Condon-Slater rules [1,40], one can restructure the sum keeping only the matrix elements with doubly excited Slater determinants. Further on, all of the Slater determinants $\varphi_{ij}^{(0)ab}$ [that are eigenfunctions of $\hat{H}_0$, Eq. (78)] in which the occupied spin-orbits i and j are replaced with the virtual spin-orbits a and b, respectively, correspond to the eigenvalue $E_m^{(0)} = \sum_{k \neq i,j}^{N} \varepsilon_k + \varepsilon_a + \varepsilon_b$, where the summation is over occupied spin-orbits. Thus, from Eq. (78) and the above expression one derives: $E_0^{(0)} - E_m^{(0)} = \sum_{i=1}^{N} \varepsilon_i - \sum_{k \neq i,j}^{N} \varepsilon_k - \varepsilon_a - \varepsilon_b = \varepsilon_i + \varepsilon_j - \varepsilon_a - \varepsilon_b$. Using the explicit expression of Eq. (79) for the perturbation Hamiltonian, one can evaluate [40] remaining non-zero matrix elements in the sum on the right of Eq. (82) containing only doubly excited Slater determinants $\varphi_{ij}^{(0)ab}$, to derive the following equations for the second-order MP correction to the energy of the ground state:

$$E_0^{(2)} = \sum_{b=a+1}^{\infty} \sum_{a=N+1}^{\infty} \sum_{i=j+1}^{N} \sum_{j=1}^{N-1} \frac{e^2 |\langle ab | r_{12}^{-1} | ij \rangle - \langle ab | r_{12}^{-1} | ji \rangle|^2}{\varepsilon_i + \varepsilon_j - \varepsilon_a + \varepsilon_b}, \tag{83}$$

where the two-electron integrals include summation over the spin variables σ_1 and σ_2 of both electrons, respectively:

$$\langle ab | r_{12}^{-1} | ij \rangle \equiv \sum_{\sigma 1} \sum_{\sigma 2} \iint d\vec{r}_1 \, d\vec{r}_2 \; u_a^*(1) u_b^*(2) r_{12}^{-1} u_i(1) u_j(2). \tag{84}$$

The MP calculations that allow obtaining the energy of the ground state of a molecule corrected up to the second order of MP theory is called MP2 theory,

$$E_0^{MP2} = E_{HF} + E_0^{(2)}. \tag{85}$$

Note, that one can use a thoughtfully chosen complete basis set of functions to expend the spin-orbits u_s of Eq. (76), and then reduce integrals over the spin-orbits with those over the basis functions. The HF SCF calculations using the infinite basis set of functions result in the exact HF energy and an infinite number of virtual orbitals, and thus in an infinite number of terms in the sums in the corresponding equation replacing Eq. (83). Regardless of the method of calculations of the electron interaction integrals, in both cases the sums in Eq. (83) have to be truncated, leading to a truncation error. However, in the case when the infinite basis set of functions is used, one would have also to truncate the basis set to deal with a finite number of basis functions.

This would lead to a basis set truncation error, that is absent if the integrals are calculated over the spin-orbits u_s. Notably, in the vast majority of the electron correlation calculations the basis set truncation errors are larger than other errors related to truncations of correlation treatments.

The MP2 corrections to molecular energy for molecules in their excited states are calculated in a fashion similar to that specific to the ground state. A perturbation theory formula similar to that of Eq. (83) for the second-order correction to the excited state energy holds. Because the first-order correction to the molecular wavefunction, $\varphi^{(1)}$, determines the values of MP2 $E^{(2)}$ and MP3 $E^{(3)}$, calculations of MP corrections $E^{(2)}$ and $E^{(3)}$ involve summations over doubly- substituted spin-orbits only. The MP4 correction $E^{(4)}$ involves summation over all substitutions up to quadruple. The formulae for higher order MP corrections to molecular energy can be found in Refs. 41–43, and other publications.

MP calculations of orders higher than 2 are very time consuming, so an approximation called *the frozen core approximation* is usually applied in such calculations. This means that terms involving excitations from the designated occupied (core) spin-orbits are omitted in calculations. MP calculations using the frozen core approximation are faster than CI ones, and their truncation at any order is size-consistent [40]. The major disadvantage of MP calculations is that they are not variational, and thus the obtained MP energies may be below the true molecular energy. Technical details and analysis of performance of MP algorithms can be found in Ref. 44.

There are several methods used to speed up MP2 calculations. The *direct* MP2 method is a technical one: it uses an algorithm that re-calculates two-electron integrals as many times as they are needed, rather than storing them on an external carrier (disc) and retrieving them. The *semi-direct* MP2 method makes use of both external storage and recalculation of two-electron integrals.

The localized MP2 (LMP2) method speeds up calculations applied to large molecules. In this method one constructs the HF SCF Slater determinant wavefunction for the ground state using local MOs (see Ref. 1, Chapter 15.9), instead of SCF MOs. For substitution purposes, atomic orbitals (say, *a* and *b*) orthogonal to the nearby localized occupied MOs that are being substituted (say, *i* and *j*, respectively) are used, instead of virtual MOs of the canonical SCF calculations.

Many specially pre-designed nanosystems and natural molecules possess open-shell ground states. In such cases, instead of the ground state HF SCF determinant used as the zero-order approximation, one can use the unrestricted HF (UHF) SCF determinant corresponding to the open-shell ground state. Such calculations are labelled UMP2, UMP3, etc. However, one has to keep in mind that UHF SCF wavefunctions are not eigenfunctions of the spin operator $\hat{S}^2$, and resulting "spin contamination" may leads to significant calculation errors. To avoid such errors, one may consider MP calculations that are based on the restricted open shell HF (ROHF) ground state wavefunction [45]. This method, called ROMP, was developed using the spin-constrained unrestricted Hartree—Fock (SUHF) approach. In SUHR approach the unrestricted HF method was amended by a requirement that the average value of the spin operator $\left\langle \hat{S}^2 \right\rangle$ should have a prescribed value. At high spin values SUHF method produces the wavefunction that approaches the high spin ROHF wavefunction and energy. The corresponding ROMP perturbation theory uses the amended Fock operators of SUHF theory that are derived in the framework of the UHF formalism. ROMP2 and ROMP4 methods converge much faster than UMP-series methods.

In addition to being non-variational, MP method does not work very well with configurations of atoms in a molecule that are far from their equilibrium configuration. The corresponding "geometrical" errors of MP2 correlation energy calculations reach up to 20 % even for small molecules. Yet another important shortcoming of MP method is that it is not generally applicable to excited electronic states.

The above limitations of MP perturbation theory restrict its use to systems in their ground states, while CI methods are applicable to both ground state and excited electron systems. In the case of the ground-state systems at equilibrium, MP2 theory is the most popular first-principle method for its efficiency and accurate accounting for electron correlation contributions to molecular properties.

The zeroth order wavefunction used as a reference function in MP theory is a single Slater determinant obtained as a result of HF SCF calculations. Alternatively, one can chose the MCSCF wavefunction as the zeroth order function, and develop a generalized MP theory using such a reference function. A CASSCF wavefunction is the most convenient for the purpose, and an MP2 method that uses a CASSCF function as a reference wavefunction is called *complete active space second order perturbation theory* [46,47], or CASPT2. However, when a MCSCF wavefunction is used as the zeroth order MP function, a choice of the unperturbed Hamiltonian $\hat{H}_0$ is not unique, Eq. (77) is no longer applicable and is replaced by a more complicated equation. The CASPT2 method allows energy calculations up to the second order correction, $E^{(2)}$, and provides accurate results of about the same high quality as those of the MRCI method, but requires much less computational effort than the MRCI method.

There are some other possible choices of an unperturbed MCSCF function (see, for example, Ref. 48 for the role of a reference function in MP theory, but they are much less popular and produce results that are not necessarily higher in quality that those of MRCI and CASPT2.

2.6 The coupled-cluster approximation

The configuration interaction method is exact in the full CI limit, but any truncation of the configuration space makes it approximate and causes a loss of size scaling, that is, the obtained results stop scaling correctly with the system size. The coupled cluster (CC) method has been introduced as a practical realization of many-body quantum theoretical diagrammatic methods that, among other virtues, overcomes size-scaling deficiency of other advanced CI methods, such as MCSCF and MP2. This is achieved by a computationally effective resummation of MBPT diagrams that offers an infinite-order approximation included in selected electron cluster operators. Originally CC method was introduced to apply MBPT to solve nuclear physics problems [49,50], and later it was transformed to serve quantum chemistry needs [51–55]. Excellent accounts of the modern state of this method can be found in Refs. 56 and 57. Currently, CC method is the most practically accurate among *ab initio* electronic structure computational techniques when applied to small and moderate-size molecules composed of light atoms. For larger molecules and molecules composed of heavy atoms MP2 method works better. At the same time, advanced versions of the CC method that use pseudopotentials to describe relativistic effects and generalized pseudopotentials to address spin-orbit interactions are successfully applied to heavy atoms and small molecules composed of such atoms.

Applied to the same molecule, the ground state MBPT and CI wavefunctions, and the exact, non-relativistic ground-state CC one, describe the same state of the molecule. However, the latter wavefunction, Ψ_{CC}, is written in a different form:

$$\Psi_{MBPT} = \Psi_{CI} = \Psi_{CC} = \exp(\hat{T})\Phi_0, \tag{86}$$

where Φ_0 is the normalized SCF wavefunction (usually HF one), the cluster operator $\hat{T}$ is a sum of excitation operators that introduce single, double, triple, *etc.* electron excitations into the SCF wavefunction substituting spin-orbits of 1, 2, 3, etc. electrons in Φ_0 with the corresponding number of virtual spin-orbits:

$$\hat{T} = \hat{T}_1 + \hat{T}_2 + \ldots + \hat{T}_N, \tag{87}$$

N denotes the number of electrons in the molecule, and the exponential operator has the standard form:

$$\exp(\hat{T}) = 1 + \hat{T} + \frac{\hat{T}^2}{2!} + \ldots + \frac{\hat{T}^l}{l!} + \ldots = \sum_{l=0}^{N} \frac{\hat{T}^l}{l!}. \tag{88}$$

Explicitly, the *l*-particle excitation operators $\hat{T}_l$ are written in terms of the creation and annihilation operators $\hat{c}_a^+$ and $\hat{c}_i$ of electrons in virtual and occupied spin-orbits (the indices $a, b, \ldots$ are retained to denote unoccupied, or virtual, spin-orbits, and the indices i, j, *etc.* occupied ones):

$$\hat{T}_l = (l!)^{-2} \sum_{\substack{i,j,\ldots \\ a,b,\ldots}} t_{ij\ldots}^{ab\ldots} \hat{c}_a^+ \hat{c}_b^+ \ldots \hat{c}_j \hat{c}_i. \tag{89}$$

Thus, the results of application of excitation operators to the SCF ground state function Φ_0 are the corresponding linear combinations of the single excited, double-excited, *etc.*, Slater determinants $\Phi_{ij\ldots}^{ab\ldots}$ in each of which 1, 2, 3, *etc.* occupied spin-orbits are replaced by virtual ones:

$$\hat{T}_l \Phi_0 = \sum_{\substack{i,j,\ldots \\ a,b,\ldots}} t_{ij\ldots}^{ab\ldots} \Phi_{ij\ldots}^{ab\ldots}. \tag{90}$$

The coefficients $t_{ij...}^{ab...}$ are called the CC amplitudes. Each operator $\hat{T}_l$ is "connected" (or irreducible, because it cannot be simplified further). The simplest is the single excitation operator,

$$\hat{T}_1 \Phi_0 = \sum_{a=N+1}^{\infty} \sum_{i=1}^{N} t_i^a \Phi_i^a, \qquad (91)$$

that produces all possible substitutions of a single electron spin-orbit from any of the occupied N spin-orbits by any one of virtual ones. The double excitation operator,

$$\hat{T}_2 \Phi_0 = \sum_{b=a+1}^{\infty} \sum_{a=N+1}^{\infty} \sum_{j=i+1}^{N} \sum_{i=1}^{N-1} t_{ij}^{ab} \Phi_{ij}^{ab}, \qquad (92)$$

realizes all doubly excited substitutions.

The simplest approximation of the CC method is limited to consideration of only double interactions (CCD, or CC double), $\Psi_{CCD} = \exp(\hat{T}_2)\Phi_0$, yet because it includes disconnected (but linked) parts of quadruple, hextuple, *etc.* interactions, it is size-scalable. The next order CC approximation that accounts only for singly and doubly excited electron states is called CCSD (CC single double). Subsequent series of approximations include triple quadruple, etc. excitations: CCSDT, CCSDTQ, and so on. In recent decades intense studies permitted to combine MBPT and CC methods to derive non-iterative approximations, such as CCSD(T) and CCSDT-1. In the widely used CCSD approximation the CC non-relativistic ground-state wavefunction is

$$\Psi_{CC} = \left[1 + (\hat{T}_1 + \hat{T}_2) + \frac{(\hat{T}_1 + \hat{T}_2)^2}{2}\right]\Phi_0 = \left[1 + \hat{T}_1 + \hat{T}_2 + \hat{T}_1\hat{T}_2 + \frac{1}{2}\hat{T}_1^2 + \frac{1}{2}\hat{T}_2^2\right]\Phi_0, \qquad (93)$$

that includes not only single and double excitations, but also some of triple and quadruple ones via the terms $\hat{T}_1\hat{T}_2\Phi_0$ and $\frac{1}{2}\hat{T}_2^2\Phi_0$:

$$\hat{T}_1\hat{T}_2\Phi_0 = \sum_{\substack{a;i\ b>c;\\j>k}} t_i^a t_{jk}^{bc} \Phi_{ijk}^{abc}, \qquad (94)$$

$$\frac{1}{2}\hat{T}_2^2\Phi_0 = \sum_{\substack{i>j;k>l;\\a>b\ c>d}} t_{ij}^{ab} t_{kl}^{cd} \Phi_{ijkl}^{abcd}, \qquad (95)$$

respectively. While such excitation operators are disconnected (that is, represented as products of simpler operators), they are still linked. [Notably, inclusion of such triple-, quadruple-, *etc.* substitution series into the lower order excitation operators is the essence of the CC resummation of selected MBPT diagram series.] For example, comparing the CISD ground-state wavefunction Ψ_{CISD} obtained by accounting for only single and double substitutions (realized by operators $\hat{C}_1$ and $\hat{C}_2$),

$$\Psi_{CISD} = (1 + \hat{C}_1 + \hat{C}_2)\Phi_0, \tag{96}$$

to the corresponding second-order CCSD wavefunction of Eq. (93), one finds that

$$\hat{C}_1 = \hat{T}_1, \tag{97}$$

$$\hat{C}_2 = \hat{T}_2 + \frac{1}{2}\hat{T}_1^2, \tag{98}$$

and that Ψ_{CISD} does not include any of triple and quadruple substitutions, while Ψ_{CCSD} does. This indicates that a CC approximation of some lower order s includes contributions that are of higher order. To justify such a selection of only parts of higher order diagram series, the CC method employs *an assumption of instantaneous interactions*. This assumption suggests that a total contribution to the system energy due to interactions of electrons in a larger cluster that includes, for example, K electrons, is dominated by contribution from instantaneously interacting electrons that belong to smaller groups in different parts of the system included in the K-electron cluster. [To a degree, this assumption correlates with the Born-Oppenheimer approximation that is almost always used in calculations.]

The Schrödinger equation for the exact Ψ_{CC} is:

$$\hat{H}\Psi_{CC} = E\Psi_{CC}, \tag{99}$$

or using Eq. (86),

$$\exp(-\hat{T})\hat{H}\exp(\hat{T})\Phi_0 = E\Phi_0. \tag{100}$$

From Eq. (100) it follows that the SCF ground-state wavefunction Φ_0 is an eigenfunction of the operator $\exp(-\hat{T})\hat{H}\exp(\hat{T})$ corresponding to the same eigenvalue as that of Ψ_{CC} in the case of the undressed Hamiltonian $\hat{H}$. Finding the inner product of both sides of Eq. (100) with the excited wavefunctions, one obtains a system of coupled equations for the CC amplitudes $t_{ij...}^{ab...}$:

$$\langle \Phi_{ij...}^{ab...} | \exp(-\hat{T})\hat{H}\exp(\hat{T}) | \Phi_0 \rangle = E \langle \Phi_{ij...}^{ab...} | \Phi_0 \rangle = 0, \tag{101}$$

where completeness and orthonormality of the set of eigenfunctions of the dressed Hamiltonian operator $\exp(-\hat{T})\hat{H}\exp(\hat{T})$ were used: $\langle \Phi_{ij...}^{ab...} | \Phi_0 \rangle = 0$. From Eq. (100) one can also obtain the following formula for the ground state energy:

$$\langle \Phi_0 | \exp(-\hat{T})\hat{H}\exp(\hat{T}) | \Phi_0 \rangle = E \langle \Phi_0 | \Phi_0 \rangle = E = \langle \Phi_0 | \tilde{H} | \Phi_0 \rangle, \tag{102}$$

where the notation $\tilde{H} \equiv \exp(-\hat{T})\hat{H}\exp(\hat{T})$ was introduced. Note, that for the Hamiltonian that contains no higher than 2-electron operators the Hausdorff expansion holds:

$$\tilde{H} = \hat{H} + [\hat{H},\hat{T}] + \frac{1}{2!}\left[[\hat{H},\hat{T}],\hat{T}\right] + \frac{1}{3!}\left[\left[[\hat{H},\hat{T}],\hat{T}\right],\hat{T}\right] + \frac{1}{4!}\left[\left[\left[[\hat{H},\hat{T}],\hat{T}\right],\hat{T}\right],\hat{T}\right]. \quad (103)$$

Thus, the expansion of the operator $\tilde{H}$ in powers of $\hat{T}$ terminates after the fourth-fold commutator. As applied to CC expansions, this means that any term that does not have indices in common between $\hat{H}$ and $\hat{T}$ will be dropped, proving that all of the excitation operators in Ψ_{CC} are either connected, or linked, making Ψ_{CC} linked. Also, from Eq. (103) and the link-ness of Ψ_{CC} it follows that the contributions to the molecular energy are linked, and thus the energy calculations are size-extensive [53]. Moreover, due to resummation enabled by $\tilde{H} \equiv \exp(-\hat{T})\hat{H}\exp(\hat{T})$, CC calculations are not only scale with size of the system, but are much more effective than the corresponding CI or MP calculations. Another notable advantage of the CC method is that CC calculations evolve around individual Slater determinants $\Phi_{ij...}^{ab...}$, rather than the configuration state function (CSF), and each of CC CSFs is composed of one or several such determinants, similar to that of the CI method.

The major errors of CC approximations of higher orders come from basis set truncations. Similar to CI method, CC approach uses sophisticated basis sets, such as Gaussian basis set, cc-pVXZ or atomic natural orbitals, but necessary truncations even of such advanced basis sets introduce inevitable errors in solutions of the Schrödinger equation. A thoughtful choice of the basis set, and a way of its truncation, help minimize such errors, but cannot eliminate them entirely.

2.7 Basis function sets

Quantum chemical calculations begin with a choice of a complete set $\{\zeta_\alpha\}_1^\infty$ of basis functions ζ_α that are used to write MOs as linear combinations of such functions, $\psi_i = \sum_{\alpha=1}^\infty c_{\alpha i} \zeta_\alpha$, Eq. (53). A thoughtful choice of the basis functions, and truncations of the basis sets, contribute significantly to success of quantum chemical computations. In particular, as discussed in the previous sections, all first-principle electronic structure calculations are extremely sensitive to truncation of a basis set used and carry on the related truncation errors through all steps of such calculations.

A choice of a basis set depends on the size and nature of a studied system. Slater-type atomic orbitals (STOs) ζ_α are centered on participating atoms and have the form well-known for the majority of atomic species. Complete sets of such STOs are a standard choice of a complete basis set for small molecules. In this case, MOs $\psi_i = \sum_{\alpha=1}^\infty c_{\alpha i} \zeta_\alpha$ are represented as linear combinations of the atomic STOs and are called LC-ATO MOs. Despite seemingly straightforward advantages of the STO basis, this basis is not convenient in electronic structure calculations for polyatomic systems, because it creates a necessity to evaluate a great number of electron-repulsion integrals $\langle\alpha\beta|\gamma\delta\rangle$, Eq. (58), and H_{ii}^{core} integrals, Eq. (61), included in the HF-Roothaan equation,

Eq. (54). Accurate calculations applied to small and medium size molecules involve several hundreds of STO basis functions. If all such functions are centered on different nuclei, then 4-center electron-repulsion integrals $\langle\alpha\beta|\gamma\delta\rangle$ have to be evaluated. Denoting the number of the basis functions in a truncated STO basis as B, one can see that the number of 4-centered integrals $\langle\alpha\beta|\gamma\delta\rangle$ to be calculated is proportional to B^4, and the total number I of different 4-centered electron-repulsion integrals, in a general case, reaches $B^4/8$. For example, if $B = 600$, then $I = 1.62 \times 10^{10}$. Such calculations require enormous computation time even when modern supercomputers are used. In early days of quantum chemical computations, the STO basis choice and small RAM of computers seriously limited capabilities of the first-principle electronic structure calculations.

There exists a number of STO-based basis sets, and the minimal one (or the *minimum STO basis set*) consists of one STO for each inner shell and valence shell AO of each atom. The *double-zeta* (DZ) STO basis set is formed by replacing each STO in the minimum basis set by two STOs that differ by their orbital zeta-exponents. In a *triple-zeta* (TZ) basis each STO of the minimum basis set is replaced by three STOs differing in their orbital zeta-exponents. *Split-valence basis sets* are formed by replacement of one valence STO in the minimum basis set with two or more STOs, and keeping the inner-shell (core) STOs of the minimum basis set, or replacing some of them with one new STO each. Such basis sets are called *valence double-zeta, valence triple-zeta*, etc., in correspondence with the type of STO replacement used. In many cases polarization of AOs upon formation of a molecule requires introduction of STOs with the quantum number l greater than maximum values of the corresponding quantum numbers of the valence shells and the ground state of the constitutive atoms. Such additions to any basis sets give rise to the corresponding *polarized (P) basis sets*, the most typical of which is DZP basis set.

Inefficiency of STO-based 4-center integral calculations was overcome only in 1950's, when the atomic-centered *Cartesian Gaussian*-type functions (GTFs) g_{ijk} were introduced to represent atomic orbits (AOs) as linear combinations of GTFs (LCAOs),

$$g_{ijk} = \left(\frac{2\alpha}{\pi}\right)^{3/4} \left[\frac{(8\alpha)^{i+j+k} i!j!k!}{(2i)!(2j)!(2k)!}\right]^{1/2} x_a^i y_a^j z_a^k \exp(-\alpha r_a^2). \tag{104}$$

In Eq. (104) $i,j,k = 0,1,2,...$, α is a positive orbital exponent, x_a, y_a, and z_a are Cartesian coordinates with the origin at the nucleus a, and $r_a^2 = x_a^2 + y_a^2 + z_a^2$. The GTFs form a complete set of functions and are orthonormalized with the normalization factor $N \equiv \left(\frac{2\alpha}{\pi}\right)^{3/4} \left[\frac{(8\alpha)^{i+j+k} i!j!k!}{(2i)!(2j)!(2k)!}\right]^{1/2}$. Several types of Cartesian Gaussian functions are identified based on the value of the sum of their indices, $\Sigma = i + j + k$. Thus, when $\Sigma = 0$ the Cartesian GTFs is called *s-type Gaussian* function; it does not contain the coordinates x_a, y_a, and z_a explicitly. When $\Sigma = 1$ the three Cartesian GTFs are called *p-type Gaussian* functions; each of these functions contains the coordinate x_a, y_a, or z_a explicitly. In the case $\Sigma = 2$, the six *d-type Gaussians* contain combinations of the

Cartesian coordinates, $x_a^2, y_a^2, z_a^2, x_a y_a, y_a z_a$, and $x_a z_a$. Five linear combinations of the d-type Gaussians possessing the factors $x_a y_a, x_a z_a, y_a z_a, x_a^2 - y_a^2$, and $3z_a^2 - r_a^2$, can be used to recreate the same angular dependence as that of the 5 real $3d$ AOs. The sixth linear combination of d-type Gaussians having the factor $r_a^2 = x_a^2 + y_a^2 + z_a^2$ behaves like a real $3s$ AO and is sometimes omitted from the GTF basis set. For $\Sigma = 3$ one has 10 *f-type Gaussians* that can be arranged into linear combinations mimicking angular behavior of 7 real $4f$ AOs. Fitting the indices i, j and k, one can form linear combinations of Cartesian Gaussian functions that have the same angular behavior as real AOs:

$$\zeta_{lm} = N r_a^l \exp(-\alpha r_a^2) \frac{(Y_l^{m*} \pm Y_l^m)}{\sqrt{2}}, \tag{105}$$

where Y_l^{m*} and Y_l^m are spherical harmonics corresponding to the quantum numbers l and m. The principal quantum number n is not included in such combinations. Angular behavior of real AOs expressed by Eq. (105) happens to be the same as that of *spherical Gaussian functions*,

$$\zeta_{nlm} = N r_a^{n-1} \exp(-\alpha r_a^2) \frac{(Y_l^{m*} \pm Y_l^m)}{\sqrt{2}}. \tag{106}$$

Spherical Gaussians also form a complete set of functions that is sometimes used as an alternative to the Cartesian GTF basis set.

To mimic real AOs, one has to use linear combinations (LCs) each of which contains at least several GTFs. Further on, a great number of such LC-GTFs are involved in LC-GTF SCF MO calculations. The total number of 4-centered integrals that have to be evaluated in LC-GTF SCF MO calculations are at least an order of magnitude larger than that in the case of LC-STO SCF MO calculations. However, LC-GTF-based integrals take much less time to evaluate than LC-STO-based ones, because all three- and four-center electron repulsion integrals reduce to two-center integrals. The latter is a corollary of a useful property of products of Cartesian Gaussian functions. In particular, a product of two Gaussian functions centered at different origins equals to one Gaussian function centered at some other origin.

In addition to the computational advantage of the use of GTFs as basis functions when calculating 3- and 4-center electron repulsion integrals, there are other advantages gained from the use of GTF basis sets. In particular, when one uses carefully selected linear combinations of GTFs, called *contracted* GTFs (CGTFs),

$$\chi_r = \sum_{s=1}^{\infty} d_{sr} g_s, \tag{107}$$

instead of individual GTFs g_{ijk} of Eq. (105), the number of variational coefficients to calculate reduces, while a loss in accuracy of calculations is very little. The contraction coefficients d_{sr} are kept constant in calculations, and the Gaussian functions g_s in

Eq. (107) are GTFs of Eq. (104) selected in such a way that their indices i, j, k are the same, and they are centered on the same atoms, but their α-factors differ.

The above computational advantage of the CGTFs is widely used in all contemporary GTF-based basis sets that are composed of CGTFs of Eq. (107). Principles of selection and classification of CGTF-based basis sets are similar to those of the STO-based basis sets. Thus, a minimal CGTF basis set consists of one CGTF for each inner shell AO and each valence shell AO, a DZ CGTF basis set includes two CGTFs for each of the valence shell AOs and the inner-shell AOs, a TZ CGTF basis set has three CGTFs for each of such AOs, a DZP CGTF basis set includes additional CGTFs with higher values of $l = i + j + k$ to the corresponding DZ CGTF basis set, and so on. Once chosen, the orbital exponents and the contraction coefficients are fixed in calculations when a selected CGTF basis set is used. Because the CGTF basis functions in a minimal CGTF basis set do not change their shape in the process of calculations, extensions of such basis set to include more CGTF basis functions are necessary to reflect the actual shape of MOs.

There exist several approaches to forming minimal CGTF basis sets. Thus, choosing CGTFs in such a way that they fit STOs one can form minimal CGTF basis sets. In particular, one can approximate each STO with one or several (N) CGTFs producing the corresponding STO-(N)CGTF minimal basis set, such as a STO-3CGTF basis set.

Alternatively, one can perform atomic SCF calculations using uncontracted GTFs of Eq. (104) to determine AOs of atoms in terms of GTFs or CGTFs, and then use such AOs to form a basis set for molecular calculations. This approach was used to recover SCF AOs of the first-row atoms, thus forming the corresponding $9 \times 5p$ and DZ [$4s2p$] GTF basis sets, SV [$3s2p$], 3-21G, 6-31G, 6-31G* and other CGTF basis sets. The series of STO-3G, 3-21G, 3-21G$^{(*)}$, 3-21G*, and 3-21G$^{(**)}$ basis sets was develop by Pople and co-workers and realized in *Gaussian* software package. The CGTF basis sets of cc-pVXZ series (cc standing for "correlation consistent"), including cc-pVDZ, cc-pVTZ, cc-pVQZ and cc-pV5Z, were developed by Dunning and co-workers.

At present, there exist a number of advanced CGTF-based and other types of basis sets that accurately recover AOs of all atoms up to lanthanides. Extensive literature covers this very important field of quantum computational chemistry and addresses quality of various types of basis sets [58]. An account of developments concerning cc-pVXZ basis sets can be found in Refs. 59 and 60. The natural orbital-based basis sets are discussed in detail in Ref. 61.

The majority of *ab initio* software packaged make use of the most popular, tested and advanced basis function sets, such as SBKJC [29], cc-pVXZ [59,60], or cc-type basis set of Ref. 62. User documentation of such software includes references and discussion of the basis sets recommended and realized by the software.

2.8 Ab initio software packages and their use

The vast majority of modern quantum chemistry software packages contain *ab initio* calculation modules of HF-, CI-, MBPT- and DFT-types. In particular, complete HF-, CI (CASCF, MCSCF, MP2)-, and CC-calculation modules include all necessary codes and permit to choose an advanced basis set, a type of electronic structure calculations, some property calculations, and a type of result representation. A number of visualization software packages allow "automatic" conversion of the quantum chemistry computational results obtained by the use of the above *ab initio* modules into standardized images of molecular orbitals, molecular electrostatic potentials, and charge and spin densities. A list of the most popular and developed software packages is available from Wikipedia: http://en.wikipedia.org/wiki/List_of_quantum_chemistry_and_solid-state_physics_software. This webpage also contains detailed information on the software, including the source websites and downloads. Academic CI-, MBPT- and CC- module containing software, namely, ACES II, ACESS III, CADPAC, *Columbus, Dalton, Dirac,* and GAMESS, is free to download and use. The oldest and most developed of such *ab initio* packages are GAUSSIAN, GAMESS, NWChem, and ACES II. NWChem and *Molpro* are newer high quality commercial software available for a relatively small fee. *Molpro* is popular among chemists working with molecules composed of relatively light atoms. *Molpro* and ACES III are probably the "youngest" of the above software packages, and thus offer a choice of improved basis sets. However, the older packages are updated on a regular basis, offer advanced and well-tested basis sets and other options, not mentioning their better compatibility with existing and emerging visualization software. CI-type *ab initio* calculations discussed in the remaining chapters of this book have been completed using GAMESS software packages [63]. GAMESS is one of a few most advanced and developed *ab initio* software packages that are widely available on supercomputers.

Visualization software includes a number of codes (http://mariovalle.name/ChemViz/tools.html) that recover quality images of calculated MOs, charge and spin density distributions converting digital MOs obtained using *ab initio* software. A particular choice of visualization software is up to a user. However, practical experience shows that many of such packages generate problems converting various types of files carrying digital MOs. In the rest of this book both older and newer versions of *Molekel* visualization software (http://molekel.cscs.ch/wiki/pmwiki.php) were used. This is very flexible software compatible with almost any type of output files. An older version of *Molekel* is the only free software package that allows visualization of the spin density distributions from GAMESS output files. It is not free from some errors (for example, on many occasions *Molekel* lists some MOs as doubly occupied while they are not), but strong points of *Molekel* overweight its few disadvantages.

Detailed instructions concerning all practical steps that lead to realization of *ab initio* quantum chemistry calculations can be found in literature. A good book for beginners is Ref. 64 where all such steps are carefully and thoughtfully outlined. Notably, from a chemical standpoint, bounded states of matter include an atom, an ion and a molecule. That is, in software documentation any system, large or small (the system size is restricted only by feasibility of calculations limited primarily by supercomputer's RAM and processor time allowance), is addressed a molecule, an atom or an ion.

The majority of *ab initio* quantum chemistry software packages has been developed by leaders in the field of theoretical and computational quantum chemistry and therefore, is very sophisticated and advanced. A regular user, let alone the beginner, should not be expected to outsmart the developers by throwing away portions of codes, replacing calculations with some simpler methods, and other corner-cutting tricks, as all such tricks will have an immediate impact on the accuracy and meaningfulness of the results. This also includes a choice of a basis set. From a general standpoint, the larger is a basis set the more accurate are calculations results, but, unfortunately, the larger also are computational resources (RAM and processor time) necessary to accomplish such calculations. Users are strongly advised to consult professional literature and/or professionals in the field to optimize their computation task. Note, that for the majority of CI-, MP- and CC-type calculations access to a supercomputer(s) is necessary, because of a huge RAM required to calculate electron repulsion integrals and to complete geometry optimization, among many other resource-consuming tasks. A standard 8-processor Dell station usually cannot handle any such calculations beyond the HF level for a system containing, for example, over 10 transition metal atoms. In many cases, even relatively small molecules composed of such atoms require significant computational resources (thousands of processor hours and hundreds of GB of RAM).

The rest of this book is devoted to HF, CI and MP *ab initio* electronic structure calculations as applied to small systems of significant interest for materials science and nanotechnology. Thus, the structure and geometry of a system are of particular interest. Unfortunately, the existing *ab initio* quantum chemistry software does not emphasize or provide means to recreate a variety of confined geometries, apart from atom-by-atom recovery of the entire environment of a molecule. The *virtual synthesis method* discussed in the following section does provide a systematic approach to this problem, allows examination of various models of confined geometries and composition, at the same time saving huge computational resources.

2.9 The virtual synthesis method

The use of the projection operator methods discussed in Chapter 1 permits to predict from the first principles thermodynamic and electronic transport properties of strongly spatially inhomogeneous systems in terms of the corresponding equilibrium Green's functions (EGFs) and electronic structure, such as electron energies, charge and spin density distributions (CDDs and SDDs, respectively), and EGFs of CDD and SDD fluctuations. The latter quantities can be derived analytically, calculated numerically, obtained via analysis of experimental data sets, or recovered using molecular simulations and modeling.

Analytical derivation of EGFs and electronic structure properties are limited to some relatively simple models that allow such explicit analysis. In more complicated cases, only qualitative results may be derived. Moreover, the majority of models that allow rigorous mathematical analysis are developed to describe bulk systems where atomistic details are not very important, and thus are neglected. The reason for this is that contributions to electronic properties of bulk systems coming from atomic-scale irregularities cancel out almost everywhere in such system due to symmetries of their bulk lattices and the corresponding symmetries of their electronic structure. At the same time, the development of novel experimental methods of manipulations with nano- and atomic-scale systems allows engineering of advanced materials and media exactly tailored to meet specific practical needs, provided one can reliably describe detailed experimental and technological conditions to guide experimental synthesis to ensure desirable system properties. The latter requires precise results that cannot be obtained by using analytical models devised to predict bulk system properties.

Experimental approaches, on the other hand, cannot proceed smoothly on their own because any data set measured experimentally needs analysis and understanding that can only be achieved using appropriate analytical and computational tools. In their turn, molecular simulations and modeling cannot provide real-time characterization of realistic quantum systems composed of more than about a hundred of atoms, including those of the environment. Moreover, such simulations and modeling needs experimental data input and reliable simulations/modeling algorithms that can only be tested against results of detailed theoretical analysis, computations and experiment. Such model and algorithm testing is extremely demanding for small nano- and atomic-scale systems where very few reliable theoretical and experimental results are available.

Therefore, the first-principle, *ab initio* numerical calculations are the only method at present that allows reliable prediction of equilibrium system properties in all atomistic detail. Such calculations may recreate various experimental conditions, system geometry, structure, composition, and similar characteristics of the environment, and thus enable a possibility of comparison of the obtained results to all appropriate and available analytical and experimental data. The use of the first-principle-derived calculation methods ensures unambiguity of the results, relative generality of conclusions, and in many cases, scalability of the results with the system size. These advantages of the fist-principle calculations are very important for quantum systems where any experimental or modeling details may have a significant impact on the results.

In the remaining chapters of this book the electronic structure of about 40 systems composed of semiconductor compound atoms are studied using HF, CI, CASSCF, MCSCF and MP2 methods discussed above in sections 2.3 to 2.6. To configure a realistic system using available software packages, one has to ascertain initial coordinates and types of the system atoms, and those of the system environment. However, the available software provides a very few options useful in the case of *ab initio* computational synthesis of quantum confined systems. Standard software setups are designed to optimize geometry and minimize the total energy of a group of atoms in "vacuum", when no foreign atoms or ions are present. To overcome this limitation, *the virtual* (*i.e.*, the first principle theory-based, computational) *synthesis method* [65–68] used to obtain computational results discussed in remaining chapters of this book suggests a systematic approach that helps chose the initial system configuration and composition. This approach ensures efficiency and success of computations confirmed by fast convergence of variational procedures and meaningfulness of the obtained results.

A regular approach to geometry optimization realized in all contemporary software packages consists of (1) assigning positions to atomic centers on ad-hoc basis, (2) calculation of the total HF energy of the system, (3) changing the Cartesian coordinates one at a time of one atom only and re-calculating the HF energy of the system again, (4) comparing the re-calculated HF energy value to the one calculated previously, (5) choosing the system coordinates that provide for the lower energy value and continuation of the calculations from that system configuration, and (6) trying other configurations (with all the above steps repeated), to make sure that the actual minimum of the HF energy is found. Such computations sometimes take thousands of steps and more processor time than actual CI calculations, and not always end up with a stable atomic configuration (that is, the procedure may not converge).

The virtual synthesis method starts from a so-called "pre-designed" configuration of atoms and composition of the system that are derived from those of symmetry elements of the natural bulk lattice(s) specific to a studied group of atoms. For example, considering a group of Ga and As atoms, one may start from their positions in an fcc cube of the natural fcc lattice of GaAs and calculate its HF energy and the ground state HF wavefunction. One would obtain MOs, CDDs and data on uncompensated valence on each atom, from which one may conclude that one or more atoms house excessive electron charge (so-called dangling bonds) that is not compensated in such configuration. If such is the case, the next step is to re-configure atoms (usually by eliminating those with dangling bonds), and complete the HF calculations again. Soon enough one would arrive with a (tetrahedral pyramid, in the case of Ga-As system) configuration that does not have dangling bonds. During all these calculations the atomic centers of all remaining atoms are kept constrained to their initial positions. Once a configuration without dangling bonds is identified (usually after a few computational steps) a sequence of desirable CI, CASSCF, MCSCF, MP2, *etc.* calculations is completed and the molecule is configured. Notably, the above computational model of synthesis of such a molecule recreates a natural "in lattice", or "in quantum confinement" synthesis setup where the atoms of the molecule are embedded into a natural bulk lattice specific to such atomic group. One must realize, though, that such a model configuration reflects the confinement only partially, because polarization of AOs of the atoms in the molecule due to those of the confinement is not entirely accounted for. The atoms of confinement are absent, and the confinement is only manifested by spatial constraints applied to the center of mass of the atoms of the cluster. Thus, only excluded volume effects, and polarization of the AOs due to such effects, are properly accounted for. Nevertheless, in the majority of cases molecular geometries so obtained correspond to stable states, and the properties of molecules so pre-designed show remarkable quantum confinement effects. To compare the electronic level structure (ELS) of such a pre-designed molecule to a molecule that is synthesized without any foreign atoms or fields present (that is, in "vacuum"), the constraints applied to the atomic positions are lifted, and the geometry of the molecule optimized again. Once the optimal geometry is HF-calculated for the unconstrained cluster, the rest of the CI, CASSCF, MCSCF and MP2 studies are done. [The CI or any other procedure in the sequence may or may not converge, although it does converge for almost all studied tetrahedral pyramidal and hexagonal prismatic semiconductor compound systems of the following chapters.] The result is the so-called "vacuum" or "unconstrained" molecule, that is, the molecule synthesized of the same atoms as those in the pre-designed molecule, but without any constraints applied to the atomic positions. Comparison of ELS and the rest of the electronic properties of pairs of molecules so synthesized leads, as a rule, to an understanding of the type and strength of model quantum confinement effects. Obviously, the pre-designed molecule in such a pair realizes a stable state of the studied atomic cluster, while the unconstrained one realizes the equilibrium state of the cluster. Electronic properties of molecules so synthesized and quantum confinement effects so modelled are discussed in the remaining chapters of this book.

Among other remarkable properties of the virtual synthesis method is its ability to help identify spatial isomers of a given cluster of atoms and cluster's "universally stable" configurations, that is, configurations that change only slightly when constraints applied to atomic positions in the pre-designed molecules are lifted. Thus, the tetrahedral pyramid and hexagonal prism configurations change only slightly for almost all small virtually synthesized III-V semiconductor compound molecules when the atomic position constraints are removed. For Ni-O and Co-O systems, other types of "universally stable" configurations are realized. In particular, Ni-O small clusters tend to become as flat as possible, while Co – O clusters tend to remain 3-dimensional. Stable configurations of In-As-Co/Ni systems deviate from pyramidal shape, but this may be due to the fact that only one or two substitution Ni or Co atoms were included in clusters built of In and As atoms. At the same time, nitrogen substitution defects do not distort much the tetrahedral pyramid shape of the carrier In-As system, while In-N systems tend to realize hexagonal prisms configurations.

In addition to improving efficiency of computations and convergence of variational procedures, a creative approach to the development of initial atomic configurations offered by the virtual synthesis method helps model many important confinement, stress and composition effects on the electronic properties of the studied systems. In the following chapters of this book many of such effects are considered in detail. In particular, in a Ga – As based pre-designed molecule, one may replace all Ga atoms with In ones without changing atomic positions (thus assigning to In atoms the covalent radius of Ga ones), to study effects of tight quantum confinement on In – As structures. Such systems model small stable In – As inclusions in a GaAs bulk lattices. Similarly, replacement of As atoms in In – As pyramidal structures with nitrogen atoms while positions of the centers of mass of all atoms are kept unchanged allows detailed investigation of an In – N molecule loosely confined in a relatively large void of In-As lattices. Increasing a number of atoms in a cluster and then synthesizing computationally the corresponding pre-designed and unconstrained molecules one can "observe" formation of closely lying bunches of electronic energy levels that with further increase in the number of atoms in the cluster develop into the corresponding valence and conduction bands, and sub-bands.

These and other manipulations with atomic cluster composition and structure, and those of quantum confinement, offer a computationally effective way to model properties of molecules that are realized in various types of quantum confinement and on surfaces without computationally demanding re-creation of all atomistic details of the environment and synthesis conditions. The virtual synthesis method offers a sophisticated and computationally feasible alternative to detailed modeling of molecular synthesis in confinement. In particular, it allows building of computational "templates" of various engineered materials and systems computationally "synthesized" in quantum confinement, and manipulation with systems' parameters, such as the ratios of the covalent radii of various atoms in a molecule, its geometry and composition, and those of the confinement to establish desirable structure-property relations. As shown in the following chapters of this book and literature citations there, this approach has been proven to deliver both insightful and practical results confirmed by experimental data and molecular simulations.

While the major analytical, algorithmic, computational and software tools are available, there are some algorithms and software modules that still have to be developed to facilitate calculations of non-equilibrium electronic properties of (nano)systems from computational data specific to their ELS and other equilibrium properties. An immediate task in this area is to develop algorithms and codes to mimic "equilibrium ensemble averaging" procedure to calculate the equilibrium "charge density - charge density" and "spin density - spin density" Green's functions. Somewhat more complicated task is to develop algorithms of calculations of the "local current - local current" and "CDD/SDD fluctuation - CDD/SDD fluctuation" Green's functions from data included in the output files of the existing CI/MP and CC-type software packages. Progress in these directions will significantly help computational design of (nano) materials and media, thus providing a practical guide for experimental synthesis of advanced nanomaterials and their technological production.

References

[1] Levin, I. N. (2000). "Quantum Chemistry," 5th ed. Prentice Hall, New Jersey.
[2] Logan, J.D. (1977). "Invariant Variational Principles." Mathematics in Science and Engineering, Vol. 138. Academic Press, New York.
[3] Anderson, R. L., and Ibragimov, N. H. (1979). "Lie-Bäcklund Transforms in Applications. SIAM Studies in Applied Mathematics." SIAM, Philadelphia.
[4] Vinogradov, A. M. (1984). The ɞ-spectral sequence, Lagrangian formalism, and conservation laws. I and II. *J. Math. Analysis and Applications* **100**, 1–40; Ibid., 41–129.
[5] Kupershmidt, B. A. (1980). Geometry of jet bundles and the structure of Lagrangian and Hamiltonian formalisms. *In* "Geometric Methods in Mathematical Physics (Lowell, MA, 1979)." Lecture Notes in Mathematical Physics (G. Kaiser and J. E. Marsden, Eds.), Vol. 775, pp. 162–218. Springer-Verlag, Berlin.
[6] Olver, P. J. (1993). "Applications of Lie Groups to Differential Equations," 2nd ed. Graduate Texts in Mathematics, Vol. 107. Springer-Verlag, New York.
[7] Zuckerman, G. J. (1987). Action principles and global geometry. *In* "Mathematical Aspects of Spring Theory (University of California, San Diego, 1986)". Advanced Series in Mathematical Physics (S. T. Yau, Ed.), Vol. 1, pp. 259–284. World Scientific, Singapore.
[8] Kolář, I., Michor, P. W., and Slovak, J. (1993). "Natural Operators in Differential Geometry". Springer-Verlag, Berlin.
[9] Deligne, P., and Freed, D. S. (1999). Classical field theory. *In* "Quantum Fields and Strings: a Course for Mathematicians" (P. Deligne, P. Etingof, D. S. Freed, L. C. Jeffrey, D. Kazhdan, J. W. Morgan, D. R. Morrison, and E. Witten, Eds.), Vol. 1, pp. 137–225. American Mathematical Society/Institute for Advance Study, Providence, RI.
[10] Noether, E. (1918). Gleichungen mit vorgeschriebener Gruppe. *Mathematische Annalen* **78**, 221–229 (in German).
[11] Noether, E. (1918). Invarianten beliebiger Differentialausdrüke. *Göttinger Nachrichten*, pp. 37–44 (presented by F. Klein, Jan. 25, 1918), (in German).

[12] Noether, E. (1918). Invariante Variationsprobleme. *Göttinger Nachrichten*, pp. 235–257 (presented by F. Klein, July 26, 1918), (in German).
[13] Kosmann-Schwarzbach, Y. (2011). "The Noether Theorems". Springer, New York.
[14] Bates, D. R. (Ed.) (1961-62). "Quantum Theory". Vols. 1 to 3. Academic Press, New York.
[15] Hameka, H. F. (1981). "Quantum Mechanics". Wiley, New York.
[16] Wilson, E. B., Jr. (1933). Wave functions for the ground state of lithium and three-electron ions. *J. Chem. Phys.* **1**, 210–218.
[17] Born, M., and Oppenheimer R. (1927). *Ann. Phys.* **389**, 457–484 (in German).
[18] Szabo, A., and Oslund, N. S. (1996). "Modern Quantum Chemistry. Introduction to Advanced Electronic Structure Theory". Dover, New York.
[19] Parr, R. G. (1963). "Quantum Theory of Molecular Electronic Structure". Benjamin, New York.
[20] Helgaker, T., Olsen, J., and Jorgensen, P. (2012). "Molecular Electronic-Structure Theory". Wiley, England.
[21] Wahl, A. C., and Das., G. (1977). The multiconfiguration self consistent field method. *In* "Methods of Electronic Structure Theory." (H. F. Schaefer, Ed.), pp. 51–78. Springer, NY.
[22] Schmidt, M.W., and Gordon, M.S. (1998). The construction and interpretation of MCSCF wavefunctions. *Ann. Rev. Phys. Chem.* **49**, 233–266. DOI: 10.1146/annurev.physchem.49.1.233.
[23] Roos, B. O. (1983). The multiconfiguration SCF method. *In* "Methods in Computational Molecular Physics." (G.H.F. Diercksen and S. Wilson, Eds.), pp. 161–187. D. Reidel Publishing, Dordrecht, Netherlands.
[24] Roos, B. O. (1994). The multiconfiguration SCF method. *In* "Lecture Notes in Quantum Chemistry." (B. O. Roos, Ed.), Vol. 58, pp. 177–254. Springer-Verlag, Berlin.
[25] Olsen, J., Yeager, D.L., and Jorgensen, P. (1983). Optimization and characterization of a MCSCF state. *Adv. Chem. Phys.* **54**, 1–176. DOI: 10.1002/9780470142783.ch1.
[26] Werner, H.-J. (1987). Matrix formulated direct MCSCF and multiconfiguration reference CI methods. *Adv. Chem. Phys.* **69**, 1–62. DOI: 10.1002/9780470142943.ch1.
[27] Shepard, R. (1987). The MCSCF method. *Adv. Chem. Phys.* **69**, 63–200. DOI: 10.1002/9780470142943.ch2.
[28] Wilson, S. (1984). "Electron Correlations in Molecules." Oxford University Press, UK.
[29] Stevens, W. J., Krauss, M., Basch, H., and Jasien, P. (1992). Relativistic compact effective potentials and efficient, shared-exponent basis sets for the third-, fourth-, and fifth-row atoms. *Can. J. Chem.* **70**, 612–630.
[30] Roos, B. O. (1987). The CASSCF method and its application in electronic structure calculations. *Adv. Chem. Phys.* **69**, 339–445. DOI: 10.1002/9780470142943.ch7.
[31] Hehre, W. J. (1995). "Practical Strategies for Electronic Structure Calculations." Wavefunction, Irvine, CA.
[32] Stewart, G. W. (2001). "Matrix Algorithms." Vol. 2. SIAM, Philadelphia, PA.
[33] van Lenthe, J. H., and Pulay, P. (1990). A space-saving modification of Davidson's eigenvector algorithm. *J. Comp. Chem.* **11**, 1164–1168.
[34] Sherrill, C. D., and Schaefer III, H. F. (1999). The configuration interaction method: advances in highly correlated approaches. *Adv. Quant. Chem.* **34**, 143–269.
[35] Ivanic, J. (2003). Direct configuration interaction and multiconfiguration self-consistent field method for multiple active spaces with variable occupations. I. Method. *J. Chem. Phys.* **119**, 9364–9376.
[36] Brooks, B. R., and Schaefer III, H. F. (1979). The unitary group approach to the electron correlation problem. Methods and preliminary applications. *J. Chem. Phys.* **70**, 5092–5106.
[37] Siegbahn, P. E. M. (1980). Generalization of the direct CI method based on the graphical unitary group approach. II. Single and double replacements from any set of reference configurations. *J. Chem. Phys.* **72**, 1647–1656.

[38] Foresman, J. B., Head-Gordon, M., Pople, J. A., and Frisch, M. J. (1996). Toward a systematic molecular orbital theory for excited states. *J. Phys. Chem.* **96**, 135–149.
[39] Webb, S. P. (2006). Ab initio electronic structure theory as an aid to understanding excited state hydrogen transfer in moderate to large systems. *Theoret. Chem. Acc.* **116**, 355–372.
[40] Shao, Y., Molnar, L. F., Jung, Y., Kussmann, J., *et al.* (2006). Advances in methods and algorithms in a modern quantum chemistry program package. *Phys. Chem. Chem. Phys.* **8**, 3172–3191.
[41] He, Y., and Cremer, D. (2000). Assessment of high order correlation effects with the help of Moller-Plesset perturbation theory up to sixth order. *Mol. Phys.* **98**, 1415–1432.
[42] Krishnan, R., and Pople, J. A. (1978). Approximate fourth-order perturbation theory of the electron correlation energy. *Int. J. Quant. Chem.* **14**, 91–100.
[43] Pople, J. A., Head-Gordon, M., and Krishnan, R. (1988). Corrections to correlations energies beyond fourth order Moller-Plesset (MP4) perturbation theory. Contributions of single, double, and triple substitutions. *Int. J. Quant. Chem.* **34**, 377–382.
[44] Foresman, J. B., and Frisch, A. E. (1996). "Exploring Chemistry with Electronic Structure Methods," 2nd ed. Gaussian, Pittsburgh, PA.
[45] Amos, R. D., Andrews, J. S., Handy, N. C., and Knowles, P. J. (1991). Open-shell Møller-Plesset perturbation theory. *Chem. Phys. Lett.* **185**, 256–264. DOI: 10.1016/S0009-2614(91)85057-4.
[46] Anderson, K., Malmqvist, P.-A., and Roos, B. O. (1992). Second-order perturbation theory with a complete active space self-consistent field reference function. *J. Chem. Phys.* **96**, 1218–1226.
[47] Anderson, K., and Roos, B. O. (1995). Multiconfigurational second-order perturbation theory. *In* "Modern Electronic Structure Theory". (D. R. Yarkony, Ed), pp. 55–109. World Scientific, Singapore.
[48] Murphy, R. B., Pollard, W. T., and Friesner, R. A. (1997). Pseudospectral localized generalized Møller-Plesset methods with a generalized valence-bond reference wave-function – Theory and calculations of conformal energies. *J. Chem. Phys.* **106**, 5073–5084.
[49] Coester, F. (1958). Bound states of a many-particle system. *Nucl. Phys.* **1**, 421–424.
[50] Coester, F., and Kümmel, H. (1960). Short range correlations in nuclear wave functions. *Nucl. Phys.* **17**, 477–485.
[51] Čížek, J. (1966). On the correlation problem in atomic and molecular systems. Calculation of wavefunction components in Ursell-type expansion using quantum-field theoretical methods. *J. Chem. Phys.* **45**, 4256–4266.
[52] Paldus, J., Čížek, J., and Shavitt, I. (1972). Correlation Problems in Atomic and Molecular Systems. IV. Extended coupled-pair many-electron theory and its application to the BH_3 molecule. *Phys. Rev. A* **5**, 50–67.
[53] Bartlett, R. J., and Purvis, G. D. III. (1978). Many-body perturbation theory, coupled-pair many-electron theory, and the importance of quadruple excitations for the correlation problem. *Int. J. Quant. Chem.* **14**, 561–581.
[54] Bishop, R. F., and Luhrmann, K. H. (1978). Electron correlations: I. Ground-state results in the high-density regime. *Phys. Rev. B* **17**, 3757–3780.
[55] Bartlett, R. J., and Purvis, G. D. III. (1980). Molecular applications of coupled cluster and many-body perturbation methods. *Phys. Scr.* **21**, 225–265.
[56] Shavitt, I., and Bartlett, R. J. (2006). "Many-Body Methods in Quantum Chemistry: Many-Body Perturbation Theory and Coupled-Cluster Theory". Cambridge University Press, Cambridge, MA.
[57] Bartlett, R. J., and Musial, M. (2007). Coupled-cluster theory in quantum chemistry. *Rev. Mod. Phys.* **79**, 291–352.
[58] Tatewaki, H., Koga, T., Shimazaki, T., and Yamamoto, S. (2004). Quality of contracted Gaussian-type function basis sets. *J. Chem. Phys.* **120**, 5938–5945.

[59] Wilson, A. K., Woon, D. E., Peterson, K. A., and Dunning, T. H., Jr. (1999). Gaussian basis sets for correlated molecular calculations. IX. The atoms gallium through krypton. *J. Chem. Phys.* **110**, 7667–7676.
[60] van Mourik, T., Wilson, A. K., Peterson, K. A., Woon, D. E., and Dunning, T. H., Jr. (1999). The effect of basis set superposition error (BSSE) on the convergence of molecular properties calculated with the correlation consistent basis sets. *Adv. Quant. Chem.* **31**, 105–133.
[61] Almlöf, J., and Taylor, P. R. (1987). General contraction of Gaussian basis sets. I. Atomic natural orbitals for first- and second-row atoms. *J. Chem. Phys.* **86**, 4070–4077.
[62] Noga, J., Kutzelnigg, W., and Klopper, W. (1992). CC-R12, a correlation-cusp corrected coupled cluster method with a pilot application to the Be_2 potential curve. *Chem. Phys. Lett.* **199**, 497–501.
[63] http://www.msg.ameslab.gov/GAMESS; http://www.emsl.pnl.gov/docs/nwchem/nwchem.html.
[64] Young, D. (2001). "Computational Chemistry: A Practical Guide for Applying Techniques to Real World Problems". Wiley, New York. ISBN 0-471-33368-9.
[65] Pozhar, L. A., Yeates, A. T., Szmulowicz, F., and Mitchel, W. C. (2005). Small atomic clusters as prototypes for sub-nanoscale heterostructure units with pre-designed charge transport properties. *EuroPhys. Lett.* **71**, 380–386.
[66] Pozhar, L. A., Yeates, A. T., Szmulowicz, F., and Mitchel, W. C. (2006). Virtual synthesis of artificial molecules of In, Ga and As with pre-designed electronic properties using a self-consistent field method. *Phys. Rev. B* **74**, 085306. See also *Virtual J. Nanoscale Sci & Technol.* **14**, No. 8 (2006). http://www.vjnano.org.
[67] Pozhar, L. A., and Mitchel, W. C. (2007). Collectivization of electronic spin distributions and magneto-electronic properties of small atomic clusters of Ga and In with As, V and Mn. *IEEE Trans. Magnet.* **43**, 3037–3039.
[68] Pozhar, L. A., and Mitchel, W. C. (2009). Virtual synthesis of electronic nanomaterials: fundamentals and prospects. *In* "Toward Functional Nanomaterials. Lecture Notes in Nanoscale Science and Technology" (Z. Wang, Ed.), Vol. 5, pp. 423–474. Springer, NY. See also arXiv:cond-mat/0502476) [cond-mat.mes-hall]; ISBN: 978-0-387-77717-7.

Part Two

Applications: Electronic Structure of Small Systems at Equilibrium

Quantum Dots of Traditional III–V Semiconductor Compounds

3.1 Introduction

GaAs and InAs bulk semiconductors have been thoroughly studied since the mid-20 century (see, for example, Ref. 1), and are among the most studied systems due to their exceptional electronic properties. Thus, GaAs systems possess a wide direct band gap (1.424 eV at 300° K), relatively high electron density and mobility, and a large dielectric constant. Among other advantages of gallium arsenide, these properties enable efficient light emitters, space electronics applications, optical windows in high power applications, high frequency and low noise transistors, switches and other electronic devices. A narrow band gap (0.354 eV at 300° K) and even higher electron mobility in InAs make it a valuable material for infrared radiation sources, photodiodes and diode lasers. Smaller GaAs- and InAs-based systems (down to microns in linear dimensions) have also been studied during the recent two decades and form a materials basis of modern electronics. Yet persistent demand for a sharp increase in the efficacy of computer systems and electronic devices commands even steeper increase in the density of active elements in integrated circuits (ICs). The latter can only be achieved by a dramatic decrease in linear dimensions of the active elements, and 3-dimensional (3D) layout of ICs. Thus, during the recent decade, sub-micron quantum dot and wire systems (QDs and QWs, respectively, or QDWs) down to several hundred nanometers in linear dimensions are vigorously studied and utilized in contemporary electronic devices providing a temporal solution to the efficacy challenge.

The next step in evolution of computing principles and electronic device architecture requires the development of 3D ICs using nano- and sub-nanoscale electronic materials with QDW-based active elements where each of QDWs is of a few nanometers in characteristic linear dimensions. In particular, QDs have been proposed as building blocks of future quantum processors operating with entangled electron spin – based quantum bits (qubits), qubits realized using entangled atomic spins, or optical qubits realized by entangled photons. Various architectures (see, for example, Ref. 2) of quantum integrated circuits (QICs) include coupling of qubits to active elements of QICs at the rate in the range of 30 MHz typical for the coupling rate of small GaAs QDs. Self-assembled GaAs QD systems may be used to implement optically-controlled large-scale computing, fast initialization of spin qubits and non-invasive quantum measurements [3,4]. At the same time, strong spin-orbit interactions of electrons with nuclei in InAs QDs provide a means to optically control spin dephasing [5] and to manipulate electron spin rotations electrically using local gate electrodes [2]. Recent experimental studies and modeling confirmed several schemes of quantum computers based on optically controlled electron spins residing on charged QDs [6–8], and demonstrated high fidelity of the corresponding one - and two - qubit gates [7]. Off-resonant Raman transition measurements [9], time-resolved Kerr rotation spectroscopy [10], vacuum Rabi splitting measurements [11] and other methods have been suggested to control precision of the quantum gates and qubit dephasing. Theoretical and experimental studies of optical properties of excitons in QD systems have inspired quantum computing schemes based on polarization-entangled photons [12,13] controlled by an electric field [14], and several optical methods [15–17]. Traditional approaches to chip technologies and the development of quantum electronic devices, such as lasers [18–29], transistors [19,27,29,31,32], sensors [16,17,19,21–23,28,33–37] and photonic crystals [38–40] also undergo revolutionary transformations as novel QDW-based nanoheterostructures become available for these applications. Novel QDW-based electronic devices utilize superior electronic and optical properties of nano- and sub-nanoscale GaAs- and InAs-based electronic nanomaterials in a way similar to that in which previous generations of electronic devices relied on advantageous properties of bulk GaAs and InAs semiconductors.

Since the end of 1990's, experimental synthesis of novel GaAs and InAs QDW-based nanoheterostructures made a substantial progress toward a decrease in linear dimensions of QDW elements from thousands of nanometers to about 2 nm [4,8,12,13,19,26,41–49] while widening a range of synthesis parameters, such as confinement structure and composition, and process temperature from about 400° K to 6° K. Yet detailed experimental characterization of each of such QDW elements constitutes a strong challenge even with modern measurement systems and software. Many of experimental synthesis methods naturally produce well-defined pyramidal QDs originally observed in Refs. 41–46 that are the major subject of this chapter.

Therefore, theoretical investigations of such QDWs, in addition to being of paramount importance for fundamental science, remain a valuable source of information necessary to advance the synthesis technologies. Many of such studies (see, for example, Refs. 50–52) have been concerned with characterization of micro- to large nanoscale QDs, and use a number of approximations and methods adopted from theoretical approaches developed to characterize bulk semiconductors and their heterostructures. Yet another large group of theoretical approaches is based on density functional theory (DFT) methods equipped with semi-phenomenological models of exchange-correlation functionals (see further analysis of such DFT-based methods in Ref. 53 and references therein). Extensive and widely publicized literature exists to reflect the nature and results of such theoretical modeling. In this chapter, an emphasis is made on the first-principles, many-body theoretical methods that do not include any semi-phenomenological or otherwise poorly characterized models or approximations. Very small clusters composed of several Ga, As and In atoms were studied in the end of 1990's and beginning of 2000's [54–66] using simpler many-body theoretical methods feasible at that time. In later studies, larger systems composed of Ga, In and As atoms have been synthesized computationally [67–73] using modern powerful many-body theoretical means of the virtual synthesis method discussed in Chapter 2. In contrast to earlier studies, in these works the major attention is paid to spatial constraints that dramatically impact the structure, chemical bonding and electronic and magnetic properties of molecules composed of Ga, As and In atoms and synthesized in confinement.

3.2 Virtual synthesis setup

At present, the smallest stable cluster built of In and As atoms, and the largest stable atomic clusters composed of Ga, In and As atoms computationally have been reported in Refs. 67, 68, and 72. All of these clusters have been derived using tetrahedral symmetry elements of the GaAs and InAs zincblende bulk lattices and the covalent radii 1.26 Å, 1.44 Å and 1.18 Å of Ga, In and As atoms, respectively, in these lattices known from experiment at 300° K for the GaAs lattice, and 298.5° K for the InAs lattice.

Optimization of the clusters to their stable molecular structure has been done using creatively total energy minimization procedures realized as a series of the first-principles, many-body quantum theoretical methods discussed in Chapter 2 in detail and implemented as the GAMESS software package [73]. Thus, the restricted Hartree-Fock (RHF) and the restricted open-shell Hartree-Fock (ROHF) method with the SBKJC basis function set has been applied to reduce the Schrödinger equation of the studied clusters and to develop the zero-order (Born-Oppenheimer) approximation of the self-consistent field (SCF) determinant wave functions that include configuration interactions (CI) only through the Pauli principle. These wave functions have been used as input for complete active space (CAS) SCF variation procedures applied to the non-reduced Schrödinger equation to improve the determinant wave functions. The resulting improved CASSCF determinant wave functions are built of linear combinations of electronic orbitals specific to various distributions of all or some of the system electrons over all or some prescribed molecular orbitals, depending on a choice of unrestricted or restricted CAS, respectively. The CASSCF procedure that uses unrestricted CAS (that is, all electrons of a systems are distributed in all possible ways over all available molecular orbits) produces the most accurate determinant wave function of the system. However, huge demands for computation time and memory make the unrestricted CASSCF procedure impractical for systems composed of multi-electron atoms, so it is usually replaced by a restricted CASSCF with a set of several valence electrons distributed over several molecular orbits (MOs). The resulting determinant wave functions of the studied systems are further improved by the multiconfiguration SCF (MCSCF) method that includes static, and partially dynamic, correlations into consideration, and produces the corresponding multi-determinant wave functions that are used to study electronic properties of the systems. The MCSCF multi-determinant wave functions can be improved to account for two- electron, 3-electron, etc. dynamic correlations using the Moller-Plesset (MP) perturbation method. Unfortunately, even the simplest realization of this method (MP2) that accounts for 2-electron dynamic correlations, is extremely resource consuming for systems built of multi-electron atoms, such as Ga, In and As, and is impractical to realize using available modern supercomputers. Thus, MP2 method has been applied thus far only to the simplest of the studied molecules, In_3As.

In Refs. 67 and 68 the standard total energy minimization procedure outlined above was generalized to include conditional total energy minimization. In this case, the total energy minimization procedure is used while some constraints are applied to motion of cluster's atoms. Such constraints may reflect the presence of (atomically-structured) quantum confinement/surfaces and/or electromagnetic fields originated outside of the cluster. In the case of Ga-As and In-As clusters discussed in this chapter, such constrains were realized as boundary conditions in the form of fixed positions of the centers of mass of clusters' atoms, while solving the Schrödinger's equation for the clusters' wave functions. The solutions (multi-determinant MCSCF wave functions) of the Schrödinger's equation so obtained are specific to stationary states of the clusters realized by local minima of the clusters' total energy. The virtual synthesis method of Refs. 67 and 68 includes both unconditional (standard) and conditional total energy minimization procedures. The unconditional energy minimization procedure is designed to optimize an atomic cluster (that is, to synthesize the corresponding molecule) in the absence of any environment, and is supposed to deliver the total energy minimum of the clusters. Below in this chapter and the book molecules so obtained are called "vacuum" or "constrained" molecules, in contrast to those called "pre-designed" molecules and obtained using the conditional energy minimization of Refs. 67 and 68. In the latter case, the structure, geometry and composition of the optimized clusters are pre-designed using symmetry considerations before application of the total energy minimization procedure. Note, that there may exist a number of local minima of the total energy of a given atomic cluster. Such minima correspond to so-called spatial isomers of a molecule, and may be realized in nature or technological processes at appropriate synthesis conditions, thus being of significant interest for applications.

In this chapter both pre-designed and vacuum molecules of Ga, In and As atoms synthesized virtually thus far are discussed in detail. Symmetry considerations and application of the conditional and unconditional total energy minimization procedures have resulted in six stable pyramidal molecules: the pre-designed and vacuum $Ga_{10}As_4$, $In_{10}As_4$ and $(In_{10}As_4)_{Ga}$ molecules. The structure and composition of the pre-designed $Ga_{10}As_4$ and $In_{10}As_4$ molecules have been derived from those of the tetrahedral symmetry elements of GaAs and InAs bulk lattices, respectively, as noted in the beginning of this section. The corresponding vacuum molecules have been obtained by lifting position constraints applied to the atoms of the pre-designed molecules, and subsequent unconditional minimization of the total energy of the "unconstrained" clusters. The pre-designed $(In_{10}As_4)_{Ga}$ molecule originates from the pre-designed $Ga_{10}As_4$ molecule where Ga atoms were replaced by In atoms without any change in their positions. That is, In atoms were assigned the covalent radii specific to that of the corresponding Ga-based molecule, and "forced" to form a molecule. Interestingly, such a molecule happened to be stable. Moreover, lifting constraints applied to positions of the centers of mass of In atoms in this molecule, and subsequent unconditional minimization of the total energy of this unconstrained molecule have not produced vacuum molecule $In_{10}As_4$. In fact, yet another stable vacuum $(In_{10}As_4)_{Ga}$ molecule with its electronic, optical and magnetic properties different from those of the vacuum $In_{10}As_4$ molecule has emerged. The pre-designed $(In_{10}As_4)_{Ga}$ molecule reflects energetics and bonding that may be expected in strained In-based clusters implanted in Ga-As lattice or nucleated on Ga-As surfaces, while the existence of the vacuum $(In_{10}As_4)_{Ga}$ molecule indicates that similar strained molecules may nucleate and exist on their own, even in the absence of quantum confinement or external electromagnetic fields.

Generally, the pre-designed synthesis procedure of Refs. 67 and 68 accounts for excluded volume effects and the major polarization effects due to quantum confinement of the cluster. At present, this procedure is the only practical way to complete many-body quantum theoretical computations for larger multi-electron atomic clusters (such as clusters composed of semiconductor compound atoms) in quantum confinement. Otherwise, hardware and software restrictions reduce such computations to about 100 atoms, including the atoms of quantum confinement.

In the case of the In_3As molecule, only the unconditional total energy minimization procedure was used, so this molecule is a vacuum one. Attempts to pre-design a similar molecule by fixing positions of its atoms using symmetry considerations did not lead to a structure different from that of the vacuum molecule. Therefore, the only form of the In_3As molecule known at present is the vacuum one, and it realizes the total energy *minimum minimorum* of the In_3As atomic cluster. Electronic spectra, optical and magnetic properties of virtually synthesized In_3As molecule, and pre-designed and vacuum $Ga_{10}As_4$, $In_{10}As_4$, and $(In_{10}As_4)_{Ga}$ molecules are discussed in sections 3 to 8 of this chapter.

3.3 The smallest 3D molecule of In and As atoms

Unconditional and conditional total energy minimization procedures applied to a cluster composed of three In and one As atoms have not produced any stable spatial isomers. Only the vacuum molecule In_3As has been obtained by unconditional total energy minimization. This molecule has been thoroughly studied using several many-body quantum field theoretical methods of increasing accuracy: RHF, ROHF, CI, MCSCF and a version of MP2 called multiconfiguration quasidegenerate perturbation method, or MC-QDP2.

The geometry of the obtained molecule resembles that of almost ideal equilateral pyramid with As atom in its apex, the In-As-In bond angle of 111.3° and the In-As bond length of 2.6794 Å (Fig. 3.1). Interestingly, the In-As bond length in this molecule is somewhat larger than the sum of In and As experimental covalent radii in the In-As bulk lattice, 2.62 Å. This increase in the bond length signifies relaxation of quantum confinement effects affecting In and As atoms arranged in the In-As bulk lattice.

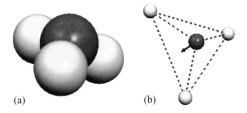

Fig. 3.1 Almost perfect pyramidal In_3As molecule: (a) general view, (b) structure. Indium atoms are yellow, As red. The arrow symbolizes the dipole moment of 1.2729 D. Atomic dimensions are reduced, other dimensions to scale. The lines provide a guide for an eye.

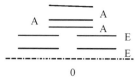

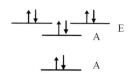

Fig. 3.2 RHF electronic level structure of the In$_3$As molecule in the HOMO – LUMO region. Letters A and E symbolize symmetry of the electron orbits. All dimensions are approximately to scale. The vacuum energy level set to 0. [Adopted from Ref. 68.]

The ground state of the In$_3$As molecule occurred to be a singlet. The RHF singlet (Fig. 3.2) features doubly degenerate highest occupied and lowest unoccupied molecular orbitals (HOMO and LUMO, respectively). Subsequent application of CI optimization procedure with RHF input orbits lifts degeneracy of RHF HOMO, but still leaves LUMO degenerate. This LUMO degeneracy is only lifted after multideterminant MCSCF optimization runs using 4×4 CAS where 4 electrons are distributed over 4 orbits (4×4 CASSCF), and 6×9 CAS with 6 electrons distributed over 9 neighbouring orbits (6×9 CASSCF) in a HOMO-LUMO region. Differences between the ground state energies obtained in various approximation runs (Table 3.1) are within 0.07 H brackets, that is, well within the error brackets of the calculations (about 1 H). The largest energy difference between the energies of the corresponding excited states is about 0.6 H, again within the error brackets of the calculations. All optimization procedures produce consistent one electron, two electron, nucleus-electron, and total molecular energies indicating good convergence of each of the optimization procedures. This convergence, however, is not uniform, which results in different shifts of the state energies dependent upon a particular optimization scheme used. The energy difference between the calculated first excited and ground state energies for each optimization procedure are collected in Table 3.2. These are computational data that correspond to the experimental direct optical transition energies (OTEs).

Given that all of the ground state energies in Table 3.1 lay within the calculation error brackets, any of these ground states can be taken as the "true" one. To ascertain the true ground state, one has to optimize the In$_3$As molecule beyond CASSCF approximation. The obtained data indicate that the lowest ground state of this molecule is produced by MC-QDPT2 optimization of the singlet ground state produced by 6×9 CASSCF approximation (see the underlined energy values in Table 3.1). While experimental data for In$_3$As molecule are not available for comparison, such data for much larger InAs nanotubes (see analysis and references in Ref. 75) and the energy gap value of

TABLE 3.1 Ground and excited state energies of the molecule In$_3$As in the studied approximations.

State	MCSCF singlet, 4×4 CASSCF, Hartree	MCSCF triplet, 4×4 CASSCF, Hartree	MCSCF singlet, 6×9 CASSCF, Hartree	MCSCF triplet, 6×9 CASSCF, Hartree	MP2 singlet (MC-QDPT2), Hartree
ground	−570.936077134	−570.841292817	<u>−570.993334060</u>	−570.892608562	<u>−571.146566455</u>
excited 1	−570.550438771	−570.820254832	−570.673128078	−570.874438424	−571.032250312
excited 2	−570.550155104	−570.820254831	−570.673128018	−570.811170365	—
excited 3	−570.454903963	−570.806850860	−570.645050865	−570.802028879	—

TABLE 3.2 Transition energies of the molecule In$_3$As in the studied approximations.

4×4 CASSCF triplet-singlet ground state transition, eV	Direct optical transition energy, 4×4 CASSCF singlet, eV	Direct optical transition energy, 4×4 CASSCF triplet, eV	6×9 CASSCF triplet-singlet ground state transition, eV	Direct optical transition energy, 6×9 CASSCF singlet, eV	Direct optical transition energy, 6×9 CASSCF triplet, eV	Direct optical transition energy, MC-QDPT2 singlet, eV
2.5692	10.3863	0.5725	2.7409	8.7132	0.4944	3.1107

0.354 eV for the InAs bulk lattice hint that the optical transition energies (OTEs) of the 6×9 CASSCF and 4×4 CASSCF singlets (that is, the energy differences between the corresponding first excited and ground states) seem to be too large. Note, that the energy differences between the ground states of the 4×4 CASSCF singlet and triplet, and 6×9 CASSCF singlet and triplet (see the columns 1 and 4 in Table 3.2, respectively) are much smaller than the OTEs of the corresponding singlets (the columns 2 and 5 in Table 3.2, respectively), being also much larger than those specific to the triplets (the columns 3 and 6, Table 3.2, respectively). The OTE of the MC-QDPT2 singlet seems reasonable and consistent with an experimentally proven tendency of an increase in OTE within an order of magnitude with a decrease in the number of atoms in a system. Further optimization in the framework of higher-order MP perturbation theory may ascertain the "true" OTE value of In$_3$As molecule.

Analysis of the electron charge density distribution (CDD) and molecular electrostatic potential (MEP) shows that the "surface" of In$_3$As molecule is negatively charged (Fig. 3.3), with electron charge of the upper-shell electrons stretching out beyond the space occupied by the molecule (as defined by the covalent radii of the atoms). Significant portion of the electron charge is accumulated in the vicinity of the As atom (the light blue regions of the MEP isosurface in Fig. 3.3c), both "inside" and "outside" of the molecular space. Remaining electron charge is re-distributed between In atoms to ensure stability of this non-stoichiometric molecule where there is only one As atom to accommodate the electron charge of 3 In atoms. This charge re-distribution

Fig. 3.3 Pyramidal MP2 In$_3$As singlet: (a) the isosurface (grey) of the electron charge density distribution (CDD) corresponding to the isovalue 0.0488; (b) and (c) color-coded surfaces of the molecular electrostatic potential [MEP; values range from negative (greenish) to positive (deep blue)] corresponding to the isovalues 0.0385 and 0.073 of CDD. In and As atoms are yellow and red spheres, respectively. In (a) and (b) all dimensions are to scale; in (c) atomic dimensions are reduced to show the MEP surface "inside" of the molecule. The radius of As atom is larger than that of In due to a bug in Molekel visualization software used.

is facilitated by ligand bonding specific to non-stoichiometric molecules [54]. Two highest occupied and two lowest unoccupied molecular orbits of this molecule are depicted in Fig. 3.4. All four MOs in Fig. 3.4 are delocalized π-bonding orbits generated by 3p valence electrons of 3 ligand In atoms. These MOs are similar to MOs observed experimentally and obtained by DFT theoretical means (see Refs. 54, 76–78 and references therein) for In$_4^2$ gas phase clusters and AlIn$_4^-$ molecules. However, in contrast to those molecules, while the 2n+2 electron counting rule still holds for the total number of valence p-electrons, the In$_3$As molecule is not planar, does not possess significant mediating π-bonding contributions, and d-electrons of In atoms do not participate noticeably in ligand bonding. Thus, more theoretical and experimental investigations are required before specific chemical properties of this molecule, such as aromaticity, can be ascertained. If some quasi-aromatic properties of this molecule exist indeed, there may be a possibility to synthesize experimentally solid nanostructures built of In$_3$As building blocks. Such nanostructures should possess very useful physical properties. In particular, in the ground state they may absorb light in the near infrared band (NIR,

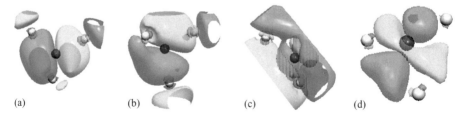

Fig. 3.4 Isosurfaces of the positive (green) and negative (red) components of molecular orbits of the pyramidal MP2 In$_3$As singlet in the HOMO-LUMO region: (a) HOMO 33, isovalue 0.015; (b) HOMO 34, isovalue 0.012; (c) LUMO 35, isovalue 0.010; (d) LUMO 36, isovalue 0.038. In atoms are yellow, As red. Atomic dimensions are reduced to reveal the isosurface structure. Other dimensions are to scale.

of about 790 ×10⁻¹² Hz). However, they may also possess stable excited states (indicated by the 6×9 CASSCF and 4×4 CASSCF triplets of Tables 3.1 and 3.2), where such nanostructures could absorb NIR band light, but lase at frequencies of about an order of magnitude smaller, that correspond to the OTEs of the 6×9 CASSCF and 4×4 CASSCF triplets.

3.4 Pre-designed and vacuum In$_{10}$As$_4$ molecules

The pre-designed and vacuum pyramidal In$_{10}$As$_4$ molecules have been virtually synthesized using ROHF, CI and MCSCF methods. The pre-designed equilateral pyramid (Fig. 3.5a and 3.5b) has been built using the covalent radii 1.44 Å and 1.26 Å of In and As atoms, respectively, in the In-As bulk zincblende lattice. This structure has emerged as a result of the conditional ROHF optimization procedure (the atomic centers of mass were fixed at their locations in the InAs bulk lattice) applied to a cubic zincblende cell with 4 As atoms positioned at ¼ of the cube body diagonals [67,68]. The obtained electroneutral tetrahedral pyramid structure emerged after removing all In atoms of the cube contributing dangling bonds. The corresponding vacuum In$_{10}$As$_4$ pyramid (Fig. 3.5c and 3.5d) has been obtained using the unconditional ROHF optimization procedure where the positioning constrains of the In and As atoms were relaxed (the atomic centers of mass were permitted to "move"). Visually, these two molecules look the same: in the vacuum pyramid In and As atoms moved only slightly (a few hundredths of an Angstrom, on average) with respect to their former positions in the pre-designed pyramid. Despite such an insignificant shift in atomic positions, electronic properties of these molecules differ dramatically. "Surfaces" of both molecules are negatively charged, but the CDD of the pre-designed one reaches into a space outside of that occupied by the molecule's atoms much farther than the CDD of the corresponding vacuum molecule. The ground state energies and OTEs of these molecules (Table 3.3 and Table 3.4) also differ more than those specific for other

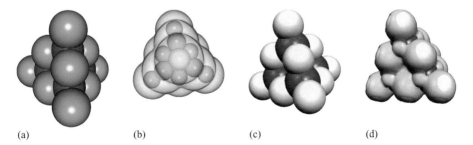

(a)   (b)   (c)   (d)

Fig. 3.5 Pyramidal pre-designed 6×9 CASSCF and vacuum 8×8 CASSCF In$_{10}$As$_4$ triplets. Pre-designed molecule (In atoms are blue, As brown): (a) structure; (b) isosurface (transparent yellow) of the electron charge density distribution (CDD) corresponding to the isovalue 0.01. Vacuum molecule (In atoms are yellow, As red): (c) structure; (d) isosurface (blue) of CDD corresponding to the isovalue 0.05. Atomic dimensions in (b) are reduced, dimensions of As atoms are significantly enlarged due to a bug in the visualization software (Molekel). All other dimensions are to scale.

TABLE 3.3 Ground state energies of the pre-designed and vacuum $In_{10}As_4$ molecules.

Molecule	ROHF triplet ground state energy, Hartree	CI triplet ground state energy, Hartree	CASCF triplet ground state energy, Hartree	CASSCF (number of electrons × number of MOs)
Pre-designed $In_{10}As_4$	−1907.035049	−1907.077252167	−1907.077255735	6×9
Vacuum $In_{10}As_4$	−1906.961402	−1906.851563399	−1906.859967977	8×8

TABLE 3.4 Direct optical transition energies (OTEs) of the pre-designed and vacuum $In_{10}As_4$ molecules.

Molecules	ROHF OTE, eV	CI OTE, eV	CASSCF OTE, eV	CASSCF (number of electrons × number of MOs)
Pre-designed $In_{10}As_4$	3.4286	1.0967	2.9964	6×9
Vacuum $In_{10}As_4$	3.4232	0.8649	0.8684	8×8

14-atomic molecules considered below. As predicted theoretically, convergence of the CI and CASSCF methods is not uniform. In particular, the ground state energy of the pre-designed $In_{10}As_4$ molecule decreases consistently with each subsequent optimization from its ROHF to CI, and to CASSCF value. However, for the corresponding vacuum molecule, ROHF optimization delivers the deepest ground state energy minimum, and the CASSCF one lies between that minimum and the one produced by CI optimization. Similar situation takes place for the OTEs (Table 3.4): there is no uniform tendency of increase or decrease in OTE with the use of progressively accurate optimization. Interestingly, the ground state energy minimum of the pre-designed molecule is deeper than that of the vacuum molecule, although the unconditional energy optimization is supposed to recover the *minimum minimorum* of the total energy. The reason for this is that the dipole moment of the pre-designed molecule is much larger (5.858489 D) than the almost zero (0.000005 D) dipole moment of the vacuum molecule. Thus, the lower conditional total energy minimum is reached at the expense of larger electron charge re-distribution leading to appearance of a large dipole. Still, both minima of the total energy are within the calculation error brackets from each other, and thus formally are indistinguishable.

MEPs of the virtually synthesized $In_{10}As_4$ molecules are depicted in Fig. 3.6 and 3.7, and the ROHF electronic level structure (ELS) in Fig. 3.8 adopted from

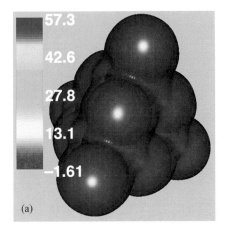 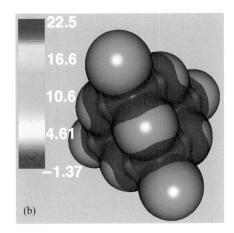

Fig. 3.6 MEP surfaces of the pre-designed 6×9 CASSCF $In_{10}As_4$ triplet corresponding to isovalues (a) 0.03 and (b) 0.05 of CDD. In both cases MEP values are negative (red). Electron charge is spread evenly over the "surface" of the molecule. All dimensions are to scale. Linear dimensions of As atoms are enlarged due to a bug in the visualization software (Molekel). In atoms are blue, As brown.

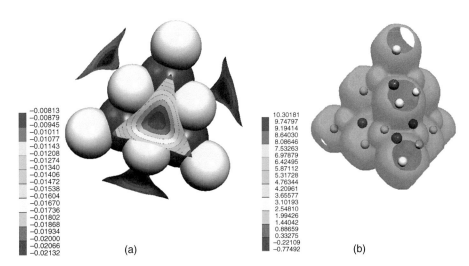

Fig. 3.7 MEP surfaces of the vacuum 8×8 CASSCF $In_{10}As_4$ triplet corresponding to isovalues (a) 0.0001 and (b) 0.05 of CDD. MEP values vary from negative (red) to almost zero (blue). Near the "surface" of the molecule (b) electron charge is spread evenly over the "surface". In (a) all dimensions are to scale. In (b) atomic dimensions are reduced. Linear dimensions of As atoms are enlarged in (a) due to a bug in the visualization software (Molekel). In atoms are yellow, As red.

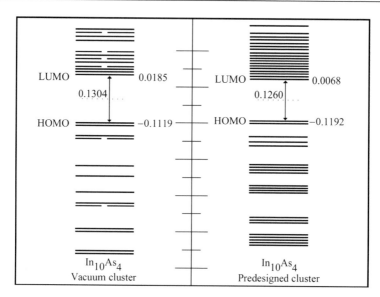

Fig. 3.8 ELSs of the vacuum (left) and pre-designed (right) ROHF $In_{10}As_4$ triplets, respectively, in their HOMO-LUMO regions. The electronic energy level values and OTEs are in Hartree units. All dimensions are approximately to scale. [Adopted from Ref. 67.]

Ref. 67. MEP pictures confirm that the electronic charge distribution in the case of the pre-designed molecule reaches into space much farther than that of the vacuum molecule. Thus, for the same isovalue 0.05, the MEP isosurface of the pre-designed molecule is far outside of the volume occupied by the molecule (Fig. 3.6b), while in the case of the vacuum molecule the corresponding isosurface is deeply inside of the molecular volume (Fig. 3.7b). This signifies that in the pre-designed molecule many more electrons are "pushed" out of the molecule's volume to minimize 2-electron repulsion. These electrons contribute strongly to ligand bonding in the pre- designed $In_{10}As_4$ molecule (see the corresponding discussion in section 3.7 below).

Despite of the visually observed tetrahedral symmetry, ROHF, CI and CASSCF ELSs of the pre-designed $In_{10}As_4$ molecule (Fig. 3.8, right) do not possess degenerate electron orbits. This is due to the fact that the atomic positions of this molecule are not fixed in exact positions corresponding to those of the numerically perfect pyramid. Rather, positions of atomic centers of mass in the vacuum molecule are much closer to those of the numerically perfect pyramid, so that the corresponding ELS (Fig. 3.8, left) exhibits a significant number of degenerate electron orbits. Once again, it can be observed that even a very small change in atomic positions has a profound effect on electronic properties of a molecule.

The ground state energies of both molecules (Table 3.3) are within the calculation error brackets from each other. The pre-designed molecule with its large dipole moment and negative surface charge models effects of possible quantum confinement on a cluster of 10 In and 4 As atoms nucleating in such confinement, as opposed to a cluster (the vacuum molecule) nucleating in a free space. Note, that such effects may include deepening of the ground state minimum and therefore, further stabilization of the molecule, at the expense of polarization of the molecule in the electromagnetic fields exerted by its confinement. Thus, one may stabilize an excited cluster by having it nucleated in quantum confinement, and then manipulate with its dipole moments using external electric fields. Such a scenario makes nucleation in quantum confinement a very attractive proposition for the development of novel materials for quantum electronics, spintronics and quantum information processing.

3.5 "Artificial" molecules $[In_{10}As_4]_{Ga}$

Tuning up electronic properties of $In_{10}As_4$ and other molecules may progress to the use of excluded volume effects of quantum confinement for the development of stressed and overstressed molecules. To understand the outcome of such tuning, two "artificial" molecules (as opposed to the two "natural" ones discussed in the previous section) denoted as $[In_{10}As_4]_{Ga}$ have been synthesized virtually [67,68]. The pre-designed $[In_{10}As_4]_{Ga}$ molecule (Fig. 3.9) has been developed by replacing Ga atoms in the pre-designed $Ga_{10}As_4$ pyramid (discussed in the following section) with In atoms, while keeping all interatomic distances unchanged, and subsequent conditional minimization (the atoms fixed in their positions) of the structure's total energy. Thus, In atoms were assigned the covalent radii of Ga atoms to model an In-based molecule perfectly inbuilt in the Ga-As zincblende lattice (In atoms replacing Ga ones in the lattice). The corresponding vacuum $[In_{10}As_4]_{Ga}$ structure (Fig. 3.10) has been obtained by lifting spatial constraints applied to the positions of atoms in the pre-designed $[In_{10}As_4]_{Ga}$ cluster and subsequent unconditional minimization of the total energy of that structure. The two major stunning results are that (i) the over-stressed pre-designed $[In_{10}As_4]_{Ga}$ pyramid did not disintegrate after the stress (in the form of fixed atomic positions) was eliminated: it keeps the pyramidal shape with its atoms only slightly moved from

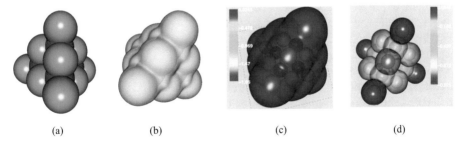

(a) (b) (c) (d)

Fig. 3.9 The pre-designed ROHF $[In_{10}As_4]_{Ga}$ triplet: (a) structure and (b) CDD isosurface (semi-transparent yellow) corresponding to the isovalue 0.036; (c) and (d): MEP surfaces corresponding to the isovalues 0.01 and 0.1 of CDD, respectively. MEP values vary from negative (blue) to almost zero (deep blue). In the space outside the molecule, electron charge (b) is spread evenly over the "surface" of the molecule. In (a) to (c) all dimensions are to scale. In (d) atomic dimensions are reduced. Linear dimensions of As atoms are enlarged in (a) to (c) due to a bug in the visualization software (Molekel). In atoms are blue, As brown.

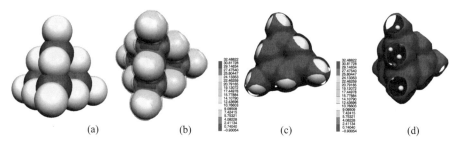

Fig. 3.10 The vacuum 8×10 CASSCF $[In_{10}As_4]_{Ga}$ triplet: (a) structure, (b) CDD isosurface (semi-transparent blue) corresponding to the isovalue 0.054; (c) and (d): MEP surface (red) corresponding to the CDD isovalue 0.036. In the vicinity of the molecular "surface" electron charge (b) is spread evenly over the "surface". In (a) to (c) all dimensions are to scale. In (d) atomic dimensions are reduced. Linear dimensions of As atoms are enlarged in (a) and (b) due to a bug in the visualization software (Molekel). In atoms are yellow, As red.

their formerly fixed positions in the pre-designed $[In_{10}As_4]_{Ga}$ pyramid, and that (ii) the covalent radius of In atoms in the new vacuum $[In_{10}As_4]_{Ga}$ pyramid differs from that specific for the "natural" vacuum $In_{10}As_4$ pyramid of section 3.4. Compared to the "surface" of the pre-designed $[In_{10}As_4]_{Ga}$ molecule, the "surface" of the vacuum one is almost an order of magnitude more negatively charged, but its dipole moment is almost zero (0.000001 D), while that of the pre-designed molecule is 5.269254 D. Atoms of the vacuum molecule "moved" by a few hundredths of Angstrom farther from each other compared to the corresponding separations in the pre-designed pyramid. The closest neighbour separation of In – In atoms is rather rigid, so the pyramidal shape has been hold.

The ground state data of the "artificial" $[In_{10}As_4]_{Ga}$ molecules are collected in Tables 3.5 and 3.6. Once again, the use of subsequently accurate approximations does not necessarily results in progressively deepening or shallowing ground state energy minimum, or progressive increase or decrease in OTE values. This means that the convergence of the used optimization procedures is not uniform. More interesting is the fact, that similar to the case of the "natural" $In_{10}As_4$ molecules, the ground state energy of the "artificial" pre-designed 8×10 CASSCF triplet $[In_{10}As_4]_{Ga}$ is smaller than that of the corresponding vacuum triplet, but the difference in those energies is within the calculation error brackets. Also in similarity to the case of the "natural" $In_{10}As_4$ molecules, the deeper ground state energy minimum of the pre-designed $[In_{10}As_4]_{Ga}$ molecule is reached at the expense of a significant electron charge re-distribution leading to the development of a large dipole moment. This point is illustrated by a comparison of the ground state interaction energies of the pre-designed and vacuum molecules (Table 3.7). Thus, in the case of the vacuum molecule, the unconditional total energy minimization results in a huge decrease (about 2000 H) in electron repulsion and a small decrease in nuclear repulsion energies at the expense of a small decrease in one-electron and nucleus-electron energies stabilizing the system. That huge decrease in the electronic repulsion energy signifies charge re-distribution that also results in almost zero dipole moment. At the same time, the total ground state energy

TABLE 3.5 Ground state energies of the pre-designed and vacuum $[In_{10}As_4]_{Ga}$ molecules.

Molecule	ROHF triplet ground state energy, Hartree	CI triplet ground state energy, Hartree	CASCF triplet ground state energy, Hartree	CASSCF (number of electrons × number of MOs)
Pre-designed $[In_{10}As_4]_{Ga}$	−1906.830250	−1906.883591070	−1906.883611124	8×10
Vacuum $[In_{10}As_4]_{Ga}$	−1906.845323	−1906.848043668	−1906.872776363	8×10

TABLE 3.6 Direct optical transition energies (OTEs) of the pre-designed and vacuum $[In_{10}As_4]_{Ga}$ molecules.

Molecules	ROHF OTE, eV	CI OTE, eV	CASSCF OTE, eV	CASSCF (number of electrons × number of MOs)
Pre-designed $[In_{10}As_4]_{Ga}$	3.5252	1.7679	1.7803	8×10
Vacuum $[In_{10}As_4]_{Ga}$	3.3334	0.4194	3.1813	8×10

of the vacuum molecule becomes larger than that of the pre-designed one, although very slightly and within the calculation error brackets. Given that the used 8×10 CAS is already rather large, these results indicate that even more accurate total energy minimization procedure, such as MP2, is necessary to finalize the ground state energies (and also OTEs) of the pre-designed and vacuum In-based molecules discussed here. This conclusion also is supported by a large number of CAS choices (2×4, 4×7, 8×7, 8×10) that lead to very closely lying unconditional total energy minima within the

TABLE 3.7 Ground state interaction energies for the pre-designed and vacuum $[In_{10}As_4]_{Ga}$ CASSCF triplets.

Molecule	One-electron Energy, Hartree	Two-electron repulsion energy, Hartree	Nuclear repulsion energy, Hartree	Nucleus-electron interaction energy, Hartree	CASSCF (number of electrons × number of MOs)
Pre-designed $[In_{10}As_4]_{Ga}$	−8845.3212179948	6107.2810355163	2738.0401824786	−9565.8857637225	8×10
Vacuum $[In_{10}As_4]_{Ga}$	−8778.1270476891	4167.5349591208	2703.7193122057	−9498.8083122813	8×10

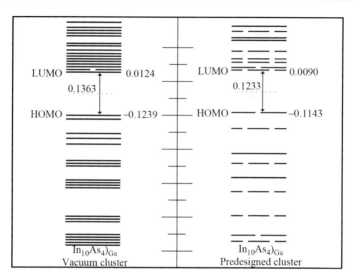

Fig. 3.11 ELSs of the vacuum (left) and pre-designed (right) ROHF $[In_{10}As_4]_{Ga}$ triplets, respectively, in the HOMO-LUMO regions. The electronic energy level values and the OTEs are in Hartree units. All dimensions are approximately to scale. [Adopted from Ref. 67.]

interval from −1906.840516 H to −1906.872776 H, respectively, and indicating that the total energy surface is rather flat near the expected ground state minimum. While one may argue [67] that the existence of the vacuum $[In_{10}As_4]_{Ga}$ molecule is likely due to shortcomings of the used computational procedures, the flatness of the total energy surface in the close proximity of its *minimum minimorum* hints that the vacuum $[In_{10}As_4]_{Ga}$ molecule may indeed be an experimentally realizable spatial isomer of the "natural" vacuum $In_{10}As_4$ molecule. The virtual synthesis method introduced in Refs. 67 and 68, and discussed in Chapter 2 of this book, uses pre-design setups based on symmetry considerations, and thus provides a powerful and independent mean to distinguish between "computational" spatial isomer candidates identifying those that can be realized in laboratory.

Both "artificial" $[In_{10}As_4]_{Ga}$, molecules do not closely resemble the corresponding "natural" $In_{10}As$ molecules. Their CDDs, while keeping appearance of tetrahedral symmetry, reach into space outside the molecules much farther than those of the corresponding "natural" molecules. At the same time, in contrast with ELS of the "natural" $In_{10}As_4$ molecule, ELS of the pre-designed $[In_{10}As_4]_{Ga}$ triplet possess degenerate energy levels typical for a numerically perfect tetrahedral pyramid. ELS of the vacuum $[In_{10}As_4]_{Ga}$ triplet exhibit only a very few degenerate MOs indicating a loss of tetrahedral symmetry. Considering that the structure of the pre-designed $[In_{10}As_4]_{Ga}$ triplet is the same as that of the pre-designed $Ga_{10}As_4$ triplet (which is a numerically perfect pyramid, as described in the following section), one comes to a conclusion that replacement of Ga atoms by In ones in the pre-designed $Ga_{10}As_4$ pyramid does not destroy impact of tetrahedral symmetry on electronic properties, despite the atom replacement.

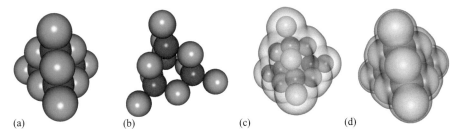

Fig. 3.12 The pre-designed 8×13 CASSCF $Ga_{10}As_4$ triplet: (a) and (b) structure, (c) and (d) CDD isosurface (semi-transparent yellow) corresponding to the isovalue 0.02. Electron charge in (c) and (d) is spread evenly over the "surface" of the molecule. In (a) and (d) all dimensions are to scale. In (b) and (c) atomic dimensions are reduced to show the structure. Linear dimensions of As atoms are enlarged in (a) and (d) due to a bug in the visualization software (Molekel). Ga atoms are blue, As brown.

3.6 $Ga_{10}As_4$ molecules

The pre-designed $Ga_{10}As_4$ molecule has been built using tetrahedral symmetry elements of the GaAs zincblende bulk lattice (Figs. 3.12a and 3.12b) with the experimental covalent radii of Ga and As atoms of 1.26 Å and 1.18 Å, respectively. Similar to the case of the pre-designed In-As molecules, coordinates of the centers of mass of 10 Ga atoms were fixed in their places at the vertexes and edges of the equilateral pre-designed pyramid, and 4 As atoms were placed at ¼ of the fcc cube body diagonals of the original lattice, thus being embedded inside of the pyramid formed by Ga atoms. The conditional total energy optimization routine included ROHF and CI optimizations, and required a rather large 8×13 CASSCF (MCSCF) to ascertain the minimum value of the total energy.

The "surface" of this molecule is almost neutral. Its electronic "surface" charge is uniformly spread in the space about the molecular "surface", but it is much smaller than that of the corresponding In-based molecules of the previous sections. CDD isosurfaces of the pre-designed $Ga_{10}As_4$ molecule for similar isovalues are much closer to the "surface" of the molecule than those of the In-based molecules. The major reason for this is that both pre-designed and vacuum $Ga_{10}As_4$ molecules have 180 electrons less than the In-based molecules of the previous sections, so As and Ga atoms accommodate the electronic charge predominantly "inside" of the molecule. Therefore, the "surface" of this molecule is significantly less negative than those of the In-based molecules.

The vacuum $Ga_{10}As_4$ molecule has been obtained by lifting the spatial constraints applied to atomic positions in the pre-designed molecule, and subsequent unconditional minimization of the total energy of the structure. Visually, the emerged molecule has kept the tetrahedral pyramidal shape (Fig. 3.13a and 3.13b), because its atoms moved only by a few hundredths of an Angstrom from their former positions in the pre-designed pyramid. The electronic "surface charge" of the vacuum molecule is about the same as that of its pre-designed counterpart ("surfaces" of both molecules are almost neutral), and is uniformly spread about the "surface" of the molecule. MEPs of the two molecules (Fig. 3.14) are also very similar, and MEP isosurfaces confirm

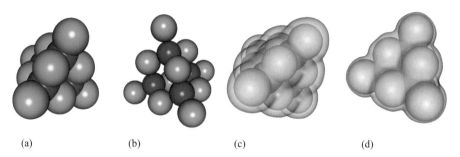

Fig. 3.13 The vacuum 8×8 CASSCF $Ga_{10}As_4$ triplet: (a) and (b): structure, (c) and (d): CDD isosurfaces (semi-transparent yellow) corresponding to the isovalues 0.01 and 0.02, respectively. Electronic charge in (c) and (d) is spread evenly over the "surface" of the molecule. In (a), (c) and (d) all dimensions are to scale. In (b) atomic dimensions are reduced to show the structure. Linear dimensions of As atoms are enlarged in (a), (c) and (d) due to a bug in the visualization software (Molekel). Ga atoms are blue, As brown.

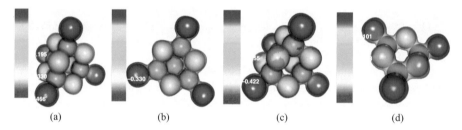

Fig. 3.14 MEP surfaces (semi-transparent light blue) of the pre-designed 8×13 CASSCF and vacuum 8×8 CASSCF $Ga_{10}As_4$ triplets [(a) and (b), and (c) and (d), respectively] corresponding to the CDD isovalues 0.08 [(a), (b) and (c)], and 0.07, (d). Atomic dimensions are reduced. Ga atoms are blue, As brown.

that almost all of the electronic charge is contained "inside" the molecules. However, ROHF ELSs of the Ga-based molecules (Fig. 3.15) differ significantly between themselves. The pre-designed $Ga_{10}As_4$ pyramid is a numerically perfect pyramid, so almost all of its MOs in the HOMO – LUMO region and beyond are double- and triple- degenerate. This numerically exact tetrahedral symmetry was lost when the atomic position constraints were lifted and the vacuum molecule was optimized. Consequently, MOs of the vacuum molecule are not degenerate. The ROHF ground state energies and OTE values of these two Ga-based molecules are also very similar (Tables 3.8 and 3.9), but their CASSCF OTEs differ about 2.5 times.

Similar to the case of the In-based molecules, the total energy surfaces of the Ga-based molecules are very flat near the expected energy minima, so the pre-designed and vacuum molecules have very close total energy minima lying within the computational error brackets from each other. Yet these molecules demonstrate very different ELSs, OTEs and dipole moments of 1.582702 D and 5.143718 D for the pre-designed and vacuum molecules, respectively, and thus are different molecules. Electronic charge of

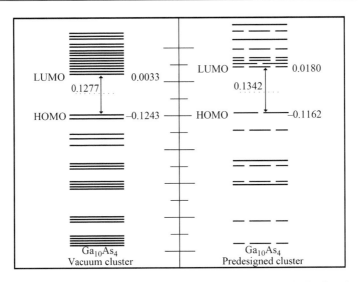

Fig. 3.15 ELSs of the vacuum (left) and pre-designed (right) ROHF $Ga_{10}As_4$ triplets, respectively, in the HOMO–LUMO regions. The electronic energy level values and the OTEs are in Hartree units. All dimensions are approximately to scale. [Adopted from Ref. 67.]

TABLE 3.8 Ground state energies of the pre-designed and vacuum $Ga_{10}As_4$ molecules.

Molecule	ROHF triplet ground state energy, Hartree	CI triplet ground state energy, Hartree	CASCF triplet ground state energy, Hartree	CASSCF (number of electrons × number of MOs)
Pre-designed $Ga_{10}As_4$	−2595.964256	−2596.024302140	−2596.026856504	8×13
Vacuum $Ga_{10}As_4$	−2595.987503	−2595.993339966	−2596.025905839	8×8

TABLE 3.9 Direct optical transition energies (OTEs) of the pre-designed and vacuum $Ga_{10}As_4$ molecules.

Molecules	ROHF OTE, eV	CI OTE, eV	CASSCF OTE, eV	CASSCF (number of electrons × number of MOs)
Pre-designed $Ga_{10}As_4$	3.6517	1.0956	0.9312	8×13
Vacuum $Ga_{10}As_4$	3.4749	1.1555	2.3508	8×8

TABLE 3.10 Ground state interaction energies for the pre-designed and vacuum $Ga_{10}As_4$ CASSCF triplets.

Molecule	One-electron Energy, Hartree	Two-electron repulsion energy, Hartree	Nuclear repulsion energy, Hartree	Nucleus-electron interaction energy, Hartree	CASSCF (number of electrons × number of MOs)
Pre-designed $Ga_{10}As_4$	−10127.0730601809	7389.0328777023	2738.0401824786	−11883.9341940530	8×10
Vacuum $Ga_{10}As_4$	−10068.7068479336	4763.8903770933	2708.7905650017	−11825.5382329680	8×10

the vacuum molecule is re-distributed signifying a loss of tetrahedral symmetry, and the molecule acquired a large dipole moment. The ground state data collected in Table 3.10 demonstrate that the vacuum $Ga_{10}As_4$ molecule has much lesser 2-electron repulsion energy than the pre-designed one, while the one-electron energy, nuclear repulsion and nucleus-electron attraction energies of these molecules are almost the same. The electron charge re-distribution that leads to a dramatic decrease of the two-electron repulsion in the vacuum molecule is penalized by acquisition of a large dipole moment.

From data of Tables 3.3, 3.5 and 3.8 it can be seen that the total energy minima of the $Ga_{10}As_4$ molecules are about 1.5 times deeper than those of the "natural" and "artificial" $In_{10}As_4$ molecules. This indicates that having a significantly smaller number of electrons, the Ga-based molecules are much more stable than the In-based ones. This conclusion is also supported by the fact that "surfaces" of the Ga-based molecules are almost neutral, while those of the In-based ones carry a significant negative "surface" charge. These observations are in a good agreement with findings of Quek et al. [63], and also with a conclusion of Ref. 63 that Ga_nAs_4 clusters with more than 10 (n>10) Ga atoms are unstable. Indeed, attempts of the authors of Refs. 67 and 68 to optimize clusters with more than 10 Ga atoms and only 4 As atoms failed.

Yet another important observation reported in Ref. 67 follows from a comparison of ROHF ELSs of the In-based and Ga-based molecules (Figs. 3.8, 3.11 and 3.15). This comparison reveals that ELS (Fig. 3.11) of the pre-designed ROHF $[In_{10}As_4]_{Ga}$ triplet (where the In atoms are enforced to have the Ga covalent radius in GaAs zincblende bulk lattice) is very similar to that of the pre-designed ROHF $Ga_{10}As_4$ triplet (Fig. 3.15). At the same time, the ground state energy characteristics (the ground state energy, interaction energies, etc.) of the pre-designed ROHF, CI and MCSCF $[In_{10}As_4]_{Ga}$ triplets (Tables 3.5 and 3.7) are close in value to those of the pre-designed ROHF, CI and MCSCF $In_{10}As_4$ triplets, respectively (Table 3.3), where the covalent radii of In and As atoms correspond to those in the InAs zincblende bulk lattice. This observation leads to a conclusion that ROHF ELS of the studied molecules is governed by the covalent radii of the atoms, while the energy values are defined by physical nature of the atoms (and in particular, by the number of electrons in the atoms).

It is also important to note that the ROHF OTE values of all of the 14-atomic molecules discussed in this chapter lie between 3.3 eV and 3.7 eV, and do not depend directly on either ELS, or the nature of the atoms (Tables 3.4, 3.6 and 3.9). Similarly, CI and MCSCF OTE values of these molecules do not correlate directly with their ELSs or physical nature of their atoms. However, a spread of CI and MCSCF OTE values is much wider than that of the corresponding ROHF OTEs: CI OTE values lie between 0.4 eV and 1.7 eV, while MCSCF OTEs fall between 0.86 eV and 3.19 eV.

3.7 Spin density distributions of the studied molecules

In recent years magnetic properties of QDWs and QDW-based nanomaterials attract significant attention due to demand for a dramatic increase in the density of active elements of electronic devices and the development of quantum information processing concept. In one of the following chapters of this book the nature of magnetism in and magnetic properties of QDWs of diluted magnetic semiconductors is related to their electron spin density distributions (SDDs). In this chapter SDDs and magnetic properties of the studied In-As and Ga-As small pyramidal QDs are analyzed.

CDDs of the considered In-As and Ga-As molecules indicate a uniform distribution of their electronic charge over the "surface" regions of the molecules. As pointed out in the following section, such a distribution is achieved through collectivization of up to 230 of the electrons populating uppermost occupied MOs in each molecule. This collectivization becomes possible due to the development of delocalized In-In or Ga-Ga ligand bonds. Such delocalization of the electronic charge also leads to spin delocalization characterized by SDDs of the virtually synthesized molecules of this chapter depicted in Figs. 3.16 to 3.18. The SDD values of all molecules discussed here are very small, with the maximum values of SDDs of In-based molecules being about half an order of magnitude larger than those of the Ga-based ones. The major contributions to SDDs of the In-based molecules are due to d electrons of In atoms

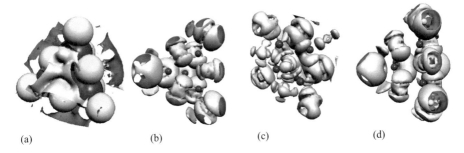

(a) (b) (c) (d)

Fig. 3.16 The SDD isosurfaces (gray) of the "natural" pre-designed 6×9 CASSCF [(a) and (b)] and vacuum 8×8 CASSCF [(c) and (d)] $In_{10}As_4$ triplets corresponding to the following isovalues: (a) 0.00005, (b) 0.0001, (c) 0.001, and (d) 0.0001. Atomic dimensions are reduced to show the structure of the surfaces. Other dimensions are to scale. In atoms are yellow and As red.

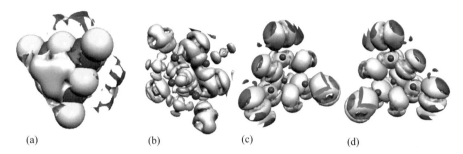

Fig. 3.17 The SDD isosurfaces (gray) of the "artificial" pre-designed 8×10 CASSCF [(a) and (b)] and vacuum 8×10 CASSCF [(c) and (d)] [In$_{10}$As$_4$]$_{Ga}$ molecules corresponding to the following isovalues: (a) 0.00005, (b) 0.0001, (c) and (d) 0.0001. Atomic dimensions are reduced to show the structure of the surfaces. Other dimensions are to scale. In atoms are yellow and As red.

(that also possess a large nucleus spin of 9/2). The maximum value of the SDD of the "natural" pre-designed 6×9 CASSCF In$_{10}$As$_4$ triplet is about 3 times smaller than that of the corresponding vacuum 8×8 CASSCF In$_{10}$As$_4$ triplet (Fig. 3.16). The reason for is likely to be that that the two-electron repulsion energy in the vacuum triplet is about 2 time smaller than that of the corresponding pre-designed triplet, and that optimization came at the expense of polarization of the vacuum molecule and spin re-distribution leading to a gain in its SDD values. In the case of the "artificial" CASSCF [In$_{10}$As$_4$]$_{Ga}$ triplets (Fig. 3.17), the maximum SDD value of the pre-designed one also is about 4 times smaller than that of the vacuum one, and again it correlates with the 2-electron repulsion energy being much smaller in the vacuum triplet. Given that the total energy minima of the In-based pre-designed molecules lie within the calculation error brackets from those of their vacuum counterparts, it is likely that electron charge re-distribution leading to minimization of the 2-electron repulsion in both In-based vacuum molecules comes at the expense of an increase in their SDD values. In view

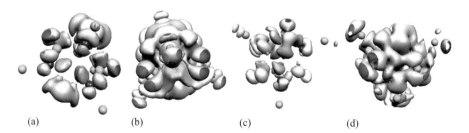

Fig. 3.18 The SDD isosurfaces (gray) of the "artificial" pre-designed 8×13 CASSCF [(a) and (b)] and vacuum 8×8 CASSCF [(c) and (d)] Ga$_{10}$As$_4$ molecules corresponding to the following isovalues: (a) 0.014, (b) 0.0004, (c) 0.0016, and (d) 0.0005. Atomic dimensions are reduced to show the structure of the surfaces. Other dimensions are to scale. Ga atoms are blue and As brown.

of importance of magnetic proper ies of the In-based molecules for applications, more targeted studies are necessary to fully understand their SDDs.

The Ga-based molecules possess extremely small SDDs (Fig. 3.18) contributed to by $3d$ electrons of Ga atoms. The SDD values of these molecules are about an order of magnitude smaller than those of the In-based molecules. Accordingly, SDD isosurfaces corresponding to the same isovalues are located much deeper "inside" of Ga molecules than the corresponding SDD isosurfaces of the In-based molecules. However, in contrast to the In-based molecules, the maximum SDD value of the pre-designed 8×13 CASSCF Ga-based triplet is larger than that of the vacuum 8×8 CASSCF Ga-based triplet, despite the 2-electron repulsion energy in the vacuum molecule being about 2 times smaller than that of the pre-designed one. A possible reason for this fact may be related to $3d$ electrons of Ga atoms contributing to the SDDs that are closer to the corresponding Ga nuclei than $4d$ electrons of In atoms to their In nuclei. Thus, despite the overall electron charge re-distribution minimizing the 2-electron repulsion energy of the vacuum Ga-based molecules, it does not affect significantly their SDD, while a similar redistribution affects SDDs of the In-based molecules. Moreover, the vacuum Ga-based triplet is not exactly tetrahedrally symmetric, so that there is no symmetry requirements applied to its MOs, CDD and SDD. At the same time, almost perfect numerical tetrahedral symmetry of the pre-designed Ga-based molecule constrains the electron charge re-distribution, and thus MOs, CDD and SDD of this molecule, to those configurations that support the tetrahedral symmetry.

Comparison of SDDs of the pre-designed and vacuum molecules confirms that quantum confinement may have a profound effect on SDDs, and thus on magnetic properties of the molecules. In-based pre-designed molecules, that model those synthesized in the presence of polarization and excluded volume effects instigated by quantum confinement, feature significantly smaller absolute values of SDDs than their vacuum counterparts, while the opposite is true for Ga-based molecules. However, in view of very small SDD values, neither of molecules discussed in this chapter is likely to be of interest for synthesis of novel nanomaterials for spintronics or quantum information processing, at present.

3.8 Electron charge delocalization and bonding in the studied molecules

As pointed out by Boldyrev and Wang [54], the existence and structure of non-stoichiometric molecules and atomic clusters cannot be explained on the basis of the classical valence theory and its octet rule. This can be further confirmed by detailed analysis of CDDs of the virtually synthesized molecules of Refs. 67 and 68 discussed above and in particular, by a consideration of contributions to the overall CDDs from the topmost "valence" and deeper MOs. From the octet rule standpoint, all of the 14-atomic molecules virtually synthesized in Refs. 67 and 68 have 50 valence electrons that are supposed to provide the major contributions to the outmost regions of their CDDs, including those that reach into the space beyond that occupied by molecules' atoms (called below CDD shells). However, data analysis reveals that up to 50% of the contributions to those CDD shells originate from electrons in deeply lying MOs (with the MO numbers below 90 in the discussed computations). According to the classical valence theory, these electrons are deeply lying, and thus are not expected to contribute to the CDD shells. [Note, that for all of the discussed 14-atomic molecules, 230 electrons in 116 topmost occupied α MOs and 114 topmost occupied β MOs were kept in calculations, while the rest of the total 622 electrons of In-based molecules and 442 electrons of Ga-based ones were accounted for by a core potential.] Moreover, analysis of MOs of these molecules in their HOMO – LUMO regions prove that such MOs are highly delocalized π-type orbits specific to ligand bonding analyzed in Ref. 54. Examples of delocalized π-type ligand bonding contributions to MOs in the HOMO-LUMO region of the natural MCSCF $In_{10}As_4$ triplets are shown in Figs. 3.19 and 3.20, and examples of such bonding contributions to similar MOs of "artificial" $[In_{10}As_4]_{Ga}$ triplets are depicted in Figs. 3.21 and 3.22. Thus, the bonding HOMO 115 and 116 of Fig. 3.19a, 3.19b, and 3.19c, respectively, contain π-type contribution due to ligand bonding of 6 In atoms. In contrast to many of MOs of

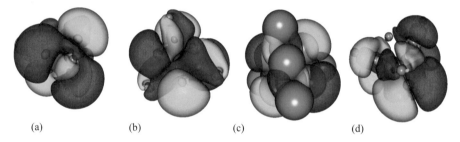

(a) (b) (c) (d)

Fig. 3.19 Isosurfaces of negative (semi-transparent red) and positive (semi-transparent green) components of MOs from the HOMO – LUMO region of the pre-designed 6×9 CASSCF $In_{10}As_4$ triplet corresponding to the isovalue 0.002: (a) HOMO 115, (b) and (c) HOMO 116, (d) LUMO 117. In atoms are blue, As brown. In (a), (b) and (d) atomic dimensions are reduced to show the structure. All other dimensions are to scale.

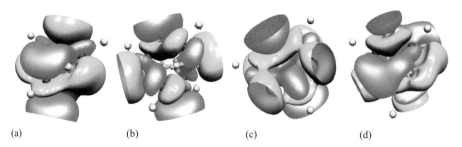

(a) (b) (c) (d)

Fig. 3.20 Isosurfaces of negative (semi-transparent orange) and positive (semi-transparent green) components of MOs from the HOMO – LUMO region of the vacuum 8×8 CASSCF $In_{10}As_4$ triplet corresponding to the isovalues 0.009 [(a) to (c)] and 0.01 (d): (a) HOMO 114, (b) HOMO 115, (c) HOMO 116, (d) LUMO 117. In atoms are yellow, As red. Atomic dimensions are reduced to show the structure. All other dimensions are to scale.

Ref. 54, these bonding MOs also have equally large contributions due to bonding of the valence *p*-electrons of In and As atoms that are also highly delocalized. LUMO 117 of Fig. 3.19d is similar in nature to the above HOMOs.

The π-type ligand bonding contributions to HOMO 116 and LUMO 117 are even more pronounced in the case of the vacuum MCSCF $In_{10}As_4$ triplet (Fig. 3.20c and 3.20d; green surfaces), while bonding of its HOMO 114 and 115 are due to shared π-electrons of In and As atoms.

It is important to note, that π-type ligand bonding is observed for the vast majority of MOs of the virtually synthesized molecules calculated in all approximation discussed here, as illustrated by the case of ROHF MOs of the pre-designed "artificial" ROHF $[In_{10}As_4]_{Ga}$ triplet in Fig. 3.21. A shape of MOs of the studied molecules vary significantly depending on what particular ligand atoms in the pyramids contribute their *p* electrons to their shared bond, and on a proportion of the ligand bonding to that of *p*-electron bonding of In and As atoms. In many cases, instead of 6-center ligand bonding one can observe 4-center one, or two 3-center ligand bonds, where

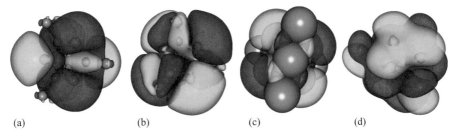

(a) (b) (c) (d)

Fig. 3.21 Isosurfaces of negative (semi-transparent red) and positive (semi-transparent green) components of MOs from the HOMO – LUMO region of the pre-designed ROHF $[In_{10}As_4]_{Ga}$ triplet: (a) HOMO 115, isovalue 0.003, (b) HOMO 116, isovalue 0.001, (c) HOMO 116, isovalue 0.002, (d) LUMO 117, isovalue 0.003. In atoms are blue, As brown. In (a), (b) and (d) atomic dimensions are reduced to show the structure. All other dimensions are to scale.

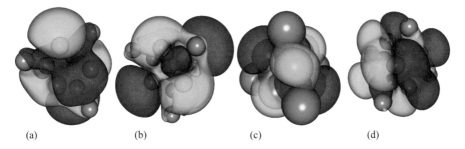

(a) (b) (c) (d)

Fig. 3.22 Isosurfaces of negative (semi-transparent red) and positive (semi-transparent green) components of MOs from the HOMO – LUMO region of the vacuum 8×10 CASSCF $[In_{10}As_4]_{Ga}$ triplet corresponding to the isovalue 0.002: (a) HOMO 115, (b) and (c) HOMO 116, (d) LUMO 117. In atoms are blue, As brown. In (a), (b) and (d) atomic dimensions are reduced to show the structure. All other dimensions are to scale.

the bonding between those 3-center ligand bonds is mediated by In – As delocalized bonding. The latter case is exemplified by the case of bonding in ROHF LUMO 117 of the "artificial" pre-designed $[In_{10}As_4]_{Ga}$ molecule shown in Fig. 3.21d. All of the other MOs in Fig. 3.21 show the "usual" 6-ligand bonding contributions. As a rule, all *p*-electrons of As atoms bond with *p*-electrons of In atoms either directly, or as mediators of the ligand bond contributions, and never produce As – As ligand bonding. Also as a rule, *p*-electrons of In atoms participate simultaneously in both ligand bonding, and that with *p*-electrons of As atoms. MOs of the vacuum $[In_{10}As_4]_{Ga}$ molecule shown in Fig. 3.22 provid e more examples of such bonding. Thus, 4-center and 2-center ligand bonds contribute to HOMOs 115 and 116 of this molecule (Figs. 3.22a to 3.22c, respectively), and *p*-electrons of its As atoms mediate bonding between those two ligand bonds. LUMO 117 of this molecule (Fig. 3.22d) features the major contribution from a 6-center ligand bond mediated by *p*-electrons of As atoms.

In the pre-designed 8×13 CASSCF Ga-based triplet ligand π-bonds are primarily 4-centered (Fig. 3.23), and all MOs contain a significant contribution due to bonding of *p*-electrons of Ga and As atoms. A similar MO structure is also typical for the vacuum 8×13 CASSCF $Ga_{10}As_4$ triplet (Fig. 3.24).

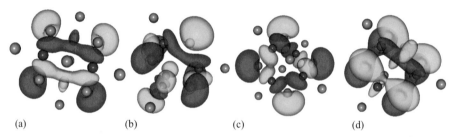

Fig. 3.23 Isosurfaces of negative (semi-transparent red) and positive (semi-transparent green) components of MOs from the HOMO – LUMO region of the pre-designed 8×13 CASSCF $Ga_{10}As_4$ triplet: (a) HOMO 114, isovalue 0.03; (b) HOMO 115, (c) HOMO 116, and (d) LUMO 117, isovalue 0.02, respectively. In atoms are blue, As brown. Atomic dimensions are reduced to show the structure. All other dimensions are to scale.

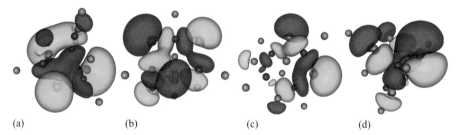

Fig. 3.24 Isosurfaces of negative (semi-transparent red) and positive (semi-transparent green) components of MOs from the HOMO – LUMO region of the vacuum 8×8 CASSCF $Ga_{10}As_4$ triplet corresponding to the isovalue 0.02: (a) HOMO 114, (b) HOMO 115, (c) HOMO 116, and (d) LUMO 117. In atoms are blue, As brown. Atomic dimensions are reduced to show the structure. All other dimensions are to scale.

Comparing MOs of the In-based and Ga-based molecules, such as those shown in Figs. 3.19 to 3.22, and 3.23, 3.24, one can observe that MOs of Ga-based molecules are contained inside of the molecular space as defined by the covalent radii of Ga and As atoms, while MOs of the In-based molecules reach outside of the corresponding molecular space. Comparison of CDDs of these groups of molecules shows that electronic charge of the Ga-based molecules is delocalized primarily inside of the space occupied by those molecules, while a significant portion of electronic charge of the In-based molecules is delocalized outside and near the surfaces of these molecules. These data indicate that the binding of electrons in the topmost valence MOs to their In-based molecules should be significantly weaker than the corresponding binding in the Ga-based molecules.

3.9 Conclusions

In this chapter small non-stoichiometric molecules of traditional III-V semiconductor compounds – In, Ga and As – have been discussed. These overall electroneutral molecules have been obtained in Refs. 67 and 68 by the use of the virtual synthesis methods developed by the authors. Linear dimensions of the 14-atomic molecules are about 1 nm, and therefore one can expect that similar systems will be synthesized in laboratory in the near future. These molecules, and their smallest sibling In_3As, appeared to be equilateral tetrahedral pyramids similar to those larger pyramids already synthesized experimentally (see, for example, Refs. 41–43).

The virtual synthesis method allows manipulation with the structure and composition of these molecules, and also covalent radii of their atoms, to better understand relations between these parameters, and physical and chemical properties of these systems. It also provides an independent means of evaluation of the properties of the molecules so synthesized using symmetry considerations. The major facts of this chapter can be summarized as follows.

1. "Surfaces" of all considered molecules are negatively charged, and their MCSCF OTEs lie in the interval from 0.86 eV to 3.18 eV. Thus, these molecules may be used as building blocks of wide-gap, n-type semiconductor nanosystems.

2. ELS of these molecules are primarily defined by the covalent radii of their atoms, while their ground state and electronic orbit energies are defined by the nature of their atoms, that is, by the number of electrons in their atoms.

3. Quantum confinement (excluded volume effects and to a degree, polarization) modeled by constraining atomic motion in the pre-designed molecules has a profound effect on all electronic, opto-electronic and magnetic properties of the systems.

4. In particular, ELSs and OTEs of the molecules are very sensitive to atomic positions, and thus can be manipulated in wide intervals (from below 1 eV to about 3 eV, in the case of OTEs of the In-based molecules) by atomic position shifts of the order of 10^{-2} Å.

5. Due to quantum confinement effects, tetrahedral symmetry of the molecules can be lost and degeneracy of their MOs lifted, so that some optical transitions forbidden in tetrahedral symmetry systems become available.

6. Binding of electrons in topmost valence MOs of the In-based molecules is significantly weaker than that in their Ga-based counterparts. Thus, charge carriers in nanomaterials that may use the In-based molecules as their building blocks may have higher charge mobility than that in nanomaterials using the Ga-based molecules.

7. Due to high mobility of charge carriers and weak binding of electrons in topmost valence MOs, the studied In-based molecules may be used as building blocks of nanomaterials for sensor and source applications.

8. SDDs of the considered molecules are small, with those of the In-based molecules several times larger than those of the Ga-based ones. Thus, neither of the molecular groups seems to be of interest for spintronics and quantum information processing applications, at least at present.

9. The octet rule of the classical valence theory does not hold for the discussed non-stoichiometric molecules. To a large degree, bonding in these non-stoichiometric molecules is realized as delocalized, π-type ligand bonding of $5p$ electrons of 3, 4, or 6 In atoms in the In-based molecules, or $4p$ electrons of 4 Ga atoms in the Ga-based ones, and is mediated by $4p$ electrons of As atoms.

10. While the studied molecules are not flat, and their π-type ligand bonding is not always similar to that of aromatic molecules, the presence of large π-type ligand bonding components in topmost valence MOs of the studied molecules indicates that there may exist a possibility to synthesize experimentally non-stoichiometric solid materials $In_{10}A_4$ and $Ga_{10}As_4$.

11. All studied molecules have deep ground state energy minima, with those of the Ga-based molecules being about 1.5 times deeper than those of the In-based molecules. All molecules are very stable, and thus may be synthesized experimentally both individually (individual QDs), and as arrays (QD arrays).

12. Due to the fact that ground state energies of the studied In-based and Ga-based molecules, respectively, lie within calculation error brackets from each other, at realistic experimental synthesis conditions arrays of such QDs may possess several types of such molecules. Therefore, such arrays may be used as nanomaterials for multi-band sensors.

All electronic, opto-electronic and magnetic properties of the molecules discussed in this chapter are in a good agreement with experimental, theoretical and computational facts available in literature. Refs. 67 and 68 provide more details on such available data, and the corresponding discussions.

References

[1] Ehrenreich, H. (1960). Band structure and electron transport in GaAs. *Phys. Rev.* **120**, 1951–1963.
[2] Petersson, K. D., McFaul, L.W., Schroer, M. D., Jung, M., Taylor, J. M., Houck, A. A., Petta, J. R. (2012). Circuit quantum electrodynamics with a spin qubit. *Nature* **490**, 380–383.
[3] Liu, R.-B., Yao, W., Sham, L. J. (2010). Quantum computing by optical control of electron spins. *Adv. Physics* **59**, 703–802.
[4] Saiki, T. (2008). Optical interaction of light with semiconductor quantum confined states at the nanoscale. *Springer Series in Optical Sciences* **139**, 1–39.
[5] Urbaszek, B., Marie, X., Amand, T., Krebs, O., Voisin, P., Malentinski, P., Högele, A., and Imamoglu, A. (2012). Nuclear spin physics in quantum dots: An optical investigation. *Rev, Mod. Phys.* **85**, 79–133.
[6] Mikkelsen, M. H., Berezovsky, J., Stoltz, N. G., Coldren, L. A., and Awschalom, D. D. (2007). Optically detected coherent spin dynamics of a single electron in a quantum dot. *Nature Physics* **3**, 770–773.
[7] Clark, S. M., Fu, K.-M. C., Ladd, T. D., and Yamamoto, Y. (2007). Quantum computers based on electron spins controlled by ultrafast off-resonant single optical pulses. *Phys. Rev. Lett.* **99**, 040501.
[8] Day, C. (2006). Semiconductor quantum dots take first steps toward spin-based quantum computations. *Physics Today* **59**, 16–18.
[9] Caillet., X., and Simon, C. (2007). Precision of single-qubit gates based on Raman transitions. *Euro. Phys. J. D* **42**, 341–348.
[10] Berezovsky, J., Mikkelsen, M. H., Gywat, O., Stoltz, W. G., Coldren, L. A., and Awschalom, D. D. (2006). *Science* **314**, 1916–1920.
[11] Khitrova, G., Gibbs, H. M., Kira, M., Koch, S. W., and Scherer, A. (2006). Vacuum Rabi splitting in semiconductors. *Nature Physics*, **2**, 81–89.
[12] Xie, W. (2013). Optical properties of an exciton in a two-dimensional quantum ring with an applied magnetic field. *Optics Comm.* **291**, 386–389.
[13] Ghali, M., Ohtani, K., Ohno, Y., and Ohno, H. (2012). Vertical electrical field-induced control of the exciton fine structure splitting in GaAs island quantum dots for the generation of polarization-entangled photons. *Japan. J. App. Phys.* **51** (II), 06FE14.
[14] Ghali, M., Ohtani, K., Ohno, Y., and Ohno, H. (2012). Generation and control of polarization-entangled photons from GaAs island quantum dots by an electric field. *Nature Communications* **3**, 661.
[15] Mathew, R., Pryor, C. E., Flatte, M. E., and Hall, K. C. (2011). Optimal quantum control for conditional rotation of exciton qubits in semiconductor quantum dots. *Phys. Rev. B*, **84**, 205322.
[16] Flagg, E. B., Robertson, J. W., Founta, S., Ma, W., Xiao, M., Salamo, G. J., and Shih, C.-K. (2009). Experimental evidence of single-phonon mediated inter-level excitonic transitions in a semiconductor quantum dot. *In:* "Proc. 2009 Conference on Lasers and Electro-Optics, and 2009 Conference on Quantum Electronics and Laser Science." CLEO/QELS, 5225836.
[17] Dou, X. M., Sun, B. Q., Xiong, Y. H., Niu, Z. C., Ni, H. Q., and Xu, Z. Y. (2009). Fine structural splitting and exciton spin relaxation in single InAs quantum dot. *J. App. Phys.* **105**, 103516.
[18] O'Driscoll, I., Blood, P., Sobiesierski, A., Gwilliam, R., and Smowton, P.M. (2012). Evaluating InAs QD lasers for space borne applications. *In:* "Conference Digest - IEEE International Semiconductor Laser Conference." No. 6348344.

[19] Bennour, M., Bouzaïene, L., Saidi, F., Sfaxi, L., and Maaref, H. (2011) Temperature dependence of optical properties of InAs quantum dots grown on GaAs(113)A and (115)A substrates. *J. Nanoparticle Res.* **13**, 6527–6535.
[20] Xie, W. (2011). Optical absorption and refractive index of an exciton quantum dot under intense laser radiation. *Physica E*, **43**, 1704–1707.
[21] Moody, G., Siemens, M. E., Bristow, A. D., Dai, X., Karaiskaj, D., Bracker, A. S., Gammon, D., and Cundiff, S. T. (2011). Spectral broadening and population relaxation in a GaAs interfacial quantum dot ensemble and quantum well nanostructure. *Physica Status Solidi (B)* **248**, 829–832.
[22] Karaiskaj, D. D., Moody, G., Bristow, A. D., Siemens, M. E., Dai, X., Bracker, A. S., Gammon, D., and Cundiff, S. T. (2010). Homogeneous linewidth temperature dependence of interfacial GaAs quantum dots studied with optical 2D Fourier-transform spectroscopy. *In:* "Proc. Lasers and Electro-Optics/Quantum Electronics and Laser Science Conference: 2010 Laser Science to Photonic Applications." *CLEO/QELS*, No. 5499665.
[23] Puangmali, T., Califano, M., and Harrison, P. (2010). Monotonic evolution of the optical properties in the transition from three- to quasi-two-dimensional quantum confinement in InAs nanorods. *J. Phys. Chem. C* **114**, 6901–6908.
[24] Yin, Z., Tang, X., Zhang, J., Zhao, J., Deny, S., and Gong, H. (2009). Photoluminescence of InAs quantum dots embedded in graded InGaAs barriers. *J. Nanoparticle Res.* **11**, 1947–1955.
[25] Mano, T., Abbarchi, M., Kuroda, T., Mastrandrea, C.A., Vinattieri, A., Sanguinetti, S., Sakoda, K., and Gurioli, M. (2009). Ultra-narrow emission from single GaAs self-assembled quantum dots grown by droplet epitaxy. *Nanotechnology* **20**, 395601.
[26] Wang, F., Yu, H., Jeong, S., Pietryga, J. M., Hollingsworth, J. A., Gibbons, P. C., and Buhro, W. E. (2008). The scaling of the effective band gaps in indium-arsenide quantum dots and wires. *ACS Nano* **2**, 1903–1913.
[27] Berezovsky, J., Mikkelsen, M. H., Stoltz, N. G., Coldren, L. A., and Awschalom, D. D. (2008). Picosecond coherent optical manipulation of a single electron spin in a quantum dot. *Science* **320** (5874), 349–352.
[28] Li, L., Liu, G., Li, Z., Li, M., and Wang, X. (2008). Growth and characterization of InAs quantum dots with low-density and long emission wavelength. *Chinese Optics Lett.* **6**, 71–73.
[29] Mikkelsen, M. H., Berezovsky, J., Stoltz, N. G., Coldren, L. A., and Awschalom, D. D. (2007). Optically detected coherent spin dynamics of a single electron in a quantum dot. *Nature Physics* **3**, 770–773.
[30] Lin, S. W., Song, A. M., Missous, M., Hawkins, I. D., Hamilton, B., Engström, O., and Peaker, A. R. (2006). Carrier emission from the electronic states of self-assembled indium arsenide quantum dots. *Mater. Sci. Eng. C* **26**, 760–765.
[31] Chan, P.-Y., Suarez, E., Gogna, M., Miller, B. I., Heller, E. K., Ayers, J. E., and Jain, F. C. (2012). Indium gallium arsenide quantum dot gate field-effect transistor using II-VI tunnel insulators showing three-state behavior. *J. Electronic Mater.* **41**, 2810–2815.
[32] Lin, S. W., Song, A. M., and Peaker, A. R. (2009). Laplace-transform deep-level spectroscopy characterization of the intrinsic and deep-level states in self-assembled InAs quantum-dot structures. *AIP Conference Proceedings* **1199**, 313–314.
[33] Shu, H., Liang, P., Wang, L., Chen, X., and Lu, W. (2011). Tailoring electronic properties of InAs nanowires by surface functionalization. *J. App. Phys.* **110**, 103713.
[34] Moody, G., Siemens, M. E., Bristow, A. D., Dai, X., Bracker, A. S., Gammon, D., and Cundiff, S. T. (2010). Temperature-dependent coupling of GaAs quantum well and interfacial quantum dots studied with optical 2D Fourier-transform spectroscopy. *In:* "Proc.

Lasers and Electro-Optics/Quantum Electronics and Laser Science Conference: 2010 Laser Science to Photonic Applications." *CLEO/QELS*, No. 5499701.
[35] Dos Santos, C. L., and Piquini, P. (2010). Diameter dependence of mechanical, electronic, and structural properties of InAs and InP nanowires: A first principles study. *Phys. Rev. B* **81**, 075408.
[36] Mukhopadhyaya, S., Boyacioglu, B., Saglam, M., and Chatterjee, A. (2008). Quantum size effect on the phonon-induced Zeeman splitting in a GaAs quantum dot with Gaussian and parabolic confining potentials. *Physica E* **40**, 2776–2782.
[37] Tackeuchi, A., Kuroda, T., Yamaguchi, K., Nakata, Y., Yokoyama, N., and Takagahara, T. (2006). Spin relaxation and antiferromagnetic coupling in semiconductor quantum dots. *Physica E* **32**, 354–358.
[38] Yang, S. Y. (2007). Analysis of the contributions of magnetic susceptibility to effective refractive indices of photonic crystals at long-wavelength limits. *Optics Express* **15**, 2669–2676.
[39] Prather, D. W., Shi, S., Murakowski, J., Schneider, G. J., Sharkawy, A., Chen, C., and Miao, B. (2006). Photonic crystal structures and applications: Perspective, overview, and development. *IEEE J. Selected Topics in Quantum Electronics* **12**, 1416–1436.
[40] Iwamoto, S., and Arakawa, Y. (2006). Advances in photonic crystals with MEMS and with semiconductor quantum dots. *Laser Physics* **16**, 223–231.
[41] Kapon, E., Pelucchi, E., Watanabe, S., Malko, A., M. H. Baier, M. H., Leifer, K., Dwir, B., Michelini, F., and Dupertuis, M.-A. (2004). Site- and energy-controlled pyramidal quantum dot heterostructures, *Physica E* **25**, 288–297.
[42] Zhe, Q., Karlsson, K. F., Pelucchi, E., and Kapon, E. (2007). Transition from two-dimensional to three-dimensional quantum confinement in semiconductor quantum wires/quantum dots. *Nano Lett.* **7**, 2227–2233.
[43] Koppens, F. H. L., Buizert, C., Tielrooij, K. J., Vink, I. T., Nowack, K. C., Meunier, T., Kouwenhoven, L. P., and Vandersypen, L. M. K. (2006). Driven coherent oscillation of a single electron spin in a quantum dot. *Nature* **442**, 766–771.
[44] Watanabe, S., Pelucchi, E., Dwir, B., Baier, M. H., Leifer, K., and Kapon, E. (2004). Dense uniform arrays of site-controlled quantum dots grown in inverted pyramids. *Appl. Phys. Lett.* **84**, 2907–2909.
[45] Baier, M. H., Watanabe, S., Pelucchi, E., and Kapon, E. (2004). High uniformity of site-controlled quantum dots grown on prepared substrates. *Appl. Phys. Lett.* **84**, 1943–1945.
[46] Pelucchi, E., Watanabe, S., Leifer, K., Dwir, B., and Kapon, E. (2004). Site-controlled quantum dots grown in inverted pyramids for photonic crystal applications. *Physica E* **23**, 476–481.
[47] Koppens, F. H. L., Nowack, K. C., and Vandersypen, L. M. K. (2008). Spin echo in a quantum dot. *Phys. Rev. Lett.* **100**, 236802.
[48] Koppens, F. H. L., Folk, J. A., Elzerman, J. M., Hanson, R., Willems van Beveren, L. H., Vink, I. T., Tranitz, H. P., Wegscheider, W., Kouwenhoven, L. P., and Vandersypen, L. M. K. Control and detection of singlet-triplet mixing in a random nuclear field. (2005). *Science* **309**, 1346–1350.
[49] Nowack, K. C., Koppens, F. H. L., Nazarov, Yu. V., Vandersypen, L. M. K. (2007). Coherent control of a single electron spin with electric fields. *Science* **318**, 1430–1433.
[50] Wang, L.-W., Kim, J., and Zunger, A. (1999). Electronic structure of [110]-faceted self-assembled pyramidal InAs/GaAs quantum dots. *Phys. Rev. B* **59**, 5678–5687.
[51] Kim, J., Wang, L.-W. and Zunger, A. (1998). Comparison of the electronic structure of InAs/GaAs pyramidal quantum dots with different facet orientations. *Phys. Rev. B* **57**, R9408–R9411.

[52] Zunger, A. and Wang, L.-W. (1996). Theory of silicon nanostructures. *Appl. Surf. Sci.* **102**, 350–359.
[53] Pozhar, L. A. (2010). Small InAsN and InN clusters: electronic properties and nitrogen stability belt, *EuroPhys. J. D* **57**, 343–354.
[54] Boldyrev, A. I., and Wang, L.-S. (2001). Beyond classical stoichiometry: experiment and theory. *J. Phys. Chem. A* **105**, 10759–10775.
[55] Singh, A. K., Briere, T. M., Kumar, V., and Kamazoe, Y. (2003). Magnetism in transition-metal-doped silicon nanotubes. *Phys. Rev. Lett.* **91**, 146802.
[56] Kumar, V., and Kawazoe, Y. (2003). Hydrogenated silicon fullerenes: effects of H on the stability of metal-encapsulated silicon clusters. *Phys. Rev. Lett.* **90**, 055502.
[57] Lu, J., and Nagase, S. (2003). Structural and electronic properties of metal-encapsulated silicon clusters in a large size range. *Phys. Rev. Lett.* **90**, 115506.
[58] Archibong, E. F., St-Amant, A., Goh, S. K., and Marynick, D. S. (2002). On the structure and electronic photodetachment spectra of Ga_3P^- and Ga_3As^-. *Chem. Phys. Lett.* **361**, 411–420.
[59] Costales, A., Kandalam, A. K., Franko, R., and Pandey, R. (2002). Theoretical study of structural and vibrational properties of $(AlP)_n$, $(AlAs)_n$, $(GaP)_n$, $(InP)_n$, and $(InAs)_n$ clusters with n=1, 2, 3. *J. Phys. Chem. B* **106**, 1940–1944.
[60] Korambath, P. P., and Karna, S. (2000). Hyper polarizabilities of GaN, GaP, and GaAs clusters: an *ab initio* time-dependent Hartree-Fock study. *J. Phys. Chem. A* **104**, 4801–4804.
[61] Karna, S. P., and Yeates, A. T. (1996). Nonlinear optical materials: theory and modelling. *In: ACS Symposium Series* **628**, Ch. 1, pp. 1–22.
[62] Piquini, P., Canuto, S., and Fazzio, A. (2004). Structural and electronic studies of Ga_3As_3, Ga_4As_3, and Ga_3As_4. *Int. J. Quant. Chem.* **52**, 571–577.
[63] Quek, H. K., Feng, Y. P., and Ong, Y. P. (1997). Tight binding molecular dynamics of Ga_mAs_n and Al_mAs_n clusters. *Z. Phys. D* **42**, 309–317.
[64] Erkoç, S., and Türker, L. (1999). Energetics and stability of GamAsn and GamAsn microclusters: PM3 calculations. *Physica E* **5**, 7–15.
[65] Yi, J.-Y. (2000). Stability, structural transformation, and reactivity of Ga13 clusters. *Phys. Rev. B* **61**, 7277–7279; Yi, J.-Y. (2000). Atomic and electronic structure of small GaAs clusters. *Chem. Phys. Lett.* **325**, 269–274.
[66] Kwong, H. H., Feng, Y. P., and Boo, T. P. (2001). Composition dependent properties of GaAs clusters. *Comp. Phys. Commun.* **142**, 290–294.
[67] Pozhar, L. A., Yeates, A. T., Szmulowicz, F., and Mitchel, W. C. (2006). Virtual synthesis of artificial molecules of In, Ga and As with pre-designed electronic properties using a self-consistent field method. *Phys. Rev. B* **74**, 085306. See also *Virtual J. Nanoscale Sci & Technol.* **14**, No. 8 (2006): http://www.vjnano.org
[68] Pozhar, L. A., Yeates, A. T., Szmulowicz, F., and Mitchel, W. C. (2005). Small atomic clusters as prototypes for sub-nanoscale heterostructure units with pre-designed charge transport properties. *EuroPhys. Lett.* **71**, 380–386.
[69] Pozhar, L. A. and Mitchel, W. C. (2007). Collectivization of electronic spin distributions and magneto-electronic properties of small atomic clusters of Ga and In with As, V and Mn. *IEEE Trans. Magnet.* **43**, 3037–3039.
[70] Pozhar, L. A., Yeates, A.T., Szmulowicz, F., and Mitchel, W.C. (2005). Virtual fabrication of small Ga-As/P and In-As/P clusters with pre-designed electronic energy level structure. *Mater. Res. Soc. Proc.* **829**, 49–54; MRS Outstanding Paper (Blue Ribbon) award.
[71] Pozhar, L. A., Yeates, A. T., Szmulowicz, F., and Mitchel, W. C. (2005). Small "magnetic" clusters of Ga and In with As and V. *Mater. Res. Soc. Proc.* **830**, D6.12 (6).

[72] Pozhar, L. A., Yeates, A. T., Szmulowicz, F., and Mitchel, W. C. (2004). Virtual synthesis of sub-nanoscale materials with prescribed physical properties. *In*: "Continuous Nanophase and Nanostructured Materials", *Mater Res. Soc. Proc.* **788**, L11.40 (6).

[73] Gordon, M. S., and Schmidt, M. W. (2005). Advances in electronic structure theory: GAMESS a decade later. *In* "Theory and Applications of Computational Chemistry: the First Forty Years" (C. E. Dykstra, G. Frenking, K. S. Kim, G. E. Scuseria, Eds.), pp. 1167–1189. Elsevier, Amsterdam. http://www.msg.ameslab.gov/GAMESS

[74] Stevens, W. J., Krauss, M., Basch, H., and Jasien, P. (1992). Relativistic compact effective potentials and efficient, shared-exponent basis sets for the third-, fourth-, and fifth-row atoms. *Can. J. Chem.* **70**, 612–630.

[75] Durgun, E., Tongay, S., and Ciraci, S. (2005). Silicon and III-V compound nanotubes: Structural and electronic properties. *Phys. Rev. B* **72**, 075420.

[76] Kuznetsov, A. E., Birch, K. A., Boldyrev, A. I., Li, X., Zhai, H.-J., and Wang, L.-S. (2003). All-metal aromatic molecules: rectangular Al_4^{4-} in the Li_3Al^{4-} anion. *Science* **300**, 622–625.

[77] Li, X., Kuznetsov, A. E., Zhang, H.-F., Boldyrev, A. I., and Wang, L.-S. (2001). Observation of all-metal aromatic molecules. *Science* **291**, 859–861.

[78] Seo, D.-K., and Cobett, J. D. (2001). Aromatic metal clusters. *Science* **291**, 841–842.

Quantum Dots of Gallium and Indium Arsenide Phosphides: Opto-electronic Properties, Spin Polarization and a Composition Effect of Quantum Confinement

4.1 Introduction

Since 1970s, a remarkable progress in the development of current-through light emitting diodes (LEDs) of very high quantum efficiency (defined as a number of photons emitted per a number of charge carriers) in a wide spectral range has been largely defined by synthesis of novel GaP-, GaAsP-, InGaAsP-, InGaP- and InP – based low-dimensional structures, such as quantum wells, wires (QWs), dots (QDs) and heterostructures [1,2]. Important problems typical for gallium arsenide phosphide materials, such as (i) emission of light in a 700 nm wavelength spectral region, where human eye sensitivity is very low, and (ii) high quantum efficiency achievable only in low current regimes, were successfully solved by inclusion of atoms of other elements (such as As, Al and N) into the materials, and the development of intricate heterostructures that included barrier layers. Additional colors and wavelengths became available due to introduction of GaP-based high efficiency red and green, and GaAsP-based orange and yellow LEDs. The development of metal-organic chemical vapor deposition (MOCVD) growth processes in 1980s permitted to synthesize high brightness and high reliability GaAlAs and InGaAlP luminescent materials with an adjustable band gap, and thus adjustable color output, using the same technology. This technology enabled two-fold increase in luminescence of the gain materials, and made it possible to overcome the "green death valley" effect by the development of LEDs with high-brightness 518 nm green light output. Further introduction of a current-blocking layer into the LED structure enhanced their luminescence even more. Synthesis of InGaAlP, GaN and SiC materials in the end of 1990s permitted a significant improvement in the light-output degradation with time and reliability of LEDs at high humidity conditions, thus enabling the use of such LEDs in solid-state traffic control and road signs, and visual message signs. At the same time, the LED structure evolved to include multiple quantum well-based waveguide layers (InGaAs and/or InGaP) and tensile strain compensation GaAsP layers [3] that enabled the development of multiple quantum well lasers [3] and LEDs [4–7].

During the recent decade, single and multiple quantum wells and wires (QWWs) of gallium and indium phosphides, and their heterostructures, continued to be thoroughly investigated for applications that permit miniaturization of optical devices and integration of optics with micro- and nano- electronics [8–13]. Meanwhile, contemporary electronic circuits and devices reached their ultimate performance speed limited by inherent resistance capacitance delay times. Many of such electronic circuits and devices can be replaced by photonic devices that do not have the above limitations, if the size of such photonic devices is reduced to nanoscale dimensions. Thus, a focus of synthesis and research of semiconductor opto-electronic materials has been shifted toward semiconductor QWWs, QD, nanoribbon and nanobelt structures as active materials for solid-state lasers, LEDs and photodetectors [14,15]. Synthesized using bottom-up chemical growth methods, such low-dimensional semiconductor nanosystems possess single-crystalline structure, superior geometric uniformity, sub-wavelength diameters, the direct band gap, relatively high refraction indices, and are a few micrometers in length. A semiconductor nanowire behaves like a Fabry-Perot cavity, so that QW nanostructures show lasing oscillations with wavelengths from ultraviolet (UV) to near infrared (IR), and also are almost ideal one-dimensional optical nanowavegiudes with tight optical confinement and low scattering losses. Very small footprints and scalability of such QW-based nanostructures permit them to be integrated with nanoelectronic circuits for on-chip communications, optical sensing, detection, signal processing and quantum optics applications. Their advanced optical characteristics include near-field mode coupling, endface reflection, and substrate-induced effects. Nanolasers based on QW nanostructures can be optically or electrically pumped, wavelength-tunable, single-mode operated, fiber-coupled and metal incorporated. However, guided modes in semiconductor nanowires are evanescent and become progressively less confined as the wire diameter decreases. This introduces a severe size restriction on the nanowire diameters (and thus the laser dimensions) to ensure a low lasing threshold. To reduce the optical mode size and the laser cavity dimensions further, especially in the case of IR nanolasers, metal-dielectric cavities are used [16–20] to generate surface plasmons and amplify them by simulated emission of radiation (so-called SPASER concept). The metal of the cavities also serves as an electric contact and heat sink. In such SPASER lasers lasing is realized by highly confined and localized surface plasmons (LSPs) that are excited in arrays of nanostructured metals by coupling of the metal to a semiconductor gain medium. LSPs are amplified through emission in the semiconductor and generate localized intense optical fields. SPASER lasers often utilize rectangular InP/InGaAs and n-InP/InGaAs/p-InP nanopillars encapsulated by silver with a thin insulating SiN layer, where InGaAs or InGaAsP is the semiconductor gain material.

Among many other distinctive features, optical properties of semiconductor Ga- and In- arsenide phosphide nanostructures exhibit nonlinearities due to multiphoton excitations [11,21], similar to those in bulk semiconductors materials that are used to create high power lasers tunable from mid-UV to near-IR ranges [22]. Nonlinear optical devices utilize these properties, and in particular higher-order transitions between electron energy levels, to realize ultrafast signal processing. In the case of electronic devices interactions between electrons, holes, and electrons and holes, cause fast decoherence of the particles' entangled quantum states. In contrast, weak photon interactions with the environment allows preservation of the entangled quantum states of the photons during long time periods, and thus make such photon states suitable for long-range interconnection and quantum communication applications. In nonlinear optical nanosystems electron transitions between energy levels occur by emission or absorption of two or more photons. In the case of non-centrosymmetric systems, sum multi-photon transmission processes lead to second and higher harmonics generation carried by so-called sum frequency photons, and to excited states of charge carriers. In the case of two-photon absorption, such excited charge carriers are used to realize all-optical logic. Among other applications, ultrafast optical multiplication based on two-photon absorption is the major physical mechanism used in modern devices for unique quantum measurements. Two-photon emission is also the basic physical process in InGaAsP Fabry-Perot laser diodes and avalanche photodiodes, InGaAs/InP single-photon quantum detectors, and GaInP/AlGaInP QW electron spin current sources. At the same time, multiphoton emission is the driving force of frequency upconversion imaging, near-field microscopy, 3D data storage and microfabrication. An alternative process of multiphoton absorption drives multiphoton–pumped frequency upconversion lasing, multiphoton absorption-based optical limiting, stabilization reshaping, stimulated scattering, multiphoton-induced surface photoelectric effects and photochemical reactions [23].

High-efficiency light harvesting is yet another field where low-dimensional Ga- and In-based arsenide phosphides gained widespread applications [24–32]. Electron transport and optical transitions in nanoscale photovoltaic devices differ markedly from those in conventional bulk solar cells and allow manipulations using size, structure and composition parameters. In particular, multiple InGaAs/GaAsP quantum well-based solar cells provide for a wide absorption range and reduction in collection time of photo-generated carriers to a few picoseconds, as opposed to several nanoseconds in micro-scale solar cells. Multiple-transition strained $InAs_{0.40}P_{0.60}$/GaAs/InP and InP/$GaAs_{0.70}P_{0.30}$/GaAs solar cells studied in Ref. 26 show a promise to reach conversion efficiency exceeding 50%. In quantum well-based solar cells and similar structures sub-wavelength metal and dielectric scattering elements are integrated for light trapping to enable improved absorption of IR radiation [27] that accounts for a half of the power in the solar spectrum. Theoretical efficiency of quantum well-based multi-junction solar cells may reach 71% due to carrier multiplication, where a single photon generates more than one electron-hole pair by avalanche-type process. Quantum well-based solar cells realize simultaneously an increase in efficiency and a decrease in costs of manufacturing, being developed using a single device/module technology and well-established manufacturing processes.

In recent years, further progress toward higher density of elements in and efficiency of electronic circuits resulted in the development of semiconductor quantum wire- and QD-based optoelectronics and photovoltaics [33–38]. In addition to advantages similar to those of quantum well-based devices, planar and radial semiconductor nanowire heterostructures can be used to engineer patterned ferromagnetic or ferroelectric domains with the pre-designed average local dielectric permittivity and magnetic permeability. Semiconductor QW-based resonant tunneling diodes, photodetectors and LEDs have dimensions commensurate with those of nanoscale components of electronic circuits, which allow high density integration of optical and electronic elements on the same platform. Furthermore, problems related to a reduction in size of optical devices, such as reduced absorption and lower carrier generation rates for photodetectors, and weaker light emission of LEDs, are absent in confined nanostructures due to optical resonances. In particular, radially confined geometry of QWs supports optical resonant modes. In the case of small diameter QW-based LEDs, efficient coupling between spontaneous emission and free-space modes lead to high light extraction efficiency. In addition, adjusting QW diameters and composition, one can engineer suitable bandgaps to achieve wavelength-selective functionality. The use of "zero-dimensional" nanostructures, such as InAsP QDs embedded in charge multiplication regions of p-n junction InP QWs, further enhances light output of the QW-based devices allowing realization of avalanche photodiodes [33]. In such structures a single incident photon with the energy equal to that of the QD optical transition creates a single exciton that dissociates into an electron and a hole pair both of which tunnel into the nanowire depletion region and undergo multiple ionization collisions in high internal and external fields. This avalanche process leads to drastic increase, up to 4 orders of magnitude, in photoelectric gain values [39,40].

As sources of coherent light, semiconductor QW (called also nanowire, NW) lasers are important elements of integrated optoelectronic circuits (IOEC). In addition to combining the gain medium and "cavity" in one element, QWs have high tolerance to lattice mismatch, and thus allow the development of high-quality heterostructures with band gaps tunable in a wide spectral range. GaAs/GaAsP core/shell QWs with lasing in near-IR region [41] serve as a good example of such structures. QW geometry provides strain relaxation mechanism to accommodate a range of lattice mismatch values between different materials, and thus allows synthesis of metastable alloys that cannot be stabilized in the case of other geometries. In addition, QW growth and assembly processes support nanoscale composition control along at least two orthogonal axes of the QWs. At the same time, QW geometry allows manipulations with the QW opto-electronic properties using confinement, and exploitation of plasmonic effects, while providing one axis to support charge carrier transport and integration with other elements of IOEC. This is especially important for applications of superconductor nanowire-based optoelectronic devices as single-photon devices and sources for emerging quantum information processing.

Further shrinking of active system dimensions to "zero" and working with QD systems may advance optoelectronics even more. Importantly, it can provide a solution to a problem of the development of OEICs based on monolithic integration of LEDs on silicon substrates. In this respect, especially interesting are highly homogeneous self-assembled (In, Ga)As QD systems grown on GaP substrates [42]. Incorporation of In atoms enhances band lineup and direct bandgap properties of large QDs composed of thousands of atoms, while GaP substrate provides for the closest-to-silicon lattice constant among all III-V superconductor materials. Further manipulations with the composition of QDs (such as addition of 2% of nitrogen) and substrates provides a means for simultaneous tuning of direct band properties and reduction in the band gap. In this respect, GaAs/GaP, InP/GaP, and (In, Ga)As/GaP QD-based systems have a strong advantage against nanowire-based ones, as they allow growing larger lattice-mismatch structures, and thus reaching smaller band gaps.

One of the most interesting properties of QD and nanowire systems with GaAsP barriers is their ability to generate a high yield of spin-polarized electrons with polarization up to 92% and 94% [43–45]. Such materials are used to build photocathodes of various devices for studies of the spin-dependent structure of nuclei, electro-weak interactions, surfaces of magnetic domains, and electronic structure of metal and semiconductor surfaces. In recent years, sources of uniformly spin-polarized electrons gained attention in spintronics and quantum electronics, where entangled electron spin states are used as qubits. Electrically-detected magnetic resonances, longitudinal magneto-transport through, and edge-magneto-plasmons in GaP- and GaAsP-based n-p-diodes have been currently investigated [46] for similar applications. Addition of "magnetic" dopants, such as Fe, Ni, Co, or Mn and Cr, to GaP structures [47] leads to ferromagnetism at or above room temperature that provides added functionality of microelectronic devices, and generates new classes of such devices and circuits, including spin transistors, ultradense nonvolatile semiconductor memory [48] and optical emitters with polarized output [49].

Multifaceted, practically important and scientifically intriguing opto-electronic and magnetic properties of nanoscale Ga and In arsenide phosphides are in need of systematic studies and detailed description, beginning from theoretical investigation of their smallest building elements, such as tetrahedral symmetry pyramids of Ga-As-P and In-As-P zincblende lattices discussed in this chapter.

4.2 Virtual synthesis procedure

Bottom-up experimental methods, such as MOCVD, self-assembly, and patterned growth helped by photolithography and etching, permitted synthesis of QDs composed of several hundreds of "traditional" semiconductor compound atoms, including Ga, In, As and P, on planar and non-planar GaAs, InP and InGaAs substrates, or embedded in these and other materials [50–59]. Such QDs exhibit tetrahedral pyramidal shape. Further perfection of experimental synthesis techniques by E. Kapon and co-workers allowed self-assembly of even smaller clusters built of Ga, In and As atoms both on specially prepared, non-planar substrates and in confinement offered by nanometer-size V-shaped grooves or pyramidal recesses [60–64]. Once again, a natural shape of such QDs of a few-nanometer in linear dimensions was that of tetrahedral pyramids. Theoretical studies of L. A. Pozhar et al. [65,66] demonstrated that the minimum of the total energy of 14-atomic clusters composed of 10 Ga or In atoms and 4 As atoms is realized when such clusters form tetrahedral pyramids. Earlier theoretical studies of these authors [67] also shown that tetrahedral pyramidal structures realize the minimum of the total energy of similar 14-atomic clusters containing phosphorus atoms.

In this chapter a current state of the art and results of virtual synthesis of 14-atomic quantum dots composed of Ga, In, As and P atoms is discussed in detail. Tetrahedral pyramidal fragments of zincblende cubes of GaAs and InAs bulk crystalline lattices optimized to become the pre-designed $Ga_{10}As_4$ and $In_{10}As_4$ molecules and discussed in Chapter 3 have provided initial structural templates. In these molecules 10 Ga or In atoms provide the fcc-pyramidal scaffolds, and 4 As atoms are placed inside of the pyramids in positions corresponding to ¼ of the body diagonals of the fcc cubes of their parental fcc lattices. The initial covalent radii 1.26 Å and 1.44 Å of Ga and In atoms, respectively, have been adopted from experiment completed at 300 K and known from literature. The covalent radius of As atoms was set to experimental value of 1.18 Å obtained at 298.5 K.

To build initial pre-designed pyramidal structures composed of 10 Ga or In atoms and containing one or two P atoms, in the pre-designed $Ga_{10}As_4$ and $In_{10}As_4$ molecules of Chapter 3 one or two As atoms were substituted with P atoms without any change in positions of the centers of mass of all participating atoms. That is, in such pre-designed clusters phosphorus atoms have been assigned As covalent radii. Therefore, considering that the As atom is larger than the P atom, four of the pre-designed clusters of this chapter – $Ga_{10}As_3P$, $Ga_{10}As_2P_2$, $In_{10}As_3P$ and $In_{10}As_2P_2$ – are rather loose. Such Ga- and In-based pre-designed clusters are useful as models of the corresponding molecules that may be synthesized in roomy voids of zincblende GaAs and InAs bulk lattices.

The third type of pre-designed pyramidal clusters studied in this chapter originates from a pre-designed $Ga_{10}As_3P$ and $Ga_{10}As_2P_2$ pre-designed clusters mentioned above, where all Ga atoms have been replaced with In atoms without any change in positions of the centers of mass of all participating atoms. In other words, in the pre-designed $[In_{10}As_3P]_{Ga}$ and $[In_{10}As_2P_2]_{Ga}$ clusters In atoms are assigned Ga covalent radii, to model significantly stressed molecules that may be synthesized in a tight quantum confinement provided by voids in zincblrnde GaAs bulk lattice or on fcc surfaces of GaAs.

The electronic energy level structure (ELS), charge and spin density distributions (CDD and SDD, respectively), and the molecular electrostatic potential (MEP) of all pre-designed atomic clusters have been investigated by means of the virtual synthesis methods [65,66] using progressively more accurate, quantum field-theoretical approximations discussed in Chapter 2. Initially, the Hartree-Fock (HF) and restricted open shell HF (ROHF) methods have been applied to minimize the total energy of the clusters and obtain their ground state structure, ELS, CDD, SDD, MEP and molecular orbits (MOs). Atomic coordinates and MOs of HF/ROHF-optimized molecules so obtained were further used as input for progressively more accurate calculations of the structure and properties of the pre-designed molecules applying configuration interactions (CI), complete active space self-consistent field (CASSCF) and MCSCF approximations as realized by the GAMESS software package [68] with SBKJC basis set. In doing these calculations, spatial constraints were applied to all atoms of the emerging pre-designed molecules. Thus, the centers of mass of the atoms in the pre-designed clusters were fixed at their initial positions in the clusters (the atoms were not permitted to "move") to model molecular synthesis in quantum confinement that is either somewhat loose or tight, or on surfaces.

At the next step of the virtual synthesis studies the spatial constraints applied to atomic positions in the pre-designed clusters were lifted to allow atomic displacement, and then another round of HF/ROHF/CI/CASCF/MCSCF optimization was applied to such unconstrained clusters. The results were unconstrained molecules that model those synthesized in the absence of any foreign atoms and external electromagnetic fields. Sometimes such molecules are denoted by the star * and called "vacuum" molecules in the text below to stress that those molecules were virtually synthesized in the absence of any mentioned constraints. As discussed below, even very small displacements (in the range of 10^{-2} Å) of the atoms in vacuum molecules from their former positions in the pre-designed molecules lead to significant, and sometimes dramatic, changes in the electronic, optical and magnetic properties of the vacuum molecules compared to those of their pre-designed counterparts. In agreement with previous results [65–67] and other data discussed in this book and elsewhere in literature, the ROHF/MCSCF direct optical transition energy (OTE) of these molecules (save for the pre-designed $In_{10}As_3P$ one) falls in the range from 1 eV to 5 eV, and can be manipulated up to 100% of its value using the composition and covalent radii of the atoms as defined by the quantum confinement. At the same time, the majority of these molecules possess ROHF/MCSCF OTEs close to 3 eV.

4.3 Ga-As molecules with one and two phosphorus atoms

The pre-designed molecule $Ga_{10}As_3P$ has been derived from the pre-designed molecule $Ga_{10}As_4$ discussed in Chapter 3 by replacement of one of its As atoms with that of phosphorus without changing the positions of all of the atoms, and subsequent conditional minimization of the total energy of the $Ga_{10}As_3P$ atomic cluster in the presence of the position constraints. This minimization procedure allows the development of molecular models to demonstrate effects of quantum confinement on the structure and properties of the molecules so synthesized. At the next step of the virtual synthesis procedure, the position constraints applied to the atoms of the pre-designed $Ga_{10}As_3P$ cluster were lifted, and the total energy of the atomic cluster was minimized again to deliver the corresponding unconstrained, or vacuum, $Ga_{10}As_3P$ molecule. Differences in ELS of these molecules are obvious already at the ROHF stage of calculations (see Fig. 4.1 and Table 4.1).

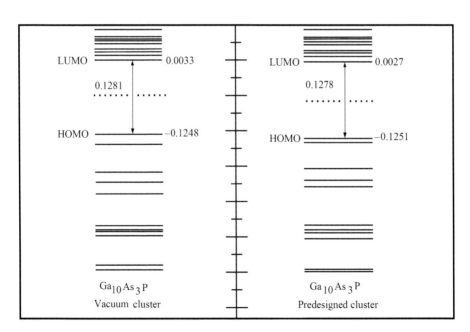

Fig. 4.1 ROHF electronic energy level structure of the vacuum (left) and pre-designed (right) $Ga_{10}As_3P$ molecules. The electronic energy level values and the OTEs are in Hartree units. All dimensions are approximately to scale.

TABLE 4.1 Ground state energies of the $Ga_{10}As_3P$ and $Ga_{10}As_2P_2$ molecules.

Molecule	ROHF ground state energy, Hartree	ROHF optical transition energy, Hartree	MCSCF ground state energy, Hartree	MCSCF optical transition energy, Hartree	CAS (number of electrons × number of MOs)
$Ga_{10}As_4$	−2595.964256423	3.6317	−2596.026856504	0.9312	8 × 13
$Ga_{10}As_4$*	−2595.987503014	3.4749	−2596.025905839	2.3508	8 × 8
$Ga_{10}As_3P$	−2595.270613817	3.4776	−2596.312563811	1.9430	6 × 9
$Ga_{10}As_3P$*	−2596.289492119	3.4858	−2596.325927751	4.7202	8 × 8
$Ga_{10}As_2P_2$	−2596.572125648	3.5102	−2596.609992140	1.9983	6 × 8
$Ga_{10}As_2P_2$*	−2596.696658512	6.5034	−2596.697688143	3.2463	CI data

* - denotes vacuum molecule

Thus, the vacuum molecule has a deeper ground state energy minimum and almost two times larger MCSCF OTE (Table 4.1) than those of its pre-designed counterpart. The reason for this, of course, is that the unconstrained minimization of the total energy of a molecular cluster results in the global minimum of the total energy, while a conditional minimization of the total energy in the presence of spatial constraints applied to atomic positions provides a local minimum of the total energy of the molecular cluster. Because of the P atom, ELSs of both $Ga_{10}As_3P$ molecules do not exhibit any signs of tetrahedral symmetry (Fig. 4.1), despite the pyramidal shape (Figs. 4.2 and 4.3) of these molecules inherent from their pre-designed $Ga_{10}As_4$ parent. In the pre-designed $Ga_{10}As_3P$ molecule several electronic energy levels moved closer to each other (Fig. 4.1) indicating that one of the major effects of quantum confinement is formation of bundles of closely lying electronic energy levels. [With an increase in the number of atoms in a cluster, such bundles of electronic energy levels lying below the highest occupied molecular orbit (HOMO) give rise to sub-bands of the valence band of the large clusters, while such bundles lying above the lowest unoccupied molecular orbit (LUMO) form sub-bands of the conductance band.] Comparing more accurate MCSCF ground state data (Table 4.1) for the $Ga_{10}As_3P$ molecules to those specific to the pre-designed and vacuum $Ga_{10}As_4$ molecules, also included in Table 4.1 for convenience, one can see that replacement of an As atom with a P atom in the parent pre-designed $Ga_{10}As_4$ molecule leads only to a small decrease of about 0.3 H in the ground state energy. Yet the MCSCF OTEs of the P-containing molecules are about two times larger than those of the corresponding $Ga_{10}As_4$ molecules. Similar to the $Ga_{10}As_4$ molecules, both $Ga_{10}As_3P$ molecules are ROHF and MCSCF triplets, and a minimum configuration of their complete active space (CAS) requires 6 to 8 electrons distributed over 8 MOs.

Not surprisingly, the pre-designed $Ga_{10}As_3P$ molecule exhibits an almost perfect tetrahedral pyramidal form (Fig. 4.2a) inherent from its pre-designed $Ga_{10}As_4$ parent. The vacuum $Ga_{10}As_3P$ molecule also keeps appearance of a tetrahedral pyramid (Fig. 4.3a), but only visually, because very small displacements of its atoms in the

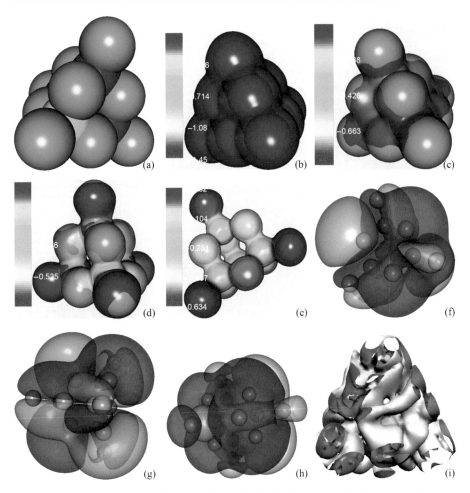

Fig. 4.2 Pre-designed CAS 6×9 MCSCF triplet $Ga_{10}As_3P$. Structure: (a). Molecular electrostatic potential (MEP) surfaces (color codding schemes are shown in the figures) corresponding to the CDD isovalues (b) 0.01, (c) 0.03, (d) 0.05 and (e) 0.07 of the charge density distribution (CDD, in arbitrary units). Isosurfaces of positive (gray) and negative (red) parts of the highest occupied (HO) and lowest unoccupied (LU) molecular orbits (MOs): (f) HOMO 115, (g) HOMO 116, and (h) LUMO 117 corresponding to the isovalue 0.001 (in arbitrary units). An isosurface of the spin density distribution (SDD) corresponding to the SDD isovalue (i) 0.0001. Ga atoms are blue, As brown and P green. In (a) to (c) dimensions of As and P atoms are somewhat enlarged; in (d) to (i) all atomic dimensions are reduced, and other dimensions are to scale.

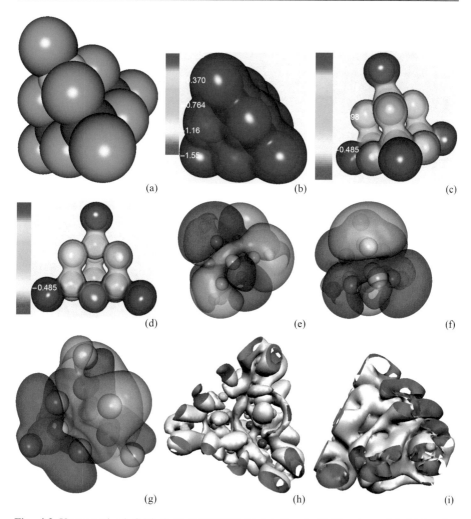

Fig. 4.3 Unconstrained CAS 8×8 MCSCF triplet $Ga_{10}As_3P$. Structure: (a). Molecular electrostatic potential (MEP) surfaces (color codding schemes are shown in the figures) corresponding to the CDD isovalues (b) 0.01, (c) and (d): 0.07 of the charge density distribution (CDD, in arbitrary units). Isosurfaces of positive (gray) and negative (red) parts of the highest occupied (HO) and lowest unoccupied (LU) molecular orbits (MOs): (e) HOMO 115, (f) HOMO 116, and (g) LUMO 117 corresponding to the isovalue 0.001 (in arbitrary units). Isosurfaces of the spin density distribution (SDD) corresponding to the SDD isovalues (h) 0.0005 and (i) 0.0001. Ga atoms are blue, As brown and P green. In (a) and (b) dimensions of As and P atoms are somewhat enlarged; in (c) to (i) all atomic dimensions are reduced, and other dimensions are to scale.

range of a few hundredths of Angstrom cannot be detected by a human eye. Such small atomic displacements, however, facilitate a dramatic increase in MCSCF OTE of this molecule as compared to that of its pre-designed counterpart (Table 4.1). [Notably, ROHF OTE values for these molecules are almost the same, due to a relative roughness of the ROHF approximation.] The most intriguing features of the $Ga_{10}As_3P$ molecules are reflected by their MEP isosurfaces (Figs. 4.2b to 4.2e and 4.3b to 4.3d). It appears that near the molecular "surfaces" (as defined by Ga atoms) outside the molecules MEPs of both molecules are slightly positive (Figs. 4.2b, 4.2c and 4.3b), indicating regions of electron charge deficit. Thus, some of the electron charge of Ga atoms is re-distributed toward As and P atoms inside the molecules providing for their remarkable stability reflected by significant depths of their ground state potential wells (Table 4.1). The corresponding "belts" of electron charge are clearly visible in Figs. 4.2d, 4.2e, 4.3c and 4.3d. Therefore, despite both molecules being electroneutral, in close proximity to their surfaces there exist electron charge deficit shells embracing the molecules. In the language of solid state physics such regions are called "holes". Thus, being imbedded into some environment or lattice, these molecules may be seen as housing holes similar to those carried by QDs of diluted magnetic semiconductors and nitrides, or those of Ni-O nanowires discussed in Chapters 5, 6 and 8, respectively. Both $Ga_{10}As_3P$ molecules possess large dipole moments. Interestingly, the dipole moment of the pre-designed molecule (4.8392 D) is smaller than that of the vacuum one (5.1552D) indicating that the deeper ground state of the vacuum molecule is achieved at the expense of an increase of its dipole moment.

Chemical properties of the $Ga_{10}As_3P$ triplets are typical for non-stoichiometric molecules where the octet rule does not hold, as discussed throughout this book and elsewhere (see, for example, Refs. 65 to 72). Highly hybridized MOs in the HOMO-LUMO regions of these molecules (Figs. 4.2f to 4.2h and 4.3e to 4.3g) are combinations of various contributions of $4p$ atomic orbits (AOs) of Ga and As atoms, and $3p$ AOs of P atoms. Thus, HOMO 115 and 116 (Fig. 4.2f and 4.2g) of the pre-designed molecule features the major contributions from $4p$ AOs of 6 Ga and 3As atoms, and a $3p$ AO of the P atom, while its LUMO 117 (Fig. 4.2h) is built primarily of $4p$ AOs of 5 Ga and 3 As atoms. In the case of the unconstrained molecule, HOMO 115 (Fig. 4.3e) features the major contributions from $4p$ AOs of 7 Ga atoms and a $3p$ AO of P, but only 2 As atoms contribute through their $4p$ AOs to this MO. Similarly, HOMO 116 of this molecule (Fig. 4.3f) is composed of $4p$ AOs of 8 Ga and 2 As, and a $3p$ AO of P atoms, while its LUMO 117 (Fig. 4.3g) is hybridized primarily of $4p$ AOs of only 4 Ga and one As atoms. Abundance of p-electrons in upper electronic levels of the constitutive atoms provides for very stable triplet ground states of both molecules and hints at a possibility of self-assembly of such molecules into larger structures. Due to its pre-designed nature, the bond length of the ligand Ga – Ga bonds is 3.997 Å, and the lengths of the Ga – As and Ga – P bonds are both equal to 2.228 Å. In the unconstrained molecule the Ga – Ga bond lengths of weak ligand Ga – Ga bonds are much longer and fall in the range from 4.031 Å to 4.064 Å, with the majority of them being about 4.064 Å. The lengths of the Ga – As and Ga – P bonds in this molecule are very close in value, with the majority of them being about 2.450 Å, 2.474 Å, 2.480 Å, and 2.518 Å. The P atoms in these molecules play a role very similar to that of As atoms. Namely, they are electron charge sinks, although they can accommodate only a few more of Ga electrons than those accommodated by each of the As atom. This is the major reason behind the fact that the replacement of an As atom with a P one in $Ga_{10}As_4$ molecules leads to only a slight decrease of about 0.3 H in the total energy. The major contributions to this loss comes from a decrease in electronic repulsion due to much smaller number of electrons in a P atom than that in an As atom, and an increase in attraction of Ga electrons to As and P nuclei that drives Ga electrons toward inner parts of the molecules.

Both $Ga_{10}As_3P$ molecules are "ferromagnetic" triplets, whose 2 parallel uncompensated spins are due to $3d$ electrons of non-neighboring Ga atoms. Thus, similar to the case of small DMS QDs discussed in Chapter 5, the delocalized charge deficit shells (holes) housed by $Ga_{10}As_3P$ molecules are spin-polarized and embrace two $3d$-electron spins accommodated inside the molecules. At the same time, completed $3d$ electron shells of Ga and As atoms provide for rather smooth SDD isosurfaces (Figs. 4.2i, 4.3h and 4.3i) of these molecules for small SDD isovalues.

Replacement of yet another As atom in the pre-designed $Ga_{10}As_3P$ molecule by a P atom, while keeping all positions of the centers of mass of all atoms fixed, and subsequent minimization of the total energy of the atomic cluster so obtained, result in the pre-designed $Ga_{10}As_2P_2$ molecule. The corresponding unconstrained counterpart of this molecule emerges after lifting the spatial constraints applied to the atomic positions, and unconditional minimization of the total energy of the atomic cluster $Ga_{10}As_2P_2$. Simultaneous replacement of 2 As atoms by P ones brings about much larger changes to ROHF ELSs (Fig. 4.4) and OTEs (Table 4.1) of these molecules

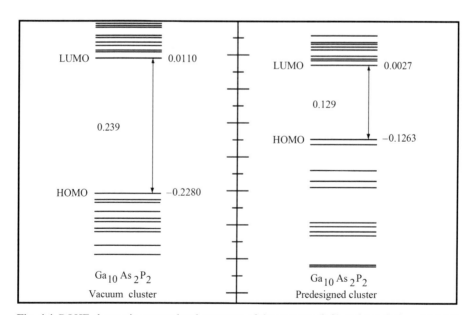

Fig. 4.4 ROHF electronic energy level structure of the vacuum (left) and pre-designed (right) $Ga_{10}As_2P_2$ molecules. The electronic energy level values and the OTEs are in Hartree units. All dimensions are approximately to scale.

compared to those caused by replacement of just one As atom in the pre-designed $Ga_{10}As_4$ molecule. However, results of more accurate CI and MCSCF calculations are less dramatic. In agreement with a general tendency, replacement of yet another As atom with a P atom causes a decrease in the MCSCF ground state energy of the pre-designed $Ga_{10}As_2P_2$ molecule by about 0.3 H compared to that of the pre-designed $Ga_{10}As_3P$ one. The CI ground state energy of the corresponding vacuum molecule is only about 0.1 H lesser than that of the pre-designed $Ga_{10}As_2P_2$ one. CI/MCSCF OTEs of these molecules differs between themselves less than those specific to the $Ga_{10}As_3P$ molecules. [Notably, a large, overestimated value of RHF OTE of the vacuum $Ga_{10}As_2P_2$ molecule is slashed by almost the factor 2 by more accurate CI calculations.] Similar to all, but one, of the gallium arsenide – based molecules of the current chapter and Chapter 3, the pre-designed $Ga_{10}As_2P_2$ molecule is a ROHF/MCSCF triplet of almost perfect pyramidal shape (Fig. 4.5a). At the same time, the vacuum $Ga_{10}As_2P_2$ molecule is the only RHF/CI/CASCF singlet among these molecules. The reason for the latter is a dramatic change in the structure of the vacuum molecule that only remotely resembles a tetrahedral pyramid (Figs. 4.6a and 4.6b). Because P atoms are much smaller and carry much smaller number of electrons than As atoms, replacement of 2 As atoms with 2 P atoms creates enough extra space inside of the pre-designed $Ga_{10}As_2P_2$ cluster to accommodate more of electron charge of Ga atoms in inner parts of the molecule, thus bringing down the total energy of this molecule. However, the centers of mass of atoms in this molecule were not allowed displacement from their averaged positions in the pre-designed parental cluster $Ga_{10}As_3P$, thus modeling excluded volume effects of quantum confinement that restricts atomic motion. Thus, the pre-designed $Ga_{10}As_2P_2$ molecule inherited the pyramidal shape of its pre-designed $Ga_{10}As_3P$ parent. As soon as the centers of mass of its atoms were allowed motion, several Ga atoms of the emerging vacuum $Ga_{10}As_2P_2$ molecule changed their positions by several tenth of Angstrom to move closer to As and P atoms (Fig. 4.6a and 4.6b). One of Ga atoms actually moved inside of the space of the former pyramid somewhat separating the As-bound part of the molecule from the P-bound one.

Similar to the $Ga_{10}As_3P$ duo, both $Ga_{10}As_2P_2$ molecules house electron charge deficit shells, but in the latter case these shells reach much farther out into the space outside of the molecules (Figs. 4.5b and 4.6c). At the same time, the corresponding electron charge "belts" stabilizing the $Ga_{10}As_2P_2$ molecules become more uniform and encompass more of inner molecular space (Figs. 4.5c, 4.6d and 4.6e). For the charge density isovalues of about 0.07 (in arbitrary units) and more, these charge belts become thinner (Figs. 4.5d and 4.6f), and their shape becomes closer to that of the corresponding charge belts of the $Ga_{10}As_3P$ molecules specific to similar CDD isovalues. The above analysis of MEP isosurfaces indicates that the $Ga_{10}As_2P_2$ molecules house thicker shells of electron charge deficit than the molecules with only one P atom, which signifies that more of electron charge of Ga atoms is re-distributed toward inner parts of the molecules. Thus, the holes of the Ga-based molecules with 2 P atoms are large in linear dimensions than those of the Ga-based molecules with one P atom. At the same time, only the pre-designed $Ga_{10}As_2P_2$ molecule is a "ferromagnetic" ROHF/MCSCF triplet with its two uncompensated electron spins accommodated in a vicinity

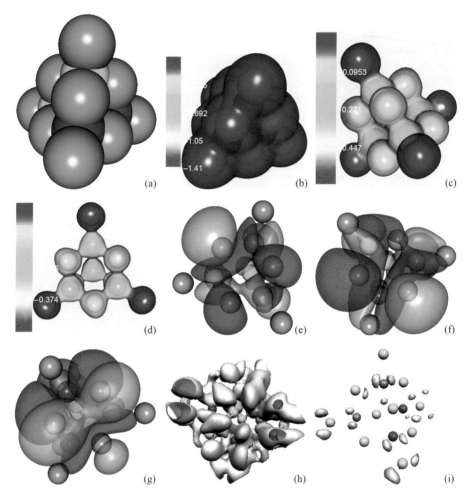

Fig. 4.5 Pre-designed CAS 6 × 8 MCSCF triplet $Ga_{10}As_2P_2$. Structure: (a). Molecular electrostatic potential (MEP) surfaces (color codding schemes are shown in the figures) corresponding to the CDD isovalues (b) 0.01, (c) 0.07 and (d) 0.1 of the charge density distribution (CDD, in arbitrary units). Isosurfaces of positive (gray) and negative (red) parts of the highest occupied (HO) and lowest unoccupied (LU) molecular orbits (MOs): (e) HOMO 115, (f) HOMO 116, and (g) LUMO 117 corresponding to the isovalues 0.01, 0.005 and 0.005 (in arbitrary units), respectively. Isosurfaces of the spin density distribution (SDD) corresponding to the SDD isovalues (h) 0.0005 and (i) 0.0029. Ga atoms are blue, As brown and P green. In (a) dimensions of As and P atoms are somewhat enlarged; in (b) to (i) all atomic dimensions are reduced, and other dimensions are to scale.

of two non-neighboring Ga atoms. The vacuum $Ga_{10}As_2P_2$ molecule is an "antiferromagnetic" HF/CI singlet with all electron spins compensated. Therefore, while both $Ga_{10}As_2P_2$ molecules house relatively large delocalized holes, only the hole of the pre-designed $Ga_{10}As_2P_2$ molecule is spin-polarized.

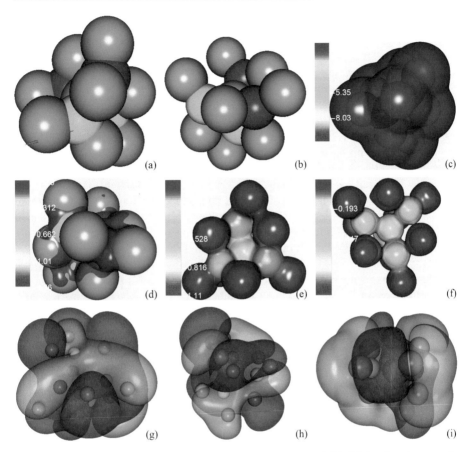

Fig. 4.6 Unconstrained HF singlet $Ga_{10}As_2P_2$. Structure: (a) and (b). Molecular electrostatic potential (MEP) surfaces (color codding schemes are shown in the figures) corresponding to the CDD isovalues (c) 0.0005, (d) 0.03, (e) 0.05 and (f) 0.07 of the charge density distribution (CDD, in arbitrary units). Isosurfaces of positive (gray) and negative (red) parts of the highest occupied (HO) and lowest unoccupied (LU) molecular orbits (MOs): (g) HOMO 114, (h) HOMO 115, and (i) LUMO 116 corresponding to the isovalue 0.002 (in arbitrary units). Ga atoms are blue, As brown and P green. In (d) dimensions of Ga atoms are to scale, and those of As and P atoms are somewhat enlarged; in all other cases atomic dimensions are reduced and other dimensions to scale.

Similar to the case of the $Ga_{10}As_3P$ molecules, hybridization of $4p$ and $3p$ AOs of Ga, As and P atoms, respectively, define the structure of MOs of these molecules in their HOMO-LUMO regions. Thus, HOMO 115 and 116 of the pre-designed $Ga_{10}As_2P_2$ molecule are built primarily of $4p$ AOs of 4 Ga and 2 As atoms, and $3p$ AOs of both P atoms (Figs. 4.5e and 4.5f), while only one $3p$ AO of a P atom contributes to the major components of LUMO 117 of this molecule (Fig. 4.5g). In the case of the vacuum $Ga_{10}As_2P_2$ molecule, p-type AOs of all atoms contribute almost equally to HOMO 114, HOMO 115 and LUMO 116 (Figs. 4.6g to 4.6h) creating all-embracing sandwich-like MOs typical for π-bonding of ligand Ga atoms mediated by $4p$ and $3p$ AOs of As and P atoms, respectively. This bonding indicates a possibility to assemble larger structures using the $Ga_{10}As_2P_2$ molecules as structural elements.

Due to the original design, the Ga – Ga, Ga – As and Ga – P bondlengths in the pre-designed $Ga_{10}As_2P_2$ molecule are exactly the same as those in the pre-designed $Ga_{10}As_3P$ molecule, and are very different from those of the vacuum $Ga_{10}As_2P_2$ molecule. In the latter case, some of the ligand Ga – Ga bonds become shorter, and the majority of such bonds falls in a wider range of values from 3.327 Å to 4.312 Å. One Ga – Ga bond in this case reaches the value of 4.745 Å, which is significantly larger than a typical 4.064 Å bondlength of the ligand Ga – Ga bonds in the vacuum $Ga_{10}As_3P$ molecule. The length of several longest Ga – As bonds in the vacuum $Ga_{10}As_2P_2$ molecule exceeds 3 Å, while the longest Ga – As bond in the vacuum $Ga_{10}As_3P$ molecule is 2.480 Å in length, with the rest of such bonds in both molecules being similar in length. The major reason for the development of longer Ga – As bonds is that in the vacuum $Ga_{10}As_2P_2$ molecule some Ga atoms moved closer to P atoms and therefore, somewhat further from As atoms. This is also reflected by several shorter Ga – P bondlengths in this molecule, where the shortest Ga – P bond is 2.402 Å, while in the case of the vacuum $Ga_{10}As_3P$ molecule the shortest Ga – As bond is 2.541 Å in length.

SDD of the pre-designed $Ga_{10}As_2P_2$ molecule (Figs. 4.5h and 4.5i) is similar to those of the $Ga_{10}As_3P$ molecules, and is relatively uniform for smaller SDD isovalues (Fig. 4.5i). Of course, in all cases, the major contributions to this SDD are defined by electrons of Ga atoms.

4.4 In – As molecules with one and two atoms of phosphorus

The pre-designed $In_{10}As_3P$ molecule has been synthesized using the standard virtual synthesis method. In particular, in the tetrahedral pyramidal symmetry element $In_{10}As_4$ of the bulk InAs lattice (see section 3.2 and Chapter 3 for more details) one of As atoms was replaced with a P atom, and the centers of mass of all of the atoms were constrained to the same positions they had in the $In_{10}As_4$ pyramid. Thus, the center of mass of the P atom was fixed in the position of the center of mass of the As atom it replaced. Therefore, the inner part of this cluster is somewhat loose, because the size of the P atom is smaller than that of the As atom. At the next step, the atomic cluster $In_{10}As_3P$ so obtained was optimized using HF, ROHF, CI, CASSCF and MCSCF methods upon a condition that all positions of the atomic centers of mass remained fixed in the corresponding positions in the pre-designed $In_{10}As_4$ cluster. This 2-step procedure produced the pre-designed molecule $In_{10}As_3P$. Its unconstrained counterpart was developed using the same pre-designed atomic cluster $In_{10}As_3P$ and applying unconditional minimization of the total energy of this cluster, while the position constraints were lifted to accommodate atomic motion. The unconstrained $In_{10}As_3P$ molecule is sometimes called a vacuum molecule, because it is virtually synthesized in the absence of any foreign atoms or external fields.

TABLE 4.2 Ground state energies of the $In_{10}As_3P$ and $In_{10}As_2P_2$ molecules.

Molecule	ROHF ground state energy, Hartree	ROHF optical transition energy, Hartree	MCSCF ground state energy, Hartree	MCSCF optical transition energy, Hartree	CAS (number of electrons × number of MOs)
$In_{10}As_4$	−1907.035041356	3.4286	−1907.077255735	2.9964	6 × 9
$In_{10}As_4$*	−1906.961402713	3.4232	−1906.859967977	0.8684	8 × 8
$In_{10}As_3P$	−1907.239502470	0.3102	not available	ROHF	triplet
$In_{10}As_3P$*	−1907.306787825	3.4041	−1907.356449778	3.2931	8 × 10
$In_{10}As_2P_2$	−1907.536235257	3.1538	−1907.589111726	3.3845	8 × 10
$In_{10}As_2P_2$*	−1907.603453104	3.4449	−1907.654457297	3.3084	8 × 10

* - denotes vacuum molecule

Both $In_{10}As_3P$ molecules are ROHF triplets, and the vacuum molecule is also an MCSCF triplet. MCSCF data for the pre-designed molecule are not yet available. The ground state data of both molecules are collected in Table 4.2, where the ground state data for the pre-designed and vacuum $In_{10}As_4$ molecules of Chapter 3 also included for conveniences of further analysis.

Comparison of the ROHF ground state energies shows that the replacement of an As atom with a P one brings about 0.2 H loss in the ground state energy of the pre-designed $In_{10}As_3P$ molecule compared to that of the pre-designed $In_{10}As_4$ one. At the same time, similar comparison for the vacuum $In_{10}As_3P$ and $In_{10}As_4$ molecules reveals a somewhat larger loss of about 0.35 H in the ground state energy of the vacuum $In_{10}As_3P$ molecule. More accurate evaluation of the energy loss in the case of the pre-designed $In_{10}As_3P$ is not possible at this time, because of the absence of MCSCF data for this molecule. MCSCF data indicate even larger loss of about 0.5 H in the ground state energy of the vacuum $In_{10}As_3P$ molecule compared to that of the vacuum $In_{10}As_4$ molecule. This loss is also significantly larger than the loss of about 0.3 H specific to the case of the gallium arsenide phosphide molecules of the previous section. The reason for such an increase in the value of the ground state energy loss in the case of the vacuum $In_{10}As_3P$ molecule is that a replacement of an As atom with a P atom makes the structure of the $In_{10}As_3P$ molecule somewhat loose, thus allowing re-distribution of some more of electron charge of In atoms toward As and P atoms inside the molecular scaffold. Such re-distribution helps decrease the electron-electron repulsion energy and at the same time, increase electron-nucleus attraction energy. In the case of the vacuum $In_{10}As_3P$ molecule, the latter effect is larger than in the case of the vacuum $Ga_{10}As_3P$ phosphide, because of a larger number of In electrons produce a larger electron repulsion energy, so any allowable re-distribution of that electron charge toward As and P atoms creates a larger effect that a similar re-distribution in the vacuum $G_{10}As_3P$ molecule.

Notably, due to a large number of indium electrons, and thus much larger electron-electron repulsion, the ROHF/MCSCF ground state energies of the $In_{10}As_3P$

molecules (Table 4.2) are much higher than those of the $Ga_{10}As_3P$ molecules of Table 4.1. At the same time, the MCSCF OTE energies of these molecules are similar in value to those of the $Ga_{10}As_3P$ one, signifying that As and P atoms effectively re-distribute a large electron charge of In atoms providing for stabilization of the In-based molecules. The ROHF OTE of the pre-designed $In_{10}As_3P$ stands out as too small for such a small molecule, indicating that ROHF approximation may significantly underestimate OTE of this molecule. Similar to ELS of the $Ga_{10}As_3P$ molecules, ELS of the $In_{10}As_3P$ ones (Fig. 4.7) reveals a total loss of the tetrahedral symmetry and a "sub-band" formation effect of quantum confinement on ELS of the pre-designed molecules. The latter leads to formation of bundles of closely lying electron energy levels in the HOMO-LUMO regions of the pre-designed molecules that with an increase in a number of $In_{10}As_3P$ elements by self-assembly will develop into sub-bands of the valence and conduction bands of $In_{10}As_3P$-based structures.

Similar to the case of the studied $Ga_{10}As_3P$ molecules, lifting of spatial constraints applied to the atomic centers of mass in the pre-designed molecule and subsequent unconditional minimization of the total energy of the $In_{10}As_3P$ cluster does not result in destruction of the tetrahedral pyramidal geometry of this cluster. Thus, the shape of the vacuum $In_{10}As_3P$ molecule visually resembles the same tetrahedral pyramid as that of the pre-designed molecule (Figs. 4.8a and 4.9a). In reality, due to unconstrained optimization procedure and rather loosely packed interior of the used pre-designed $In_{10}As_3P$ cluster, atoms in the vacuum $In_{10}As_3P$ molecule moved a few hundredths of Angstrom from their respective positions in the pre-designed $In_{10}As_3P$

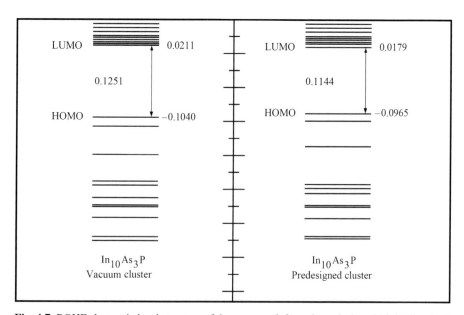

Fig. 4.7 ROHF electronic level structure of the vacuum (left) and pre-designed (right) $In_{10}As_3P$ molecules. The electronic energy level values and the OTEs are in Hartree units. All dimensions are approximately to scale.

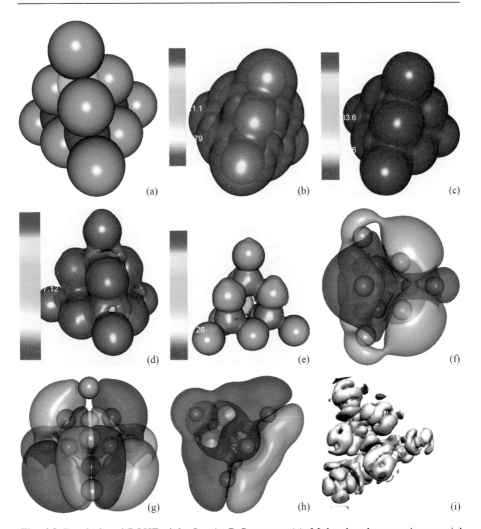

Fig. 4.8 Pre-designed ROHF triplet $In_{10}As_3P$. Structure: (a). Molecular electrostatic potential (MEP) surfaces (color codding schemes are shown in the figures) corresponding to the CDD isovalues (b) 0.005, (c) 0.03, (d) 0.07 and (e) 0.09 of the charge density distribution (CDD, in arbitrary units). Isosurfaces of positive (gray) and negative (red) parts of the highest occupied (HO) and lowest unoccupied (LU) molecular orbits (MOs): (f) HOMO 115, (g) HOMO 116, and (h) LUMO 117 corresponding to the isovalue 0.001 (in arbitrary units). An isosurface of the spin density distribution (SDD) corresponding to the SDD isovalue (i) 0.0005. Indium atoms are blue, As brown and P green. In (a) to (c) dimensions of In atoms are to scale, and As and P atoms are somewhat enlarged; in (d) to (i) all atomic dimensions are reduced and other dimensions are to scale.

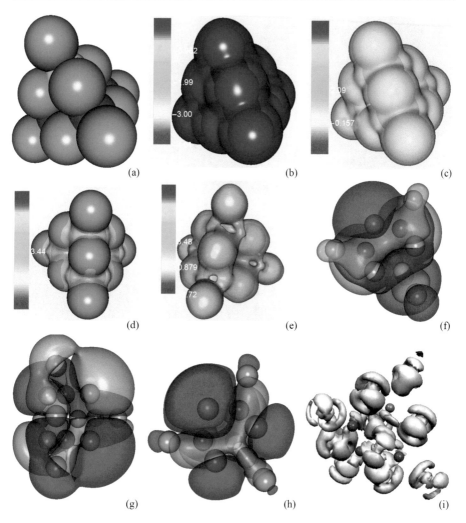

Fig. 4.9 Unconstrained CAS 8 × 10 MCSCF triplet $In_{10}As_3P$. Structure: (a). Molecular electrostatic potential (MEP) surfaces (color codding schemes are shown in the figures) corresponding to the CDD isovalues (b) 0.01, (c) 0.03, (d) 0.05 and (e) 0.07 of the charge density distribution (CDD, in arbitrary units). Isosurfaces of positive (gray) and negative (red) parts of the highest occupied (HO) and lowest unoccupied (LU) molecular orbits (MOs): (f) HOMO 115, (g) HOMO 116, and (h) LUMO 117 corresponding to the isovalue 0.001 (in arbitrary units). An isosurface of the spin density distribution (SDD) corresponding to the SDD isovalue (i) 0.001. Indium atoms are blue, As brown and P green. In (a) to (d) dimensions of In atoms are to scale, and As and P atoms are somewhat enlarged; in (e) to (i) all atomic dimensions are reduced and other dimensions are to scale.

molecule. Meanwhile, MEPs of the two $In_{10}As_3P$ molecules are dramatically different. The pre-designed $In_{10}As_3P$ molecule does not show any sign of an electron charge deficit shell embracing the molecule in close proximity of the "surface" (Figs. 4.8b and 4.8c), where its MEP remains slightly negative. In contrast, MEP of the vacuum molecule (Figs. 4.9b and 4.9c) resembles those of the $Ga_{10}As_3P$ molecules considered in section 4.3 and is slightly positive outside and in a close proximity of the molecular "surface" indicating the presence of a shell of electron charge deficit, or "hole". Due to the fact that some of the electron charge of In atoms is redistributed toward inner parts of the molecules, both $In_{10}As_3P$ molecules exhibit electron charge stability "belts" (Figs. 4.8d, 4.8e, 4.9d and 4.9e). Again, similar to those of the $Ga_{10}As_3P$ molecules, MOs of both $In_{10}As_3P$ molecules in their HOMO-LUMO regions are highly hybridized arrangements of various p-type AOs of In, As and P atoms. In particular, HOMO 115 of the pre-designed molecule (Fig. 4.8f) features the major contributions from $5p$ AOs of 5 In and $4p$ AOs of all As atoms, and a $3p$ AO of the phosphorus atom. HOMO 116 of this molecule (Fig. 4.8g) is similar in nature, but phosphorus does not contribute significantly to this MO. LUMO 117 of the pre-designed $In_{10}As_3P$ molecule (Fig. 4.8h) is hybridized primarily of $5p$ AOs of 8 In atoms and $4p$ AOs of all As atoms, while a $3p$ AO of P and $5p$ AOs of two other In atoms provide smaller contributions to this MO. HOMO 115 of the vacuum $In_{10}As_3P$ molecule (Fig. 4.9f) exhibits the major contributions of the same type as those of HOMO 115 of the pre-designed molecule, but particular $5p$, $4p$ and $3p$ AOs of the constitutive In, As and P atoms, respectively, in the vacuum molecule case differ from those contributing to HOMO 115 of the pre-designed molecule. Four In, all As and P atoms contribute via their respective $5p$, $4p$ and $3p$ AOs to form the major parts of HOMO 116 (Fig. 4.9g) of the vacuum $In_{10}As_3P$ molecule, while only $5p$ AOs of 3 In atoms and $4p$ AOs of 2 As ones contribute to formation of LUMO 117 (Fig. 4.9h) of this molecule.

Similar to many other cases of non-stoichiometric molecules discussed throughout this book, the octet rule of the standard valence theory does not hold for $In_{10}As_3P$ molecules. All of MOs in HOMO-LUMO regions of these molecules are well-developed π-type MOs developed due π-type ligand bonding of In atoms mitigated by their simultaneous bonding to As or P atoms. The In – In bonds in the pre-designed molecule are rather long, reaching 4.284 Å in length. Strong In – As and In – P bonds in this molecule are somewhat longer than those in the pre-designed $Ga_{10}As_3P$ molecule, with their bondlengths ranging from 2.220 Å to 2.934 Å. In the vacuum molecule, the length of the In – In ligand bonds is longer than that in the pre-designed molecule, and falls in the range from 4.291 Å to 4.347 Å. The bondlengths of In – As bonds in the vacuum molecule lie in a somewhat tighter range from 2.339 Å to 2.832 Å, and the In – P bondlengths adopt only 3 values of 2.344 Å, 2.689 Å and 2.892 Å, all of which are somewhat different from those in the pre-designed $In_{10}As_3P$ molecule.

For small spin density isovalues, SDD isosurfaces of both $In_{10}As_3P$ molecules are relatively uniform. The maximum SDD values of these molecules are 0.0304 (in arbitrary units) for the pre-designed molecule and 0.0207 in the case of the vacuum one. The dipole moment of the pre-designed $In_{10}As_3P$ molecule is 5.2933 D, and that of the vacuum one 3.5427 D. Thus, unconstrained optimization of the $In_{10}As_3P$ atomic cluster leads to a simultaneous decrease in the maximum SDD and dipole moment compared to those of the pre-designed molecule. In this respect, the studied $In_{10}As_3P$ molecules differ from the $Ga_{10}As_3P$ molecules of section 4.3, where unconstrained optimization leads to a significant increase in the dipole moment and a slight decrease in the maximum SDD value of the vacuum molecule with regard to those of the pre-designed one. Being ROHF/ MCSCF triplets, each of the $In_{10}As_3P$ molecules carries 2 uncompensated electron spins accommodated in a vicinity of 2 In atoms. Thus, similar to $Ga_{10}As_3P$ molecules, these molecules are "ferromagnetic" in nature. However, only one of them, the vacuum $In_{10}As_3P$ molecule, houses an electron charge deficit shell (a hole), while the pre-designed molecule does not. The electron charge deficit shell of the vacuum $In_{10}As_3P$ molecule is somewhat thinner than those of the $Ga_{10}As_3P$ molecules, but still embraces the entire molecule with its two uncompensated electron spins. Thus, similar to $Ga_{10}As_3P$ molecules, the vacuum $In_{10}As_3P$ molecule houses a delocalized, spin-polarized hole.

Replacement of another As atom with a P one in the pre-designed atomic cluster $In_{10}As_3P$ and conditional minimization of the total energy of the $In_{10}As_2P_2$ atomic cluster, with the centers of mass of the clusters atoms kept in the same positions as they were in the pre-designed $In_{10}As_3P$ cluster, brings about the pre-designed $In_{10}As_2P_2$ molecule. The corresponding vacuum molecule is derived from the pre-designed one upon lifting the spatial constraints applied to the atomic positions, and subsequent unconditional total energy minimization.

The ROHF ground state energy of the $In_{10}As_2P_2$ molecules, and the MCSCF ground state energy of the vacuum $In_{10}As_2P_2$ molecule, are about 0.3 H smaller than the ROHF and MCSCF ground state energy values specific to the corresponding $In_{10}As_3P$ molecules (Table 4.2). ROHF OTE of the pre-designed $In_{10}As_2P_2$ molecule is about an order of magnitude larger than that of the pre-designed $In_{10}As_3P$ molecule, indicating once again, that the ROHF approximation underestimates OTE of the pre-designed $In_{10}As_3P$ molecule. At the same time, ROHF OTE of the vacuum $In_{10}As_2P_2$ molecule is only about 0.04 eV larger than that of the vacuum $In_{10}As_3P$ molecule, and the MCSCF OTEs specific to the pre-designed and vacuum $In_{10}As_2P_2$ molecules exceed that of the vacuum $In_{10}As_3P$ one only by about 0.09 eV and 0.015 eV, respectively.

Similar to other cases studied in this chapter, ELS of the pre-designed $In_{10}As_2P_2$ molecule in the HOMO-LUMO region (Fig. 4.10) reflects the presence of a model quantum confinement. In particular, ROHF OTE and separation between electron energy levels in the HOMO-LUMO region of this molecule are somewhat smaller than those of its vacuum counterpart. The dipole moment 3.3845 D of this molecule is somewhat smaller than that of the vacuum $In_{10}As_2P_2$ molecule (3.7121 D), indicating that a decrease in the ground state energy of the vacuum molecule compared to that of the pre-designed one is due to such re-distribution of electron charge that leads to an increase in the dipole moment of the molecule. This mechanism of electron charge redistribution is similar to the one specific to small gallium arsenide phosphide molecules of section 4.3, and is different from the one specific to $In_{10}As_3P$ molecules.

By its nature, the pre-designed $In_{10}As_2P_2$ molecule is a tetrahedral pyramid in shape (Fig. 4.11a). The vacuum $In_{10}As_2P_2$ molecule also occurs pyramidal in shape (Fig. 4.12a), but this is due to the fact that atomic displacement at sub-Angstrom scale cannot be noticed by a human eye. As a result of unconstrained energy minimization, atoms of the vacuum $In_{10}As_2P_2$ molecule moved from their respective positions in the

Quantum Dots of Gallium and Indium Arsenide Phosphides

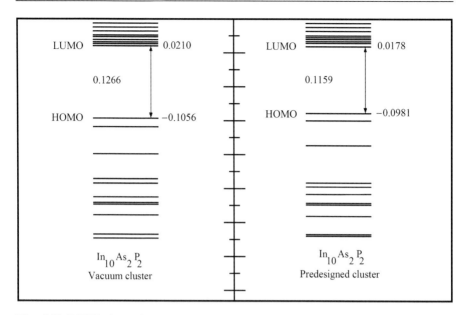

Fig. 4.10 ROHF electronic level structure of the vacuum (left) and pre-designed (right) $In_{10}As_2P_2$ molecules. The electronic energy level values and the OTEs are in Hartree units. All dimensions are approximately to scale.

pre-designed $In_{10}As_2P_2$ cluster by only a few hundredths of Angstrom. Larger atomic displacements, such as those in the vacuum $Ga_{10}As_2P_2$ molecule of sec. 3, were not possible in this case, because an additional space provided by replacement of two As atoms with two P ones was not enough to accommodate significant displacements of much larger (than Ga) In atoms toward As and P atoms.

In correspondence with their structural similarity, the two $In_{10}As_2P_2$ molecules possess similar MEPs, and chemical and magnetic properties. Both molecules do not develop any electron charge deficit shells, as their MEPs are almost zero immediately outside of their "surfaces" (Figs. 4.11b and 4.12b), and become progressively negative toward the central regions of the molecular structures (Figs. 4.11c and 4.12c). Due to re-distribution of electron charge of In atoms toward As and P atoms, there are electron charge accumulation belts in the inner parts of the molecules (Figs. 4.11d, 4.11e, 4.12d, and 4.12e) playing a stabilizing role.

In the HOMO-LUMO regions MOs of the $In_{10}As_2P_2$ molecules are hybridized of $5p$, $4p$ and $3p$ AOs of their constitutive In, As and P atoms, respectively, included in various proportions. Thus, HOMO 115 of the pre-designed $In_{10}As_2P_2$ molecule (Fig. 4.11f) contains the major contributions from $5p$ AOs of 5 In atoms, $4p$ AOs of 2 As atoms and a $3p$ AO of one P atom. The major contributions to HOMO 116 of this molecule (Fig. 4.11g) come from $5p$ AOs of only 3 In, $4p$ AOs of 2 As and a $3p$ AO of one P atoms, while $5p$ AOs of two other In atoms contribute significantly less. The major contributions to LUMO 117 of this molecule (Fig. 4.11h) come from a 5p AO of only one In atom and 4p AOs of both As atoms. Somewhat smaller contributions to

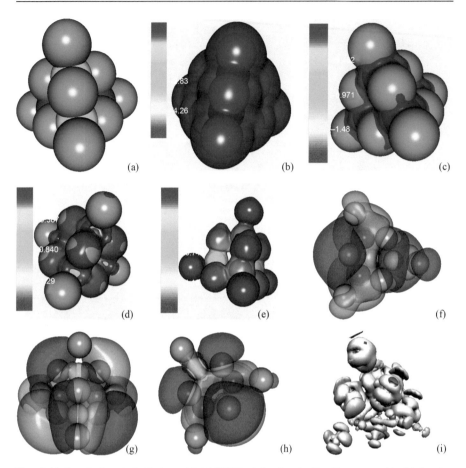

Fig. 4.11 Pre-designed CAS 8 × 10 MCSCF triplet $In_{10}As_2P_2$. Structure: (a). Molecular electrostatic potential (MEP) surfaces (color codding schemes are shown in the figures) corresponding to the CDD isovalues (b) 0.005, (c) 0.05, (d) 0.07 and (e) 0.09 of the charge density distribution (CDD, in arbitrary units). Isosurfaces of positive (gray) and negative (red) parts of the highest occupied (HO) and lowest unoccupied (LU) molecular orbits (MOs): (f) HOMO 115, (g) HOMO 116, and (h) LUMO 117 corresponding to the isovalue 0.001 (in arbitrary units). An isosurface of the spin density distribution (SDD) corresponding to the SDD isovalue (i) 0.0005. Indium atoms are blue, As brown and P green. In (a) to (c) dimensions of In atoms are to scale, and As and P atoms are somewhat enlarged; in (d) to (i) all atomic dimensions are reduced and other dimensions are to scale.

this MO are due to 5p AOs of four other In atoms, while 3p AOs of P atoms do not contribute at all. In the case of the vacuum $In_{10}As_2P_2$ molecule, HOMO 115 (Fig. 4.12f) reflects a larger role of P atoms, as the major contributions to this MO come from 5p AOs of 4 In, 4p AOs of 2 As and 3p AOs of both P atoms. The proper HOMO 116 of this molecule (Fig. 4.12g) is almost a mirror image of the proper HOMO 116 of the pre-designed molecule (Fig. 4.11g), with the major contributions coming from 5p

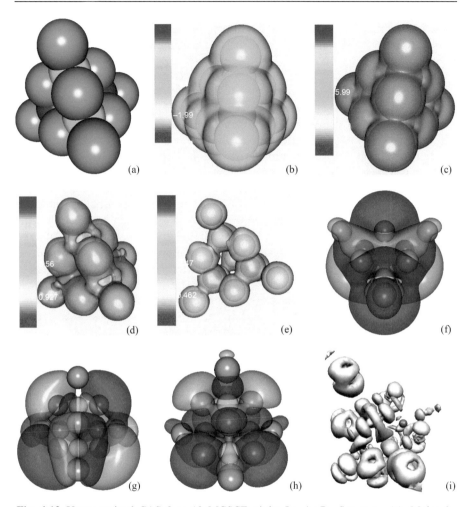

Fig. 4.12 Unconstrained CAS 8 × 10 MCSCF triplet $In_{10}As_2P_2$. Structure: (a). Molecular electrostatic potential (MEP) surfaces (color codding schemes are shown in the figures) corresponding to the CDD isovalues (b) 0.005, (c) 0.03, (d) 0.07 and (e) 0.09 of the charge density distribution (CDD, in arbitrary units). Isosurfaces of positive (gray) and negative (red) parts of the highest occupied (HO) and lowest unoccupied (LU) molecular orbits (MOs): (f) HOMO 115, (g) HOMO 116, and (h) LUMO 117 corresponding to the isovalue 0.001 (in arbitrary units). An isosurface of the spin density distribution (SDD) corresponding to the SDD isovalue (i) 0.001. Indium atoms are blue, As brown and P green. In (a) to (c) dimensions of In atoms are to scale, and As and P atoms are somewhat enlarged; in (d) to (i) all atomic dimensions are reduced and other dimensions are to scale.

AOs of 4 In, 4p AOs of As and a 3p AO of one P atoms, and a significantly smaller contribution due to a 5p AO of yet another In atom. The proper LUMO 117 of the vacuum $In_{10}As_2P_2$ molecule (Fig. 4.12h) is hybridized in a way similar to that of the pre-designed $In_{10}As_2P_2$ one (Fig. 4.11c). Thus, the major contributions to this MO come from a $5p$ AO of one In and $4p$ AOs of both As atoms, with smaller contributions to this MO from $5p$ AOs of 4 other In atoms, and no significant contributions from 3p AOs of P atoms.

Obviously, the bonding in the $In_{10}As_2P_2$ molecules breaks the standard octet rule, and is defined by the π-type In – In ligand bonding mediated by As and P atoms at almost equal grounds. By the π-nature of the bonding, larger nanostructures may be developed using $In_{10}As_2P_2$ molecules as structural elements. All In – In bonds in the pre-designed $In_{10}As_2P_2$ molecule are the same in length, 4.284 Å, equal to that of In – In bonds in the pre-designed $In_{10}As_3P$ molecule, as defined by the same model quantum confinement of this molecules. However, In – As and In – P bondlengths in the pre-designed $In_{10}As_2P_2$ molecule mirror the bondlengths of In – P and In – As bonds, respectively, of the pre-designed $In_{10}As_3P$ molecule. In particular, there are 3 types of In – As bonds in the pre-designed $In_{10}As_2P_2$ molecule with their lengths 2.220 Å, 2.717 Å and 2.934 Å. These bondlengths are exactly equal to those of three types of In – P bonds in the pre-designed $In_{10}As_3P$ molecule. At the same time, all 5 types of In – P bonds in the pre-designed $In_{10}As_2P_2$ molecule are exactly equal in length to those of the 5 types of In – As bonds in the pre-designed $In_{10}As_3P$ molecule. Thus, bonding-wise, in the pre-designed $In_{10}As_2P_2$ molecule P atoms seem assuming the role played by As atoms in the pre-designed $In_{10}As_3P$ molecule, while its two As atoms play a role similar to that of one P atom in the pre-designed $In_{10}As_3P$ molecule.

This very interesting and important phenomenon is entirely due to quantum confinement, and as such, can be called **a quantum confinement composition effect**. That is, dependent on a particular nature of quantum confinement, polarization of the confined molecules due to quantum confinement may affect atomic bonding in a molecule to a degree that it becomes similar to that of some other atoms in similar quantum confinements. Due to similar geometries of model quantum confinements, $4p$ AOs of the two As atoms in the pre-designed $In_{10}As_2P_2$ molecule play the same role as that of $3p$ AOs of just one P atom in the pre-designed $In_{10}As_3P$ molecule. Similarly, $3p$ AOs of two P atoms of the pre-designed $In_{10}As_2P_2$ molecule play the same role in bonding in this molecule as that of $4p$ AOs of the two As atoms in the pre-designed $In_{10}As_3P$ molecule. Thus, one may expect that these molecules may be interchangeable, or at least equally probable, as reactants or products of chemical reactions taking place in quantum confinement or on surfaces.

The above effect is also manifested by results of comparison of bondlengths in the vacuum $In_{10}As_2P_2$ and $In_{10}As_3P$ molecules. Due to its higher symmetry, the vacuum $In_{10}As_2P_2$ molecule develops only 3 types of ligand In – In bonds of 4.302 Å, 4.343 Å and 4.347 Å in length. At the same time, in the asymmetric vacuum $In_{10}As_3P$ molecule there develop 5 types of ligand In – In bonds of 4.291 Å, 4.304 Å, 4.309 Å, 4.343 Å and 4.347 Å in length. Similarly, in the symmetric vacuum $In_{10}As_2P_2$ molecule there exist 4 types of In – As bonds of 2.341 Å, 2.702 Å, 2.709 Å and 2.892 Å in length, while in the asymmetric vacuum $In_{10}As_3P$ molecule there are 7 types of such bonds with their lengths 2.339 Å, 2.492 Å, 2.650 Å, 2.702 Å, 2.704 Å, 2.709 Å and 2.892 Å. Moreover, in the symmetric vacuum $In_{10}As_2P_2$ molecule there exist 6 types of In – P bonds of 2.347 Å, 2.487 Å, 2.647 Å, 2.689 Å, 2.701 Å and 2.891 Å in length, while in the asymmetric vacuum $In_{10}As_3P$ molecule there are only 3 types of In – P bonds of 2.344 Å, 2.689 Å and 2.892 Å in length. At the same time, several In – As bondlengths in the vacuum $In_{10}As_2P_2$ molecule are equal to In – P bondlengths in the vacuum $In_{10}As_3P$ molecule, and *vice versa*. This latter phenomenon provides a physical mechanism behind the composition effect of quantum confinement.

Both $In_{10}As_2P_2$ molecules are "ferromagnetic" ROHF/MCSCF triplets, each with two uncompensated $4d$ electron spins accommodated near 2 In atoms, and small maximum values of about 0.02 (in arbitrary units) of their SDDs. None of these molecules houses an electron charge deficit hole. Thus, among 4 studied small In – As phosphide molecules, all of which are "ferromagnetic" triplets, only one houses a delocalized, spin polarized hole. In contrast, all 4 small Ga – As phosphide molecules studied in section 4.3 house delocalized holes, and three of such holes are spin-polarized.

4.5 More about composition effects of quantum confinement: small molecules of In-As–based phosphides "imbedded" into a model Ga-As confinement

One of the most attractive features of the virtual synthesis method is its ability to realize computationally almost any constrained or unconstrained system, regardless of whether or not such a system has been studied experimentally. In this section the virtual synthesis method is used to model small pyramidal molecules whose composition is the same as that of the $In_{10}As_3P$ and $In_{10}As_2P_2$ molecules of section 4.4, but their geometry is defined by covalent radii of Ga and As specific to those in the zincblende GaAs bulk lattice. The simplest way to build pre-designed pyramidal models of such molecules is to replace Ga atoms in the pre-designed $Ga_{10}As_3P$ and $Ga_{10}As_2P_2$ pyramidal molecules of section 4.3 with In atoms while keeping all interatomic distances unchanged. This means, that In atoms are assigned the value of Ga covalent radius specific to the zincblende GaAs bulk lattice, and therefore, the In-based atomic clusters so designed are significantly strained. Conditional optimization of these clusters uses HF, ROHF, CI, CASSCF and MCSCF methods applied at the condition that the atomic centers of mass remain fixed at their positions in space. This optimization produces the pre-designed molecules denoted $[In_{10}As_3P]_{Ga}$ and $[In_{10}As_2P_2]_{Ga}$, where the subscript Ga indicates that the covalent radii of In atoms in these molecules are equal to that of Ga in the zincblende bulk GaAs lattice. These two molecules model likely results of molecular synthesis of small indium arsenide phosphide molecules in tetrahedral voids of the zincblende bulk GaAs lattice. The corresponding unconstrained molecules have been obtained after lifting spatial constrained applied to the centers of mass of the $[In_{10}As_3P]_{Ga}$ and $[In_{10}As_2P_2]_{Ga}$ cluster atoms, and subsequent unconditional minimization of the total energy of the clusters. The unconstrained molecules so obtained are denoted $[In_{10}As_3P]*_{Ga}$ and $[In_{10}As_2P_2]*_{Ga}$. These vacuum molecules are developed to investigate relaxation of the pre-designed molecules synthesized in quantum confinement after such a confinement is removed (or the molecules being removed from their quantum confinement). The simplest prediction of the results of such relaxation is that the pre-designed $[In_{10}As_3P]_{Ga}$ and $[In_{10}As_2P_2]_{Ga}$, molecules should relax to become the vacuum $In_{10}As_3P$ and $In_{10}As_2P_2$ molecules of the previous section that realize the global minima of the total energy of the $In_{10}As_3P$ and $In_{10}As_2P_2$ clusters, respectively. This expectation, however, is wrong. It is shown below in this section, that there exist metastable and long-lived states of these strained atomic clusters whose total energies are close to that of the global minima of the total energies of these clusters. Such long-lived metastable states are realized at the first step of relaxation of the pre-designed molecules toward the equilibrium vacuum $In_{10}As_3P$ and $In_{10}As_2P_2$ molecules, respectively, studied in section 4.4. The ground state data specific to all 4 molecules modelled in this section are collected in Table 4.3, where the ground state data of similarly developed pre-designed and vacuum $[In_{10}As_4]_{Ga}$ and $[In_{10}As_4]^*_{Ga}$, molecules, respectively, studied in Chapter 3 are also included for comparison. All molecules of Table 4.3 are ROHF/MCSCF triplets. Note, that only ROHF data for the pre-designed $[In_{10}As_3P]_{Ga}$ and vacuum $[In_{10}As_3P]^*_{Ga}$ molecules are available at present.

The ROHF ground state energy of the pre-designed $[In_{10}As_3P]_{Ga}$ molecule is only slightly larger than that of its vacuum $[In_{10}As_3P]^*_{Ga}$ counterpart. Accordingly, ROHF OTE of this "artificial" molecule is somewhat smaller than that of the vacuum $[In_{10}As_3P]^*_{Ga}$ molecule. At the same time, the ground state energies of the pre-designed $[In_{10}As_3P]_{Ga}$ and vacuum $[In_{10}As_3P]^*_{Ga}$ molecules both are smaller than those of the pre-designed $[In_{10}As_4]_{Ga}$ and vacuum $[In_{10}As_4]^*_{Ga}$ molecules, respectively. Moreover, the ROHF ground state energy of the pre-designed $[In_{10}As_3P]_{Ga}$ triplet is about 0.3 H smaller than that of the pre-designed $[In_{10}As_4]_{Ga}$ triplet. Thus, for all models of quantum-confined molecules considered in this chapter and regardless of whether such molecules are loose or strained, a replacement of an As atom with a P one leads to a decrease by about 0.3 H in the ROHF ground state energy. In the case of the vacuum molecules, the corresponding decrease in the ROHF ground state energy is about 0.3 H for the molecules of Tables 4.1 and 4.2, and about 0.26 H for the molecules of Table 4.3. The latter result confirms, once

TABLE 4.3 Ground state energies of the $In_{10}As_3P$ and $In_{10}As_2P_2$ molecules derived from Ga-based clusters. In the pre-designed molecules In atoms are assigned Ga covalent radius.

Molecule	ROHF ground state energy, Hartree	ROHF optical transition energy, Hartree	MCSCF ground state energy, Hartree	MCSCF optical transition energy, Hartree	CAS (number of electrons × number of MOs)
$[In_{10}As_4]_{Ga}$	−1906.830250473	3.5252	−1906.883611124	1.7803	8 × 10
$[In_{10}As_4]^*_{Ga}$	−1906.845323382	3.3334	−1906.872776363	3.1813	8 × 10
$[In_{10}As_3P]_{Ga}$	−1907.149620456	3.2681	not available	ROHF	triplet
$[In_{10}As_3P]^*_{Ga}$	−1907.205148269	3.3007	not available	ROHF	triplet
$[In_{10}As_2P_2]_{Ga}$	−1907.464429502	3.0313	−1907.501145979	1.0374	8 × 9
$[In_{10}As_2P_2]^*_{Ga}$	−1907.463364082	3.4858	−1907.512980892	1.3436	8 × 10

* - denotes vacuum molecule

again, that molecules of Table 4.3 are realization of metastable states of the corresponding atomic clusters rather than the equilibrium states of these clusters.

ROHF OTEs of the vacuum $[In_{10}As_3P]*_{Ga}$ triplet is close in value to that of the vacuum $In_{10}As_3P$ molecule of Table 4.2, while ROHF OTE of the pre-designed $[In_{10}As_3P]_{Ga}$ triplet is about an order of magnitude larger than that of the pre-designed $In_{10}As_3P$ triplet of Table 4.2, in agreement with already mentioned observation that the ROHF approximation underestimates the OTE value of the latter triplet. Note, that with the exceptions of the mentioned triplet and the vacuum $Ga_{10}As_2P_2$ molecule of Table 4.2, ROHF OTEs of all In-based molecules of this chapter, and those of the pre-designed $[In_{10}As_4]_{Ga}$ and vacuum $[In_{10}As_4]*_{Ga}$ molecules, fall in the range from about 3 eV to slightly over 3.5 eV.

Both pre-designed $[In_{10}As_3P]_{Ga}$ and vacuum $[In_{10}As_3P]*_{Ga}$ molecules retain tetrahedral pyramidal geometry (Figs. 4.13a and 4.14a), although in the vacuum molecule case the pyramidal structure is only approximate, as the atoms of this molecule moved by a few hundredths of Angstrom from their former places in the parent pre-designed molecule due to relaxation of the special constraints. MEPs isosurfaces of the $[In_{10}As_3P]_{Ga}$ molecule outside of the space occupied by its atoms do not indicate any region of electron charge deficit. MEP values of this molecule change from zero at some distance from the molecular "surface" (Fig. 4.13b) to small negative values in the immediate vicinity of the surface (Fig. 4.13c), and to progressively larger negative values in the inner regions of the molecule (Figs. 4.13d and 4.13e). In contrast, MEP values of the vacuum $[In_{10}As_3P]*_{Ga}$ molecule are slightly positive in the immediate vicinity of the molecular "surface" (Figs. 4.14b and 4.14c) outside the molecule, indicating that there exists a thin shell of electron charge deficit embracing this molecule. In this respect, the vacuum $[In_{10}As_3P]*_{Ga}$ molecule resembles the vacuum $In_{10}As_3P$ molecule of section 4.4, which is the only molecule among "natural" In-based molecules of section 4.4 that houses a delocalized, spin-polarized hole. The electron charge of In atoms redistributed toward inner region of the pre-designed $[In_{10}As_3P]_{Ga}$ and vacuum $[In_{10}As_3P]*_{Ga}$ molecules is accumulated in the "belts" stabilizing these molecules (Figs. 4.13e, 4.14d and 4.14e). The electron charge stability belt is less distinctive in the pre-designed $[In_{10}As_3P]_{Ga}$ molecule than in the vacuum $[In_{10}As_3P]*_{Ga}$ molecule.

Occupied MOs in the HOMO-LUMO region of both pre-designed $[In_{10}As_3P]_{Ga}$ and vacuum $[In_{10}As_3P]*_{Ga}$ molecules are very similar. Thus, HOMOs 115 and HOMOs 116 of both molecules are hybridized of $5p$ AOs of 4 In, $4p$ AOs of all As and a $3p$ AO of P atoms, respectively (Figs. 4.13f, 4.13g, 4.14f and 4.14g). The proper LUMOs 117 of these molecules are somewhat different. In the case of the pre-designed molecule, this LUMO is composed primarily of $5p$ AOs of 6 In atoms, $4p$ AOs of all As atoms and a $3p$ AO of the P atom, with a smaller contributions due to $5p$ AOs of yet another two In atoms. In the case of the vacuum molecule, a $5p$ AO contribution due to yet another In atom replaces a contribution from a $3p$ AO of the P atom, which becomes very small.

Similar to all other cases discussed in this chapter, bonding in the "artificial" pre-designed $[In_{10}As_3P]_{Ga}$ and vacuum $[In_{10}As_3P]*_{Ga}$ molecules does not confirm to the standard octet rule. By its design, the lengths of all π-type ligand In – In bonds in the pre-designed molecule are the same and equal to 3.997 Å, and those of all In – As and In – P bonds are equal to 2.448 Å. These bondlengths are defined by the GaAs-based pyramidal geometry of the quantum confinement and Ga covalent radius

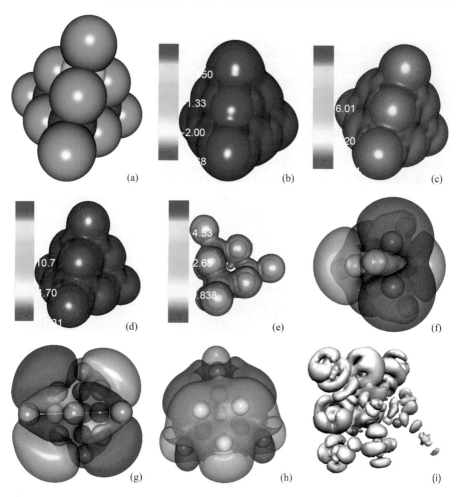

Fig. 4.13 Pre-designed ROHF triplet $[In_{10}As_3P]_{Ga}$. Structure: (a). Molecular electrostatic potential (MEP) surfaces (color codding schemes are shown in the figures) corresponding to the CDD isovalues (b) 0.01, (c) 0.03, (d) 0.05 and (e) 0.1 of the charge density distribution (CDD, in arbitrary units). Isosurfaces of positive (gray) and negative (red) parts of the highest occupied (HO) and lowest unoccupied (LU) molecular orbits (MOs): (f) HOMO 115, (g) HOMO 116, and (h) LUMO 117 corresponding to the isovalue 0.001 (in arbitrary units). An isosurface of the spin density distribution (SDD) corresponding to the SDD isovalue (i) 0.001. Indium atoms are blue, As brown and P green. In (a) to (c) dimensions of In atoms are to scale, and As and P atoms are somewhat enlarged; in (d) to (i) all atomic dimensions are reduced and other dimensions are to scale.

enforced on In atoms. In the vacuum molecule, where the geometrical constraints are relaxed, 5π-type ligand In – In bonding is realized via 11 bonds with their bondlengths very close in value and ranging from 4.033 Å to 4.089 Å. Notably, these values of In – In bondlengths are almost the same as those of Ga – Ga bonds in the vacuum

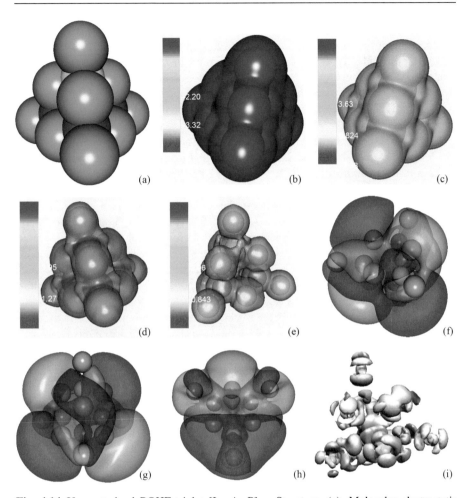

Fig. 4.14 Unconstrained ROHF triplet $[In_{10}As_3P]_{Ga}$. Structure: (a). Molecular electrostatic potential (MEP) surfaces (color codding schemes are shown in the figures) corresponding to the CDD isovalues (b) 0.005, (c) 0.03, (d) 0.07 and (e) 0.09 of the charge density distribution (CDD, in arbitrary units). Isosurfaces of positive (gray) and negative (red) parts of the highest occupied (HO) and lowest unoccupied (LU) molecular orbits (MOs): (f) HOMO 115, (g) HOMO 116, and (h) LUMO 117 corresponding to the isovalue 0.001 (in arbitrary units). An isosurface of the spin density distribution (SDD) corresponding to the SDD isovalue (i) 0.001. Indium atoms are blue, As brown and P green. In (a) to (c) dimensions of In atoms are to scale, and As and P atoms are somewhat enlarged; in (d) to (i) all atomic dimensions are reduced and other dimensions are to scale.

$Ga_{10}As_3P$ molecule of section 4.3, and are much shorter than the 5π-type ligand In – In bondlengths in the vacuum $In_{10}As_3P$ molecule of section 4.4. Similarly, the $5p$-$4p$-type In – As bonding in the vacuum $[In_{10}As_3P]^*_{Ga}$ molecule is represented by 7 types of bonds with 4 bonds of 2.457 Å, 2.465 Å, 2.488 Å and 2.490 Å in length, and 3 other bondlengths close to 2.496 Å. Two types of the 5p-3p In – P bonds in this molecule are of the same length as some of those of the In – As bonds (2.457 Å and 2.490 Å), and 3 other types of 5p-3p In – P bonds are of 2.449 Å, 2.477 Å and 2.482 Å in length. The majority of In – As and In – P bonds in the vacuum [In$_{10}$As$_3$P]*$_{Ga}$ molecule are shorter in length than the corresponding bonds in In$_{10}$As$_3$P molecule of section 4.4, but somewhat longer than the corresponding bonds in the Ga$_{10}$As$_3$P molecule of section 4.3. While these data are obtained using a rather rough ROHF approximation, the bondlength values so calculated are supported by MCSCF data for pre-designed [In$_{10}$As$_2$P$_2$]$_{Ga}$ and vacuum [In$_{10}$As$_2$P$_2$]*$_{Ga}$ molecules discussed below.

As already mentioned, both "artificial" pre-designed [In$_{10}$As$_3$P]$_{Ga}$ and vacuum [In$_{10}$As$_3$P]*$_{Ga}$ molecules are "ferromagnetic" triplets, with their uncompensated electron spins accommodated near two non-neighboring In atoms in each case, and the maximum SDD values being about 0.046 and 0.034 (in arbitrary units), respectively. Examples of SDD isosurfaces of these molecules corresponding to the SDD isovalue 0.001 are shown in Figs. 4.13i and 4.14i. Of these two molecules, only the vacuum [In$_{10}$As$_3$P]*$_{Ga}$ one houses a delocalized, spin-polarized hole. The ROHF dipole moment of the pre-designed [In$_{10}$As$_3$P]$_{Ga}$ molecule is 5.7078 D, and that of the vacuum [In$_{10}$As$_3$P]*$_{Ga}$ molecule is 5.7689 D. Thus, a decrease in the total energy of the vacuum molecule compared to that of the pre-designed one is accompanied by an increase in its dipole moment. This indicates that the electron charge re-distribution mechanism leading to synthesis of the vacuum [In$_{10}$As$_3$P]*$_{Ga}$ molecule resembles the one that takes place in the case of all small gallium arsenide phosphide molecules of section 4.3 and In$_{10}$As$_2$P$_2$ molecules of section 4.4, while it differs from the charge re-distribution mechanism specific to In$_{10}$As$_3$P molecules of section 4.4. Therefore, as a result of applied strain due to the tight tetrahedral GaAs-defined quantum confinement, the electron charge redistribution mechanism in the vacuum [In$_{10}$As$_3$P]*$_{Ga}$ molecule changes and remains similar to the one specific to Ga-As-based small phosphide molecules of section 4.3, even after removal of the model quantum confinement. In general, opto-electronic properties of the "artificial" pre-designed [In$_{10}$As$_3$P]$_{Ga}$ and vacuum [In$_{10}$As$_3$P]*$_{Ga}$ molecules are closer to those of the corresponding Ga-As-based molecules with one P atom from section 4.3, while magnetic properties of these artificial molecules are closer to those of the corresponding In$_{10}$As$_3$P molecules of section 4.4.

Replacement of yet another As atom in the pre-designed [In$_{10}$As$_3$P]$_{Ga}$ molecule with a P atom and subsequent conditional optimization (that is, the total energy minimization completed on a condition that the centers of mass of the atoms in the [In$_{10}$As$_2$P$_2$]$_{Ga}$ cluster were fixed at their original positions in space) brought about the pre-designed "artificial" molecule [In$_{10}$As$_2$P$_2$]$_{Ga}$. The corresponding unconstrained, or vacuum, molecule [In$_{10}$As$_2$P$_2$]*$_{Ga}$ was obtained by lifting the position constraints applied to the atoms of the [In$_{10}$As$_2$P$_2$]$_{Ga}$ cluster, and subsequent unconditional minimization of its total energy. ROHF and MCSCF ground state data specific to these two molecules are collected in Table 4.3.

Once again, replacement of another As atom with a P one leads to a decrease in the ROHF ground state energies by about 0.35 H and 0.26 H in the case of the pre-designed [In$_{10}$As$_2$P$_2$]$_{Ga}$ and vacuum [In$_{10}$As$_2$P$_2$]*$_{Ga}$ molecules compared to those of the pre-designed [In$_{10}$As$_3$P]$_{Ga}$ and vacuum [In$_{10}$As$_3$P]*$_{Ga}$, respectively. Comparison of the MCSCF ground state energies of the pre-designed $[In_{10}As_2P_2]_{Ga}$ and vacuum $[In_{10}As_2P_2]*_{Ga}$ molecules to those of the pre-designed $[In_{10}As_4]_{Ga}$ and vacuum $[In_{10}As_4]*_{Ga}$ molecules (Table 4.3) produces similar results for the ground state energy loss per P atom: about 0.31 H and 0.32 H, respectively. At the same time, MCSCF OTEs of the "artificial" molecules with two phosphorus atoms (Table 4.3) are about 2 times smaller than both ROHF and MCSCF OTEs of all other molecules with one or two P atom discuss in this section and section 4.4, with an exception of the pre-designed $In_{10}As_3P$ molecule of section 4.4, whose ROHF OTE is severely underestimated by the ROHF calculations. This result is due to the fact that by design of these molecules, In atoms are still too close to As and P ones (the molecules are strained), so that to minimize the total energy, In electrons are accommodated in closer lying MOs. This strain effect is similar in nature to the excluded volume effect of quantum confinement, and adds to the excluded volume effect of the quantum confinement in the case of the pre-designed $[In_{10}As_2P_2]_{Ga}$ molecule. In the case of the vacuum $[In_{10}As_2P_2]*_{Ga}$ molecule, the remaining strain effect is the only one that is responsible for a decrease in MCSCF OTE of these molecule as compared to that of the vacuum molecule $In_{10}As_2P_2$ of Table 4.3 and other molecules of this section. Note, that similar to almost all other In-based molecule cases considered in this chapter, the minimum CAS necessary to complete CASSCF and MCSCF calculations for these molecules is large and requires 8 electrons distributed over 9 or 10 occupied and unoccupied MOs.

Despite of replacement of two As atoms with P ones that provides for accommodation of a larger number of In electrons in the inner parts of the molecule, additional space created by such replacement is still too small to allow In atoms moving closer to As and/or P atoms in a way Ga atoms moved in the vacuum $Ga_{10}As_2P_2$ vacuum molecule. Therefore, the shape of both molecules remains pyramidal. In the case of the pre-designed molecule, it is a tetrahedral pyramid by design (Fig. 4.15a), and in the case of vacuum molecule it is a visual pyramid (Fig. 4.16a), as the atoms of this molecule are slightly displaced from their former positions in the corresponding "true" tetrahedral pyramid. At the same time, that additional space is sufficient for significant redistribution of electron charge of In atoms toward inner parts of the molecules, which brings about electron charge deficit regions (holes), as demonstrated by isosurfaces of small positive values of MEPs of these molecules corresponding to such regions. The delocalized hole residing on the pre-designed $[In_{10}As_2P_2]_{Ga}$ molecule is a rather thick shell that embraces the molecule and embeds parts of In atoms' cores defined by their covalent radius (Figs. 4.15b and 4.15c). A similar hole residing on the vacuum $[In_{10}As_2P_2]*_{Ga}$ molecule is thinner, and embraces the molecule on outside of its "surface", so that only small parts of cores of In atoms are embedded into the shell (Figs. 4.16b and 4.16c). Similar to all other stable molecules of semiconductor compound atoms, both molecules possess stabilization charge belts, that is, regions of electron charge accumulation surrounding As and P atoms (Figs. 4.15d, 4.15e, 4.16d and 4.16e).

In common with other cases considered in this chapter, MOs of the "artificial" $[In_{10}As_2P_2]_{Ga}$ and $[In_{10}As_2P_2]*_{Ga}$ molecules in their respective HOMO-LUMO regions are hybridized of only p-type AOs, which provides for strong π-type bonding with its typical sandwich-like MO shape. HOMO 115 (Fig. 4.15f) of the pre-designed

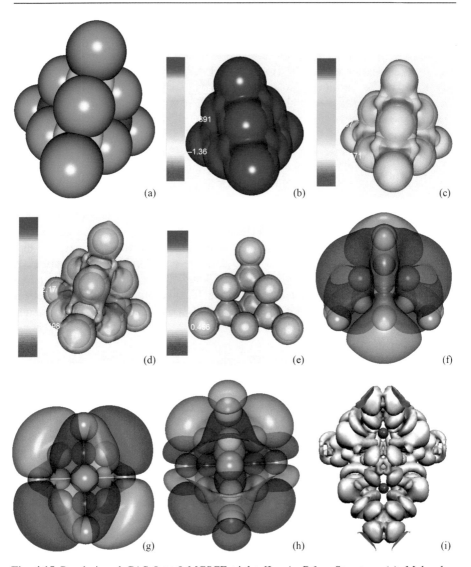

Fig. 4.15 Pre-designed CAS 8 × 9 MCSCF triplet [$In_{10}As_2P_2$]$_{Ga}$. Structure: (a). Molecular electrostatic potential (MEP) surfaces (color codding schemes are shown in the figures) corresponding to the CDD isovalues (b) 0.03, (c) 0.07, (d) 0.09 and (e) 0.11 of the charge density distribution (CDD, in arbitrary units). Isosurfaces of positive (gray) and negative (red) parts of the highest occupied (HO) and lowest unoccupied (LU) molecular orbits (MOs): (f) HOMO 115, (g) HOMO 116, and (h) LUMO 117 corresponding to the isovalue 0.001 (in arbitrary units). An isosurface of the spin density distribution (SDD) corresponding to the SDD isovalue (i) 0.0005. Indium atoms are blue, As brown and P green. In (a) and (b) dimensions of In atoms are to scale, and As and P atoms are somewhat enlarged; in (c) to (i) all atomic dimensions are reduced and other dimensions are to scale.

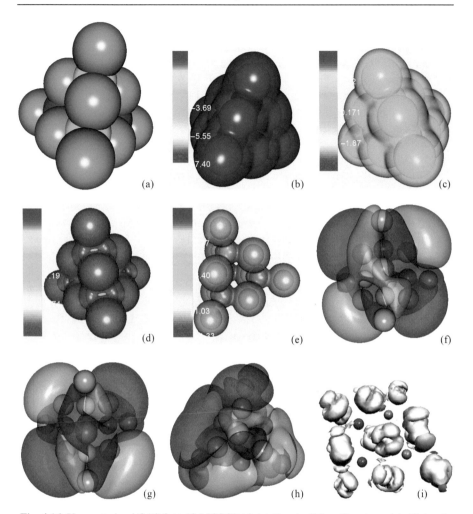

Fig. 4.16 Unconstrained CAS 8 × 10 MCSCF triplet $[In_{10}As_2P_2]_{Ga}$. Structure: (a). Molecular electrostatic potential (MEP) surfaces (color codding schemes are shown in the figures) corresponding to the CDD isovalues (b) 0.005, (c) 0.01, (d) 0.05 and (e) 0.09 of the charge density distribution (CDD, in arbitrary units). Isosurfaces of positive (gray) and negative (red) parts of the highest occupied (HO) and lowest unoccupied (LU) molecular orbits (MOs): (f) HOMO 115, (g) HOMO 116, and (h) LUMO 117 corresponding to the isovalue 0.001 (in arbitrary units). An isosurface of the spin density distribution (SDD) corresponding to the SDD isovalue (i) 0.001. Indium atoms are blue, As brown and P green. In (a) to (c) dimensions of In atoms are to scale, and As and P atoms are somewhat enlarged; in (d) to (i) all atomic dimensions are reduced and other dimensions are to scale.

molecule is built primarily of $5p$ AOs of 6 In and $4p$ AOs of both As atoms, with small contributions due to $3p$ AOs of both P atoms. HOMO 115 (Fig. 4.16f) of the vacuum molecule is similar to that of the pre-designed one, but in this case only $5p$ AOs of 4 In atoms contribute significantly, while a contribution from a $5p$ AO of the 5-th In atom becomes small. The proper HOMOs 116 (Figs. 4.15g and 4.16g) of both molecules are alike, with the major contributions to them coming from $5p$ AOs of 4 In atoms and $4p$ AOs of both As atoms, while $3p$ AOs of P atoms contribute little. At the same time, HOMO 116 of the vacuum molecule has yet another large contribution due to a $5p$ AO of the 5-th In atom, while a similar contribution is absent in the case of HOMO 116 of the pre-designed molecule. The proper LUMOs 117 (Figs. 4.15h and 4.16h) of these molecules are also alike and hybridized primarily of $5p$ AOs of 4 In atoms and $4p$ AOs of both As atoms. This time, LUMO 117 of the pre-designed molecule (Fig. 4.15h) features a large additional contribution from a $5p$ AO of the 5-th In atom, while a similar contribution is absent in the case of LUMO 117 of the vacuum molecule (Fig. 4.16h). In both cases, contributions to LUMOs 117 from $3p$ AOs of P atoms are very small.

In similarity to other cases considered in this chapter, the atoms of "artificial" $[In_{10}As_2P_2]_{Ga}$ and $[In_{10}As_2P_2]*_{Ga}$ molecules of this section are held together primarily due to 5π-type ligand In – In bonding mediated by $4p$ and $3p$ AOs of As and P atoms, respectively. Such bonding violates the standard octet rule, and is typical for small non-stoichiometric molecules. By design, the majority of In – In bonds in the pre-designed molecule are 3.997 Å in length, and the length of all In – As and In – P bonds is 2.448 Å. There exist also several of In – In bonds in this molecules with the bond-length equal to 5.653 Å. The ligand bonds of 3.997 Å in length are a typical property of the pre-designed $Ga_{10}As_3P$ and $Ga_{10}As_2P_2$ molecules of sec. 3 due to their design. No other types of ligand bonds develop in those Ga-based molecules, because their quantum confinement is rather loose. The development of the longer In – In bonds in the pre-designed $[In_{10}As_2P_2]_{Ga}$ molecule signifies a response of the system to strain due to the tight quantum confinement.

Once this strain is somewhat relaxed in the vacuum $[In_{10}As_2P_2]*_{Ga}$ molecule, 3 other types of ligand In – In bonds of 4.046 Å, 4.069 Å and 4.088 Å in length are developed. These bondlengths are very close to those of ligand Ga – Ga bonds in the vacuum $Ga_{10}As_3P$ molecule of section 4.3, indicating that the metastable state of the atomic cluster $[In_{10}As_2P_2]_{Ga}$ realized as the "artificial" vacuum $[In_{10}As_2P_2]*_{Ga}$ molecule is chemically very similar to the ground state of the natural vacuum $Ga_{10}As_3P$ molecule. This example may serve as yet another manifestation of the quantum confinement composition effect, due to which one group of atoms may be strained by a quantum confinement into a metastable state that is chemically similar to either ground state, or a metastable state, of some other unconstrained group of atoms.

Due to some strain relaxation, 3 types of In – As and In – P bonds also are developed in the vacuum $[In_{10}As_2P_2]*_{Ga}$ molecule. The bondlengths of such In – As bonds take values of 2.474 Å, 2.499 Å and 2.505 Å, while those of the In – P bonds are equal to 2.456 Å, 2.483 Å and 2.498 Å. All of these bondlengths either coincide with, or are very close in value to those specific to Ga – As and Ga – P bonds in the vacuum $Ga_{10}As_3P$ molecule of section 4.3. These findings are yet another realization of composition effects of quantum confinement. Indeed, not only the pre-designed $[In_{10}As_2P_2]_{Ga}$ molecule embedded into a tight confinement provided by a tetrahedral void in a model GaAs bulk lattice exhibits chemical properties, such as bondlengths, similar to those specific to bonding between atoms of the confinement, but also metastable states of molecules originally synthesized in quantum confinement may exhibit chemical properties inherent from those acquired due to the confinement.

Similar to other In-based molecules of this chapter, the pre-designed $[In_{10}As_2P_2]_{Ga}$ and vacuum $[In_{10}As_2P_2]^*_{Ga}$ molecules are "ferromagnetic" ROHF/MCSCF triplets. One of the two uncompensated electron spins of the pre-designed molecule is localized near an In atom, and the second spin is accommodated in an inner region of the molecule between 4 In and 2 As atoms. SDD of the pre-designed molecule is primarily due to $4d$ electrons of In atoms. In the vacuum $[In_{10}As_2P_2]^*_{Ga}$ molecule both uncompensated spins are accommodated in the regions between one In and one As atoms, one spin per atomic pair. The maximum value of SDD of the pre-designed $[In_{10}As_2P_2]_{Ga}$ molecule, 0.0314 (in arbitrary units), is close to those of the "natural" In-based molecule with 2 P atoms described in section 4.4, with SDD relatively smooth for small spin density isovalues (Fig. 4.15i). The dipole moment of the pre-designed molecule also is suitably mediocre and equal to 2.2078 D.

In contrast, the maximum spin density distribution value specific to the vacuum $[In_{10}As_2P_2]^*_{Ga}$ molecule, 0.1155, is about an order of magnitude larger than any of such values specific to molecules considered in this chapter. Regions of relatively large SDD values surround In atoms of this molecule and are prominent at larger SDD isovalues (Fig. 4.16i). Referring again to a metastable nature of the state of the $[In_{10}As_2P_2]^*_{Ga}$ molecule discussed above, one comes to a conclusion that this metastable state corresponds to large SDD values near In atoms, and the large dipole moment equal to 5.7970 D. Notably, the vacuum $Ga_{10}As_3P$ molecule also possess a large dipole moment exceeding 5 D. Therefore, the large dipole moment of the vacuum $[In_{10}As_2P_2]^*_{Ga}$ molecule similar to that of the vacuum $Ga_{10}As_3P$ molecule is yet another manifestation of the quantum confinement composition effect that reveals itself even in the studied metastable state represented by the vacuum $[In_{10}As_2P_2]^*_{Ga}$ molecule and obtained after relaxation of the pre-designed $[In_{10}As_2P_2]_{Ga}$ molecule strained by the presence of the quantum confinement.

Considering that both $[In_{10}As_2P_2]_{Ga}$ and $[In_{10}As_2P_2]^*_{Ga}$ molecules are embraced by shells of electron charge deficit (holes), and each of them carries 2 uncompensated electron spins, one observes that these molecules house delocalized, spin-polarized holes. Thus, being implanted into a solid lattice they may become sources of magnetization, and in external electric fields they may provide hole charge carriers. Taking into account the fact that the "natural" $In_{10}As_2P_2$ molecules of section 4.4 do not house charge holes, one concludes that the development of the holes housed by the $[In_{10}As_2P_2]_{Ga}$ and $[In_{10}As_2P_2]^*_{Ga}$ molecules is almost exclusively due to either existing strain (in the case of the pre-designed molecule), or partially relaxed strain (in the case of the vacuum molecule) caused by quantum confinement in the process of synthesis of the pre-designed molecule.

4.6 Conclusions

Opto-electronic and magnetic properties of 12 small QDs of gallium and indium arsenide phosphides have been studied in this chapter using the first-principle, quantum many body- theoretical methods. The obtained results provide an insight into fundamental physical and chemical mechanisms defining electronic, optical and magnetic properties of gallium and indium arsenide phosphides known from experimental observations, and also predict new phenomena that have not been observed yet. Such information is necessary to develop new technologies of experimental synthesis of nanoscale gallium and indium arsenide phosphide systems for applications in quantum communications, quantum electronics, spintronics and photovoltaics.

1. All gallium and indium arsenide phosphide systems of about 1 nanometer in linear dimensions studied in this chapter exhibit electronic, optical and magnetic properties strikingly different from those of their bulk system counterparts. By their design and properties, these systems are molecules either in their ground or metastable states. Their ELSs consist of bundles of well-separated energy levels, their MOs are results of strong hybridization of upper shell AOs of their constitutive atoms, and their magnetization is defined by up to two uncompensated $3d$ or $4d$ electron spins of Ga or In atoms, respectively. Such strongly correlated systems cannot be properly described in the framework of theoretical approaches developed for bulk gallium and indium arsenide phosphides, or other bulk systems, and require application of the first-principle, quantum many body-theoretical methods.

2. Re-distribution of electron charge from Ga or In atoms toward As and P atoms in the inner parts of the molecules is due to integral Coulomb interaction effects. This is the major physical mechanism that provides for exceptional stability of the studied molecules. Replacement of one or two As atoms with P ones in the parent gallium or indium arsenide molecules of Chapter 3 somewhat enhances stability of the derived phosphide molecules compared to that of their parent molecules due to (i) a decrease in the total number of electrons in the P-containing molecules that leads to a decrease in Coulomb repulsion between electrons of Ga or In atoms in the pyramidal "scaffolds" and electrons that belong to As and P atoms in the body of the molecules, and (ii) an increase in Coulomb attraction of electrons of the Ga and In atoms in the pyramidal "scaffolds" of the molecules to the nuclei of As and P atoms inside of the pyramids. These two major Coulomb effects lead to the development of stability "belts" of electron charge in the inner regions of the molecule that lead to a decrease by about 0.3 H per P atom in the ground state energy of the molecules where one or two As atoms are replaced with one or two P atoms.

3. Despite of a small decrease in the ground state energy, the opto-electronic properties of the P-containing molecules differ dramatically from those of "pure" gallium or indium arsenide molecules of Chapter 3. All studied "natural" Ga-based arsenide phosphide molecules, quantum – confined or otherwise, develop shells of electron charge deficit embracing the molecules, called delocalized charge holes, similar to those in DMS materials [73,74]. In contrast, only one of the

studied In-based "natural" arsenide phosphide molecules (the unconstrained $In_{10}As_3P$ triplet) develops such a delocalized hole.

4. At the same time, 3 of the 4 studied strained (or "artificial") molecules of In-based arsenide phosphides that model effects of quantum confinement provided by tight tetrahedral symmetry voids in the zincblende GaAs bulk lattice also develop such delocalized charge holes.

5. All but one molecules housing holes are "ferromagnetic" triplets carrying the uncompensated magnetic moment of two parallel electron spins. Thus, the delocalized holes of these molecules are spin-polarized. While magnetization of these molecules (and thus spin-polarization of their delocalized holes) is small, it is still equal to that of the majority of similar small molecules modeling nanometer-size QDs of DMS compounds and studied in Chapter 5. Again, similar to some DMS QDs of Chapter 5, one of the Ga arsenide phosphide molecules of section 4.3 (the vacuum $Ga_{10}As_2P_2$ singlet) is "antiferromagnetic", and therefore, its delocalized charge hole is not spin-polarized

6. Notably, mechanisms of the charge hole development in the studied Ga- and In-based arsenide phosphides are strikingly different from those in DMS QDs of Chapter 5. In contrast to the case of DMS molecules of Chapter 5 and bulk DMS systems [73], the *pd*-type hybrid bonding is almost negligible in the small Ga and In arsenide phosphide molecules, and the development of the magnetization is entirely due to hybrid $5p$-$4p$ and $5p$-$3p$ hybrid bonding realized in these non-stoichiometric molecules that leaves behind two uncompensated spins of *d*-electrons of In or Ga atoms.

7. The development of complex 5π-type ligand bonds mediated by $4p$- and $3p$-electrons of As and P atoms, respectively, in all of the studied molecules hints at a possibility of self-assembly of such molecules into larger nanostructures, with such molecules playing a role of building blocks. Nanomaterials so synthesized may combine magnetic properties typical for DMS nanomaterials with advanced opto-electronic properties of GaAs or InAs nanosystems, free of known problems typical for DMS nanomaterials. Such nanomaterials are in high demand for the use in spin transistors, non-volatile memory devices, optical emitters with polarized output, and sources of entangled electron spin state qubits.

8. Hybrid bonding in all of the molecules studied in this chapter is not described by the standard octet rule. In this respect, chemical properties of the studied molecules are similar to those of other non-stoichiometric molecules discussed throughout this book and in literature (see, for example, Refs. 69 to 72).

9. In addition to quantum confinement effects on the structure and properties of molecules synthesized in such confinement described throughout this book and in literature, results of virtual synthesis of Ga and In arsenide phosphide molecules discussed in this chapter reveal a novel type of quantum confinement effects that may be called **a composition effect of quantum confinement**. For example, chemical properties of the "artificial" pre-designed $[In_{10}As_2P_2]_{Ga}$ and

vacuum $[In_{10}As_2P_2]*_{Ga}$ triplets that model synthesis of the indium arsenide phosphide molecules in tight tetrahedral voids of the zincblende GaAs bulk lattice, and relaxation of the synthesized molecules, respectively, are strongly affected by the strain that sets up in these molecules due to the presence of quantum confinement. In particular, in the pre-designed $[In_{10}As_2P_2]_{Ga}$ molecule there develops two kinds of 5π-bonds of ligand In atoms with two strikingly different values of their bondlengths, one of which is equal to that of $4p$-type ligand Ga bonds in the pre-designed $Ga_{10}As_3P$ molecule with only one P atom, and the other is almost twice as long. Similarly, the In – As and In – P bondlengths in the $[In_{10}As_2P_2]_{Ga}$ molecule are equal to those in the pre-designed $Ga_{10}As_3P$ molecule.

10. Moreover, after the tight quantum confinement is lifted, the pre-designed $[In_{10}As_2P_2]_{Ga}$ molecule relaxes to the ground state of the vacuum $[In_{10}As_2P_2]*_{Ga}$ molecule that is electrochemically different from that of the "natural" vacuum $In_{10}As_2P_2$ molecule synthesized in the absence of any quantum confinement or fields. Rather, bonding in the vacuum $[In_{10}As_2P_2]*_{Ga}$ molecule still is similar to that in the pre-designed $Ga_{10}As_3P$ molecule, with many bondlengths of ligand In – In bonds, and those of In – As and In – P bonds, being either equal, or close to the corresponding bondlengths in that molecule. Thus, chemical properties of these strained molecules carry on an impact of quantum confinement, and by design of that confinement, are close to chemical properties of the pre-designed $Ga_{10}As_3P$ molecule than to those of the "natural" In-based molecules of section 4.4. This quantum composition effect also contributes to OTEs of the "artificial" $[In_{10}As_2P_2]_{Ga}$ and $[In_{10}As_2P_2]*_{Ga}$ molecules making them closer in value to that of the pre-designed $Ga_{10}As_3P$ molecule (Table 4.1), which is about 50 % smaller than those of the "natural" In-based molecules.

11. At the same time, quantum composition effect does not bear much on the ground state energies of the "artificial" $[In_{10}As_2P_2]_{Ga}$ and $[In_{10}As_2P_2]*_{Ga}$ molecules that remain very close to those specific to the "natural" In-based molecules of Table 4.2.

12. The molecules of this chapter are virtually synthesized at conditions corresponding to zero temperature. Because of all of them possess OTEs above 1 eV, with the majority of them being above 3 eV, the major results and conclusions reached in this chapter and concerning ground and steady states of the studied molecules are likely to remain either quantitatively, or qualitatively correct, or both, even at room temperature. At the same time, excited states of these molecules described by many closely lying MOs in the LUMO-bound portions of their ELSs may change significantly.

References

[1] Kramers, M. R., Shchekin, O. B., Mueller-Mach, R., Zhou, L., Harber, G., Craford, M. G. (2007). Status and future of high-power light emitting diodes for solid-state lighting. *J. Display Technol.* **3**, 160–175.
[2] Martin, D. (1997). Material advances light full-color LED displays. *Laser Focus World* **33**, 3.

[3] Han, H., Freeman, P. N., Hobson, W. S., Dutta, N. K., Lopata, J., Wynn, J. D., and Chu, S. N. G. (1996). High-speed modulation of strain-compensated InGaAs-GaAsP-InGaP multiple quantum well lasers. *IEEE Photonics Tech. Lett.* **8**, 1133–1134.

[4] Akinaga, F., Kuniyasu, T., Matsumoto, K., Fukunaga, T., and Hayakawa, T. (2003). 350 mW reliable operation in fundamental transverse-mode InGaAs (λ=1.05 μm)/GaAsP strain-compensated laser diodes. *Electronics Lett.* **39**, 55–57.

[5] Arimoto, K., Usami, N., and Shiraki, Y. (2000). Correlation between electronic states and optical properties in indirect GaAsP/GaP quantum wells with insertion of an ultrathin AlP layer. *Physica E* **8**, 323–327.

[6] Takanohashi, T., and Ozeki, M. (1992). Luminescence characteristics of the (GaP)n (GaAs)n/GaAs atomic layer short-period structure. *J. App. Phys.* **71**, 5614–5618.

[7] Tanaka, Y., and Toyama, T. (1994). Analysis of current-temperature-light characteristics of GaAsP light-emitting diodes. *IEEE Trans. Electron Devices* **41**, 1475–1477.

[8] Ma, S., Sodabanlu, H., Watanabe, K., Sugiyama, M., and Nakano, Y. (2011). Strain-compensation measurement and simulation of InGaAs/GaAsP multiple quantum wells by metal organic vapour phase epitaxy using wafer-curvature. *J. App. Phys.* **110**, 113501.

[9] Mori, M. J., and Fitzgerald, E. A. (2009). Microstructure and luminescent properties of novel InGaP alloys on relaxed GaAsP substrates. *J. Appl. Phys.* **105**, 013107.

[10] Lerner, E. J. (2001). The photodiode is a workhorse of detection. *Laser Focus World* **11**, 133.

[11] Hayat, A., Nevet, A., Ginzburg, P., and Orenstein, M. (2011). Application of two-photon processes in semiconductor photonic devices: invited review. *Semicond. Sci. Technol.* **26**, 083001.

[12] Noorma, M., Karha, P., Lamminpaa, A., Nevas, S., *et al.* (2005). Characterization of GaAsP trap detector for radiometric measurements in ultraviolet wavelength region. *Rev. Sci. Inst.* **76**, 033110.

[13] Zhang, P., Yanrong, S., Tian, J., and Zhang, X. (2009). Gain characteristics of the InGaAs strained quantum wells with GaAs, AlGaAs and GaAsP barriers in vertical-external-cavity surface-emitting lasers. *J. Appl. Phys.* **105**, 053103.

[14] Ma, Y., Guo, X., Wu, X., Dai, L., and Tong, L. (2013). Semiconductor nanowire lasers. *Adv. in Optics and Photonics* **5**, 216–273.

[15] Saxena, D., Mokkapati, S., and Jagadish, C. (2012). Semiconductor nanolasers. *IEEE Photonics J.* **4**, 582–585.

[16] Bergman, D. J., and Stockman, M. I. (2003). Surface plasmon amplification by simulated emission of radiation: quantum generation of coherent surface plasmons in nanosystems. *Phys. Rev. Lett.* **90**, 027402.

[17] Hill, M. T., Oei, Y.-S., Smalbrugge, B., Zhu, Y., de Vries, T., van Veldhoven, P. J., van Offen, F. W. M., Eijkemans, T. J., Turkiewicz, J. P., de Waardt, H., Geluk, E. J., Kwon, S.-H., Lee, Y.-H., Notzel, R., and Smith, M. K. (2007). Lasing in metallic-coated nanocavities. *Nature Photonics* **1**, 589–594.

[18] Luk'yanchuk, B., Zheludev, N. I., Maier, S. A., Halas, N. J., Nordlander, P., Giessen, H., Chong, C. T. (2010). The Fano-resonance in plasmonic nanostructures and metamaterials. *Nature Materials* **9**, 707–715.

[19] Noginov, M. A., Zhu, G., Belgrave, A. M., Bakker, R., Shalaev, V. M., Narimanov, E. E., Stout, S., Herz, E., Suteewong, T., and Wiesner, U. (2009). Demonstration of spacer-based nanolaser. *Nature* **460**, 1110–1112.

[20] Oulton, R. F., Sorger, V. J., Zentgraf, T., Ma, R.-M., Gladden C., Dai, L., Bartal, G., and Zhang, X. (2009). Plasmon lasers at deep subwavelength scale. *Nature* **461**, 629–632.

[21] Pal, S., Singh, S. D., Porwal, S., Sarma, T. K., Khan, S., Jayabalan, J., Chari, R., and Oak, S. M. (2013). Effect of light hole tunnelling on the excitonic properties of GaAsP/AlGaAs near-surface quantum wells. *Semicond. Sci. Technol.* **28**, 035016.

[22] Hopkins, K. (1995). Nonlinear materials extend the range of high-power lasers. *Laser Focus World*, **31**, 87–93.
[23] He, G. S., Tan, L.-S., Zheng, Q., and Prasad, P. N. (2008). Multiphoton absorbing materials: molecular designs, characterizations, and applications. *Chem. Rev.* **108**, 1245–1330.
[24] Aeberhard, U. (2013). Simulation of nanostructure-based high-efficiency solar cells: challenges, existing approaches, and future directions. *IEEE J. Selected Topics in Quantum Electronics* **19**, 4000411.
[25] Lapierre, R. R., Chia, A. C. E., Gibson, S. J., Haapamaki, C. M., Boulanger, J., Yee, R., Kuyanov, P., Zhang, J., Tajik, N., Jewell, N., and Rahman, K. M. A. (2013). III-V nanowire photovoltaics: review of design for high efficiency. *Phys. Status Solidi* **7**, 815–830.
[26] Levy, M. Y., and Honsberg, C. (2008). Nanostructured absorbers for multiple transition solar cells. *IEEE Trans. Electron Devices* **55**, 706–711.
[27] McPheeters, C. O., Hu, D., Schoadt, D. M., and Yu, E. T. (2012). Semiconductor heterostructures and optimization of light-trapping structures for efficient thin-film solar cells. *J. Optics* **14**, 024007.
[28] Phillips, S. P., Dimroth, F., and Belt, A. W. (2012). High-efficiency III-V multijunction solar cells. In "Practical Handbook of Photovoltaics. Solar Cells", 2nd Ed. (A. McEvoy, L. Castaner and T. Markvart, Eds.) Elsevier, New York, pp. 417–448. ISBN 9780123869647.
[29] Rault, F. K., and Zahedi, A. (2004). Idealized quantum solar cell design. *Physica E* **21**, 61–70.
[30] Tsakalakos, L. (2008). Nanostructures for photovoltaics. *Materials Sci. Eng.* R **62**, 175–189.
[31] Wen, Y., Wang, Y., Watanabe, K., Sugiyama, M., and Nakano, Y. (2013). Effect of GaAs step layer thickness in InGaAs/GaAsP stepped quantum well solar cell. *IEEE Photovoltaics* **3**, 289–295.
[32] Zhang, Y., Geng, H., Zhou, Z., Wu, J., Wang, Z., Zhang, Y, Li, Z., Zhang, L., Yang, Z., and Hwang, H.-L. (2012). Development of inorganic solar cells by nanotechnology. *Nano-Micro Lett.* **4**, 124–134.
[33] Hyun, J. K., Zhang, S., and Lauhon, L. J. (2013). Nanowire heterostructures. *Ann. Rev. Mater. Res.* **43**, 451–479.
[34] Couto, Jr., O. D. D., Sercombe, D., Puebla, J., Otubo, L., Luxmoore, I. J., Sich, M., Elliott, T. J., Chekhovich, E. A., Wilson, L. R., Skolnick, M. S., Liu, H. Y., and Tartakovskii, A. I. (2012). Effect of GaAsP shell on the optical properties of self-catalized GaAs nanowires grown on silicon. *Nano Lett.* **12**, 5269–5274.
[35] Algra, R. E., Weheijen, M. A., Borgstrom, M. T., Feiner, L.-F., Immink, G., van Enckevolt, W. J. P., Vlieg, E., and Bakkers, E. P. A. M. (2008). Twinning superlattices in indium phosphide nanowires. *Nature* **456**, 369–572.
[36] Kobayashi, T., Taira, K., Nakamura, F., and Kawai, H. (1989). Band lineup for GaIn/GaAs heterojunction measured by high-gain npn heterojunction bipolar transistor grown by metalorganic chemical vapour deposition. *J. Appl. Phys.* **65**, 4898–4902.
[37] Pan, A., Nichols, P. L., and Ning, C. Z. (2011). Semiconductor alloy nanowires and nanobelts with tunable optical properties. *IEEE J. Selected Topics in Quantum Electronics* **17**, 808–818.
[38] Yan, R., Gargas, D., and Yang, P. (2009). Nanowire photonics. *Nature Photonics* **3**, 569–576.
[39] Hayden, O., Agarwal, R., and Lieber, C. M. (2006). Nanoscale avalanche photodiodes for highly sensitive and spatially resolved photon detection. *Nature Materials* **5**, 352–356.
[40] Bulgarini, G., Reimer, M. E., Hocevar, M., Bakkers, E. P. A. M., Kouwenhoven, L. P., and Zwiller, V. (2012). Avalanche amplification of single exciton in a semiconductor nanowire. *Nature Photonics* **6**, 455–458.
[41] Hua, B., Motohisa, J., Kobayashi, Y., Hara, S., and Fukui, T. (2009). Single GaAs/GaAsP coaxial core-shell nanowire lasers. *Nano Lett.* **9**, 112–116.

[42] Thanh, T. N., Robert, C., Cornet, C., Perrin, M., Jancu, J. M., Bertu, N., Even, J., Chevalier, N., Folliot, H., Durand, O., and Le Corre, A. (2011). Room temperature photoluminescence of high density (In, Ga)As/GaP quantum dots. *Appl. Phys. Lett.* **99**, 143123.

[43] Nashitani, T., Nakanishi, T, Yamamoto, M., Okumi, S., et al. (2005). Highly polarized electrons from GaAs-GaAsP and InGaAs-AlGaAs strained-layer superlattice photocathodes. *J. Appl. Phys.* **97**, 094907.

[44] Mamaev, Yu. A., Gerchikov, L. G., and Yashin, Yu. P. (2008). On the way to perfect spin-polarized electron source. *Semicond. Sci. Technol.* **23**, 114014.

[45] Meier, F., and Zakharchenya, P. (1984). "Optical orientation in solid state physics." North Holland, Amsterdam.

[46] Brandt, M. S., Goennenwein, S. T. B., Graf, T., Huebl, H., Lauterbach, S., and Stutzmann, M. (2004). Spin-dependent transport in elemental and compound semiconductors and nanostructures. *Phys. Status Solidi*, (c) **8**, 2056–2093.

[47] Gerchikov, L. G., Mamaev, Yu. A., Yashin, Yu. P., Vasiliev, D. A., Kuz'michev, V. V., Ustinov, V. M., Zhukov, A. E., Vasiliev, A. P., and Mikhin, V. S. (2009). Resonance enhancement of spin-polarized electron emission. *Semiconductors* **43**, 463–467.

[48] Pearton, S. J., Abernathy, A. C., Overberg, M. E., Thaler, G. T., Norton, D. P., Theodoropoulos, N., Hebard, A. F., Park, Y. D., Ren, F., Kim, J., and Boatner, L. A. (2003). Wide band gap ferromagnetic semiconductors and oxides. *J. Appl. Phys.* **93**, 1–13.

[49] Subashiev, A. V., Gerchikov, L. G., and Ipatov, A. N. (2003). Optical spin orientation in strained superlattices. arXiv:cond-mat/0312719 [cond-mat.mtrl-sci]

[50] Fukui, T., Ando, S., Tokura, Y., and Toriyama, T. (1991). GaAs tetrahedral quantum dot structures fabricated using selective area metalorganic chemical vapor deposition. *Appl. Phys. Lett.* **58**, 2018–2020.

[51] Fukui, T., and Ando, S. (1992). GaAs tetrahedral quantum dots structures fabricated using selective area MOCVD. *Surfice Sci.* **1–3**, 236–240.

[52] Nagamune, Y., Nishioka, M., Tsukamoto, S., and Arakava, Y. (1994). GaAs quantum dots with lateral dimension of 25 nm fabricated by selective metalorganic chemical vapor deposition growth. *Appl. Phys. Lett.* **64**, 2495–2497.

[53] Rajkumar, K. C., Kaviani, K., Chen, J., Chen, P., and Madhukar, A. (1992). Nanofeatures on GaAs (111)B via photolithography. *Appl. Phys. Lett.* **60**, 850–852.

[54] Madhukar, A., Rajkumar, K. C., and Chen, P. (1993). In situ approach to realization of three-dimensionally confined structures via substrate encoded size reducing epitaxy on non-planar patterned substrates. *Appl. Phys. Lett.* **62**, 1547–1549.

[55] Rajkumar, K. C., Madhukar, A., Rammohan, K., Rich, D. H., Chen, P., and Chen, L. (1993). Optically active three-dimensionally confined structures realized via molecular beam epitaxy grown on GaAs (111)B. *Appl. Phys. Lett.* **63**, 2905–2907.

[56] Rajkumar, K. C., Madhukar, A., Chen, P., Konkar, A., Chen, L., Rammohan, K., and Rich, D. H. (1994). Realization of three-dimensionally confined structures via one-step in situ molecular beam epitaxy on appropriately patterned GaAs (111)B and GaAs (001). *J. Vac. Sci. Technol.* B **12**, 1071–1923.

[57] Williams, R. L., Aers, G. C., Poole, P. J., Lefebvre, J., Chithrani, D., and Lamontagne, B. (2001). Controlling the self-assembly of InAs/InP quantum dots. *J. Cryst. Growth* **223**, 321–446.

[58] Lefebvre, J., Poole, P. J., Aers, G. C., Chithrani, D., and Williams, R. L. (2002). Tunable emission from InAs quantum dots on InP nanotemplates. *J. Vac. Sci. Technol. B* **20**, 2173–2176.

[59] Chithrani, D., Williams, R. L., Lefebvre, J., Poole, P. J., Aers, G. C. (2004). Optical spectroscopy of single, site-selected, InAs/InP self-assembled quantum dots. *Appl. Phys. Lett.* **84**, 978–980.

[60] Kapon, E., Pelucchi, E., Watanabe, S., Malko, A., Baier, M. H., Leifer, K., Dwir, B., Michelini, F., and Dupertuis, M.-A. (2004). Site- and energy-controlled pyramidal quantum dot heterostructures. *Physica E* **25**, 288–297.

[61] Michelini, F., Dupertuis, M.-A., and Kapon, E. (2004). Effects of one-dimensional quantum barriers in pyramidal quantum dots. *Appl. Phys. Lett.* **84**, 4086–4088.

[62] Watanabe, S., Pelucchi, E., Dwir, B., Baier, M. H., Leifer, K., and Kapon, E. (2004). Dense uniform arrays of site-controlled quantum dots grown in inverted pyramids. *Appl. Phys. Lett.* **84**, 2907–2909.

[63] Baier, M. H., Watanabe, S., Pelucchi, E., and Kapon, E. (2004). High uniformity of site-controlled pyramidal quantum dots grown on prepatterned substrates. *Appl. Phys. Lett.* **84**, 1943–1945.

[64] Pelucchi, E., Watanabe, S., Leifer, K., Dwir, B., and Kapon, E. (2003). Site-controlled quantum dots grown in inverted pyramids for photonic crystal applications. *Physica E* **23**, 476–481.

[65] Pozhar, L. A., Yeates, A.T., Szmulowicz, F., and Mitchel, W.C. (2005). Small atomic clusters as prototypes for sub-nanoscale heterostructure units with pre-designed charge transport properties. *EuroPhys. Lett.* **71**, 380–386.

[66] Pozhar, L. A., Yeates, A.T., Szmulowicz, F., and Mitchel, W.C. (2006). Virtual synthesis of artificial molecules of In, Ga and As with pre-designed electronic properties using a self-consistent field method. *Phys. Rev. B* **74**, 085306. See also *Virtual J. Nanoscale Sci & Technol.* **14**, No. 8 (2006): http://www.vjnano.org.

[67] Pozhar, L. A., Yeates, A.T., Szmulowicz, F., and Mitchel, W.C. (2005). Virtual fabrication of small Ga-As/P and In-As/P clusters with pre-designed electronic energy level structure. *Mat. Res. Soc. Proc.* **829**, 49–54.

[68] Gordon, M. S., and Schmidt, M. W. (2005). Advances in electronic structure theory: GAMESS a decade later. *In* "Theory and Applications of Computational Chemistry: the First Forty Years" (C. E. Dykstra, G. Frenking, K. S. Kim, G. E. Scuseria, Eds.), pp. 1167–1189. Elsevier, Amsterdam; http:// www.msg.ameslab.gov/GAMESS

[69] Boldyrev, A. I., and Wang, L.-S. (2001). Beyond classical stoichiometry: experiment and theory. *J. Phys. Chem.* **105**, 10759–10775.

[70] Pozhar, L. A. (2010). Small InAsN and InN clusters: electronic properties and nitrogen stability belt. *Eur. Phys. J. D* **57**, 343–354.

[71] Pozhar, L. A. (2011). Magneto-optic properties of InAsN and InN quantum dots doped with Co or Ni atoms. *J. Appl. Phys.* **109**, 07C303.

[72] Pozhar, L. A. (2011). Electronic and magneto-optic properties of Co- and Ni-doped small quantum dots of indium nitrides. *Phys. Status Solidi C* **8**, 2261–2263.

[73] Dietl, T., Ohno, H., Matsukura, F. (2001). Hole-mediated ferromagnetism in tetrahedrally coordinated semiconductors. *Phys. Rev. B*, **63**, 195205.

[74] Galicka, M., Buczko, R., and Kacman, P. (2011). Structure-dependent ferromagnetism in Mn-doped III-V nanowires. *Nano Lett.* **11**, 3319–3323.

Quantum Dots of Diluted Magnetic Semiconductor Compounds

5.1 Introduction

Since a discovery of ferromagnetism in bulk Mn-doped GaAs and InAs bulk semiconductors [1–4], III-V- and II-VI- based diluted magnetic semiconductor (DMS) systems, sometimes called molecular magnets, have been vigorously studied by both experimental and theoretical means. Such systems exhibit a variety of electronic and magneto-optical properties defined by carrier-mediated ferromagnetism, including (i) spin polarization of holes, (ii) hole concentration-induced switching of the magnetic phase at constant temperature, (iii) hysteresis of Hall resistance as a function of an external magnetic field and gate voltage below the Curie temperature (Tc), (iv) electric-field assisted magnetization reversal, (v) magnetization enhancement by circularly polarized light, (vi) current-induced magnetization switching, (vii) a strong increase in resistance with lowering temperature, (viii) colossal negative magnetoresistance, (ix) current-induced domain wall motion, *etc.* (see Refs. 5–7 and references therein). These properties are exceptionally important in materials for quantum electronics, spintronics and quantum information processing, as they allow control of carriers' spin generation, relaxation and deflection, and magneto-optical properties, using not only external magnetic fields, but also external electric fields, current, light, carrier concentration, and quantum confinement effects.

Unfortunately, bulk DMS systems possess very low T_C heavily limiting their technological use. Thus, synthesis of DMS materials possessing room temperature and higher T_C has become one of the central issues of application of DMS-based materials in novel spintronics and quantum information processing technologies. In particular, following the existing practical experience and the *p-d* Zener model-based predictions [8,9], state-of-the-art low temperature molecular beam epitaxy (MBE) and other thin film synthesis methods, layered device development techniques, and numerous methods of synthesis of quantum wire (QW)- and QD- based nanostructures have been established over the years (see, for example, Refs. 10–16) to realize low dimensional nanostructures where DMS components possess enhanced T_C up to 173 K [17–26]. Recently, room temperature ferromagnetism was reported in Mn-doped InGaAs and GaAsSb nanowires [27], and MBE-based formation of (Ga,Mn)As crystalline nanowires with Tc = 190 K was achieved [28]. However, reached values of T_C for ingenious layered structures based on Mn-doped (Ga,Mn)As are still significantly lower than those predicted by the existing theoretical models of hole-mediated ferromagnetism in bulk, spatially homogeneous Mn-doped GaAs and InAs systems that are supposed to exhibit the highest T_C among bulk III-V DMS candidate materials. In particular, the *p-d* Zener model [8,9] predicts T_C = 345 K for a spatially homogeneous sample

$Ga_{0.8}Mn_{0.2}As$ with high Mn concentration from Ref. 11, while the observed T_C for that material is 118 K. Because the p-d Zener model has been developed to describe DMSs in the low impurity concentration limit, it fails to predict (even qualitatively) an increase in T_C of samples with high Mn concentration [11]. Similarly, an improved approach [29] that uses a semi-phenomenological model predicts unrealistic values of T_C above 500 K in structures with remote doping, such as heterostructures of Ref. 24. [In such heterostructures impurities are introduced as a so-called δ layer on one side of a heterojunction to separate impurity and carrier channels. This provides for the carrier mobility up to three orders of magnitude higher than that of heterostructures where impurities and carriers share the same channel.]

It is important to note that the exiting theoretical approaches have been developed to describe ferromagnetism (FM) in bulk, spatially homogeneous DMS materials with low impurity concentration [30,31]. Correspondingly, they incorporate assumptions that are not applicable to quantum-confined systems and nanoscale heterostructures, such as those composed of layered films, QWs or QDs of a few nanometers in linear dimensions. In such systems carrier motion is quantized in the direction(s) of confinement, which is not incorporated into the models' formalism. Moreover, a small number of atoms aligned in one direction (2 to 4, in the case of a confinement with at least one linear dimension of about 1 nm) requires solving the many-particle quantum problem. In the absence of a rigorous, first principle quantum theory of solids, and in particular semiconductors, necessary to fully account for quantum effects, the semi-phenomenological band theory of semiconductors was the first theoretical approach subjected to modifications that produced numerous semi-phenomenological models designed to quantify observed electronic structure and properties of semiconductor films, quantum wells, thick QWs, large QDs, and their heterostructures [30,32,33].

One of such modifications due to Zener [34], called the p-d Zener model, has been very successful in the case of bulk DMS materials [8,9], and with some additional *ad hoc* fitting has also been successfully applied to thick DMS films and layered structures (see, for example, Ref. 31 and references therein). This model makes use of the fact that in DMSs the Fermi energy E_F lies, as a rule, within the majority t-band of the "magnetic" impurity (such as Mn atoms) which is t_{2g} symmetric. Moreover, d-states of magnetic impurity atoms lie below the valence p-states of the host semiconductor structure, such as (Ga,Mn)As. These p- and d-states hybridize pushing the majority valence states to high energies, and the minority valence states to lower energies. This mechanism, called kinetic p-d exchange, causes the development of holes in the valence band, and thus to generation of a significant magnetic moment μ per an impurity atom (μ = 5 $μ_B$ in the case of Mn impurity, where $μ_B$ is the Bohr magneton) leading to hole-mediated ferromagnetism. At the same time, the majority valence band becomes spin-polarized with a much smaller magnetic moment per an impurity atom (equal to -$μ_B$ per a Mn atom) due to weak d-d interactions between impurity electrons. Thus, the majority (d) - majority (p) orbit hybridization (the p-d exchange) produces FM coupling that leads to a decrease in the total energy of the system.

Importantly, the p-d Zener model uses the virtual crystal and molecular field uncontrolled approximations, the limit of low hole and impurity concentrations (with the hole concentration much smaller than that of the impurities), and an assumption of weak p-d coupling [35]. When the p-d hybridization is large the coupling becomes strong, and both approximations become invalid [36,37]. Also, the p-d Zener model is expected to fail in the case of nanometer-thin DMS films, thin QWs and small QDs where the concentration of impurity atoms may reach tens of percent. Moreover, for DMS systems possessing large characteristic dimensions, but heterogeneous in nature (that is, featuring regions of the higher and smaller impurity concentration, or different magnetic phases) this model is also invalid.

The p-d Zener model predicts the Curie temperature $T_C \approx 345°$ K in the case of a FM-homogeneous $Ga_{1-x}Mn_xAs$ film with the Mn concentration x over 10%, while the experimental value is approximately $118°$ K [12]. According to this model, T_C increases with the increasing hole concentration. However, when the hole concentration exceeds that of the impurities, experimentally observed T_C of 5 nm thin $Ga_{1-x}Mn_xAs$ films from Ref. 11 tends to decrease with an increase in the hole concentration. Moreover, the p-d Zener model does not account for many other experimental data [14,38–41] describing magnetic properties of DMS films. There were attempts to modify this model to reflect new experimental observations [18,42], but they did not accommodate emerging experimental evidence indicating that FM mechanisms other than the p-d Zener exchange, significantly contributed to magnetic properties of DMS systems.

Thus, other semi-phenomenological models, such as an impurity band (IB) model, have been developed [43]. Yet, even in the case of relatively thick DMS films of Ref. 14, the major issue concerning the hole-mediated FM mechanism remains unresolved. In particular, it is not quite clear where the hole carriers reside – in a disordered valance band (VB), as stated in Refs. 6 and 8, or in the impurity band that may be detachable from the host VB and retaining d-orbit properties of the impurities, such as Mn [43–50].

More sophisticated theoretical approaches steaming from the first-principle theoretical basis have been suggested to explain non-conventional origin of ferromagnetism in and complex magnetic properties of DMS systems. One of the earliest of them was an approach that used theory of scattering to solve the Schrödinger equation [51–54] for electrons in a periodic system. However, another approach due to Korringa, Kohn and Rostoker (KKR) introduced in Refs. 54–56 became more popular due to its simplicity and mathematical transparency. This approach uses the density functional theory's (DFT) mathematical foundation [54] and a smart choice of a reference system to simplify a general DFT procedure of band structure calculations. Numerous KKR-based models were developed and applied to periodic systems with localized perturbations [57–65]. The KKR method itself was further improved when the problem of determining the charge density $n(\mathbf{r})$ was re-formulated [66,67] in terms of the single particle Green's function $G(\mathbf{r}_1, \mathbf{r}_2)$ introduced as the solution of the inhomogeneous Kohn-Sham equation of DFT with the δ-function source, and the algebraic Dyson equation was used to relate the Green's function of structural defects to the Green's function of the ideal crystal [68]. Further advance of the KKR method is related to inclusion of non-spherical interaction potentials using a shape function technique to model non-spherical parts of the potential, and to the use of the Lippmann-Schwinger equation to develop an iterative scheme of determination of the Green's functions from the Dyson equation [68–72]. A concept of a coherent potential [73,74] was incorporated to make KKR-based methods applicable to disordered systems, giving rise to the so-called coherent potential approximation (CPA). The use of the local force "theorem" [75] enabled further successful applications of KKR-CPA models to disordered magnetic systems. According to this "theorem", in the case of a frozen ground state potential and small perturbations in the charge and magnetization densities of a system, a variation of the total energy of the system is equal to the sum of the single particle energies over all occupied energy states. This "theorem" simplified evaluation of the exchange interaction energy between two magnetic atoms [76].

Simplification of the exchange-correlation functional by the use of the local density approximation (LDA) made it practical to apply DFT-based methods to calculate the electronic energy level structure (ELS) of large molecules, and the band structure of solids. However, LDA failed in the case of transition metal compounds and DMS systems, where it predicted a partially filled d-band with metallic character of ELS [77]. Thus, other adjustments of DFT-based methods were introduced, including the most popular self-interaction correction LDA (SIC-LDA) and LDA + U [78,79] methods, with U denoting the Hubbard potential, to achieve at least partial success in predicting the band structure and T_C of strongly correlated systems. In GaAs systems with Mn impurities the Hubbard potential accounts for Coulomb correlation effects that have to be incorporated to calculate more accurately the band structure of such DMS systems [79], and also leads to an increase in hole delocalization and a decrease in the p–d Zener exchange [78]. It also inhibits the double exchange and pushes the d-states to lower energies [80]. As a result, in LDA + U approximation the mean-field Curie temperature is a linear function of the carrier concentration [81]. Despite important results, it is not clear whether further improvement in prediction of the band structure and magnetic properties of DMS is possible in the framework of LDA and its modifications.

In recent years, a more general, spin-polarized DFT approach, called generalized gradient approximation (σGGA), and its modification by inclusion of the Hubbard potential (σGGA + U) were used [82] to calculate electronic and magnetic properties of Mn impurities incorporated in bulk GaAs. The electron-electron interactions were described using the Perdew–Burke–Ernzerhof exchange–correlation potential [83], and electron-ion interactions were accounted for using ultrasoft pseudopotentials [84]. It was proved that inclusion of the Hubbard potential into calculations (σGGA + U) leads to contraction of geometrical parameters by about 2% as compared to those calculated without the Hubbard potential by the σGGA method. The σGGA + U calculations of Ref. 78 have used the value 4.0 eV for the Hubbard potential, which is usually adopted in LDA + U calculations, and which is usually chosen close to the value U = 3.5 eV obtained on the basis of experimental data for photoemission of $Ga_{1-x}Mn_xAs$ systems with small values of the impurity concentrations x [85]. In both cases of σGGA and σGGA + U, a Mn impurity atom introduces a d-hole, and the majority spin state of the impurity lies at 0.25 eV above the Fermi level. These theoretical findings correlate with experimental observations for bulk GaAs systems with low concentration of Mn, and constitute improvement of the previous LDA + U results.

It has become obvious, that being a non-variational theory by its nature [86], DFT cannot be applied successfully to predict properties of transition metal semiconductor [87] and magnetic systems without *ad hoc* schemes (such as KKR), uncontrolled approximations (such as LDA, GGA or σGGA), phenomenological pseudopotentials, and adjustable parameters, such as the Hubbard potential U. All these improvements and adjustments are incorporated to fit experimental observations, and are not derivable from the first principles. They are necessary because DFT is a "non-variational" theory, that is, in the framework of DFT one cannot introduce a self-consistent scheme of controlled approximations converging to the "real" exchange-correlation functional. Even more such *ad hoc* manipulations are required to apply DFT-based methods to strongly correlated nanoscale systems, such as layered DMS systems, or DMS QW and QD assemblies. Generally, one cannot expect realistic results from any of DFT-based theoretical methods, unless phenomenological theoretical models for the exchange-correlation functionals and adjustable parameters steaming from experimental data are incorporated into such theoretical schemes.

During two recent decades, another class of theoretical approaches to the electronic structure-property correlations in spatially inhomogeneous and nanoscale magnetic systems has emerged. These approaches have been developed originally [88–94] to describe quantum phenomena, such as Coulomb blockade, in large nanoscale systems, and are focused on electron dynamics. They use a rigorous quantum statistical mechanical basis (see, for example, Refs. 95 and 96), but include numerous uncontrolled approximations, *ad hoc* assumptions and intuitive models, such as those explored in Refs. 97 and 98. These approaches have received significant attention in literature in conjunction with emerging quantum computing technologies [99], because spin states of electrons residing on magnetic QDs are considered as the most practical, and at the same time, long-lived realization of quanta of information qubits [100–102]. Being derived semi-phenomenologically from a theoretical basis developed either for bulk systems or for simple quantum mechanical models, these approaches [103–111] tend to provide physically incorrect predictions even for mesoscopic systems, such as tunneling junctions, because they do not include adequate consideration of electron spin interactions, quantum confinement effects, QD-to-QD and QD-to-environment coupling, *etc*. Yet such interactions, effects and coupling are responsible for the major physical mechanisms of ferromagnetism in DMSs. By their nature, such models do not allow for the first-principle predictions of the electronic structure and magneto-electronic properties of layered or nanoscale systems. Fortunately, in the case of small nanosystems one can exploit rigorous methods of quantum statistical mechanics (QSM) directly using QSM theory-based "quantum chemistry" software packages, such as GAMESS, GAUSSIAN or Molpro, as described in the following section.

5.2 Virtual synthesis of small quantum dots of diluted magnetic semiconductor compounds

Latest resonant tunneling spectroscopy studies of a variety of GaMnAs surface layers [112] have proven that GaAs band structure is not significantly affected by the presence of Mn substitution defects for a range of Mn concentrations from 6% to 15%. It appears that the GaAs valence band does not merge with the impurity band, and that the exchange splitting of the valence band is in the range of several millielectronvolts even for the layers with T_C as high as 154° K. At the same time, in such systems the ferromagnetic state is more pronounced than that of bulk (Ga,Mn)As and (Ga,Mn)As [26]. Moreover, new model studies [113] show that similar to atoms and nuclei DMS QDs exhibit formation of electron shells, and that their magnetism depends on shell occupancy. In (Ga,Mn)As/GaAs heterostructures there exist nanoscale potential fluctuations that lead to formation of electrostatic QDs [114]. Moreover, transversal Kerr effect spectroscopy studies [115] identified that MnAs inclusions of 10 nm to 40 nm in linear dimensions were responsible for a strong resonant band in the energy range from 0.5 eV to 2.7 eV in GaMnAs and InMnAs layers formed by laser ablation on GaAs and InAs surfaces. It has been also observed [116] that transitions from quasi-two dimensional (2D) to three dimensional (3D) $In_{0.85}Mn_{0.15}As$ structures develop gradually and are rather slow even at 270° C [116]. At the same time, the electron spin relaxation time in MnAs nanoparticles formed in GaAs lattice was found to be as long as 10 μs at 2° K [117], in contrast to known relaxation times of the order of 100 ns in other QD structures, and several picoseconds in bulk semiconductor systems [102].

These latest experimental data are in contradiction with the existing conventional models of mechanisms responsible for magnetism in and the band structure of nanoscale DMS systems, as described in section 5.1. In particular, the major assumption that holes occupy GaAs-like valence band used to obtain successful predictions [118] of magnetization of GaMnAs as a function of the magnetic field at T_C contradicts to experimental results of Ref. 112 and modeling results of Ref. 113. Thus, with advance of spintronics and quantum computing, more accurate QSM-based methods are required to evaluate spin entanglement and decoherence, and spin transport properties in small DMS QDs and their heterostructures.

In this chapter the first principle QSM methods free from any assumptions concerning FM mechanisms and ELS of the studied systems have been used to synthesize virtually a range of 14-atomic QDs of GaAs and InAs containing one or two Mn or V dopant atoms, and to study their electronic and magnetic properties. The virtual synthesis method of Refs. 119–121 (see also section 2.9 of Chapter 2) realized using GAMESS software has been applied to minimize the total energy of several atomic clusters composed of tetrahedral symmetry elements (14 - atomic pyramids) of the GaAs and InAs bulk zincblende lattices to obtain $Ga_{10}As_4$ and $In_{10}As_4$ molecules. Originally, the pyramidal frames of these clusters have been built of 10 Ga or In atoms, with four As atoms placed at ¼ of the cube body diagonals in the corresponding bulk zincblende lattices (that is, inside of the pyramidal scaffolds composed of 14 Ga or In atoms; see Chapter 3 for details). The initial covalent radii of Ga, In and As atoms in these clusters have been adopted from experiment: 1.26, 1.44 and 1.18, respectively. Such clusters have been synthesized at conditions mimicking quantum confinement (modeled by spatial constraints applied to the centers of mass of clusters' atoms) and in "vacuum", when the spatial constraints have been relaxed (that is, in the absence of any constraints, such as external electromagnetic fields and "foreign" atoms interacting with the clusters' atoms). At the next step, one or two As atoms have been replaced by one or two vanadium or manganese atoms, without any changes to the positions of the remaining Ga or In and As atoms. The total energy of the developed "diluted magnetic semiconductor" clusters has been minimized in the presence and in the absence of spatial constraints

TABLE 5.1 Ground states of semiconductor compound atoms.

Atom	Electronic configuration	Atomic term	Nuclear spin	Nuclear magnetic moment in μ_P units
^{31}Ga	[Ar]3d^{10}4s^{2}4p	$^2P_{1/2}$	3/2	2.0166
^{49}In	[Kr]4d^{10}5s^{2}5p	$^2P_{1/2}$	9/2	5.5340
^{33}As	[Ar]3d^{10}4s^{2}4p^3	$^4S_{3/2}$	3/2	1.4395
^{25}Mn	[Ar]3d^{5}4s^2	$^6S_{5/2}$	5/2	3.4687
^{23}V	[Ar]3d^{3}4s^2	$^4F_{3/2}$	7/2	5.1574

applied to the positions of the clusters' atoms to obtain new pre-designed and vacuum molecules, respectively. Note again, that spatial constraints applied to the centers of mass of the clusters' atoms have been incorporated to model effects of quantum confinement on the properties of molecules synthesized in such confinement. The total energy minimization procedures have used Hartree-Fock (HF) and restricted open shell Hartree-Fock (ROHF) approximations, resulting in the pre-designed molecules Ga$_{10}$As$_3$V, Ga$_{10}$As$_2$V$_2$, In$_{10}$As$_3$V, In$_{10}$As$_2$V$_2$, and In$_{10}$As$_3$Mn. That is, the Schrödinger equation has been solved numerically in the first-order approximation using GAMESS software to obtain the ground state of the above molecules in the presence of the boundary conditions realized as spatial constraints applied to positions of the centers of mass of atoms of the initial atomic clusters. Once the pre-designed molecules were obtained, the corresponding unconstrained (or "vacuum") molecules have been obtained by relaxation of the spatial constraints applied to the atomic positions in the pre-designed clusters, and subsequent optimization of the clusters (that is, solving the corresponding Schrödinger equations for the clusters in the absence of the spatial constraints). Electronic and magnetic properties of the virtually synthesized molecules are discussed in this chapter. Ground state data for the constitutive atoms are listed in Table 5.1.

5.3 Pre-designed and vacuum In$_{10}$As$_3$Mn molecules

The case of the In$_{10}$As$_3$Mn molecules is very special for both fundamentals and applications. In particular, findings obtained in the process of virtual synthesis of such molecules may help shed light of the origin of some controversies between the existing semi-phenomenological theoretical predictions and experimental observations specific to Mn-doped InAs nanoscale systems.

In contrast to many of zincblende-derived molecules described in this book, the shape of the unconstrained In$_{10}$As$_3$Mn molecule visibly deviates from pyramidal shape of its pre-designed counterpart (Figs. 5.1 and 5.2, respectively). As compared to the pre-designed molecule of Fig. 5.1, in the vacuum molecule In atoms moved somewhat closer toward each other and Mn atom (Fig. 5.2a). This is enough for the vacuum molecule to lose the pyramidal form. Position adjustment accommodates stabilization of the vacuum In$_{10}$As$_3$Mn molecule in the absence of any spatial constraints and external fields. The electron charge density distribution

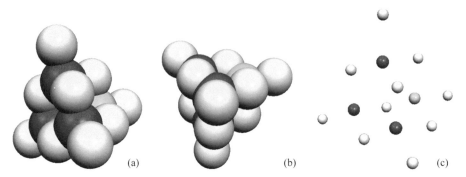

Fig. 5.1 Pre-designed pyramidal molecule $In_{10}As_3Mn$: (a) front view; (b) top view; (c) atomic positions. Indium atoms are yellow, As red, and Mn blue. In (a) and (b) atomic dimensions of In atoms are reduced to show the structure. In (c) all atomic dimensions are reduced.

(CDD) and molecular electronic potential (MEP) of these molecules are shown in Figs. 5.3 and 5.4.

Both $In_{10}As_3Mn$ molecules have similarly shaped CDD and MEP surfaces, and almost identical maximum and minimum values of CDD. However, the CDD and MEP surfaces of the vacuum molecule reflect significant loss of tetrahedral symmetry (compare the surfaces in Fig. 5.3a and Fig. 5.4c, for example). In both cases electron charge is delocalized, spread through the space occupied by the molecules (as defined by the covalent radii of their atoms) and reaches beyond it. However, this charge delocalization is not spatially uniform. Both molecules exhibit regions of electron charge accumulation and deficit. In particular, there are regions of electron charge deficit near the "surfaces" of the molecules on both "sides" of the surfaces, excluding for the "cores" of In atoms framing the molecules. [The molecular "surfaces" are defined as surfaces drown through the centers of mass of In atoms.] In the case of the pre-designed molecule, MEP values at the distances over two In covalent radii from the surface of the molecule are more negative (Fig. 5.3a) than those closer to the "surface" of the molecule on both "sides" of the surface (Figs. 5.3b to 5.3d). Only well "inside" of the molecule MEP values become more

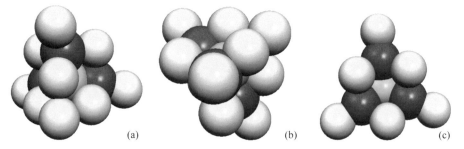

Fig. 5.2 Vacuum molecule $In_{10}As_3Mn$. (a) and (b): side views; (c) front view. In atoms are yellow, As red, and Mn blue. In (a) and (b) atomic dimensions of In atoms are reduced to show the structure. In (c) all atomic dimensions are reduced.

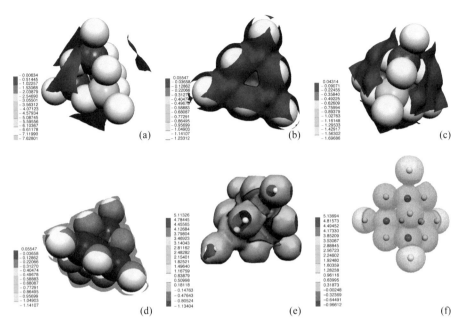

Fig. 5.3 Molecular electrostatic potential (MEP) of the pre-designed molecule $In_{10}As_3Mn$ calculated for several isosurfaces of the electron charge density distribution (CDD): (a) 0.001; (b) 0.02; (c) and (d) 0.05; (e) 0.08; (f) 0.1. The CDD maximum value 4.25145 is in arbitrary units. The color coding scheme is shown in each figure with red corresponding to negative and blue to positive MEP values. Indium atoms are yellow, As red and Mn blue. In (a) to (d) atomic dimensions of In atoms are reduced to show the structure. In (e) and (f) all atomic dimensions are reduced.

negative than those in the space outside the molecule. Thus, despite of its total charge equal to zero, the pre-designed molecule may be characterized experimentally as carrying a "shell" of delocalized electron charge surrounding the molecule at separations of about 2 In covalent radii from the surface of the molecule, and a "shell" of electron charge deficit near the "surface" on its both sides, with In atomic cores (featuring large localized electron charge) immersed in the charge deficit shell. The thickness of the electron charge deficit shell is roughly 3 In covalent radii. In the case of the vacuum molecule, the delocalized electron charge is closer to the molecular surface on its both "sides" signifying that more electron charge is kept closer to the molecular surface and more of it is kept in the "bulk" of the molecule. Correspondingly, the shell of electron charge deficit of the vacuum molecule is thinner than that of the pre-designed one. Such shells of electron charge deficit may be viewed as delocalized, positively charged holes of roughly pyramidal shape embracing the pyramid-like frames of the $In_{10}As_3Mn$ molecules.

Due to quantum confinement (modeled by means of application of spatial constraints to atomic positions), the pre-designed molecule is significantly polarized, with its dipole moment almost twice as large as that of the vacuum molecule (Table 5.2). The vacuum molecule is a ROHF septet whose uncompensated spin magnetic moment

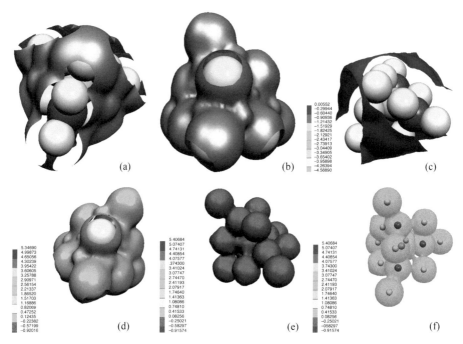

Fig. 5.4 Electron charge density distribution (CDD) and molecular electrostatic potential (MEP) of the vacuum molecule $In_{10}As_3Mn$. The CDD maximum value is 4.61805 (in arbitrary units). CDD isosurfaces correspond to the isovalues (a) 0.01 and (b) 0.05, and MEP surfaces to CDD isovalues: (c) 0.001, (d) 0.05, (e) 0.1 and (f) 0.1. The color coding scheme for MEP surfaces is shown in each figure with red corresponding to negative and blue to positive MEP values. Indium atoms are yellow, As red and Mn blue. In (a) to (d) atomic dimensions of In atoms are reduced. In (e) and (f) all atomic dimensions are reduced to show the MEP surface structure.

is equal to 6 μ_B (where μ_B denotes Bohr's magneton) and uncompensated electron spins localized in $4d$ orbits of In atoms (Fig. 5.5). At the same time, the pre-designed molecule is a HF singlet with all electron spins compensated and the total spin magnetic moment equal to zero. The spin density distribution (SDD) of the vacuum molecule reaches into the space far outside of the molecule, in agreement with the CDD and MEP data (Fig. 5.4). Thus, the vacuum molecule $In_{10}As_3Mn$ is a nanoscale ferromagnet possessing a large spin magnetic moment brought about by re-distribution of electron charge of In atoms due to d-electrons of the Mn atom and spatial symmetry breaking specific for low dimensional systems. Such integrated quantum Coulomb effect is typical for nanoscale objects where spatial symmetry breaking causes violation of the octet rule of the standard valence theory. To stabilize such systems, the electron charge must be delocalized throughout the system and even in the space surrounding the system. During such delocalization the electron spin components may or may not be fully compensated giving rise to molecules with non-zero, and sometimes large, total magnetic moments.

TABLE 5.2 Ground state data for the studied RHF/ROHF InAs- and GaAs-based molecules with one and two manganese and vanadium atoms.

Molecule	Spin multiplicity	RHF/ROHF ground state energy (Hartree)	RHF/ROHF direct optical transition energy (eV)	Dipole moment (Debye)
$In_{10}As_3Mn$	1	−2003.6919187179	3.9076	7.93980
$In_{10}As_3Mn*$	7	−2004.2390957056	1.2436	3.46941
$In_{10}As_3V$	9	−1971.4803877865	0.0571	4.12400
$In_{10}As_3V*$	5	−1971.4252752330	1.2653	3.26400
$Ga_{10}As_3V$	3	−2660.4273232664	1.2626	1.38721
$Ga_{10}As_3V*$	3	−2660.5087787590	1.0585	4.56927
$In_{10}As_2V_2$	11	−2035.9308103524	0.1551	3.44417
$In_{10}As_2V_2*$	1	−2036.0936312586	3.6082	3.27030
$Ga_{10}As_2V_2$	7	−2724.8716270248	0.8735	3.58697
$Ga_{10}As_2V_2*$	9	−2724.9370304183	1.5837	2.37091

* Vacuum molecules.

ELS of both molecules reflects the loss of tetrahedral symmetry due to the presence of the substitution Mn atom in the pre-designed molecule and also a loss of tetrahedral symmetry in the case of the vacuum molecule (Fig. 5.6). The vacuum molecule features bunches of closely lying molecular orbits (MOs) both near the highest occupied and the lowest unoccupied orbits (HOMO and LUMO, respectively). Nevertheless, the ground states of both molecules correspond to deep minima of the total energy, and their ELSs exhibit large OTEs (Table 5.2). About 240 electrons contribute to formation of the outer MOs of these molecules. The nature of these MOs confirms the violation of the octet rule, and some of the MOs are similar to those studied in Ref. 122 in the case of small non-stoichiometric atomic clusters. For the pre-designed molecule, the bonding HOMO 118, which is the third

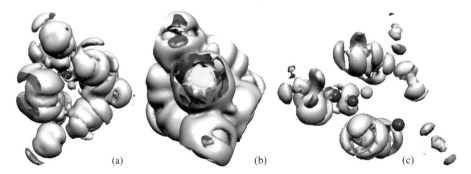

(a) (b) (c)

Fig. 5.5 Isosurfaces (gray) of the spin density distribution (SDD) of the vacuum molecule $In_{10}As_3Mn$ corresponding to SDD isovalues (a) 0.0002, (b) 0.001 and (c) 0.005. Indium atoms are yellow, As red and Mn blue. In (a) dimensions of In atoms are reduced. In (b) and (c) all atomic dimensions are reduced to show the SDD isosurface structure.

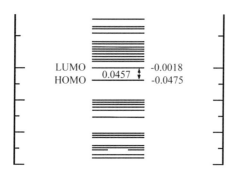

Fig. 5.6 The electronic energy level structure (ELS) in the HOMO-LUMO region of the vacuum $In_{10}As_3Mn$ molecule. The HOMO and LUMO energy values, and the optical transition energy (OTE), are in Hartree (H) units.

occupied MO counting from the uppermost HOMO 120, is a *pd*-type MO derived from several major contributions: (1) $5p$ atomic orbits (AOs) of In atoms hybridized with $3d$ AOs of Mn, (2) $4p$ AOs of As atoms hybridized with $5p$ AOs of In atoms and $3d$ AOs of Mn atom, and (3) large *pp*- and small *pd*-ligand bonding of In atoms between themselves (Figs. 5.7a and 5.7b). AOs of this type also contribute to the rest of MOs in the near HOMO – LUMO region, including HOMO 119 and the proper HOMO 120 (Figs. 5.7c to 5.7e). In the latter case, contributions from $4d$ AOs of In atoms are significant. In – In π-type ligand bonding similar to that studied in Ref. 122, can be observed in all cases. This ligand bonding is mediated by As and Mn atoms. All three HOMOs in Figs. 5.7a to 5.7f are bonding MOs signifying stability of the pre-designed singlet. The LUMO 122 also is a bonding MO similar in nature to the HOMOs, but contributions from $4d$ AOs of In atoms to this MO are negligibly small. <u>In all cased In atoms bond to Mn via hybridization of their $5p$ AOs and $3d$ AOs of Mn, in agreement with Zener's assumption [34]</u>. Mn and As atoms in the pre-designed $In_{10}As_3Mn$ molecule do not bond directly: their bonding is always mediated by $5p$ AOs of In atoms, which is somewhat different from As-Mn *p-d* bonding suggested in Ref. 79.

Isosurfaces of several MOs in the HOMO – LUMO region of the vacuum $In_{10}As_3Mn$ molecule are depicted in Fig. 5.8. In this case, all MOs in the immediate HOMO – LUMO region are essentially shaped by hybridization of $3d$ AOs of the Mn atom and $4p$ AOs of As atoms, in a perfect agreement with suggestion of Ref. 79. The lower HOMO 121 (Figs. 5.8a and 5.8b) exhibits bonding of In atoms mediated by As atoms in two equal parts of the molecule. The HOMO 122 (Figs. 5.8c and 5.8d) is a bonding MO with the major contributions from the $3d_z^2$ AO of the Mn atom and $4p$ AOs of As atoms, and very small contributions from $4d$ AOs of In atoms. In contrast, the proper HOMO 123 (Figs. 5.8e and 5.8f) has significant contributions from $4d$ orbits of two In atoms, in addition to the major contributions from $3d_{xy}$ AOs of the Mn atom and $4p$ AOs of As atoms. Similar to HOMO 121 and 122, the proper HOMO 123 is a bonding MO. The LUMOs 124, 125 and 126 (Figs. 5.8g to 5.8i) are also derived from the above mentioned AOs of In, As and Mn atoms, but these MOs feature significant ligand (In to In) bonding contributions. In particular, the major contribution to LUMO 124

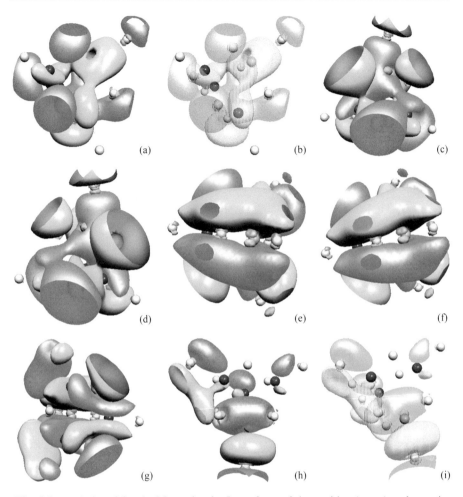

Fig. 5.7 Pre-designed $In_{10}As_3Mn$ molecule. Isosurfaces of the positive (green) and negative (orange) parts of the highest occupied and lowest unoccupied molecular orbits (HOMOs and LUMOs, respectively) corresponding to the isovalues 0.01 [(a) to (e), and (g) and (h)] and 0.1 [(f) and (i)]; (a) and (b): HOMO 118; (c) HOMO 119; (d) to (f): HOMO 120; (g) to (h): LUMO 122. Indium atoms are yellow, As red and Mn blue. Atomic dimensions are reduced to show the surface structure; other dimensions are to scale. In (b) and (i) MO surfaces are transparent to show contributing atoms.

(Fig. 5.8g) comes from bonding of 4 In atoms between themselves through their $5p$ AOs (a "sandwich" in the lower left part of Fig. 5.8g that consists of a large negative "football" surface in the lower left part of Fig. 5.8g and the corresponding positive "lid" on the top), and with 3 other In atoms whose bonding is mediated by $3d$ AOs of Mn and $4p$ AOs of As atoms. The two higher MOs in the LUMO region, LUMO 125 and LUMO 126, are also bonding MOs, but In ligand bonding in this cases is mediated by Mn and As atoms. The discussed properties of MOs of $In_{10}As_3Mn$

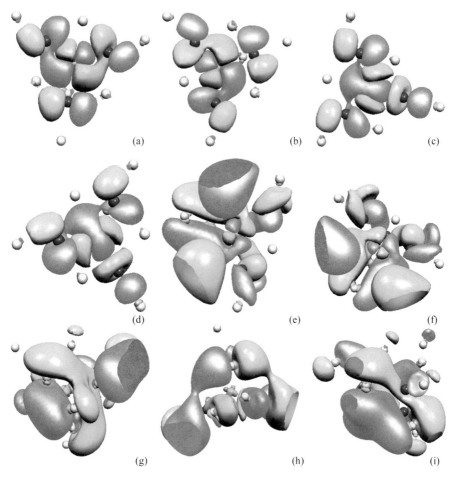

Fig. 5.8 Vacuum $In_{10}As_3Mn$ molecule. Isosurfaces of the positive (green) and negative (orange) parts of the highest occupied and lowest unoccupied molecular orbits (HOMOs and LUMOs, respectively) corresponding to several isovalues. (a) and (b): HOMO 121, isovalue 0.005; (c) and (d): HOMO 122, isovalue 0.005; (e) and (f): HOMO 123, isovalue 0.01; (g), (h) and (i): LUMOs 124, 125 and 126, respectively, isovalue 0.015. Indium atoms are yellow, As red and Mn blue. Atomic dimensions are reduced to show the isosurface structure; other dimensions are to scale.

molecules manifest violation of the octet rule of the standard valence theory, in agreement with other available observations [119–122], and indicate that a wide range of non-stoichiometric molecules may exist that are stable either on their own, or can be stabilized by their environment, such as quantum confinement.

The RHF and ROHF MOs are not very accurate, and the corresponding OTEs of the two In-based molecules are rather large (Table 5.2), which is typical for HF approximation (see a discussion in Chapter 2). However, these MOs constitute the major contributions to the "true" MOs of the studied systems. The obtained results reveal the existing opportunities to manipulate electronic and magnetic properties of small QDs composed of semiconductor compound atoms using the synthesis environment, such as quantum confinement, and adjusting the composition. The case of the pre-designed $In_{10}As_3Mn$ molecule demonstrates rich prospects opened by the use of quantum confinement as an element of molecular synthesis conditions. Virtually synthesized at conditions mimicking quantum confinement, this molecule is a HF singlet ("antiferromagnetic" electron spin arrangement in the molecule) with the total magnetic moment equal to zero and RHF OTE about 3.9 eV. At the same time, in the absence of the spatial constraints (quantum confinement) the same atoms self-assemble into a different molecule with a relatively large magnetic moment and ROHF OTE about 1.24 eV.

While hardware restrictions do not allow virtual synthesis of much larger atomic systems (see Chapter 2 for more information), the discussed results in the case of small $In_{10}As_3Mn$ molecules lead to several important conclusions concerning electronic and magnetic properties of thin films of, and possibly bulk, InAsMn. [Note, that the composition of the studied molecules corresponds to about 7% of Mn in the zincblende InAs structure, that is, to the upper limit of the case of "diluted magnetic semiconductors".] Based on the analysis of CDDs and SDDs of the $In_{10}As_3Mn$ molecules and those of $In_{10}As_4$ molecules of Chapter 3, one can observe that electron charge deficit shells embracing the "surfaces" of the molecules are orchestrated, to a large degree, by the Mn atom. In the language of the semi-phenomenological theory of semiconductors this charge deficit is called "hole". When average linear dimensions of a system are orders of magnitude larger than 1 nm (the case of bulk solids, thick films, large QDs and QWs, *etc.*) such a "hole" may be considered roughly localized in the vicinity of a Mn atom. However, in the case of nanometer-thin films that are widely investigated at present (see section 5.1) such a hole cannot be treated as a localized object, because its "localization dimensions" are of the order of the thickness of the film. Therefore, in the case of thin films, small QDs and QWs of InAsMn with a few percent of substitution Mn impurities and characteristic dimensions of several nanometers, such a hole must be considered as a region of electron charge deficit delocalized about the "surface" of the pyramidal element of the InAs zincblende lattice composed of 10 to 14 In and As atoms and containing a Mn atom. Moreover, even in the pre-designed case (that is, the case reflecting conditions of quantum confinement such as that provided by the zinkblende bulk lattice of InAs) the region of electron charge deficit is not uniform, and is not centered on Mn atom. Also, there exist smaller electron charge deficit regions in the "bulk" of the molecule. These findings are further supported by the obtained data on SDD of the vacuum molecule (Fig. 5.5). In particular, for this molecule one can observe a large spin density values near In atoms embraced by the electron charge deficit region surrounding the molecular "surface". That is, the vacuum $In_{10}As_3Mn$ molecule houses a delocalized and spin-polarized hole, while in the case of the pre-designed $In_{10}As_3Mn$ molecule a similar hole is delocalized, but not spin-polarized.

Using the virtual synthesis data discussed above one can finally answer the question concerning "the place of residence" of delocalized and spin-polarized holes in DMS systems. In particular, in the case when a nanometer-thick InAsMn layer is sandwiched between layers of other materials (such systems are widely studied at present, see section 5.1), such holes do not reside exclusively inside of the InAsMn leyer. Rather, they include portions of the confining systems.

ELS of the studied $In_{10}As_3Mn$ molecules differs significantly, especially in the HOMO-LUMO regions. This means, that for systems with one or more linear dimensions in the range of a few nanometers, the "band" structure (if such a structure still can be properly identified) is not derivable directly from that of InAs "valence band" and Mn "impurity band". Instead, ELS of nanoscale DMS systems should be treated as a new band structure shaped by quantum confinement and Coulomb effects governing quantum motion of strongly correlated electrons in such broken symmetry systems. Thus, reasoning about the "impurity band" and its position related to the "valence band" of the host lattice is meaningless for DMS systems with one or more linear dimensions in the range of a few nanometers.

5.4 Pre-designed and vacuum $In_{10}As_3V$ molecules

Virtual synthesis studies of $In_{10}As_3V$ molecules are of significant importance for understanding a role of $3d$ AOs of V and Mn atoms in the development of positive MEP regions (delocalized charge deficit regions, or holes) in InAs and GaAs zincblende bulk lattices and low dimensional structures containing a few percent of Mn or V atoms. Similar to the pre-designed $In_{10}As_3Mn$ molecule, the pre-designed $In_{10}As_3V$ molecule has been developed by substitution of As with V atom in the pyramidal symmetry element of the zincblende InAs lattice, and subsequent RHF/ROHF minimization of the total energy of the atomic cluster so obtained in the presence of spatial constraints applied to the centers of mass of the cluster's atoms. The corresponding $In_{10}As_3V$ vacuum molecule has been obtained upon the total energy minimization of the pre-designed cluster when the spatial constraints were lifted The structure of the $In_{10}As_3V$ molecules so obtained is detailed in Fig. 5.9. Visually these two molecules are almost indistinguishable, because in the vacuum molecule all As atoms have moved only slightly from their original positions in the pre-designed molecule. This motion also resulted in small changes to the angles between V or As and In atoms. For example, in the case of tetrafold-coordinated As atoms in the pre-designed molecule the In-As-In angles with the closest 4 In atoms are 113.0°, 113.0°, 105.7° and 105.7°, while in the case of the vacuum molecule those angles are 118.7°, 110.6°, 105.6° and 99.7° for two of the As atoms on the zincblende cube body diagonal opposite to that of the V atom, and 111.8°, 111.1°, 107.7°, and 107.0° for the As atom sharing the zincblende cube body diagonal with V atom. The vanadium atom has also been slightly displaced from its original position in the pre-designed molecule. Thus, the In-V-In angles in the small, tetrafold-coordinated V-containing pyramid of the vacuum molecule are 118.4°, 111.0°, 105.4° and 99.6°. This small displacement, however, is such that it does not affect V - As atomic coordination: both the distances and angles between V and As atoms in the vacuum molecule remain equal to those in the pre-designed molecule.

CDDs (Figs. 5.10 and 5.11) and MEPs (Figs. 5.12 and 5.13) of both molecules retain a general appearance of tetrahedral symmetry. However, detailed analysis

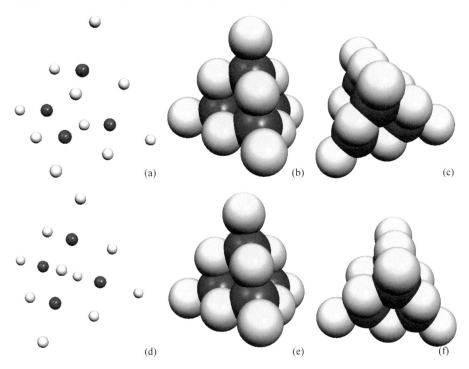

Fig. 5.9 The structure of the pre-designed [(a) to (c)] and vacuum [(d) to (f)] $In_{10}As_3V$ molecules. Indium atoms are yellow, As red and V purple. In (a) and (d) atomic dimensions are reduced to show atomic positions; other dimensions are to scale. The dimensions of In atoms in (b), (c), (e) and (f) are somewhat reduced, and those of As and V atoms somewhat enlarged.

reveals tetrahedral symmetry breaking in both cases. Comparing these CDD and MEP isosurfaces to those of $In_{10}As_3Mn$ molecules one can see that in the latter case delocalized electron charge is pushed further into the space outside the molecules with Mn atom. In the case of V-containing molecules V atoms accumulate more electron charge, and a portion of electron charge pushed outside is much closer to "surfaces" for the same CDD isovalues as those of the Mn-containing molecules. Thus, the shells of electron charge deficit of the V-containing molecules are much thinner than those of their Mn-containing counterparts. Correspondingly, the MEP values are positive near the "surfaces" of both V-containing molecules (Figs. 5.12d and 5.13d) and in the immediate vicinity of the "surfaces" on the inner side of the molecules, excepting for the cores of In atoms embedded into the charge deficit shells (Figs. 5.12e to 12f, and 5.13e to 5.13f). The shell of positive MEP values (hole) is less distinctive, and the absolute values of MEP are lesser, in the case of the pre-designed $In_{10}As_3V$ molecule (Figs. 5.10 and 5.12) than those in the case of the vacuum one. To a degree, this may be a consequence of the fact that the pre-designed molecule is a ROHF nonet. Such states are at the verge of applicability of the ROHF approximation (see a discussion in Chapter 2), indicating that this molecule may not be stable. In contrast, the

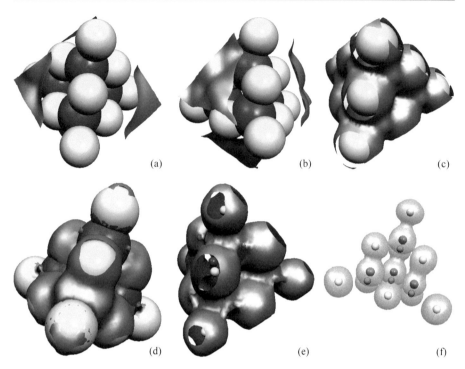

Fig. 5.10 The electron charge density distribution (CDD) of the pre-designed $In_{10}As_3V$ molecule. CDD isosurfaces (golden) correspond to the isovalues (a) 0.0005, (b) 0.01, (c) 0.05, (d) 0.075, (e) 0.075, and (f) 0.15. Indium atoms are yellow, As red and V purple. In (e) and (f) atomic dimensions are reduced to show the isosurface structure; other dimensions are to scale. The dimensions of In atoms in (a) to (d) are somewhat reduced, and those of As and V atoms somewhat enlarged.

vacuum V-containing molecule is a ROHF pentet, and its stability is more certain: the shell of positive MEP values surrounding its surface on both sides is much better defined. It contains regions in the immediate vicinity of the molecular "surface" on its outer side, and much thicker regions near the molecular "surface" on its inner side. Using the hole terminology, the holes orchestrated by V atoms are smaller than those orchestrated by Mn atoms, and most of their volume is contained "inside" of the 14-atomic clusters. Respectively, one can expect that V-facilitated holes in the case of bulk InAsV DMS materials should reside inside of the tetrahedral symmetry elements of the zincblende InAs structure with V substitution defect, while Mn-facilitated holes should be delocalized about a larger region that would contain the corresponding 14-atomic pyramidal element. The "volumes" of Mn-mediated holes in $In_{10}As_3Mn$ molecules is at least 30% larger than those in the case of V-mediated holes in $In_{10}As_3V$ molecules. Therefore, a V-mediated hole delocalized in a pyramidal symmetry element of the InAs zincblende lattice containing a vanadium atom should be more localized than a Mn-facilitated hole in that symmetry element. The major reason for this steams from the fact that the vanadium atom has smaller number

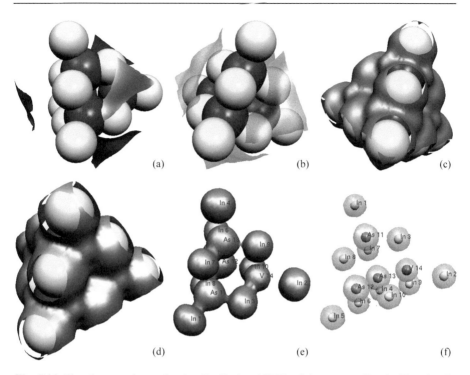

Fig. 5.11 The electron charge density distribution (CDD) of the vacuum $In_{10}As_3V$ molecule. CDD isosurfaces (golden) correspond to the isovalues (a) 0.001, (b) 0.01, (c) 0.05, (d) 0.05, (e) 0.15, and (f) 0.25. Indium atoms are yellow, As red and V purple. In (e) and (f) atomic dimensions are reduced to show the isosurface structure; other dimensions are to scale. The dimensions of In atoms in (a) to (d) are somewhat reduced, and those of As and V atoms somewhat enlarged.

of $3d$-electrons (2, as opposed to 5 $3d$-electrons of a Mn atom), and those electrons are closer to the V nucleus. Therefore, more electron charge of In atoms can be kept "inside" of $In_{10}As_3V$ molecules (Figs. 5.12e and 5.13e) compared to that of $In_{10}As_3Mn$ molecules (Figs. 5.3e and 5.4e). These findings hint at that that the semi-phenomenological modifications of the band theory of semiconductors designed to embrace a realm of DMS systems should work better for InAsV systems than for InAsMn ones.

Analysis of MOs of $In_{10}As_3V$ molecules provides further information on electron charge configuration, a role of the vanadium atom in stabilization of these molecules and the development of regions of positive MEP values ("holes") associated with V-atoms. Several such MOs in the HOMO-LUMO regions of these molecules are depicted in Figs. 5.14 and 5.15. Similar to the case of $In_{10}As_3Mn$ molecules, the major contributions to MOs of $In_{10}As_3V$ molecules in the HOMO-LUMO region come from $5p$ AOs of In atoms, $3d$ AOs of V atom and $4p$ AOs of As atoms hybridized in various proportions. Much smaller contributions due to $4d$ AOs of In atoms must be ascertained by further CI, MCSCF and MP-2 studies. In the case of the pre-designed ROHF

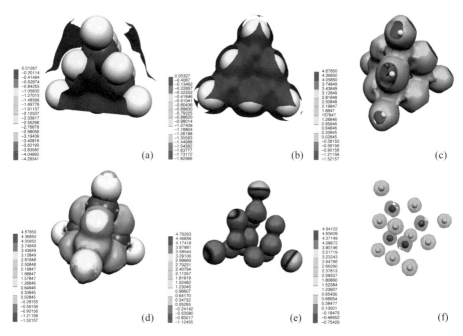

Fig. 5.12 Surfaces of the molecular electrostatic potential (MEP) of the pre-designed molecule $In_{10}As_3V$ corresponding to several isovalues of its CDD (a) 0.01, (b) 0.05, (c) and (d) 0.075, (e) 0.15, and (f) 0.3. The CDD maximum value 3.54328 (arbitrary units) is not shown. The color coding scheme for MEP surfaces (with red corresponding to negative and blue to positive MEP values) is shown in each figure. Indium atoms are yellow, As red and V purple. In (a) to (d) dimensions of In atoms are somewhat reduced and those of As and V atoms enlarged. In (d) to (f) all atomic dimensions are reduced to show the MEP surface structure.

nonet, HOMO 121 (Figs. 5.14a and 5.14b) contains both bonding and non-bonding regions mediated by $3d_{xy}$ AOs of V atom that orchestrates In ligand bonding via their $5p$ AOs, and a significant contribution from $4p$ AOs of As atoms. HOMOs 122 and 123 are bonding MOs with the major contributions from $3d_{z^2}$ AOs of the vanadium atom hybridized with both $5p$ and $4d$ AOs of several In ligand atoms (Figs. 5.14c to 5.14f). In the case of HOMO 122 (Fig. 5.14d), $4p$ AOs of only one As atom contribute to bonding, while $4p$ AOs of all As atoms significantly contribute to bonding in the case of HOMO 123 (Figs. 5.14e and 5.14f). LUMOs 124 and 126 of this molecule (Figs. 5.14g and 5.14i, respectively) are bonding MOs where the In ligand bonding is mediated by $3d_{z^2}$ AOs of the vanadium atom. LUMO 125 contains both non-bonding and bonding regions, all mediated by $3d_{xy}$ AOs of the vanadium atom. In contrast to $In_{10}As_3Mn$ molecules, all MOs in the HOMO-LUMO region of the pre-designed $In_{10}As_3V$ molecule exhibit significant contributions from ligand bonding of 4 or more participating In atoms. Both $5p$ and $4d$ AOs of In atoms contribute to this bonding. This is in agreement with the corresponding CDD results indicating that in the case of V-containing molecules electron charge of In atoms is more effectively redistributed inside the molecules, and less of this charge is pushed outside of the molecular "surfaces".

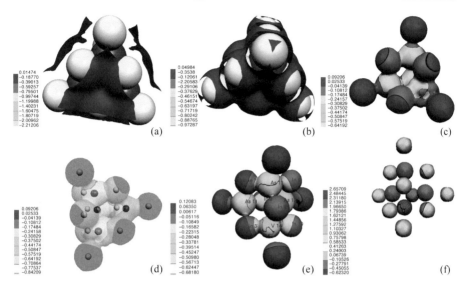

Fig. 5.13 Surfaces of the molecular electrostatic potential (MEP) of the vacuum molecule In$_{10}$As$_3$V corresponding to several isovalues of its CDD: (a) 0.01, (b) 0.05, (c) and (d) 0.1, (e) 0.15, and (f) 0.25. The CDD maximum value 3.264001 (arbitrary units) is not shown. The color coding scheme for MEP surfaces (with red corresponding to negative and blue positive MEP values) is shown in each figure. Indium atoms are yellow, As red and V purple. In (a) and (b) atomic dimensions of In atoms are somewhat reduced and those of As and V atoms enlarged. In (c) to (f) all atomic dimensions are reduced to show the MEP surface structure.

All MOs in the HOMO-LUMO region of the vacuum In$_{10}$As$_3$V molecule shown in Fig. 5.15 are bonding orbits. The $3d_{z2}$ AO of the vanadium atom provides the major contribution to all MOs except for LUMO 124. In the latter case, contributions due to this AO are small, and all bonding is mediated by $4p$ AOs of 3 As atoms. All MOs of the vacuum In$_{10}$As$_3$V molecule exhibit two large areas of shared electron charge: one is a delocalized electron charge of 4 to 6 In ligand atoms mediated by V, and the other a delocalized electron charge of 3 In atoms mediated by one to three As atoms. These two portions of the MOs are bonded to each other through one of the In ligand atoms. Once again, this type of bonding violates the octet rule and is the major means of stabilization of the studied non-stoichiometric molecules.

The In ligand bond length in the case of the pre-designed molecule is defined entirely by the pyramid geometry and is 4.284 Å, which is also the distance between any two In neighbours. In the vacuum molecule the ligand bond length is flexible taking several values between 4.272 Å and 4.999 Å. The In-As bond length is also flexible in both molecules and allows for several values in the ranges from 2.220 Å to 2.934 Å (the pre-designed molecule) and 2.346 Å to 2.895 Å (the vacuum molecule). The arsenic and vanadium atoms bond directly only to In ones. In the case of the pre-designed molecule the length of the V - In bond can take 4 values: 2.220 Å, 2.717 Å, 2.934 Å and 4.695 Å. The first 3 values are the same as those of the In - As bond in

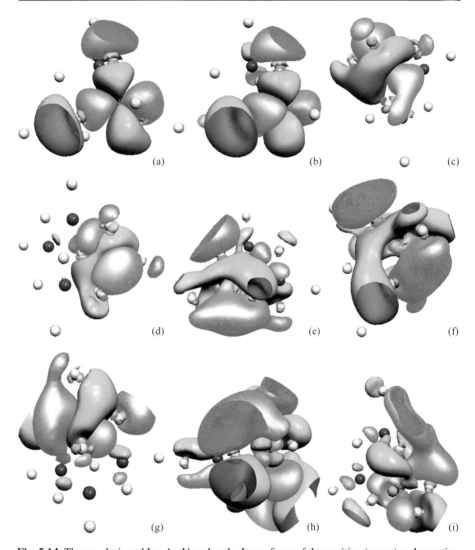

Fig. 5.14 The pre-designed $In_{10}As_3V$ molecule. Isosurfaces of the positive (green) and negative (orange) parts of the highest occupied and lowest unoccupied molecular orbits (HOMOs and LUMOs, respectively) corresponding to several isovalues; (a) and (b): HOMO 121, isovalue 0.01; (c) and (d): HOMO 122, isovalue 0.01; (e) and (f): HOMO 123, isovalue 0.015; (g), (h) and (i): LUMO 124, 125 and 126, isovalues 0.01, 0.01 and 0.02, respectively. Indium atoms are yellow, As red and V purple. Atomic dimensions are reduced to show the isosurface structure; other dimensions are to scale.

this molecule. In the case of the vacuum molecule the length of V - In bond takes only 3 values: 2.340 Å, 2.706 Å and 2.907 Å.

Both $In_{10}As_3V$ molecules are ROHF spin multiplets with "ferromagnetic" arrangement of their uncompensated electron spins (Fig. 5.16). The total spin of these

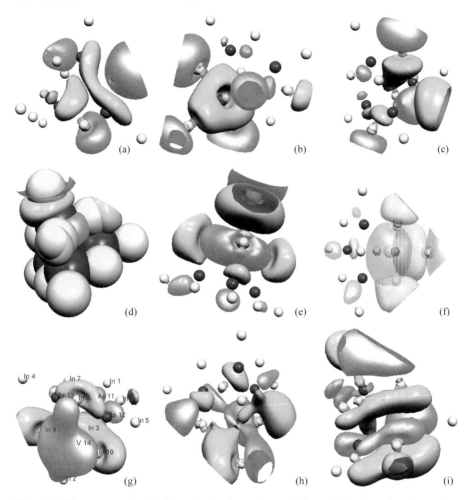

Fig. 5.15 The vacuum $In_{10}As_3V$ molecule. Isosurfaces of the positive (green) and negative (orange) parts of the highest occupied and lowest unoccupied molecular orbits (HOMOs and LUMOs, respectively) corresponding to several isovalues; (a) to (c): HOMO 118, 119 and 120, isovalue 0.01, respectively; (d) and (f): HOMO 121, isovalue 0.01; (g) and (h): LUMO 122 and 123, isovalue 0.015, respectively; and (i) LUMO 124, isovalue 0.015. Indium atoms are yellow, As red and V purple. In all figures except for (d) atomic dimensions are reduced to show the isosurface structure; other dimensions are to scale. In (d) Indium atomic dimensions are reduced and those of As and V atoms somewhat enlarged.

molecules is primarily contributed to by spin components of delocalized $4d$ electrons of In atoms. Electrons of arsenic atoms do not contribute to SDDs. SDD values of the pre-designed molecule (which is a "ferromagnetic" ROHF nonet with the uncompensated magnetic moment 8 μ_B, Figs. 5.16a to 5.16c) are about 3 times larger than those of the vacuum one, which is a "ferromagnetic" ROHF pentet with the uncompensated magnetic moment 4 μ_B (Figs. 5.16d to 5.16i). Thus, the InAs-based molecules with

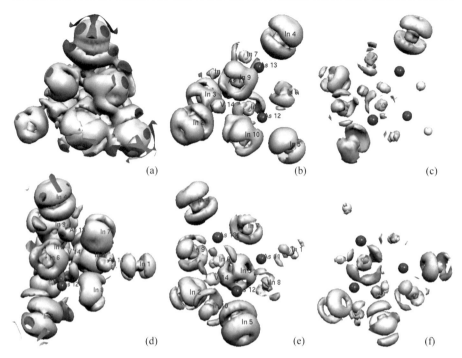

Fig. 5.16 Isosurfaces of the spin density distribution (SDD) of the pre-designed [(a) to (c)] and vacuum [(d) to (f)] $In_{10}As_3V$ molecules corresponding to SDD isovalues (a) 0.001, (b) 0.005, (c) 0.01, (d) 0.001, (e) 0.003, (f) 0.005. Indium atoms are yellow, As red and Mn blue. All atomic dimensions are reduced to show the SDD surface structure.

one vanadium atom are stronger "magnets" than those with one Mn atom, and thus may be more suitable for DMS applications. In particular, the pre-designed $In_{10}As_3V$ molecule is "ferromagnetic" and possesses the largest magnetic moment among the studied InAs-based molecules. At the same time, the pre-designed $In_{10}As_3Mn$ molecule is "antiferromagnetic" singlet with its zero uncompensated magnetic moment. [The latter finding is consistent with experimental observation that with a change in thermodynamic conditions some thin DMS films exhibit magnetic phase transitions; see section 5.1 for further details and references.] At the same time, much larger and heavier "holes" facilitated by Mn atoms in the $In_{10}As_3Mn$ structures may have their own uses for applications.

5.5 $Ga_{10}As_3V$ molecules with one vanadium atom

Interest to GaAs-based DMS materials is currently rising, because such systems have some technological advantages over InAs-based DMS ones. Among such systems, GaAsV structures may be simpler to understand than GaAsMn materials. At present, GaAsV DMS systems have not been well investigated, so virtual synthesis and

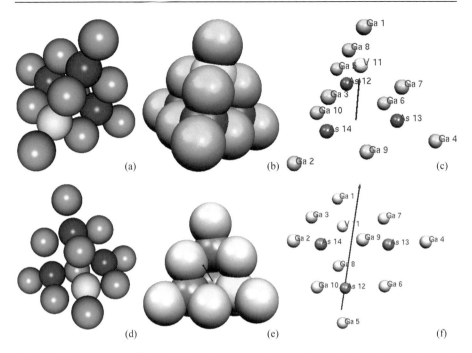

Fig. 5.17 Pre-designed [(a) to (c)] and vacuum [(d) to (f)] molecules $Ga_{10}As_3V$. In (b) dimensions of Ga atoms roughly correspond to Ga covalent radius, and in (e) are somewhat reduced, while atomic dimensions of As and V atoms are enlarged. In (a), (c), (d) and (f) all atomic dimensions are reduced to show the structure. Red arrows in (c) and (f) show the dipole moment. Gallium atoms are blue, As brown, and V yellow.

computational studies of basic GaAsV structures may provide a valuable guidance to experimentalists and engineers.

Similar to the studied $In_{10}As_3V$ molecules, two $Ga_{10}As_3V$ molecules have been virtually synthesized using computational approaches discussed in Chapter 2. Thus, the pre-designed $Ga_{10}As_3V$ molecule was obtained by the total energy minimization procedure applied to a tetrahedral symmetry element (a pyramid) of the GaAs zincblende bulk lattice described in Chapter 3, where one of As atoms was replaced by the vanadium one. During such conditional energy minimization all positions of the centers of mass of the pyramid atoms were fixed. The corresponding vacuum $Ga_{10}As_3V$ molecule was virtually synthesized by lifting the spatial constraints applied to the centers of mass of the atoms in the pre-designed molecule, and minimizing the total energy of the atomic cluster unconditionally (that is, without any constrains applied). The structure of the obtained molecules is depicted in Fig. 5.17. To a human eye, these structures seem to be the same, but analysis reveals that many atoms in the vacuum pyramidal molecule moved from their former positions in the pre-designed one. The pre-designed pyramid includes 4 smaller pyramids built on As and V atoms. All distances between Ga and As atoms in the three As-coordinated pyramids are equal to 2.448 Å, and all Ga-As-Ga angles are 109.5°. This, of course, corresponds to a separation between an

As atom and its 4 closest Ga neighbors, and related angles, respectively, in the GaAs zincblende lattice. The vanadium atom in this molecule simply substitutes an As one, so all dimension of V-coordinated small pyramid are equal to those of the small As-coordinated pyramids. All closest neighbor distances between As atoms and V atom are 3.997Å, and the corresponding angles 60°. Thus, geometrically, the pre-designed $Ga_{10}As_3V$ molecule is the perfect pyramid.

The vacuum molecule is far from being of perfect pyramidal structure. All As and V atoms in this molecule moved from their original positions in the perfect pre-designed pyramid. Thus, all distances in the small As-coordinated pyramidal arrangements have been changed by several tenths of Å, and the corresponding angles by about 2° to 7°. In particular, separations of Ga and V atoms in the V-coordinated small pyramid have become 2.505Å, 2.998Å, 2.998Å and 2.998Å, and the Ga-V-Ga angles 118.9°, 111.5°, 111.5° and 101.3°. The As-coordinated small pyramids have changed even more dramatically: the distances between Ga and As atoms in each of the pyramids have become 2.835Å, 2.519Å, 2835Å and 2.487Å. The sets of Ga-As-Ga angles in the As-coordinated small pyramids differ for each of the pyramids. For the pyramid coordinated by the As-13 atom (Figs. 5.17c and 5.17f) the set of angles includes 58.5°, 58.5°, 103.7° and 107.9°; for the As-14-coordinated pyramid the set includes 103.7°, 103,7°, 116.6° and 116.6°, and for the As-12-coordinated pyramid it is 103.7°, 106.7°, 116.6° and 107.9°. This tetrahedral symmetry breaking developed to stabilize a molecule when spatial constraints previously applied to the atoms were lifted.

The dipole moment of the perfect pre-designed pyramid $Ga_{10}As_3V$ is 1.387208 D. It is applied directly to the center of the pyramid base facing the V atom, and runs strictly along the pyramid heights toward the vertex Ga atom (Fig. 5.17c). In the case of the vacuum molecule the dipole moment is about 3 times larger: 4.569266 D; it is applied to the pyramid base closest to the V atom and runs through one of the As atoms (Figs. 5.17e and 5.17f).

The MEPs of these molecules are pictured in Figs. 5.18 and 5.19. Similar to $In_{10}As_3Mn$ and $In_{10}As_3V$ molecules of sections 5.3 and 5.4, CDD and MEP surfaces of both $Ga_{10}As_3V$ molecules retain some indication of tetrahedral symmetry. Characteristic features of CDD and MEP surfaces of the pre-designed $Ga_{10}As_3V$ molecule are close to those of $In_{10}As_3Mn$ molecules, while such characteristics in the case of the vacuum $Ga_{10}As_3V$ molecule resemble those of $In_{10}As_3V$ molecule. Indeed, electron charge of the pre-designed molecule is pushed further outside of molecule's surface (Figs. 5.18a to 5.18c), and also deeper inside of the molecule (Figs. 5.18d to 5.18f), so a "shell" of electron charge deficit surrounding the surface is thicker than that of the vacuum $Ga_{10}As_3V$ molecule. Electron charge of the vacuum molecule is distributed relatively close to the molecular surface (Figs. 5.19a and 5.19b) both on the outer side and inside of the molecular volume creating a relatively thin shell of electron charge deficit surrounding the molecular surface (Fig. 5.19a). The average thickness of the electron charge deficit shell of this molecule is about 2 covalent radii of Ga atom.

The major reason behind the fact that CDD and MEP properties of the pre-designed $Ga_{10}As_3V$ molecule resemble closer those of $In_{10}As_3Mn$ molecules, while CDD and MEP of the vacuum $Ga_{10}As_3V$ molecule exhibits more similarity with those of $In_{10}As_3V$ molecules, steams from the fact that the pre-designed $Ga_{10}As_3V$ molecule is

Quantum Dots of Diluted Magnetic Semiconductor Compounds 217

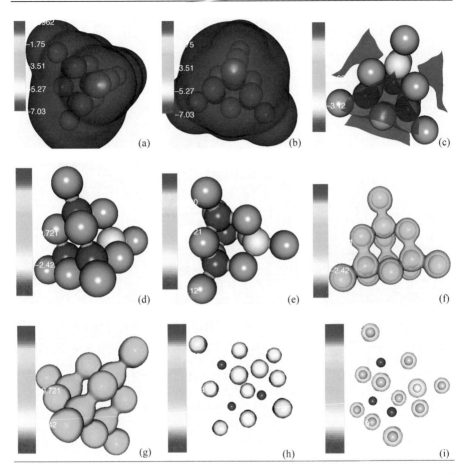

Fig. 5.18 The molecular electrostatic potential (MEP) of the pre-designed molecule $Ga_{10}As_3V$ calculated for several CDD isovalues: (a) and (b): 0.001; (c) 0.01; (d) to (g): 0.1; (h) and (i): 0.3. The color coding scheme for MEP surfaces is shown in each figure. Ga atoms are blue, As brown and V yellow. Atomic dimensions are reduced to show the MEP surface structure. In (a) to (f) and (i) MEP surfaces are semi-transparent.

strained. The covalent radius of the Ga atom is much smaller than that of the In atom (Table 5.1), and is close to that of the As atom. Thus, the partial volume occupied by V atom in the pre-designed $Ga_{10}As_3V$ molecule is larger than that in the pre-designed $In_{10}As_3V$ molecule. As a result, replacement of an As atom with the V one causes much more electron charge imbalance in the pre-designed $Ga_{10}As_3V$ molecule as compared to that in the pre-designed $In_{10}As_3V$ molecule, and therefore, the former molecule is more strained than the latter, and in this respect resembles the pre-designed $In_{10}As_3Mn$ molecule. Correspondingly, CDDs of the pre-designed and vacuum $In_{10}As_3V$ molecules differ less between themselves than CDDs of the pre-designed and vacuum $Ga_{10}As_3V$ molecules. The latter is illustrated in Figs. 5.18 and 5.19.

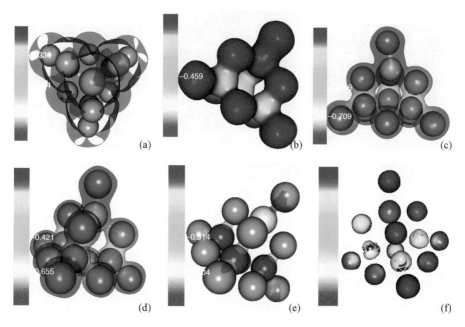

Fig. 5.19 The molecular electrostatic potential (MEP) of the vacuum molecule $Ga_{10}As_3V$ calculated for several CDD isovalues: (a) 0.02; (b) and (c) 0.05; (d) 0.06; (e) 0.1, and (f) 0.15. The color coding scheme for MEP surfaces is shown in each figure. Ga atoms are blue, As brown and V yellow. Atomic dimensions are reduced to show the MEP surface structure. In (a), (c), (d) and (e) MEP surfaces are semi-transparent to reveal the atoms.

More detailed analysis of the data obtained further confirms the above conclusion that the pre-designed $Ga_{10}As_3V$ molecule is more strained than the pre-designed $In_{10}As_3V$ molecule. Indeed, the total number of electrons (238) contributing to the upper 118 doubly occupied MOs of the pre-designed $Ga_{10}As_3V$ molecule is the same as that contributing to the upper 115 doubly occupied MOs of the pre-designed $In_{10}As_3V$ molecule. Given that the Ga-based molecule has 180 less electrons than the In-based one, the above finding signifies that much more AOs of Ga electrons in the Ga-based molecule have to be re-configured in response to a disturbance caused by the V atom than in the case of the In-based molecule. Moreover, the V atom in the pre-designed $Ga_{10}As_3V$ molecule accumulates more of re-distributed electron charge of Ga atoms than As atoms in this molecule (Figs. 5.18c to 5.18i), in contrast to that of the vacuum $Ga_{10}As_3V$ molecule (Figs. 5.19b to 5.19f). This is yet another sign that the pre-designed molecule is strained, so the V atom has to accommodate more of Ga electron charge to provide for a stable state (a ROHF triplet; Table 5.2) similar to that of the corresponding vacuum molecule. In the case of much roomier vacuum $Ga_{10}As_3V$ molecule there is less need to accumulate Ga electron charge near the V atom, or to push that charge outside the molecular "surface" to stabilize the molecule.

The above considerations lead to a conclusion that dimensions and properties of delocalized, spin-polarized holes associated with substitution V-atoms in the

Quantum Dots of Diluted Magnetic Semiconductor Compounds

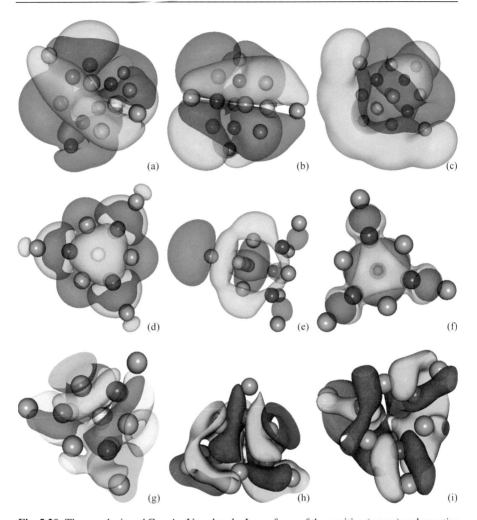

Fig. 5.20 The pre-designed $Ga_{10}As_3V$ molecule. Isosurfaces of the positive (green) and negative (orange) parts of the highest occupied and lowest unoccupied molecular orbits (HOMOs and LUMOs, respectively) corresponding to several isovalues; (a) and (b): HOMOs 117 and 118, isovalue 0.003, respectively; (c) HOMO 119, isovalue 0.001; (d) HOMO 119, isovalue 0.007; (e) and (f): HOMO 120, isovalue 0.007; (g) LUMO 121, isovalue 0.01; (h): LUMO 122, isovalue 0.007, and (i) LUMO 123, isovalue 0.01. Ga atoms are blue, As brown and V yellow. Atomic dimensions are reduced and isosurfaces made semi-transparent to show the structure.

zincblende GaAs lattice should be more sensitive to the lattice strain than those in the case of the zibcblende InAs lattice. This phenomenon can be used to develop a sensitive device to measure lattice strain by measuring the hole conductivity, or *vice versa*.

Several MOs from the HOMO-LUMO regions of the studied $Ga_{10}As_3V$ molecules, both of which are ROHF triplets (Table 5.2), are shown in Figs. 5.20 and 5.21. The

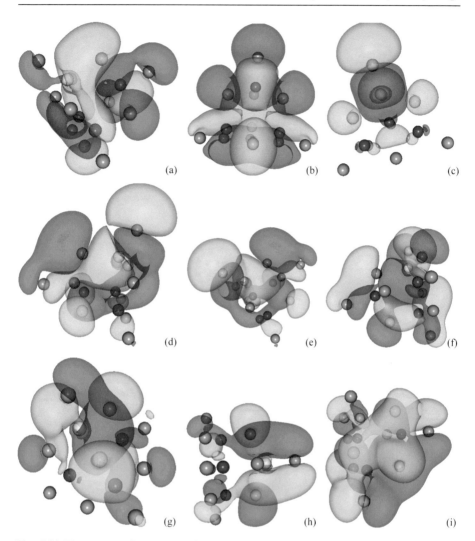

Fig. 5.21 The vacuum $Ga_{10}As_3V$ molecule. Isosurfaces of the positive (green) and negative (orange) parts of the highest occupied and lowest unoccupied molecular orbits (HOMOs and LUMOs, respectively) corresponding to several isovalues; (a) and (b): HOMO 118, isovalues 0.01 and 0.005, respectively; (c) HOMO 119, isovalue 0.01; (d) and (e): HOMO 120, isovalue 0.005; (f): LUMO 121, isovalue 0.015; (g) and (h): LUMO 122, isovalues 0.01 and 0.01, respectively, and (i) LUMO 123, isovalue 0.01. Ga atoms are blue, As brown and V yellow. Atomic dimensions are reduced and isosurfaces made semi-transparent to show the structure.

electronic level structure of the pre-designed molecule retains significant symmetry in the HOMO-LUMO region exhibiting doubly degenerate MOs, and in particular LUMO. Counting from the proper HOMO 121 (which is non-degenerate) toward the core MOs, the near HOMO-LUMO region of this molecule consists of MOs of the following symmetry: A (HOMO 120), A (MO 119), E (MOs 118 and 117), A (MO 116), E (MOs 115 and 114), E (MOs 113 and 112), A (MO 111), and so on. In the LUMO region, counting from the proper LUMO 121 and up, MOs of this molecule features E (LUMO 121 and MO 122), E (MOs 123 and 124), A (MO 125), E (MOs 126 and 127), E (MOs 128 and 129), A (MO 130), A (MO 131), T (MOs 132, 133 and 134), *etc.* types of symmetry. This is a result of constraining all centers of mass of atoms to their positions in the tetrahedral symmetry element of the zincblende GaAs lattice. The reduction in symmetry of the electronic charge distribution of this molecule is caused only by replacement of one of As atoms by the V atom.

In the pre-designed $Ga_{10}As_3V$ molecule $3d$ AOs of tetra-coordinated vanadium atom always bond it directly to 4 Ga atoms. The arsenic atoms bond 6 other Ga atoms, and the first 4 Ga atoms (bonded to V) also bond 3 Ga atoms from the "arsenic bonding triangle", thus completing an MO. [One of the 4 Ga atoms bonded to V is in a pyramid vertex, and does not contribute much to Ga-As π-type bonding.] Ga atoms bond both V and As ones via their $4p$ AOs. In contrast to the case of InAs-based molecules where some contributions to bonding come from $4d$ AOs of In atoms, $3d$ AOs of Ga atoms do not contribute to bonding in the GaAs-based molecules. Arsenic atoms bond through their $4p$ AOs only to Ga atoms, and do not bond to the vanadium atom directly. This arrangement is typical for all MOs in Fig. 5.20. The (4 + 3) Ga atom $4p$-bonding brings about a strong π-type ligand bonding MOs of this molecule in the HOMO region (see Ref. 122 for further discussion of "aromatic" π-type ligand bonding). This π-type ligand bonding is responsible for the pre-designed molecule being a stable ROHF triplet whose OTE (over 1.26 eV) is larger than that of the vacuum $Ga_{10}As_3V$ molecule (about 1.058 eV), and whose minimum of the total energy is almost as deep as that of the vacuum molecule (see Table 5.2).

ELS of the vacuum $Ga_{10}As_3V$ molecule does not exhibit any symmetry, being composed only of A-type orbits and showing only a very few spontaneously degenerate MOs of E-type in the higher LUMO region. Several MOs in the near HOMO-LUMO region are depicted in Fig. 5.21. The type of bonding in this molecule is very similar to that of the pre-designed molecule, with some small deviations. As a rule, $3d$ AOs of the V atom bond $4p$ AOs of the nearest 4 Ga atoms. Four or 3 of these Ga atoms bond another 4 or 3 Ga atoms (Ga-Ga $4p - 4p$ bonding), where the 3 Ga are bonded to all 3 As atoms (Ga-As $4p - 4p$ bonding). In some cases the Ga atom bonded to the V atom and positioned in the pyramid vertex closest to the V atom may be only weakly bonded to the Ga-As portion of the entire hybrid MO. Such MOs provide for the vacuum molecule being a robust ROHF triplet with a large OTE and a deep energy minimum of its ground state (Table 5.2). The Ga-V bond length in this molecule takes only 2 values: 2.505 Å and 2.998 Å. The bond length of the ligand Ga-Ga bonding also takes two values: 4.046 Å and 4.532 Å, while Ga-As bonding is more flexible, so the Ga-As bond may be of 2.487 Å, 2.519 Å and 2.835 Å in length. Similar to the case of the pre-designed molecule, in the vacuum molecule vanadium and arsenic atoms do not bond directly.

Compared to InAs-based molecules with one V atom that are pre-designed ROHF nonet and vacuum ROHF pentet, GaAs-based ones are more stable, being ROHF triplets. Such stability is due to a strong and large π-type Ga-ligand and Ga-As bonding contributions [122].

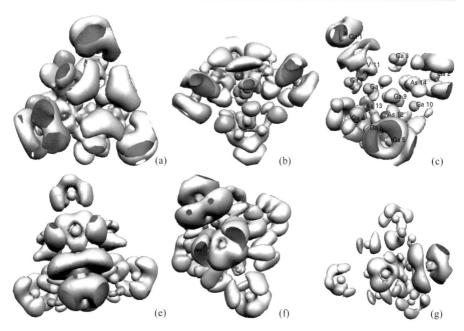

Fig. 5.22 Isosurfaces of the spin density distribution (SDD) of the pre-designed [(a) to (c)] and vacuum [(d) to (f)] $Ga_{10}As_3V$ molecules corresponding to SDD isovalues: (a) 0.0005, (b) 0.0007, (c) 0.001, and (d) 0.0007, (e) 0.0005, (f) 0.001, respectively. Indium atoms are yellow, As red and Mn blue. All atomic dimensions are reduced to show the SDD surface structure.

Being ROHF triplets, both $Ga_{10}As_3V$ molecules have "ferromagnetic" arrangement of their aligned uncompensated spins (Fig. 5.22) contributed by electrons in $3d$ AOs of Ga atoms. The SDD values of these molecules are about an order of magnitude smaller than those of the InAs-based molecules with one V atom discussed in section 5.4, that are higher ROHF spin multiplets. Correspondingly, the total magnetic moment of the GaAs-based molecules with one V atom is significantly smaller (2 μ_B) than the smallest of the magnetic moments of InAs-based molecules 4 μ_B) with one V atom. Thus, for DMS applications concerned with large magnetic moments $In_{10}As_3V$ molecules may be a better choice.

5.6 InAs - and GaAs - based molecules with two vanadium atoms

Pre-designed and vacuum $In_{10}As_2V_2$ and $Ga_{10}As_2V_2$ molecules have been virtually synthesized using the same procedures applied to virtual synthesis of $In_{10}As_2V$ and $Ga_{10}As_2V$ molecules (see sections 5.2 to 5.5 of this chapter and section 2.9 of Chapter 2), with the only difference that this time two As atoms were replaced by two V atoms in

Quantum Dots of Diluted Magnetic Semiconductor Compounds 223

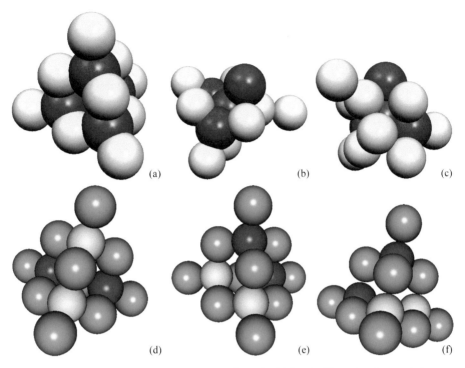

Fig. 5.23 The pre-designed (a) and vacuum [(b) and (c)] $In_{10}As_2V_2$ molecules, and the pre-designed (d) and vacuum [(e) and (f)] $Ga_{10}As_2V_2$ molecules. In (a) to (c) atomic dimensions of In atoms are reduced, while those of As and V somewhat enlarged; in (d) to (f) all atomic dimensions are reduced. In the $In_{10}As_2V_2$ molecules In atoms are yellow, As red and V purple. In the $Ga_{10}As_2V_2$ molecules Ga atoms are blue, As brown, and V yellow.

each of the pre-designed $In_{10}As_4$ and $Ga_{10}As_4$ tetrahedral pyramids of the zincblende InAs and GaAs lattices (see Chapter 3 for details). After such replacement, the $In_{10}As_2V_2$ and $Ga_{10}As_2V_2$ pyramidal structures were optimized while the positions of the centers of mass of their atoms were fixed. The result of this optimization is two pre-designed molecules $In_{10}As_2V_2$ and $Ga_{10}As_2V_2$. Then the initial structures were optimized again, this time when the constraints applied to the atomic positions were lifted. This optimization produced two vacuum molecules $In_{10}As_2V_2$ and $Ga_{10}As_2V_2$. The structure of all four molecules is pictured in Fig. 5.23. Because of the spatial constraints applied to the atomic positions, both of the pre-designed molecules (Figs. 5.23a and 5.23d) retain their pyramidal shape. When the constrains are lifted, only the vacuum $Ga_{10}As_2V_2$ molecule (Figs. 5.23e and 5.23f) appears pyramidal, while the vacuum $In_{10}As_2V_2$ molecule loses any resemblance to the tetrahedral pyramid of its parent pre-designed structure. The final structures, of course, are a consequence of relative sizes of the participating atoms and the numbers of electrons they possess, and quantum confinement effects (in the case of the pre-designed molecules). In particular, the V atom is smaller and has fewer electrons than As and Ga atoms, so after the replacement of 2 As atoms by 2 V atoms in the tetrahedral $Ga_{10}As_4$ pyramid (see Chapter 3 for detailed information on this structure) the pre-designed molecule $Ga_{10}As_2V_2$ is somewhat loose. When spatial constraints on atomic positions are lifted, the atoms move from their original positions, but only slightly. Thus, visually, the vacuum $Ga_{10}As_2V_2$ molecule appears pyramidal. In the case of $In_{10}As_2V_2$, the V atom is much smaller and has significantly fewer electrons that the In atom. After the replacement of two As atoms by two V ones the original $In_{10}As_4$ structure becomes too loose, as the absence of 20 electrons that occur after the replacement creates a large electron charge distribution imbalance. Thus, when the constraints applied to atomic positions are lifted, electron charge of atoms in the structure adjust dramatically in response to that imbalance, and integrated Coulomb effects push the atoms to new positions. As a result, the vacuum $In_{10}As_2V_2$ molecule assumes a shape entirely different from that of its parent tetrahedral pyramid, and sets in a structure realizing a stable ROHF singlet. In contrast, the pre-designed $In_{10}As_2V_2$ molecule has spin multiplicity 11, and thus is an unstable molecule (see Table 5.2) in the framework of ROHF approximation used here. The CDDs and MEPs of the studied molecules with two V atoms reflect characteristic features of their composition and shape (Figs. 5.24 and 5.25).

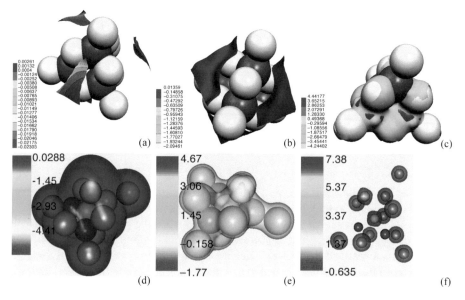

Fig. 5.24 The molecular electrostatic potential (MEP) of the pre-designed [(a) to (c)] and vacuum [(d) to (f)] $In_{10}As_2V_2$ molecules for several CDD isovalues: (a) 0.001, (b) 0.01, (c) 0.08, (d) 0.0007, (e) 0.01, and (f) 0.1. The CDD maximum values are 3.44417 and 3.01378 (in arbitrary units) for the pre-designed and vacuum molecules, respectively. The color coding scheme for MEP surfaces is shown in each figure. In the pre-designed molecule [(a) to (c)] In atoms are yellow, As red and V purple. In the vacuum molecule [(d) to (f)] In atoms are blue, As brown and V yellow. In (a) to (c) dimensions of In atoms are somewhat smaller than those corresponding to the In covalent radius, and the dimensions of V and As atoms enlarged. In (d) to (f) all atomic dimensions are reduced, and the MEP surfaces are made semi-transparent to reveal the structure.

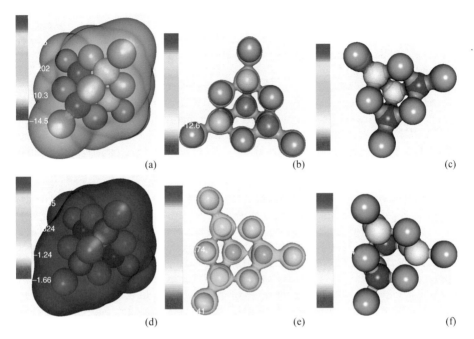

Fig. 5.25 The molecular electrostatic potential (MEP) of the pre-designed [(a) to (c)] and vacuum [(d) to (f)] $Ga_{10}As_2V_2$ molecules for several CDD isovalues (a) 0.007, (b) 0.01, (c) 0.08, (d) 0.005, (e) and (f): 0.1. The CDD maximum values are 9.46923 and 10.37640 (in arbitrary units) for the pre-designed and vacuum molecules, respectively. The color coding scheme for MEP surfaces is shown in each figure. Ga atoms are blue, As brown and V yellow. All atomic dimensions are reduced. In all cases but (c) MEP surfaces are semi-transparent to reveal the structure.

Comparison of MEP surfaces of the pre-designed and vacuum $In_{10}As_2V_2$ molecules in Fig. 5.24 obtained for similar CDD isovalues reveals strikingly different electrostatics of these molecules. Thus, the MEP surface values of the pre-designed molecule corresponding to the CDD isovalue 0.001 ran from highly negative to positive ones (Fig. 5.24a), while for the similar CDD isovalue 0.0007 the MEP values of the vacuum molecule are uniformly close to 0 (Fig. 5.24d). Thus, in the case of the pre-designed molecule, at large distances from the molecular "surface" there still exists large MEP fluctuations that signifies instability of this molecule. In contrast, the vacuum molecule is stable exhibiting text-book MEP values for large separations from the molecular "surface". Similar conclusions follow from comparison of MEP values in Figs. 5.24b and 5.24e for the same CDD isovalue 0.01, and those in Figs. 5.24c and 5.24f for similar CDD isovalues 0.08 and 0.1, respectively. Comparison of the ground state energy values of these molecules reveals that their ground state energy values differ by about 1 H (Table 5.2). This difference lies at the limit of the total energy evaluation accuracy of the RHF-ROHF approximation method. The dipole moment of the pre-designed $In_{10}As_2V_2$ molecule is somewhat larger than that of the vacuum one (Table 5.2). These facts further confirms that the pre-designed molecule, that realizes a local minimum of the total energy of the $In_{10}As_2V_2$ atomic cluster corresponding to the tetrahedral symmetry spatial constraints applied to the cluster's atoms, may be unstable and unrealizable experimentally.

MEP surfaces of $Ga_{10}As_2V_2$ molecules for several CDD isovalues are shown in Fig. 5.25. In contrast to those of the two $In_{10}As_2V_2$ molecules, MEP surfaces of the $Ga_{10}As_2V_2$ ones are very similar reflecting the fact that both $Ga_{10}As_2V_2$ molecules are close ROHF spin multiplets (see Table 5.2). On the other hand, both the pre-designed ROHF $Ga_{10}As_2V_2$ septet and vacuum $Ga_{10}As_2V_2$ nonet are not typical for the ground state of small stable molecules. The pre-designed $Ga_{10}As_2V_2$ septet is more stable than the vacuum nonet, indicating that the calculated minimum of the total energy in the case of the vacuum nonet may not be a global minimum of the total energy. Indeed, comparing for example, MEP surfaces of these molecules depicted in Figs. 5.25c and 5.25f, one can see that the pre-designed molecule has larger areas of delocalized electron charge shared between Ga, V and As atoms than the pre-designed one. Most likely, the vacuum nonet is either realization of a yet another local minimum of energy of $Ga_{10}As_2V_2$ atomic cluster, or signifies that the molecule is not stable. This consideration is further supported by comparison of the ground state energies of the $Ga_{10}As_2V_2$ molecules (Table 5.2) that are lying within the error brackets of the RHF-ROHF approximation. Indeed, the difference in the ground state energy values of these molecules is only 0.055 H, while the averaged accuracy of the total energy evaluation provided by the used calculation method is about 1 Hartree. The dipole moment of the vacuum molecule is considerably smaller than that of the pre-designed one (Table 5.2). This further supports the conclusion that the vacuum nonet is a local minimum of the total energy of the unconstrained $Ga_{10}As_2V_2$ cluster that corresponds to a less polarized electron charge distribution than that of the local minimum of the constrained pre-designed molecule.

Molecular orbits of all studied molecules doped with two V atoms are shown in Figs. 5.26 and 5.27. In support of the conclusion concerning instability of the pre-designed $In_{10}As_2V_2$ molecule discussed above, some MOs of this molecule in the near HOMO-LUMO region (Figs. 5.26a and 5.26b) are only partially bonding and feature antibonding contributions from $4d$ AOs of In and $3d$ AOs of V atoms, and contributions due to (i) sp – type As-In bonding and (ii) π-type In ligand bonding. The LUMO (Fig. 5.26c) of this molecule is a bonding MO realized via hybrid pd-type In-V, π-type In-As, and π-type In-In ligand bonding. MOs of the vacuum $In_{10}As_2V_2$ molecule in its HOMO-LUMO region (Figs. 5.26d to 5.26f) are bonding hybrid MOs with significant portions contributed by pd In-V bonding, and As-mediated π-type In ligand bonding.

In the case of $Ga_{10}As_2V_2$ molecules (Fig. 5.27), all MOs in the HOMO region and the LUMOs are bonding hybrid MOs contributed to by (i) pd-type Ga-V bonding due to $4p$ AOs of Ga and $3d$ AOs of V atoms, (ii) $4p$-type As-Ga bonding, and [3] π-type Ga ligand bonding.

Magnetic properties of the studied InAs- and GaAs-based molecules with 2 vanadium atoms are illustrated in Fig. 5.28. An important finding is that replacement of another As atom by the second vanadium one in the studied $In_{10}As_3V$ and $Ga_{10}As_3V$ molecules does not ensure an increase in the magnetic moment of the obtained molecules, because it generally destabilizes the molecules. At this point in studies, it

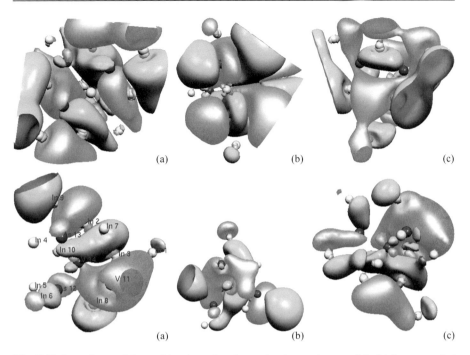

Fig. 5.26 Isosurfaces of the positive (green) and negative (orange) parts of the highest occupied and lowest unoccupied molecular orbits (HOMOs and LUMOs, respectively) corresponding to several isovalues. The pre-designed $In_{10}As_2V_2$ molecule: (a) HOMO 127, isovalue 0.01; (b) HOMO 128, isovalue 0.001; (c) LUMO 129, isovalue 0.015. The vacuum $In_{10}As_2V_2$ molecule: (d) HOMO 122, isovalue 0.015; (e): HOMO 123, isovalue 0.01; (f): LUMO 124, isovalue 0.015. Indium atoms are yellow, As red and V purple. Atomic dimensions are reduced to reveal the isosurface structure.

seems that the pre-designed $Ga_{10}As_2V_2$ septet (Figs. 5.28c and 5.28d) may be experimentally realizable in quantum confinement. Indeed, this molecule has a deep ground state energy minimum (Table 5.2) that is about 720 H deeper than that of the vacuum $In_{10}As_3Mn$ molecule, which is the only other relatively stable molecule with a large uncompensated magnetic moment 6 μ_B. At the same time, the vacuum $In_{10}As_3Mn$ molecule has been computationally synthesized in the absence of any spatial constraints applied to its atoms, and such conditions are not exactly realizable in experiment. Moreover, the pre-designed $In_{10}As_3Mn$ molecule is "antiferromagnetic" with zero total uncompensated magnetic moment. Thus, it seems likely that application of spatial constraints to positions of atoms in $In_{10}As_3Mn$ cluster atoms in the process of experimental synthesis of $In_{10}As_3Mn$ molecules (let alone films) may lead to a molecule with a smaller uncompensated magnetic moment than that of the vacuum $In_{10}As_3Mn$ one analyzed above. Therefore, among InAs- and GaAs-based molecules with two V substitution atoms only the pre-designed $Ga_{10}As_2V_2$ may be of interest for DMS applications. Indeed, as already noted, the pre-designed $In_{10}As_2V_2$ molecule is a ROHF singlet, while the vacuum $In_{10}As_2V_2$

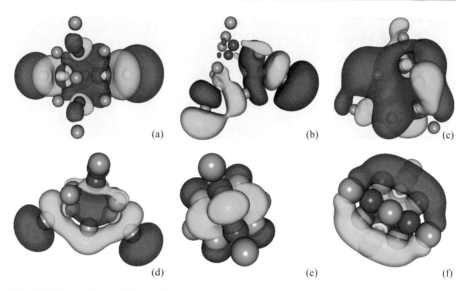

Fig. 5.27 Isosurfaces of the positive (green) and negative (orange) parts of the highest occupied and lowest unoccupied molecular orbits (HOMOs and LUMOs, respectively) corresponding to several isovalues. The pre-designed $Ga_{10}As_2V_2$ molecule: (a) HOMO 125, isovalue 0.008; (b) HOMO 126, isovalue 0.01; (c) LUMO 127, isovalue 0.008. The vacuum $Ga_{10}As_2V_2$ molecule: (d) HOMO 126, isovalue 0.01; (e): HOMO 127, isovalue 0.01; (f): LUMO 128, isovalue 0.01. Ga atoms are blue, As brown and V yellow. In (a) to (d) atomic dimensions are significantly reduced to reveal the isosurface structure. In (e) and (f) the atomic dimensions are reduced only slightly. Isosurfaces are semi-transparent to reveal the structure.

(Figs. 5.28a and 5.28b) and $Ga_{10}As_2V_2$ (Figs. 5.28e and 5.28f) molecules may not be realizable in laboratory.

5.7 Conclusions

In this chapter the first-principle, quantum many body theoretical methods, RHF and ROHF, were discussed in conjunction with virtual synthesis of 10 molecules composed of In, Ga and As atoms in arrangements that either reflect tetrahedral symmetries of the zincblende InAs and GaAs bulk lattices, or are derivable from such arrangements by the total energy minimization.

This discussion summarizes initial steps toward realization of a program of virtual synthesis studies of semiconductor QDs and QWs introduced in Refs. 119–121, 123, 124 and related publications. Much more research has to be performed before any final quantitative conclusions concerning the above systems are reached. It is well known (see Chapter 2) that RHF and ROHF approximations overestimate OTE values and provide only major contributions to molecular orbits of atomic clusters so optimized. However, it is also true that these approximations are necessary to obtain MOs and other data to enable further theoretical studies and to guide experimental

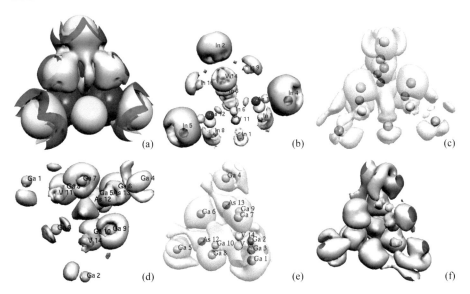

Fig. 5.28 Isosurfaces of the spin density distributions (SDDs) of the pre-designed $In_{10}As_2V_2$ [(a) and (b)] and $Ga_{10}As_2V_2$ [(c) and (d)] molecules, and the vacuum molecule $Ga_{10}As_2V_2$ [(e) and (f)] corresponding to the respective SDD isovalues (a) 0.001, (b) 0.004, (c) 0.002, (d) 0.003, (e) 0.002 and (f) 0.002. In (a) and (b) indium atoms are yellow, As red and V purple. In (c) to (f) Ga atoms are blue, As brown and V yellow. All atomic dimensions are reduced to show the SDD surface structure.

developments. In particular, progressively more accurate CI, MCSCF, MP-2 and other first principle many-body theoretical methods use RHF or ROHF MOs as input MOs, and the RHF/ROHF ground state energy, CDD and SDD data provide important information and ideas for prospective applications.

In recent years, applications of DMS systems raised several important problems concerning the nature of magnetism and band structure of DMS systems, changes from magnetic to non-magnetic phase, and the location of the delocalized, spin-polarized holes associated with the impurity atoms (see section 5.1 for references). Analysis of RHF/ROHF data provided in this chapter leads to some conclusions and insights that may help solve these problems.

1. In the case of nanoscale InAs- and GaAs-based DMS systems (such as thin films, small QD and QWs, etc.) with at least one linear dimension in the range of a few nanometers the electronic energy level structure is defined by quantum confinement and Coulomb interaction effects. Such ELSs can hardly be described in terms of impurity-driven modifications of the band level systems of the parent InAs or GaAs bulk lattices. Moreover, ELS of such systems may only remotely resemble a band structure of the parent solids. Rather, with a decrease in dimensionality and/or the number of atoms in a system, its ELS acquires similarity with that of molecules, and is contributed to by all participating atoms. Perturbation-type approaches do not work for such strongly correlated electron systems.

2. Magnetism in the studied DMS nanosystems results from Coulomb effect-driven electron charge and spin re-distribution in a broad vicinity of a substitution impurity atom possessing $3d$ electrons that may include up to 13 lattice atoms surrounding the impurity atom, and is not centered at the impurity atom. Such re-distribution is a response of the nanosystem to a disturbance of the electron charge and spin distributions of system's atoms due to $3d$ electrons of the impurity atom. In the majority of the studied cases, the non-zero total spin magnetic moment arises from uncompensated electron spins of $4d$ In or $3d$ Ga electrons in the process of such electron charge and spin re-distribution, and the $3d$ electrons of the impurity atoms do not contribute directly to this total spin magnetic moment.

3. In the studied cases, the process of the electron charge re-distribution leads to delocalization of a portion of electron charge of up to 13 atoms neighboring an impurity atom. The delocalized electron charge is accommodated in highly hybridized molecular orbits that possess three major types of contributions. In the majority of the studied cases, one of the two major contributions to such MOs come from pd-type hybrid bonding between 3 to 4 ligand In or Ga atoms and Mn or V atoms, where In or Ga atoms contribute via their hybridized $4p$ AOs, and Mn or V via their $3d$ AOs. The other major contribution comes from hybrid π-bonding of Ga or In with As atoms that mediate ligand bonding between up to 4 In or Ga atoms per one As atom.

4. While the above results confirm and enrich the major idea of the p-d Zener model, some other findings indicate that the p-d Zener exchange may not be sufficient for the development of the uncompensated spin magnetic moment in DMS structures. In particular, the pre-designed $In_{10}As_3Mn$ and vacuum $In_{10}As_2V_2$ molecules are RHF singlets with zero total uncompensated spin magnetic moments, despite of their bonding MOs being a result of hybridization of $4p$ AOs of In and $3d$ AOs of Mn or V atoms. The case of vacuum $In_{10}As_3Mn$ molecule confirms, that pd As-Mn bonding may provide other important mechanism leading to magnetization in (In,As)Mn systems, in support of suggestion of Ref. 79.

5. The total uncompensated magnetic moment in the majority (8 cases) of the studied molecules is contributed to by $4d$ electrons of In or $3d$ electrons of Ga atoms. Because RHF/ROHF MOs may not be quite accurate, especially in the case where many-electron atoms, such as In, Ga and As, are involved, further detailed studies of the considered systems by CI, MCSCF and MP2 methods are necessary to identify quantitatively mechanisms of "generation" of the uncompensated magnetic moment in the studied molecules. At this stage, it is clear that uncompensated total spin magnetic moment appears in response to charge disturbance introduced by $3d$ electrons of Mn or V atoms. However, only in two of the studied cases the total uncompensated magnetic moment is directly contributed to by $3d$ electrons of Mn or V atoms. Even in these two cases, $3d$ electrons of Mn or V atoms alone contribute only a fraction of the total uncompensated spin magnetic moment. The spin multiplicity of the virtually synthesized molecules range from 1 to 11, which signifies that many more electron spins residing on In or Ga atoms are involved. Indeed, calculations of each of the

topmost occupied RHF/ROHF MOs for the studied molecules involves about 120 electrons of their constitutive atoms.

6. The electron charge delocalization of atoms surrounding an impurity atom creates tetrahedral-symmetry regions of electron charge deficit embracing the "surfaces" of the studied molecule from outside and inside of the molecule structures. Although the total charge of the molecules is zero, such shells of electron charge deficit may be considered as positively charged "holes" that may move when external electromagnetic fields are applied. Such shells separate electron charge accumulated deeper inside of the molecular structures from that pushed outside the molecular "surfaces". These electron charge deficit shells are physical realization of delocalized, positively charged holes of the semi-phenomenological band-theoretical models. Indium or Ga atoms are embedded in these holes and contribute the uncompensated spin magnetic moment, thus providing for spin polarization of the holes.

7. Such shells of electron charge deficit surround from about 8 to 14 structure atoms in the studied cases. Holes carried by In-based molecules are larger in size than those of the Ga-based ones and approach 1 nm in linear dimensions. Thus, in thin DMS films sandwiched between other materials, and in other DMS systems with one or more linear dimensions in the range of a few nanometers confined by layers of other materials, similar holes may include atoms of the confining material layers. Moreover, if a suitably directed electromagnetic field gradient is applied, such delocalized, spin-polarized holes may move and leave the confined low dimensional DMS system of their origin.

8. With a change in thermodynamic conditions, such as temperature or pressure, structural changes in low-dimensional DMS systems may provoke thermodynamics-based electron charge re-distribution that may affect shape, linear dimensions, delocalization and spin polarization of the impurity-mediated holes, and thus affect the total impurity-mediated magnetic moment of the DMS systems. Among the studied molecules only two $Ga_{10}As_3V$ ones did not change their spin multiplicity in response to a change in conditions of their synthesis. The rest of the studied molecules do changed values of their total uncompensated spin magnetic moments in response to a change in synthesis conditions. Two of such molecules (the pre-designed $In_{10}As_3Mn$ and vacuum $In_{10}As_2V_2$ ones) are "antiferromagnetic" RHF singlets with the total magnetic moment equal to zero, while their vacuum $In_{10}As_3Mn$ and pre-designed $In_{10}As_2V_2$ counterparts virtually synthesized in model quantum confinement are ROHF spin multiplets. These results indicate that changes in synthesis conditions can be a suitable physical mechanism behind transitions from a ferromagnetic state to a non-magnetic state in some DMS thin film systems observed in recent experiments (see section 5.1 for references and a further discussion of those experimental observations).

9. All calculations of this chapter are specific for systems at zero absolute temperature. However, the OTE of all virtually synthesized, stable molecules (including even septets, Table 5.2) are several times larger than any temperature-derived contributions to the OTEs at room temperature. Thus, the obtained results are likely to be quantitatively reasonable for the studied systems even at room temperature.

References

[1] Munekata, H., Ohno, H., von Molnár, S., Segmüller, A., Chang, L. L., and Esaki, L. (1989). Diluted magnetic III–V semiconductors. *Phys. Rev. Lett.* **63**, 1849–1853.
[2] Ohno, H., Munekata, H., Penney, T., von Molnár, S., and Chang, L. L. (1992). Magnetotransport properties of p-type (In,Mn)As diluted magnetic III–V semiconductors. *Phys. Rev. Lett.* **68**, 2664–2668.
[3] Ohno, H., Shen, A., Matsukura, F., Oiwa, A., Endo, A., Katsumoto, and S., Iye, Y. (1996). (Ga,Mn)As: A new diluted magnetic semiconductor based on GaAs, *Appl. Phys. Lett.* **69**, 63–67.
[4] Ohno, H. (1998). Making nonmagnetic semiconductors ferromagnetic. *Science* **281**, 951–956.
[5] Matsukura, F., Ohno, H., and Dietl, T. (2002). III-V Ferromagnetic semiconductors. *In* "Handbook of Magnetic Materials" (K. H. J. Buschow, Ed.), Vol. 14, pp. 1–88. Elsevier, Amsterdam.
[6] Jungwirth, T., Sinova, J., Masek, J., Kucera, J., and MacDonald, A. H. (2006). Theory of ferromagnetic (III,Mn)V semiconductors. *Rev. Mod. Phys.* **78**, 809–864.
[7] Dietl, T., Ohno, H., and Matsukura, F. (2007). Ferromagnetic semiconductor heterostructures for spintronics. *IEEE Trans. Electronic Devices* **54**, 945–954.
[8] Dietl, T., Ohno, H., Matsukura, F., Gibert, J., and Ferrand, D. (2000). Zener model description of ferromagnetism in zinc-blende magnetic semiconductors. *Science* **287**, 1019–1928.
[9] Dietl, T., Ohno, H., Matsukura, F. (2001). Hole-mediated ferromagnetism in tetrahedrally coordinated semiconductors. *Phys. Rev. B* **63**, 195205.
[10] Beyer, J., Buyanova, I. A., Suraprapapich, S., Tu, C. W., and Chen, W. M. (2009). Spin injection in lateral InAs quantum dot structures by optical orientation spectroscopy. *Nanotechnology* **20**, 375401.
[11] Cho, Y. J., Liu, X., and Furdyna, J. K. (2008). Collapse of ferromagnetism in (Ga,Mn)As at high hole concentration. *Semicond. Sci. Technol.* **23**, 125010.
[12] Chiba, D., Nishitani, Y., Matsukura, F., and Ohno, H. (2007). Properties of $Ga_{1-x}Mn_xAs$ with high Mn composition (x>0.1). *Appl. Phys. Lett.* **90**, 122503.
[13] Ji, C.-J., Zhang, C.-Q., Zhao, G., Wang, W.-J., Sun, G., Yuan, H.-M., and Han, Q.-F. (2011). Preparation and properties of dilute magnetic semiconductors GaMnAs by low temperature molecular epitaxy. *Chin. Phys. Lett.* **28**, 097101.
[14] Chapler, B. S., Mack, S., Ju, L., Elson, T. W., Boudouris, B. W., Namdas, E., Yuen, J. D., Heeger, A. J., Samarth, N., Di Ventra, M., Basov, D. N. (2012). Infrared conductivity of hole accumulation and depletion layers in (Ga,Mn)As- and (Ga,Be)As- based electric field-effect devices. Arxiv: 1207.0895, v. 2 [cond-mat. mtrl-sci].
[15] Aronzon, B. L., Kovalcheck, M. V., Pashaev, E. M., Chuev, M. A., Kvardakov, V. V., Subbotin, I. A., Rylkov, V. V., Pankov, M. A., Likhachev, I. A., Zvonkov, B. N., Danilov, Yu. A., Vihrova, O. V., Lashkul, A. V., and Laiho, R. (2008). Structural and transport properties of GaAs/δ-Mn/GaAs/$In_xGa_{1-x}As$/GaAs quantum wells. *J. Phys.: Condens. Matter* **20**, 145207.
[16] Borschel, C., Messing, M. E., Borgström, M. T., Paschoal Jr., W., Wallentine, J., Kumar, S., Mergenthaler, K., Deppert, K., Canali, C. M., Petterson, H., Samuelson, L., and Ronning, C. (2011). A new route toward semiconductor nanospintronics: highly Mn-doped GaAs nanowires realized by ion-implantation under dynamic annealing conditions. *Nanoletters* **11**, 3935–3940.
[17] Chiba, D., Takamura, K., Matsukura, F., and Ohno, H. (2003). Effect of low temperature annealing on (Ga,Mn)As trilayer structures. *Appl. Phys. Lett.* **82**, 3020–3022.
[18] Jungwirth, T., Wang, K. Y., Masék, J., Edmonds, K. W., König, J., Sinova, J., Polini, M., Goncharuk, N. A., MacDonald, A. H., Sawicki, M., Rushforth, A. W., Campion, R. P.,

Zhao, L. X., Foxton, C. T., and Gallagher, B. L. (2005). Prospects of high temperature ferromagnetism in (Ga,Mn)As semiconductors. *Phys. Rev. Lett.* **72**, 165204.
[19] Nazmul, A. M., Sudahara, S., and Tanaka, M. (2003). Ferromagnetism and high Curie temperature in semiconductor heterostructures with Mn δ-doped GaAs and p-type selective doping. *Phys. Rev. B* **67**, 241308R.
[20] Yu, K. M., Walukiewicz, W., Wojtowicz, T., Kuryliszyn, I., Liu, X., Sasaki, Y., and Furdyna, J. K. (2002). Robustness of the fractional Hall effect. *Phys. Rev. B* **72**, 201303R.
[21] Wojtowicz, T., Furdyna, J. K., Liu, X., Yu, K. M., Walukiewicz, W. (2004). Electronic effects determining the formation of ferromagnetic $III_{1-x}Mn_xV$ alloys during epitaxial growth. *Physica E* **25**, 171–180.
[22] Wang, K. Y., Edmonds, K. W., Campion, R. P., Gallagher, B. L., Farley, N. R. S., Foxton, C. T., Sawicki, M., Boguslawski, P., and Dietl, T. (2004). Influence of the Mn interstitial on the magnetic and transport properties of (Ga,Mn)As. *J. Appl. Phys.* **95**, 6512.
[23] Edmonds, K. W., Farley, N. R. S., Johal, T. K., van der Laan, G., Campion, R. P., Gallagher, B. L., and Foxton, C. T. (2005). Ferromagnetic moment and antiferromagnetic coupling in (Ga,Mn)As thin films. *Phys. Rev. B* **71**, 064418.
[24] Dmitrieva, A.I., Morgunova, R. B., and Zaitsev, S. V. (2011). Electron spin resonance in InGaAs/GaAs heterostructures with a manganese δ layer. *J. Exper. Theor. Physics* **112**, 317–326.
[25] Chen, L., Yang, X., Yang, F., Zhao, J., Misuraca, J., Hong, P., von Molnar, S. (2011). Enhancing the Curie temperature of ferromagnetic semiconductor (Ga,Mn)As via nanostructure engineering. *Nano Lett.* **11**, 2584–2589.
[26] Galicka, M., Buczko, R., and Kacman, P. (2011). Structure-dependent ferromagnetism in Mn-doped III-V nanowires. *Nano Lett.* **11**, 3319–3323.
[27] Kong, K.-J., Jung, C.-S., Jung, G.-B., Cho, Y.-J., Kim, H.-S., Park, J., Yu, N.-E., and Kang, C. (2010). Room-temperature ferromagnetism and terahertz emission of Mn-doped InGaAs and GaAsSb nanowires. *Nanotechnology* **21**, 435703.
[28] Bouravleuv, A. D., Girlin, G. E., Romanov, V. V., Bagraev, N. T., Brilinskaya, E. S., Lebedeva, N. A., Novikov, S. V., Lipsanen, H., and Dubrovskii, V. G. (2012). Formation of (Ga,Mn)As nanowires and study of their magnetic properties. *Semiconductors* **46**, 178–183.
[29] Meilikhov, E. Z., and Farzetdinova, R. M. (2010). Ferromagnetism in heterostructures based on a dilute magnetic semiconductor. *J. Exper. Theor. Phys.* **110**, 794-804.
[30] Dietl, T. (2010). A ten-year perspective on diluted magnetic semiconductors and oxides. *Nature Materials* **9**, 965–974.
[31] Bonnani, A., and Dietl, T. (2010). A story of high-temperature ferromagnetism in semiconductors. *Chem. Soc. Rev.* **39**, 528–539.
[32] Lee, B., Jungwirth, T., and MacDonald, A. H. (2000). *Phys. Rev. B* **61**, 15606.
[33] Ivchenko, E. L., and Pikus, G. E. (1997). "Superlattices and Other Heterostructures," 2nd ed. (M. Cardona, Ed.), Springer Series in Solid-State Sciences, Vol. 110. Springer, Berlin.
[34] Zener, C. (1951). Interaction between the d-shells in the transition metals. *Phys. Rev.* **81**, 440–444; (1951). Interaction between the d-shells in the transition metals. II. Ferromagnetic compounds of manganese with perovskite structure. *Phys. Rev.* **82**, 403–405.
[35] Dietl, T. (2007). Origin of ferromagnetic response in diluted magnetic semiconductors and oxides. *J. Phys.: Condens. Matter* **19**, 165204.
[36] Benoit `a la Guillaume, C., Scalbert, D., and Dietl, T. (1992). Wigner-Seitz approach to spin splitting. *Phys. Rev. B* **46**, R9853-R9856.
[37] Dietl, T., Matsukura, F., and Ohno, H. (2002). Ferromagnetism of magnetic semiconductors: Zhang-Rice limit. *Phys. Rev. B* **66**, 033203.

[38] Rokhinson, L., Lyanda-Geller, Y., Ge, Z., Shen, S., Liu, X., Dobrowolska, M., and Furdyna, J. (2007). Weak localization in $Ga_{1-x}Mn_xAs$: evidence of impurity band transport. *Phys. Rev. B* **76**, 161201.
[39] Sheu, B. L., Myers, R. C., Tang, J.-M., Samarth, N., Awschalom, D. D., Schiffer, P., and Flatte, M. E. (2007). Onset of ferromagnetism in low-doped $Ga_{1-x}Mn_xAs$. *Phys. Rev. Lett.* **99**, 227205.
[40] Mack, S., Myers, R. C., Heron, J. T., Gossard, A. C., and Awschalom, D. D. (2008). Stoichiometric growth of high Curie temperature heavily alloyed GaMnAs. *Appl. Phys. Lett.* **92**, 192502.
[41] Dobrowolska, M., Tivakornsasithorn, K., Liu, X., Furdyna, J. K., Berciu, M., Yu, K. M., and Walukiewicz, W. (2012). Controlling the Curie temperature in (Ga,Mn)As through location of the Fermi level within the impurity band. *Nature Materials* **11**, 444–449.
[42] Timm, C., and MacDonald, A. H. (2005). Anisotropic exchange interactions in III-V diluted magnetic semiconductors. *Phys. Rev. B* **71**, 155206.
[43] Litvinov, V. I., and Dugaev, V. K. (2001). Ferromagnetism in magnetically doped III-V semiconductors. *Phys. Rev. Lett.* **86**, 5593–5596.
[44] Sato, K., Kudrnovsky, J., Dederichs, P. H., Ericsson, O., Turek, I., Sanyal, B., Bouserar, J., Katayama - Yoshida, H., Dinh, V. A., Fukushima, T., Kizaki, H., and Zeller, K. (2010). First-principles theory of diluted magnetic semiconductors. *Rev. Mod. Phys.* **82**, 1633–1690.
[45] Sato, K., Dederichs, P. H., and Katayama-Yoshida, H. (2003). Curie temperature of III-V magnetic semiconductors calculated from first principles. *Europhys. Lett.* **61**, 403–408.
[46] Berciu, M., and Bhatt, R. N. (2001). Effects of disorder on ferromagnetism in diluted magnetic semicondutors. *Phys. Rev. Lett.* **87**, 107203.
[47] Mahadevan, P., and Zunger, A. (2004). Trends in ferromagnetism, hole localization and acceptor level depth for Mn substitution in GaN, GaP, GaAs and GaSb. *Appl. Phys. Lett.* **85**, 2860.
[48] Chattopadhyay, A., Das Sarma, S., and Mills, A. (2001). Transition temperature in ferromagnetic semiconductors: a dynamic mean field study. *Phys. Rev. Lett.* **87**, 227202.
[49] Priour, D. J., and Das Sarma, S. (2006). Phase diagram of the disordered RKKY model in diluted magnetic semiconductors. *Phys. Rev. Lett.* **97**, 127201.
[50] Kacman, C. (2001). Spin interactions in diluted magnetic semiconductors and magnetic semiconductor structures. *Semiconductor Sci. Technol.* **16**, R25–R38.
[51] Dupree, T. H. (1961). Electron scattering in a crystal lattice. *Ann. Phys.* (New York) **15**, 63–78.
[52] Beeby, J. L. (1967). The density of electrons in a perfect and imperfect lattice. *Proc. Roy. Soc. London A* **302**, 113–136.
[53] Morgan, G. J. (1966). Bloch waves and scattering impurities. *Proc. Phys. Soc.* **89**, 365–371.
[54] Hohenberg, P., and Kohn, W. (1964). Calculation of inelastic alpha-particle scattering by Ni^{58}. *Phys. Rev.* **136**, B864–B870.
[55] Korringa, J., (1947). On the calculation of the energy of a Bloch wave in a metal. *Physica* **13**, 392–400.
[56] Kohn, W., Rostoker, N. (1954). Solution of the Schrödinger equation in periodic lattices with an application to metallic lithium. *Phys. Rev.* **94**, 1111–1120.
[57] Nonas, B., Wildberger, K., Zeller, R., Dederichs, P. H., and Georffy, B. L. (1998). Magnetic properties of 4d impurities on (001) surfaces of nickel and iron. *Phys. Rev. B* **57**, 84–87.
[58] Asato, M., Settels, A., Hoshino, T., Asada, T., Blügel, S., Zeller, R., and Dederichs, P. H. (1999). Full-potential KKY calculations for metals and semiconductors. *Phys. Rev. B* **60**, 5202–5210.
[59] Popescu, V., Ebert, H., Nonas, B., and Dederichs, P. H. (2001). Spin and orbital magnetic moments of 3d and 4d impurities on and in the (001) surface of bcc Fe. *Phys. Rev. B* **64**, 184407.
[60] Galanakis, I., Bihlmayer, G., Bellini, V., Papanikolaou, N., Zeller, R., Blügel, S., and Dederichs P. H. (2002). Brocken-bond rule for the surface energies of noble metals. *Europhys. Lett.* **58**, 751–757.

[61] Freyss, M., Papanikolaou, N., Bellini, V., Zeller, R., and Dederichs, P. H. (2002). Electronic structure of Fe/semiconductor/Fe(001) tunnel junction. *Phys. Rev. B* **66**, 014445.
[62] Mavropoulos, P., Wunnicke, O., and Dederichs, P. H. (2002). Ballistic spin injection in Fe/semiconductor/Fe junctions. *Phys. Rev. B* **66**, 024416.
[63] Papanikolaou, N., Opitz, J., Zahn, P., and Mertig, I. (2002). Spin filter in metallic nanowires. *Phys. Rev. B* **66**, 165441.
[64] Ebert H., and Mankovsky, S., (2003). Field-induced magnetic circular X-ray dichroism in paramagnetic solids: a new magneto-optical effect. *Phys. Rev. Lett.* **90**, 077404.
[65] Vernes, A., Ebert, H., and Banhart, J. (2003). Electronic conductivity in Ni_xCr_{1-x} and Ni_xCu_{1-x} fcc alloy systems. *Phys. Rev. B* **68**, 134404.
[66] Zeller, R., Deutz, J., and Dederichs, P. H. (1982). Applications of complex energy integration to self-consistent electronic structure calculations. *Solid State Commun.* **44**, 993–997.
[67] Wildberger, K., Lang, P., Zeller, R., and Dederichs, P. H. (1995). Fermi-Dirac distribution in *ab initio* Green's-function calculations. *Phys. Rev. B* **52**, 11502–11508.
[68] Zeller, R. (1987). Multiple-scattering solution of Schrödinger equation for potentials of general shape. *J. Phys. C* **20**, 2347–2351.
[69] Anderson, O. K., and Woolley, R. G. (1973). Muffin-tin orbitals and molecular calculations: general formalism. *Mol. Phys.* **26**, 905–927.
[70] Wang, Y., Stocks, G. M., and Faulkner, J. S. (1994). General method for evaluating shape truncation functions of Voronoi polyhedral. *Phys. Rev. B* **49**, 5028–5031.
[71] Stefanou, N., Aki, H., and Zeller, R. (1990). An efficient numerical method to calculate shape truncation functions for Wigner-Seitz atomic polyhedra. *Comp. Phys. Comm.* **60**, 231–238.
[72] Stefanou, N., and Zeller, R. (1991). Calculation of shape-truncation functions for Voronoi polyhedral. *J. Phys.: Condens. Matter* **3**, 7599–7606.
[73] Shiba, D. H. (1971). A reformulation of the coherent potential approximation and its application. *Progr. Theor. Phys.* **46**, 77–94.
[74] Soven, P. (1971). Coherent-potential model of substitutional disordered alloy. *Phys Rev.* **156**, 809–813.
[75] Oswald, A., Zeller, R., Braspenning, P. J., and Dederichs, P. H. (1985). Interaction of magnetic impurities in Cu and Ag. *J. Phys. F* **15**, 193–212.
[76] Lichtenstein, A. I., Katsnelson, M. I., Antropov, V. P., and Gubanov, V. A. (1987). Local spin density functional approach to the theory of exchange interactions in ferromagnetic metals and alloys. *J. Magn. Magn. Mater.* **67**, 65–74.
[77] Schulthess, T. C., Temmerman, W. M., Szotek, Z., Svane, A., and Petit, L. (2007). First-principles electronic structure of Mn-doped GaAs, GaP and GaN semiconductors, *J. Phys.: Condens. Matter* **16**, 165207.
[78] Sandratskii, L. M., Bruno, P., and Kudrnovsky, J. (2004). On-site Coulomb interaction and magnetism of (Ga,Mn)N and (Ga,Mn)As. *Phys. Rev. B* **69**, 195203.
[79] Park, J. H., Kwon, S. K., and Min, B. I. (2000). Electronic structure of III-V ferromagnetic semiconductors: half-metallic phase. *Physica B* **281/282**, 703–704.
[80] Belhadji, B., Bergqvist, L., Zeller, R., Dederichs, P. H., Sato, K., and Katayama-Yoshida, H. (2007). Trends of exchange interactions in dilute magnetic semiconductors. *J. Phys.: Condens. Matter* **19**, 436227.
[81] Bergqvist, L., Belhadji, B., Picozzi, S., and Dederichs, P. H. (2008). Volume dependence of the Curie temperatures in diluted magnetic semiconductors. *Phys. Rev. B.* **77**, 014418.
[82] AlZahrani, A. Z., Srivastava, G. P., Gard, R., and Migliorato, M. A. (2009). Ab initio study of electronic and structural properties of Mn in GaAs environment. *J. Phys.: Condens. Matter* **21**, 485504.

[83] Perdew, J. P., Burke, K., and Ernzerhof, M. (1996). Generalized gradient approximation made simple. *Phys. Rev. Lett.* **77**, 3865–3868.

[84] Vanderbilt, D. (1990). Soft self-consistent pseudopotentials in a generalized eigenvalue formalism. *Phys. Rev. B* **41**, 7892–7895.

[85] Okabayashi, J., Kimura, A., Rader, O., Mizokawa, T., Fujimori, A., Hayashi, T., and Tanaka, M. (1998). Core-level photoemission study of $Ga_{1-x}Mn_xAs$. *Phys. Rev. B* **58**, R4211–R4214.

[86] Lieb, E. H. (1985). Density functionals for Coulomb systems. *In* "Density Functional Methods in Physics. NATO Advanced Science Institute, Series B: Physics" (R.M. Dreizler, and J. da Providencia, Eds.), Vol. 123, pp. 31–80. Springer, New York; see also Lieb, E. H. (1983). Density functionals for Coulomb systems. *Int. J. Quant. Chem.* **24**, 243–277.

[87] Pozhar, L. A. (2010). Small InAsN and InN clusters: electronic properties and nitrogen stability belt. *Eur. Phys. J. D* **57**, 343–354.

[88] Imry, Y. (1986). Directions in Condensed Matter Physics, Vol. 1. World Scientific, Singapore.

[89] Beenakker, C. W. J. (1991). Theory of Coulomb-blockade oscillations in the conductance of a quantum dot. *Phys. Rev. B* **44**, 1646–1656.

[90] Y. Imry and R. Landauer. (1999). Conductance viewed as transmission. *Rev. Mod. Phys.* **71**, S306–S312.

[91] Averin, D. V., and Likharev, K. K. (1991). Theory of single electron charging of quantum wells and dots. *Phys. Rev. B* **44**, 6199–6211.

[92] Averin, D. V., and Nazarov, Yu. N. (1990). Virtual electron diffusion during quantum tunneling of the electric charge. *Phys. Rev. Lett.* **65**, 2446–2449.

[93] Glazman, L. I., and Matveev, K. A. (1990). Residual quantum conductivity under Coulomb-blockage conditions. *JETP Lett.* **51**, 484–487.

[94] Meir, Y., and Wingreen, N. S. (1992). Landauer formula for the current through an interacting electron region. *Phys. Rev. Lett.* **68**, 2512–2515.

[95] Mahan, G. D. (1993). "Many Particles Physics", 2nd ed. Plenum, New York.

[96] Fujita, S. (1986). "Introduction to Non-Equilibrium Quantum Statistical Mechanics". W.B. Saunders, Philadelphia.

[97] Datta, S. (1995). "Electronic Transport in Mesoscopic Systems". Cambridge University, Cambridge, England.

[98] Ferry, D. K., and Goodnick, S. M. (1997). "Transport in Nanostructures". Cambridge University, Cambridge, England.

[99] DiVincenzo, D. P., and Loss, D. (1999). Quantum computers and quantum coherence. *J. Magnet. Magnet. Mater.* **200**, 202–218.

[100] Koppens, F. H. L., Buizert, C., Tielrooij, K. J., Vink, I. T., Nowack, K. C., Meunier, T., Kouwenhoven, L. P., and Vandersypen, L. M. K. (2006). Driven coherent oscillations of a single electron spin in a quantum dot. *Nature* **442**, 766–771.

[101] Kitchens, D., Richardella, A., Tang, J.-M., Flatte, M. E., and Yazdani, A. (2006). Atom-by-atom substitution of Mn in GaAs and visualization of their hole-mediated interactions. *Nature* **442**, 436–439.

[102] Nogues, J., Sort, J., Langlais, V., Skumryev, V., Sarinach, S., Munoz, J. S., Baro, M. D. (2005). Exchange bias in nanostructures. *Phys. Reports* **422**, 65–117.

[103] Hu, X., and Das Sarma, S. (2000). Hilbert-space structure of a solid-state quantum computer: two-electron states of a double –quantum-dot artificial molecule. *Phys. Rev. A* **61**, 062301.

[104] Recher, P., Sukhorukov, E, V., and Loss, D. (2000). Quantum dot as spin filter and spin memory. *Phys. Rev. Lett.* **85**, 1962–1965.

[105] Meier, F., Levy, J., and Loss, D. (2003). Quantum computing with antiferromagnetic spin clusters. *Phys. Rev. B* **68**, 134417.

[106] Scrola, V. W., and Das Sarma, S. (2005). Exchange gate in solid-state quantum computation: the applicability of the Heisenberg model. *Phys. Rev. A* **71**, 032340.

[107] Mizel, A., and Lidar, D. A. (2004). Exchange interaction between three and four coupled quantum dots: theory and applications to quantum computing. *Phys. Rev. B* **70**, 115310.
[108] Meier, F., Carletti, V., Gywat, O., Loss, D., and Awschalom, D. D. (2004). Molecular spintronics: coherent spin transfer in coupled quantum dots. *Phys. Rev. B* **69**, 195315.
[109] Lehmann, J., and Loss, D. (2006). Cotunneling current through quantum dots with phonon-assisted spin-flip processes. *Phys. Rev. B* **73**, 045328.
[110] Datta, S. N. (2005). Derivation of quantum Langevin equation from an explicit molecular-medium treatment in interaction picture. *J. Phys. Chem. A* **109**, 11417–11423.
[111] Reimann, S. M., and Manninen, M. (2002). Electronic structure of quantum dots. *Rev. Mod. Phys.* **74**, 1283–1327.
[112] Ohya, S., Takata, K., Tanaka, M. (2011). Nearly non-magnetic valence band of the ferromagnetic semiconductor GaMnAs. *Nature Physics* **7**, 342–347.
[113] Oszwaldowski, R., Žutić, I., and Petukhov, A. G. (2011). Magnetism in closed-shell quantum dots: emergence of magnetic bipolars. *Phys. Rev. Lett.* **106**, 177201.
[114] Wijnheijmer, A. P., Makarovsky, O., Garleff, J. K., Eaves, L., Campion, R. P., Gallagher, B. L., and Koenraad, P. (2010). Nanoscale potential fluctuations in (Ga,Mn)As/GaAs heterostructuresfrom individual ions to charged clusters and electrostatic quantum dots. *Nano Lett.* **10**, 4874–4879.
[115] Gan'shina, E. A., Golik, L. L., Kovalev, V. I., Kun'kova, Z. E., Temiryazeva, M. P., Danilov, Y. A., Viktorova, O. V., Zvonkov, B. N., Rubacheva A. D., Tcherbak, P. N., and Vinogradov, A. N. (2011). On nature of resonant transversal Kerr effect in InMnAs and GaMnAs layers. *Solid State Phenomena* **168–169**, 35–38.
[116] Placidi, E., Zallo, E., Arciprete, F., Fantoni, M., Patella, F., and Balzarotti, A. (2011) Comparative study of low temperature growth of InAs and InMnAs quantum dots. *Nanotechnology* **22**, 195602.
[117] Hai, P. N., Ohya, S., and Tanaka, M. (2010). Long spin-relaxation time in a single metal nanoparticle. *Nature Nanotechnology* **5**, 593–596.
[118] Šliwa, C., and Dietl, T. (2011). Thermodynamic and thermoelectric properties of (Ga,Mn) As and related compounds. *Phys. Rev. B* **83**, 245210.
[119] Pozhar, L. A., Yeates, A. T., Szmulowicz, F., and Mitchel, W. C. (2006). Virtual synthesis of artificial molecules of In, Ga and As with pre-designed electronic properties using a self-consistent field method. *Phys. Rev. B* **74**, 085306. See also: (2006). *Virtual J. Nanoscale Sci & Technol.* 14, No. 8, http://www.vjnano.org.
[120] Pozhar, L. A., Yeates, A. T., Szmulowicz, F., and Mitchel, W. C. (2005). Small atomic clusters as prototypes for sub-nanoscale heterostructure units with pre-designed charge transport properties. *EuroPhys. Lett.* **71**, 380–386.
[121] Pozhar, L. P., and Mitchel, W. C. (2009). Virtual synthesis of electronic nanomaterials: fundamentals and prospects. *In* "Toward Functional Nanomaterials. Lecture Notes in Nanoscale Science and Technology" (Z. Wang, A. Waag and G. Salamo, Eds.), Vol. 5, pp. 423–474. Springer, New York.
[122] Boldyrev, A. I., and Wang, L.-S. (2001). Beyond classical stoichiometry: experiment and theory. *J. Phys. Chem.* **105**, 10759–10775.
[123] Pozhar, L. A., and Mitchel, W. C. (2007). Collectivization of electronic spin distributions and magneto-electronic properties of small atomic clusters of Ga and In with As, V and Mn. *IEEE Trans. Magn.* **43**, 3037–3039.
[124] Pozhar, L. A., Yeates, A. T., Szmulowicz, F., and Mitchel, W. C. (2006). Magneto-optical properties of small atomic clusters of Ga or In with As, V and Mn. *Mater. Res. Soc. Proc.* **906E**, 0906-HH01-05.

Quantum Dots of Indium Nitrides 6

6.1 Introduction

In recent years indium nitride semiconductors have been in focus of materials physics research and device development due to their outstanding value for optoelectronics. In particular, zincblende and wurtzite InAsN and InN semiconductor structures exhibit a unique combination of electronic properties, such as the band gap tunable from ultraviolet to mid-infrared and high carrier mobility. Such properties are indispensable for the development of new-generation optoelectronic and high-speed electronic devices [1–3], including field effect transistors [4], photoelectrodes [5], light emitting diodes [6,7], and photodetectors [8]. Yet another line of research is driven by aspirations to synthesize p-type semiconductors on the basis of indium nitrides [2], and to develop novel devices enabled by terahertz emission [9], field emission [10] and ion sensing properties [11] of indium nitrides. Overcoming numerous difficulties in preparation of good quality films, single crystals and nanosystems of well-defined structure [12], experimental efforts are concentrated on synthesis and characterization of optoelectronic properties of quantum-confined indium nitride nanowires, low-dimensional structures and quantum dots (QDs) [1,13–20] down to 100 nm in linear dimensions. However, experimental synthesis of smaller indium nitride clusters, as well as characterization of various indium nitride systems, remains a challenging task. For example, the value of the fundamental band gap in bulk wurtzite InN was ascertained only in 2005 [2,21–23], and the electronic level structure and properties of small QDs and quantum wires (QWs) in the range of several nanometers in characteristic dimensions are virtually unknown experimentally. Yet such small QDs and QWs are envisioned as active elements of superdense three-dimensional (3D) integrated circuits (IC) [24] of spintronics, quantum electronics and quantum information processing devices that essentially use quantum coherence effects to increase efficacy and signal processing capabilities by orders of magnitude. It becomes increasingly obvious that experimental methods alone cannot provide for design of such devices, because it requires exceptionally numerous, detailed and reliable data on electronic and magnetic properties of each particular element of such systems synthesized on surfaces, in quantum confinement, and self-assembled in stacks. As in other cases, the only source of such a multitude of reliable data is the first principle-based theoretical predictions realized using realistic modeling of synthesis conditions and functionality of such systems.

Because of seeming simplicity, computational methods based in the density functional theory (DFT) are usually used to serve such theoretical predictions and modeling. However, as discussed elsewhere in literature (see, for example, Ref. 25), the majority of DFT-based methods suffer from various deficiencies that impact self-consistency and predictive capacity of such metods. In particular, to alleviate for the presence of self-interactions and the absence of a variational procedure [26] that would allow approximation of the exchange-correlation functional (ECF) in a controlled manner (and thus could enable calculations of system properties with systematically improving accuracy), numerous semi-phenomenological ECFs, adjustable parameters, pseudopotentials and fitting potentials, such as the Hubbard potential [27], are used to calculate electronic energy level structure (ELS) and thermodynamic properties comparable in value to those obtained experimentally. For example, in the case of diluted magnetic semiconductors only a specifically designed spin-polarized version of a generalized gradient approximation (GGA) that includes the Perdew-Burke-Ernzerhof exchange-correlation functional [28] and the Hubbard potential (GGA + U) was able to deliver results comparable to experimental data for bulk (Ga,As)Mn systems. Further adjustment of this scheme by the addition of a screened Stoner-like exchange parameter [29] (so-called J-correction) was necessary to develop a reasonable model (GGA + U + J) of ferromagnetism in Mn-doped (In,Mn)As and (Ga,Mn)As nanowires [30]. Even with such corrections and fitting, DFT-based modeling of electronic and magnetic properties of nanoscale semiconductor systems is not sufficiently accurate, and provides results that may differ from the corresponding experimental data within an order of magnitude. Simple DFT-based approaches, such as the local density approximation (LDA), sometimes lead to erroneous results predicting for example, a negative band gap for bulk wurtzite InN.

In addition to already known intrinsic shortcomings of DFT as a non-variational method [26], studies of computational complexity of quantum algorithms [31,32] indicate that efficiently computable good approximations of ECF, even semi-phenomenological ones, may not exist. Moreover, removal of the self-interactions (responsible for yet another deficiency of DFT-based approaches) in a recent so-called exact-exchange (EXX) Kohn-Sham DFT did not improve EXX-DFT predictions of the electronic properties of materials [33,34]. Only enhancement of DFT with quantum many-body theoretical methods enables the DFT-based approaches, such as the quasiparticle self-consistent GW (QSGW) one [35], to predict the fundamental band gap in the bulk indium nitride with a reasonable accuracy. However, even in this case, the QSGW band gap values somewhat exceed those obtained experimentally [36,37].

Despite of known shortcomings and inconsistencies of DFT-based methods, the majority of available computational studies of atomic clusters of indium nitrides [38–41], as well as other studies of clusters built of carbon, metal and semiconductor compound atoms, [42–44] originate from simple DFT-based approaches, such as LDA and GGA. In some cases, this modeling includes stabilization of atomic clusters achieved by introduction of hydrogen or metal atoms to saturate so-called dangling bonds. While indeed, sometimes experimental conditions include the use of hydrogen atoms to stabilize atomic clusters and nanosystems, such conditions are not inclusive, because they do not represent nucleation of atomic clusters that are not only stable on their own (that is, are molecules), but also have remarkable electronic and magnetic properties that can be controlled by manipulations with the composition and geometry of the quantum confinement shaping formation of such molecules. In modeling clusters's nucleation in quantum confinement as a process governed by the octet rule of the standard valence theory, both such computational and experimental studies miss an opportunity to discover much more general mechanisms of the total energy minimization of small nanosystems. In particular, as follows from numerous studies cited and discussed in this book, and in this chapter in particular, any atomic cluster is exposed to its environment (which is certainly not necessarily represented by hydrogen or metal atoms) via quantum confinement/surface effects, such as Coulomb interactions with electrons of the confinement's atoms and excluded volume effects. Both types of effects lead to polarization of cluster's atoms and the development of new type of bonding that violates the octet rule and promotes chemical reactions impossible otherwise. Thus, unrealistic models of nucleation conditions lead to predictions of either unrealistic or close-to-bulk properties and structure of the ground states of such clusters [45,46], and therefore, does not mimic properties of the corresponding molecules synthesis in confinement or on surfaces. Notably, no DFT-based or other cluster nucleation models reflecting quantum confinement/surface effects have been reported in literature.

In the chapter presented below virtual synthesis methods, discussed in Chapter 2 and based in the first-principle quantum many-body theory, are applied to produce realistic models of stable 14-atomic clusters composed of In, As and N atoms, and nucleated in model quantum confinements (below called pre-designed molecules), and in in the absence of such confinement (called vacuum molecules). First, the restricted and restricted open shell Hartree-Fock (RHF and ROHF, respectively) approximations are used to obtain the Slater determinant wavefunctions (and thus the structure and electronic properties) of the studied molecules specific to their ground states in RHF/ROHF approximation. Further on, the configuration interaction (CI), complete active space self-consistent field (CASSCF) and multiconfiguration self-consistent field (MCSCF) approximations are used with the RHF/ROHF wavefunctions to subsequently improve these wavefunctions and compute more accurate molecular orbits (MOs), ELSs, the direct optical transition energies (OTEs), and charge and spin density distributions (CDDs and SDDs, respectively) of the studied moecules. As everywhere in this book, all computations use GAMESS software package and the SBKJC standard basis set [45,47].

6.2 Virtual synthesis of small indium nitride QDs

The first principle, quantum many-body theory-based optimization of the total energy of atomic systems built of many-electron atoms and nucleating in quantum confinement or on surfaces is a challenging computational task for large systems due to software and hardware restrictions. However, such modeling is feasible and practical for atomic clusters composed of a few tens of many-electron atoms and nucleating in a "free space", that is, without any other atoms or fields present. In the case of nucleation in quantum confinement, such modeling is feasible at present only if there is a way to incorporate quantum confinement without inclusion of its atomistic details. This problem has been addressed in the framework of the virtual synthesis method described in Chapter 2. In a particular case of atomic clusters built of In, As and N atoms, analysis of spatial symmetry elements of zincblende and wurtzite bulk lattices of

InAs and InN, respectively, have been used to develop pre-designed configurations of selected clusters whose atoms already occupy specific positions as defined by their atomistic quantum confinement in the bulk lattices. Such pre-designed configurations reflect major excluded volume and polarization effects due to the environment. Moreover, such symmetry-guided pre-designed configurations are in true equilibrium with the rest of their parent bulk lattices, and thus mimic the true ground state molecules formed of the atomic clusters in the presence of the parent lattices. Therefore, such equilibrium structures must be selected among sometimes numerous spatial isomers produced by the total energy minimization procedure in the process of modeling cluster nucleation in their parent lattices. Even in the cases when the composition of such pre-designed clusters is modified by doping (compared to the composition of symmetry elements of their parent lattices), such pre-designed cluster structures still include all major effects of the confinement. In the studies described in this chapter, the virtual synthesis methods exploited these ideas to produce computational templates of several pre-designed zincblende-derived, substitution-doped $In_{10}As_3N$, $In_{10}As_2N_2$ and $In_{10}N_4$ atomic clusters, and one wurtzite-derived In_6N_6 cluster nucleated in quantum confinement served by the respective zincblende InAs and wurtzite InN bulk lattices. The virtual synthesis ideas that pre-designed equilibrium cluster structures reflect the major effects of quantum confinement are supported by well-studied experimental synthesis of semiconductor compound clusters composed of Ga or In atoms with As and P atoms. In particular, experimental data confirm that tetrahedral pyramids and hexagonal prismatic structures pre-designed computationally and discussed in this book closely resemble the structure of similar (although larger) molecules synthesized experimentally at solid interfaces and in 3D confinement [48–54].

In the pre-designed structures of this chapter, atoms of the pre-designed $In_{10}As_3N$, $In_{10}As_2N_2$ and $In_{10}N_4$ atomic clusters are arranged in tetrahedral pyramids of the symmetry elements of the zincblende InAs lattice described in detail in Chapter 3. In their turn, atoms of the pre-designed wurtzite-derived In_6N_6 cluster are arranged in a hexagonal prism derived from the corresponding symmetry element of the wurtzite InN lattice with an introduced dislocation defect. The total energy of the pre-designed clusters was minimized using controlled and subsequently more accurate RHF/ROHF, CI, CASSCF and MCSCF approximations. During the minimization procedure the centers of mass of the cluster atoms were constrained to their original positions in the pyramids or prism (that is, the atoms were fixed in their original positions in the pre-designed pyramids ot the prism) to model excluded volume and polarization effects due to quantum confinement. The obtained ground states of these clusters are called below pre-designed molecules and realize local minima of the total energies of the clusters. The corresponding unconstrained clusters mimicking cluster formation in the absence of quantum confinement, and called below vacuum molecules, were virtually synthesized upon optimization of the geometry of the pre-designed clusters. In this case, the pre-designed clusters were used as initial structures, and then the spatial constraints applied previously to atomic positions were lifted to generate various unconstrained atomic configurations mimicking the absence of any confinement or electromagnetic fields. For each change in a position of an atom, the total energy of the atomic configuration was calculated. The steepest descent method was used to accelerate a search for the structure corresponding to the global minima of the total energy among hundreds of thousands of such unconstrained atomic configurations. The minimum of the total energy so obtained corresponds to the ground state of the vacuum molecule derived from the original pre-designed atomic cluster by the cluster's structure optimization in the absence of any conditions limiting the structure nucleation. Similar to the vacuum molecules $In_{10}As_4$ and $Ga_{10}As_4$ of Chapter 3, and other such molecules described in this book, in some cases the structure optimization does not significantly distort the original geometry of the pre-designed clusters, and in other cases it does. However, in all of the studied cases the structure of vacuum molecules resembles that of the pre-designed molecules indicating that the cluster structures derived from that of symmetry elements of bulk lattices are extremely robust and not easily affected by even close proximity of quantum confinement atoms (such as that provided by quantum confinement of the tetrahedral and hexagonal bulk lattice atoms). Notably, despite a symmetry-driven choice of the initial structure and composition of a cluster, the exact value of a minimum (local or global) of the total energy of a cluster may not be found due to intrinsic restrictions of mathematical methods realizing the variational procedure numerically and used in existing software. Hardware restrictions are also relevant. In particular, software restrictions do not allow a significant change in atomic positions at each particular step of computations, thus restricting a total number of achievable atomic configurations to hundreds of thousands (instead of hundreds of millions, for example) during each calculation run. At the same time, should all such configurations have been attainable, hardware restrictions would not allow to complete such computations within reasonable time limits.

Symmetry elements of the InAs zincblende and InN wurtzite lattices open rich prospects for investigation of nitrogen-containing In-As QDs with desirable optical, electronic and magnetic properties by the virtual synthesis method. In this section a range of such molecules are virtually synthesized and characterized in detail.

6.3 Pyramidal InAs-based molecules with one nitrogen atom

The pre-designed $In_{10}As_4$ pyramid of Chapter 3 has been used throughout this book as the major source of initial virtual structures of numerous nanometer size QDs. In this section three $In_{10}As_3N$ atomic clusters have been constructed by substitution of one of arsenic atoms by a nitrogen atom either in the pre-designed $In_{10}As_4$ pyramid (with the covalent radii of In and As atoms of 1.44Å and 1.18Å, respectively, known from experiment) or in clusters derived from it. Thus, one of the pre-designed $In_{10}As_3N$ atomic clusters discussed below was derived from the $In_{10}As_4$ pyramid directly, and the other one from the vacuum $In_{10}As_3P$ cluster of Chapter 4 where the phosphorus atom was replaced by a nitrogen one. In this latter cluster the original covalent radii of In and As atoms were slightly changed in the process of the total energy minimization that produced the vacuum $In_{10}As_3P$ molecule. Substitution of the P atom with N atom that lead to the pre-designed InAsP-derived $In_{10}As_3N$ atomic cluster did not involve a change in interatomic distances, so that the total energy of this cluster was minimized in the presence of the spatial constraints applied to the

TABLE 6.1 Ground state energies of the $In_{10}As_3N$ molecules.

Molecule	ROHF ground state energy, Hartree	MCSCF ground state energy, Hartree	CAS (number of electrons × number of MOs)
Pre-designed $In_{10}As_3N$	−1910.470004	Not available	ROHF triplet
[a]Pre-designed $In_{10}As_3N$	−1910.526654 −1910.526654	−1910.671094340 −1910.527331893	12 × 11 8 × 7
Vacuum $In_{10}As_3N$	−1910.637634	−1910.740085831	12 × 11

[a] The geometry of this molecule was derived from the vacuum $In_{10}As_3P$ cluster of Chapter 4.

centers of mass of the cluster atoms. At the second step of the virtual synthesis procedure, the spatial constraints applied to atomic positions in the pre-designed $In_{10}As_3N$ cluster derived from $In_{10}As_4$ pyramid were relaxed (that is, the atoms were allowed to move from their initial positions), and its total energy minimized again. This produced the vacuum $In_{10}As_3N$ molecule. Electronic properties of these two pre-designed and one vacuum molecules are summarized in Tables 6.1 and 6.2, and their electrochemical signature is depicted in Figs. 6.1 to 6.6.

Because the atoms of the $In_{10}As_4$-derived pre-designed $In_{10}As_3N$ molecule were constrained to their original positions defined by those of $In_{10}A_4$ atoms, this molecule (Fig. 6.1a) retains the tetrahedral pyramid geometry identical to that of the $In_{10}A_4$ one. The pyramidal geometry was also retained in the case of the $In_{10}As_3P$-derived pre-designed $In_{10}A_3N$ molecule (Fig. 6.5a), because atoms of its parent vacuum $In_{10}As_3P$ molecule (see Chapter 4 for details) only slightly changed their position with regard to those of the pre-designed $In_{10}As_4$ cluster in the process of the total energy

TABLE 6.2 Direct optical transition energies (OTEs) of the $In_{10}As_3N$ molecules.

Molecules	ROHF OTE, eV	MCSCF OTE, eV	CAS (number of electrons × number of MOs)
Pre-designed $In_{10}As_3N$	3.0667	Not available	ROHF triplet
[a]Pre-designed $In_{10}As_3N$	2.9769	1.8788	12 × 11
[a]Pre-designed $In_{10}As_3N$	2.9769	1.9229	8 × 7
Vacuum $In_{10}As_3N$	3.1375	0.6716	12 × 11

[a] The geometry of this molecule was derived from the vacuum $In_{10}As_3P$ cluster.

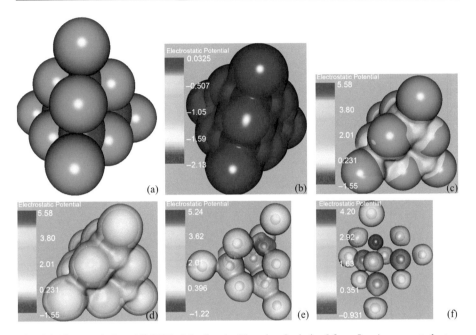

Fig. 6.1 The pre-designed ROHF triplet $In_{10}As_3N$ molecule derived from $In_{10}As_4$ parent cluster: (a) structure; MEP surfaces (color scheme) corresponding to the CDD isovalues (b) 0.02, (c) 0.05, (d) 0.05, (e) 0.1 and (f) 0.2. MEP minimum and maximum values: −5.07322 and 2.54872, respectively (in arbitrary units). In (e) negative potential "belt" due to electron charge accumulated by nitrogen atom is prominent. In (a) to (c) all dimensions are to scale. In (d) to (f) atomic dimensions are reduced. Linear dimensions of As atoms are somewhat enlarged in (a) to (c) due to inaccuracy of the visualization software (Molekel). Indium atoms are blue, As brown, N deep blue.

minimization. Such small atomic position changes in the range of a few hundredths of Angstrom are unnoticeable to a human eye. As a result, both pre-designed $In_{10}As_3N$ molecules visually resemble the original $In_{10}As_4$ tetrahedral pyramid. In the case of the vacuum molecule $In_{10}As_3N$, the total energy minimization affected the shape of the molecule profoundly (Figs. 6.3a and 6.3c). While it still significantly resembles a pyramid, a distortion of the pyramidal geometry is significant. The covalent radius of As atoms in InAs zincblende lattice is significantly larger than that of the N atom. Therefore, replacement of an As atom with the N one in the original $In_{10}As_4$ pyramid created a somewhat loose atomic cluster. In the process of the total energy minimization, In atoms had to move closer to the N atom to provide for the total energy minimum, and re-shaped the original pyramid in the process. Correspondingly, the ground state energy of both pre-designed $In_{10}As_3N$ molecules realizing local minima of the total energy of the $In_{10}As_3N$ cluster is larger than that of their vacuum counterpart (Table 6.1) that realizes the global minimum of the total energy of the $In_{10}As_3N$ cluster. Interestingly, the ROHF ground state energy (−1910.526654 H) of the pre-designed $In_{10}As_3N$ molecule derived from the vacuum $In_{10}As_3P$ cluster is closer in value to that of the vacuum molecule than the ROHF ground state energy (−1910.470004 H) of the

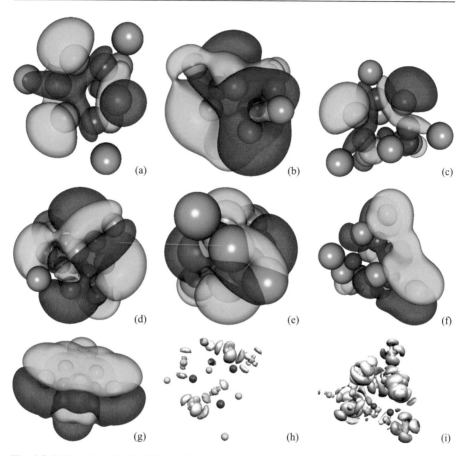

Fig. 6.2 Molecular orbitals (MOs) and the spin density distribution (SDD) of the pre-designed ROHF triplet $In_{10}As_3N$ molecule derived from $In_{10}As_4$ parent cluster. Isosurfaces of negative (semi-transparent red) and positive (semi-transparent green) components of MOs in the HOMO –LUMO region: (a) HOMO 115, isovalue 0.01; (b) HOMO 115, isovalue 0.001, (c) HOMO 116, isovalue 0.01, (d) HOMO116, isovalue 0.001, (e) HOMO 116, isovalue 0.001, (f) LUMO 117, isovalue 0.01, and (g) LUMO 117, isovalue 0.001. SDD isosurfaces (gray) corresponding to the isovalues (h) 0.005 and (i) 0.001, respectively. Indium atoms are blue, As brown or red [in (h) and (i)], N deep blue. Atomic dimensions are reduced everywhere, but in (e), to show the structure.

pre-designed $In_{10}As_3N$ molecule derived directly from the pre-designed $In_{10}As_4$ pyramid. The reason behind this finding is that in the parent vacuum $In_{10}As_3P$ molecule In atoms already moved closer to P atom in the course of the total energy minimization of the $In_{10}As_3P$ cluster (the covalent radius of a P atom in the InAs zincblende lattice is smaller than that of As). Thus, taking the vacuum $In_{10}As_3P$ molecule as a template for pre-designing the $In_{10}As_3N$ molecule meant that the ground state energy of the pre-designed $In_{10}As_3N$ molecule so obtained would be closer to the global minimum of the total energy realized by the ground state of the vacuum $In_{10}As_3N$ molecule. Refining the

Quantum Dots of Indium Nitrides

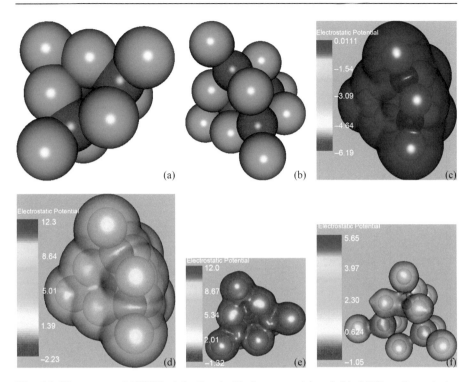

Fig. 6.3 The vacuum MCSCF triplet $In_{10}As_3N$. Structure: (a) and (b). MEP surfaces (color scheme) corresponding to the CDD isovalues (c) 0.001, (d) 0.01, (e) 0.05 and (f) 0.1. MEP minimum and maximum values: −4.51069 and 1.59942, respectively (in arbitrary units). In (e) and (f) negative potential "belt" due to electron charge accumulated by nitrogen atom is prominent. In (a) and (b) all dimensions are to scale. In (c) to (f) atomic dimensions are reduced. Linear dimensions of As atoms are somewhat enlarged in (a) and (b) due to inaccuracy of the visualization software (Molekel). Indium atoms are blue, As brown, N deep blue.

ROHF data by the use of CI, CASSCF and MCSCF total energy minimization procedures reveals a strong correlation between the local minimum values of the total energy and the size of the chosen complete active space (CAS). In particular, when the same 12×11 CAS (12 electrons distributed over 11 orbits) is used, the difference between the ground state energy of the $In_{10}As_3P$-derived pre-designed molecule and that of the vacuum one is about 3 times smaller than in the case of 8×7 CAS. Further on, in accord with theoretical considerations discussed in Chapter 2, the use of CASSCF and MCSCF approximations significantly changes the RHF/ROHF optical transition energy (OTE) results, while the ground state energies are not affected nearly that much. This is especially obvious in the case of the vacuum molecule where the ROHF OTE of 3.1375 eV decreases almost by the factor of 5 to 12×11 CAS MCSCF OTE 0.6716 eV (Table 6.2). However, this latter OTE value seems to be too small in view of a recent experimental evaluation of the fundamental band gap (0.9 eV) of the hexagonal InN [55].

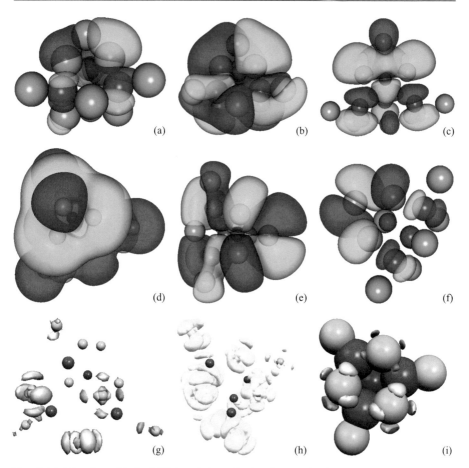

Fig. 6.4 Molecular orbitals (MOs) and the spin density distribution (SDD) of the vacuum MCSCF triplet $In_{10}As_3N$. Isosurfaces of negative (semi-transparent red) and positive (semi-transparent green) components of MOs in the HOMO – LUMO region: (a) HOMO 115, isovalue 0.01; (b) HOMO 115, isovalue 0.001, (c) HOMO 116, isovalue 0.01, (d) HOMO116, isovalue 0.001, (e) LUMO 117, isovalue 0.001 and (f) LUMO 117, isovalue 0.01. SDD isosurfaces (gray and yellow) corresponding to the isovalues (g) 0.01, (h) 0.001 and (i) 0.01, respectively. Indium atoms are blue, As brown or red [in (g) to (i)], N deep blue. Atomic dimensions are reduced everywhere, to show the structure.

In contrast, the OTE results for the $In_{10}As_3P$-derived pre-designed molecule look realistic, but point out again, that larger CASs may produce smaller OTE values.

All virtually synthesized $In_{10}As_3N$ molecules are electrostatic dipoles with the dipole moments equal to 5.1181 D, 5.0269 D and 2.4796 D for the pre-designed $In_{10}As_3N$, pre-designed InAsP-derived $In_{10}As_3N$, and vacuum $In_{10}As_3N$ molecules, respectively. The polar nature of these molecules is further manifested by their molecular electrostatic potentials (MEPs). Analysis of MEPs of these molecules (Figs. 6.1, 6.3 and 6.5) reveals that the molecular surfaces (as defined by outbound

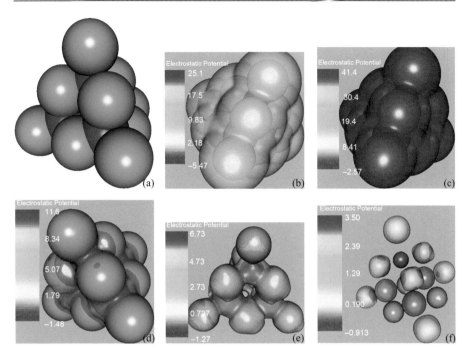

Fig. 6.5 The pre-designed MCSCF triplet $In_{10}As_3N$ derived from the vacuum $In_{10}As_3P$ cluster. Structure: (a). MEP surfaces (color scheme) corresponding to the CDD isovalues (b) 0.02, (c) 0.01, (d) 0.05, (e) 0.08 and (f) 0.2. MEP minimum and maximum values: -3.34631 and 2.73188, respectively (in arbitrary units). In (e) negative potential "belt" due to electron charge accumulated by nitrogen atom is prominent. In (a) to (d) all dimensions are to scale. In (e) and (f) atomic dimensions are reduced. Linear dimensions of As atoms are somewhat enlarged in (a) to (d) due to inaccuracy of the visualization software (Molekel). Indium atoms are blue, As brown, N deep blue.

portions of atomic spherical cores of In atoms and defined by In covalent radii) carry small delocalized electron charge. The pre-designed MCSCF triplet $In_{10}As_3N$ derived from the vacuum $In_{10}As_3P$ molecule (Figs. 6.5b to 6.5d) carries more surface charge than the other two $In_{10}As_3N$ molecules (Figs. 6.1b, 6.1c and 6.3c). As expected, electron charge is the most evenly distributed over the surface and volume in the vacuum MCSCF triplet (Figs. 6.3d to 6.3f). In all cases, the nitrogen atoms accumulate electronic charge of In atoms creating a "belt" of electronic charge, and help As atoms to stabilize the molecules. Such electron charge belts are especially prominent in Figs. 6.1e, 6.3e, 6.3f, and 6.5e corresponding to larger negative values of MEPs. In all molecules, a significant portion of the electronic charge is redistributed toward surface from the volume of the molecules. These findings agree very well with numerous experimental observations [16,17,56–61] indicating that electron charge in larger nanostructures and bulk lattices of the group III-V nitrides is re-distributed to the surface from the bulk to allow such broken symmetry systems minimize their total energy.

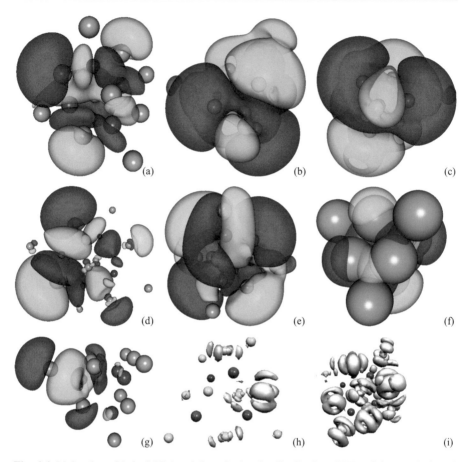

Fig. 6.6 Molecular orbitals (MOs) and the spin density distribution (SDD) of the pre-designed MCSCF triplet $In_{10}As_3N$ derived from the vacuum $In_{10}As_3P$ cluster. Isosurfaces of negative (semi-transparent red) and positive (semi-transparent green) components of MOs in the HOMO – LUMO region: (a) HOMO 115, isovalue 0.01; (b) HOMO 115, isovalue 0.001, (c) HOMO 115, isovalue 0.001, (d) HOMO116, isovalue 0.01, (e) HOMO 116, isovalue 0.001, (f) HOMO 116, isovalue 0.003, and (g) LUMO 117, isovalue 0.01. SDD isosurfaces (gray) corresponding to the isovalues (h) 0.005 and (i) 0.001, respectively. Indium atoms are blue, As brown or red [in (h) and (i)], N deep blue. Atomic dimensions are reduced everywhere, but in (f), to show the structure.

This charge re-distribution is facilitated by the development of a new type of chemical bonding between ligand and other atoms in small atomic clusters resulting in formation of both non-stoichiometric and some stoichiometric molecules [25,62–65]. The octet rule of the standard valence theory is violated in the case of non-stoichiometric molecules, as many atomic orbits (AOs) of the participating atoms contribute to the development of highly hybridized and delocalized molecular orbits (MOs) of such molecules. Thus, MOs in the highest occupied molecular orbital (HOMO) – lowest unoccupied molecular orbital (LUMO) regions of the studied $In_{10}As_3N$ molecules are a result of hybridization from 12 to about 100 AOs. The topmost occupied MOs of these molecules feature the major contributions from $4d$ AOs of several In atoms, $4p$ AOs of all As atoms, and $2p$ AOs of the N atom (Figs. 6.2, 6.4 and 6.6) hybridized in various proportions. The MOs in the HOMO-LUMO region of the pre-designed ROHF triplet $In_{10}As_3N$ also have a significant contribution from $5p$ AOs of In atoms. Such are HOMO 115 (Figs. 6.2a and 6.2b), HOMO 116 (Figs. 6.2c to 6.2d) and LUMO (Figs. 6.2f and 6.2g) of this molecule developed of highly hybridized and delocalized $4p$ AOs of 3 As atoms, $5p$ AOs of 2 In atoms, and $2p$ AOs of the N atom. Two of these MOs (HOMO 116 and LUMO 117) also include contributions from $4d$ AOs of several In atoms. In the case of the pre-designed MCSCF triplet $In_{10}As_3N$ derived from $In_{10}As_3P$ molecule, the major contributions to HOMO 115 (Figs. 6.6a to 6.6c) come from $5p$ AOs of 2 In atoms, $4d$ AOs of 2 other In atoms, $4p$ AOs of 3 As atoms and $2p$ AO of nitrogen. The proper HOMO 116 of this molecule (Figs. 6.6d to 6.6f) possesses the major contributions from $5p$ AOs of 3 In atoms, somewhat smaller contributions from $4p$ AOs of 3 As and $2p$ AOs of N atoms, and $4d$ AOs of 4 other In atoms. In the case of LUMO of this molecule (Fig. 6.6g), contributions from $5p$ AOs of 3 In atoms and $2p$ AO of the N atom are the largest. The major contributions to the proper HOMO 116 of the vacuum MCSCF triplet $In_{10}As_3N$ come from $4d$ AOs of 4 In atoms, $4p$ AOs of all As atoms and $2p$ AO of the N atom (Figs. 6.4c and 6.4d). HOMO 115 (Figs. 6.4a and 6.4b) and LUMO (Figs. 6.4e and 6.4f) of this molecule have a significant contribution due to $5p$ AOs of four and three In atoms, respectively.

Almost all MOs in the near HOMO-LUMO region of the studied $In_{10}As_3N$ molecules include a large part formed by a delocalized and hybridized $5p$ ligand bond of In atoms (for example, large sandwich-like regions of MOs in Figs. 6.2, 6.4 and 6.6) mediated by $4p$ AOs of As atoms and $2p$ AO of nitrogen. Contributions from $4d$ In ligand bonding mediated by $4p$ arsenic AOs or $2p$ nitrogen AO convey to MOs into which they are included a distinctive bow-like shape. Both $5p$, $4d$, and $5p$ - $4d$ ligand bonding of In atoms mediated by As and N is responsible for remarkable stability of the synthesized molecules and an opportunity to easily optimize structures of various geometry composed of an appropriate number of In, As and/or N atoms, as will be shown further in this chapter. A specific feature of bonding in MOs of the studied $In_{10}As_3N$ molecules is the presence of bonding groups, such as: 3 In + N (with the N atom on the top of the pyramid and 3 In atoms in the verges), As + (2 In + N) + As, and 4 In + N (where 4 In atoms are arranged in a pyramid with the N atom in its center). Thus, the nitrogen atom in the studied $In_{10}As_3N$ molecules tends to bond 3 to 4 nearby In and other atoms, and In atoms in N-mediated groups also form bonds with each other.

Magnetic properties of the studied molecules are not especially interesting for spintronics. All studied $In_{10}As_3N$ molecules are "ferromagnetic" triplets with uncompensated z-components of the spin magnetic moment equal to 2 Bohr magnetons contributed by $4d$ electrons of In atoms. The corresponding spin density distributions (SDDs) of the $In_{10}As_3N$ molecules are highly delocalized with small maximum values of similar magnitude: 0.0272235 (the pre-designed ROHF triplet, Figs. 6.2h and 6.2i), 0.0322926 (the $In_{10}As_3P$-derived MCSCF triplet, Figs. 6.6h and 6.6i), and 0.0262703 (the vacuum MCSCF triplet, Fig. 6.4g to 6.4i), in arbitrary units.

6.4 Pyramidal InAs-based molecules with two nitrogen atoms

Similar to $In_{10}As_3N$ molecules of the previous section, three molecules of $In_{10}As_2N_2$ have been derived by substitution of two of arsenic atoms in the pre-designed $In_{10}As_4$ pyramid of Chapter 3, or two P atoms in the vacuum $In_{10}As_2P_2$ cluster of Chapter 4 with two nitrogen atoms. As described in Chapter 4, the molecule $In_{10}As_2P_2$ itself was derived from the pre-designed $In_{10}As_4$ one after replacement of 2 As atoms with 2 P atoms without changing interatomic distances, followed by relaxation of spatial constraints applied to the atoms of the pre-designed $In_{10}As_2P_2$ cluster so obtained, and subsequent minimization of its total energy. During this latter step, the atoms of the pre-designed $In_{10}As_2P_2$ cluster were permitted to move, so that the original covalent radii of In and As atoms were slightly changed, and the covalent radii of P atoms in this structure ascertained, in the process of the total energy minimization that produced the vacuum $In_{10}As_2P_2$ molecule. Thus, substitution of the two P atoms with N atoms in the vacuum $In_{10}As_2P_2$ molecule without changing interatomic distances leads to the development of the pre-designed InAsP-derived $In_{10}As_2N_2$ cluster that does not possess tetrahedral symmetry. Minimization of the total energy of this cluster with spatial constraints still applied to the atomic positions produced the pre-designed InAsP-derived $In_{10}As_2N_2$ molecule. Another pre-designed $In_{10}As_2N_2$ molecule was obtained directly from the pre-designed $In_{10}As_4$ molecule of Chapter 3 upon substitution of 2 of its As atoms with 2 N ones (interatomic distances were kept unchanged). Finally, the vacuum $In_{10}As_2N_2$ molecule was derived from the latter pre-designed molecule upon lifting the spatial constraints and subsequent total energy minimization of the "unconstrained" cluster so obtained.

Electronic properties and electrochemical characteristics of these virtually synthesized $In_{10}As_2N_2$ molecules are summarized in Tables 6.3 and 6.4, and Figs. 6.7 to 6.12. According to the virtual synthesis setup, the ground states of the two pre-designed molecules realizes two local minima of the total energy of the $In_{10}As_2N_2$ cluster, and the vacuum molecule realizes the global minimum of the total energy of this cluster. Data of Table 6.3 readily support this conclusion. Interestingly, replacement of the second

TABLE 6.3 Ground state energies of the $In_{10}As_2N_2$ molecules.

Molecule	ROHF ground state energy, Hartree	MCSCF ground state energy, Hartree	CAS (number of electrons × number of MOs)
Pre-designed $In_{10}As_2N_2$	−1914.017418	Not available	ROHF triplet
[a]Pre-designed $In_{10}As_2N_2$	−1914.057876	−1914.201967222	12 × 11
Vacuum $In_{10}As_2N_2$	−1914.312056	Not available	RHF singlet

[a] The geometry of this molecule was derived from the vacuum $In_{10}As_2P_2$ cluster.

TABLE 6.4 Direct optical transition energies (OTEs) of the $In_{10}As_2N_2$ molecules.

Molecule	ROHF OTE, eV	MCSCF OTE, eV	CAS (number of electrons × number of MOs)
Pre-designed $In_{10}As_2N_2$	3.8014	Not available	ROHF triplet
[a]Pre-designed $In_{10}As_2N_2$	2.3756	2.0491	12 × 11
Vacuum $In_{10}As_2N_2$	6.2831	Not available	RHF singlet

[a] The geometry of this molecule was derived from the vacuum $In_{10}As_2P_2$ cluster.

As atom with the nitrogen one leads to a significant decrease in the ground state energies of all $In_{10}As_2N_2$ molecules compared to those of the $In_{10}As_3N$ molecules by about 3.4 H, which is much larger than the computation error brackets (about 1 H). Therefore, the pyramidal molecules with 2 nitrogen atoms appear to be more stable than the corresponding molecules with one nitrogen atom. This result is further supported

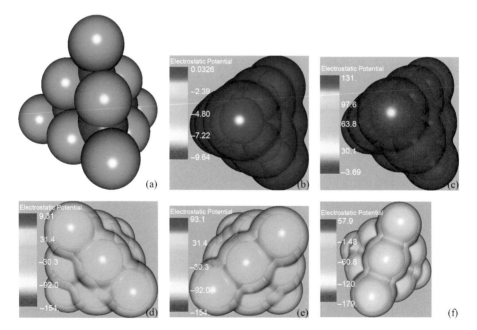

Fig. 6.7 The pre-designed ROHF triplet $In_{10}As_2N_2$. Structure: (a). MEP surfaces (color scheme) corresponding to the CDD isovalues (b) 0.001, (c) 0.005, (d) 0.01, (e) 0.01 and (f) 0.02. MEP minimum and maximum values: −5.87588 and 2.08240, respectively (in arbitrary units). In (a) to (c) all dimensions are to scale. In (d) to (f) atomic dimensions are reduced. Linear dimensions of As atoms are somewhat enlarged in (a) to (c) due to inaccuracy of the visualization software (Molekel). Indium atoms are blue, As brown, N deep blue.

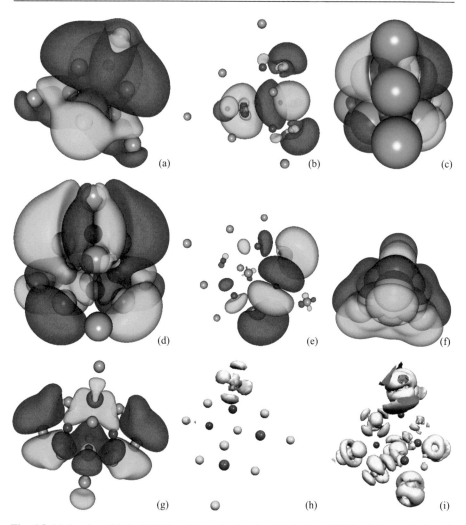

Fig. 6.8 Molecular orbitals (MOs) and the spin density distribution (SDD) of the pre-designed ROHF triplet $In_{10}As_2N_2$. Isosurfaces of negative (semi-transparent red) and positive (semi-transparent green) components of MOs in the HOMO – LUMO region: (a) HOMO 115, isovalue 0.001; (b) HOMO 115, isovalue 0.01, (c) HOMO 116, isovalue 0.001, (d) HOMO116, isovalue 0.001, (e) HOMO 116, isovalue 0.01, (f) LUMO 117, isovalue 0.001, and (g) LUMO 117, isovalue 0.01. SDD isosurfaces (yellow) corresponding to the isovalues (h) 0.01 and (i) 0.001, respectively. Indium atoms are blue, As brown or red [in (h) and (i)], N deep blue. Atomic dimensions are reduced everywhere, but in (c) and (d), to show the structure.

by the fact that ROHF OTEs of the pre-designed and vacuum $In_{10}As_2N_2$ molecules are larger than those of the corresponding pre-designed and vacuum $In_{10}As_3N$ ones. Moreover, the vacuum $In_{10}As_2N_2$ molecule is a RHF singlet, as opposed to the vacuum $In_{10}As_3N$ molecule being a ROHF triplet. However, ROHF OTEs of the pre-designed

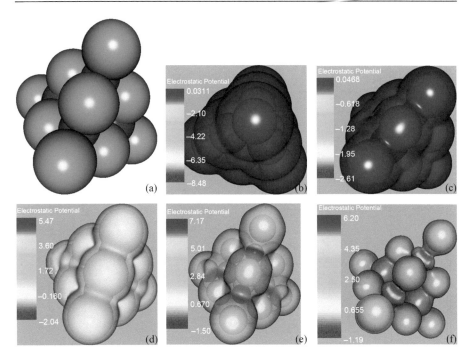

Fig. 6.9 The pre-designed MCSCF triplet $In_{10}As_2N_2$ derived from the vacuum $In_{10}As_2P_2$ cluster. Structure: (a). MEP surfaces (color scheme) corresponding to the CDD isovalues (b) 0.001, (c) 0.01, (d) 0.02, (e) 0.05 and (f) 0.1. MEP minimum and maximum values: −3.73601 and 2.36422, respectively (in arbitrary units). In (a) to (c) all dimensions are to scale. In (d) to (f) atomic dimensions are reduced. Linear dimensions of As atoms are somewhat enlarged in (a) to (c) due to inaccuracy of the visualization software (Molekel). Indium atoms are blue, As brown, N deep blue.

InAsP-derived $In_{10}As_2N_2$ molecule is smaller than that of the corresponding $In_{10}As_3N$ molecule. At the same time, 12 × 11 CAS MCSCF OTE of the pre-designed InAsP-derived $In_{10}As_2N_2$ molecule is larger than both 12 × 11 CAS MCSCF and 8 × 7 CAS MCSCF OTEs of the pre-designed InAsP-derived $In_{10}As_3N$ molecule, in line with expectatins. A decline in value of ROHF OTE of this molecule by about 0.6eV as compared to ROHF OTE of the the pre-designed InAsP-derived $In_{10}As_3N$ molecule is explained by the nature of HF approximation that is relatively rough, especially for transition metal atoms.

The above conclusion that $In_{10}As_2N_2$ molecules are more stable than their counterparts with one nitrogen atom is also confirmed by data on their dipole moments. Similar to $In_{10}As_3N$ molecules, the virtually synthesized $In_{10}As_2N_2$ molecules are polar, but their dipole moments are smaller than those of the corresponding $In_{10}As_3N$ molecules. In particular, the pre-designed, the pre-designed InAsP-derived, and the vacuum $In_{10}As_2N_2$ molecules possess the dipole moments 1.5455 D, 1.7128 D and 2.1791 D, respectively. These values are over 3.5 D lesser than the values of the dipole moments of the corresponding pre-designed $In_{10}As_3N$ molecules.

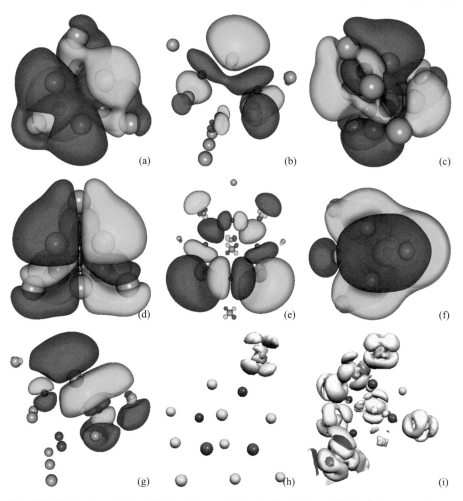

Fig. 6.10 Molecular orbitals (MOs) and the spin density distribution (SDD) of the pre-designed MCSCF triplet $In_{10}As_2N_2$ derived from the vacuum $In_{10}As_2P_2$ cluster. Isosurfaces of negative (semi-transparent red) and positive (semi-transparent green) components of MOs in the HOMO – LUMO region: (a) HOMO 115, isovalue 0.001; (b) HOMO 115, isovalue 0.01, (c) HOMO 116, isovalue 0.001, (d) HOMO116, isovalue 0.001, (e) HOMO 116, isovalue 0.01, (f) LUMO 117, isovalue 0.001, and (g) LUMO 117, isovalue 0.01. SDD isosurfaces (yellow) corresponding to the isovalues (h) 0.01 and (i) 0.001, respectively. Indium atoms are blue, As brown or red [in (h) and (i)], N deep blue. Atomic dimensions are reduced to show the structure.

Because of its synthesis setup, the geometry of the pre-designed $In_{10}As_2N_2$ molecule (Fig. 6.7a) is inherent from that of its parent tetrahedral pyramid - the pre-designed $In_{10}As_4$ molecule. However, the replacement of 2 As atoms by 2 N ones lead to breaking of tetrahedral symmetries of the electronic energy level structure (ELS), MOs, MEP and SDD of this molecule. Similarly, the pre-designed InAsP-derived $In_{10}As_2N_2$ molecule retains visual appearance of tetrahedral geometry (Fig. 6.9a) of

Quantum Dots of Indium Nitrides

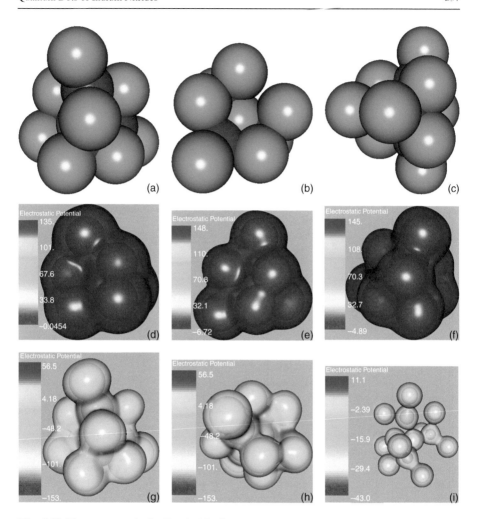

Fig. 6.11 The vacuum singlet $In_{10}As_2N_2$. Structure: (a) to (c). MEP surfaces (color scheme) corresponding to the CDD isovalues (d) 0.0001, (e) 0.0005, (f) 0.001, (g) 0.01, (h) 0.01 and (i) 0.05. MEP minimum and maximum values: −4.56018 and 3.23418, respectively (in arbitrary units). In (a) to (f) all dimensions are to scale. In (g) to (i) atomic dimensions are reduced. Linear dimensions of As atoms are somewhat enlarged in (a) to (f) due to inaccuracy of the visualization software (Molekel). Indium atoms are blue, As brown, N deep blue.

its parent vacuum $In_{10}As_2P_2$ molecule, whose atoms moved only slightly from the positions once occupied by In and As atoms in the pre-designed $In_{10}As_4$ molecule. At the same time, ELS, MOs, MEP and SDD of the pre-designed InAsP-derived $In_{10}As_2N_2$ molecule retain only distant resemblance to the tetrahedral pyramid. As expected, in the process of unconstrained total energy minimization 6 In atoms of the vacuum $In_{10}As_2N_2$ molecule moved closer to their closest nitrogen neighbors causing a loss of tetrahedral geometry of this molecule (Figs. 6.11a to 6.11c).

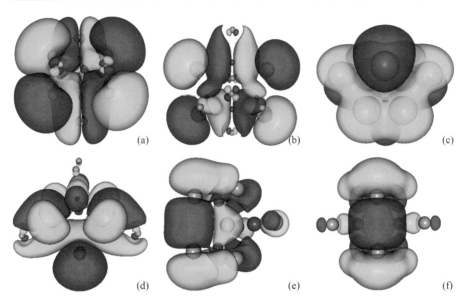

Fig. 6.12 Molecular orbitals (MOs) of the vacuum RHF singlet $In_{10}As_2N_2$. Isosurfaces of negative (semi-transparent red) and positive (semi-transparent green) components of MOs in the HOMO – LUMO region: (a) HOMO 114, isovalue 0.001; (b) HOMO 114, isovalue 0.01, (c) HOMO 115, isovalue 0.001, (d) HOMO 115, isovalue 0.01, (e) LUMO 116, isovalue 0.01, (f) LUMO 116, isovalue 0.01. Indium atoms are blue, As brown, N deep blue. Atomic dimensions are reduced to show the structure.

Outside of the molecular volume, MEP surfaces of both pre-designed $In_{10}As_2N_2$ molecules retain tetrahedral symmetry (Figs. 6.7b, 6.7c, 6.9b and 6.9c). Electron charge of In atoms in these molecules is re-distributed toward the surfaces causing an increase in the electron charge density distribution (CDD) values at the surfaces. In some areas of the surfaces CDD values are larger than those in the case of the corresponding pre-designed $In_{10}As_3N$ molecules. This effect is due to nitrogen atoms whose "belt" of electron charge attracted from the neighbouring In atoms is wider and CDD values larger than those of the corresponding pre-designed $In_{10}As_3N$ molecules. This "belt" reaches further into space encompassing parts of the molecular surfaces, and beyond. The electron charge of In atoms attracted by nitrogen also forces additional re-distribution of the remaining electron charge in these molecules some of which becomes squeezed near the molecular surfaces and beyond due to higher CDD values inside of the molecules in the vicinity of the nitrogen atoms (Figs. 6.7d to 6.7f, and 6.9d to 6.9f).

MEP surfaces of the vacuum $In_{10}As_2N_2$ molecule exhibit only traces of tetrahedral symmetry (Figs. 6.11d to 6.11i). The surface of this molecule (Figs. 6.11d to 6.11f) is much less negatively charged than those of the pre-designed $In_{10}As_2N_2$ molecules, and is almost neutral. In the volume of this molecule a significant electron charge re-distribution toward nitrogen atoms produces a prominent "belt" of electron charge that causes higher negative CDD values exhibited by portions of MEP surfaces in the vicinity of the negative charge belt (Figs. 6.11g to 6.11i).

Similar to the case of $In_{10}As_3N$ and many other non-stoichiometric molecules, charge re-distribution toward surfaces of $In_{10}As_2N_2$ molecules is facilitated by the development of N-and As- mediated In ligand bonding that violates the traditional octet rule. MOs of these molecules are suitably highly delocalized and hybridized, and feature up to 50 contributions from AOs of various atoms. In some contrast to the corresponding MOs of $In_{10}As_3N$ molecules, MOs of $In_{10}As_2N_2$ molecules in the HOMO-LUMO region do not exhibit significant contributions from $5p$ AOs of In atoms. Thus, HOMO 115 of the pre-designed $In_{10}As_2N_2$ molecule (Figs. 6.8a and 6.8b) includes the major contributions coming from $4d$ AOs of 3 In atoms, 2p AO of a nitrogen atom and $4p$ AOs of 2 As atoms (named in the order of the magnitude of the contributions from the largest to smaller ones). The major contributions to the proper HOMO 116 of this molecule (Figs. 6.8c to 6.8e) come from $4d$ AOs of 2 In atoms and a $2p$ AO of one nitrogen atom, while 2 As atoms and another nitrogen contribute less, and $4d$ AOs of 2 other In atoms contribute even less. In the near HOMO-LUMO region of this molecule, only LUMO 117 (Figs. 6.8f and 6.8g) exhibits a large contribution from $5p$ AOs of 3 In atoms. At the same time, 2 N and other 2 In atoms contribute to this MO through their $2p$ and $4d$ AOs, respectively.

The structure of MOs of the pre-designed InAsP-derived $In_{10}As_2N_2$ molecule in the near HOMO-LUMO region differs only slightly from that of the pre-designed $In_{10}As_2N_2$ molecule described above. Thus, HOMO 115 of this molecule (Figs. 6.10a and 6.10b) exhibits the major contributions from $5p$ AO of one In atom, $4d$ AOs of 2 In atoms and $2p$ AOs of 2 N atoms, and $4p$ AOs of both As atoms. The proper HOMO 116 of this molecule (Figs. 6.10c to 6.10e) includes the major contributions from $4d$ AOs of 2 In atoms, $2p$ AO of a nitrogen and $4p$ AOs of 2 As atoms, $4d$ AOs of 3 other In atoms, and a small contribution from $5p$ AOs of 2 In atoms. LUMO 117 of this molecule resembles LUMO 117 of the pre-designed $In_{10}As_2N_2$ molecule in that that the major contribution to it comes from $5p$ AO of an In atom, while $2p$ AOs of 2 N and $4d$ AOs of 2 other In atoms contribute a little less.

In the case of the vacuum $In_{10}As_2N_2$ molecule, MOs in its near HOMO-LUMO region demonstrate somewhat larger signature of $5p$ AOs of In atoms. In particular, the HOMO 114 (Figs. 6.12a and 6.12b) of this molecule features the major contributions coming from $4p$ AOs of both As atoms, $5p$ AOs of 4 In atoms, $2p$ AOs of both N atoms, and $4d$ AOs of another 4 In atoms. The proper HOMO 115 (Figs. 6.12c and 6.12d) of this molecule includes $2p$ AOs of 2 As and $5p$ AO of one In atoms, somewhat smaller contributions from $5p$ AOs of other 3 In atoms, $4d$ AO of yet another indium atom, $2p$ AOs of 2 N atoms, and $4d$ AOs of 4 other In atoms. LUMO 116 of this molecule (Figs. 6.12e and 6.12f) is defined by $5p$ AOs of 8 In atoms, $2p$ AOs of both N atoms and $4p$ AOs of both As atoms.

As a rule, MOs in the near HOMO-LUMO region of all $In_{10}As_2N_2$ molecules contain significant portions formed by bonding of 3 In atoms to one of the nitrogen atoms on the top of such a pyramid (3 In + N), and/or the group bonding of the form (2 In + N) + In +(N + 2 In), and/or As + (In + N + In) +As. Therefore, a nitrogen atom in the $In_{10}As_2N_2$ molecules exhibit a tendency to bond 2 or 3 In atoms and one or two As or N atoms at the same time. All MOs of these molecules in the near HOMO-LUMO region are bonding and exhibit the presence of significant contributions from $4d$ AOs of In atoms. The latter leads to a specific "knotted sandwich" (or bow-like) structure of these MOs where portions of $5p$ In – $4p$ As, $4d$ In – $2p$ N and $4p$ In – $2p$ N are hybridized by portions of $4d$ AOs of In atoms (Figs. 6.7, 6.9 and 6.10). Ligand bonding of In atoms (via their 4d or $5p$ AOs) are usually mediated by $4p$ AOs of As and $2p$ AOs of N atoms.

Magnetic properties of $In_{10}As_2N_2$ molecules are not remarkable. The pre-designed molecules are "ferromagnetic" triplets with uncompensated z-component of the spin magnetic moment aligned and totaling to 2 Bohr magnetons. The vacuum $In_{10}As_2N_2$ molecule is an "antiferromagnetic" singlet with the uncompensated spin magnetic moment equal to zero. To be useful for spintronics, magnetic properties of these QDs have to be enhanced by doping with "magnetic" atoms, such as Ni or Co.

6.5 Pyramidal molecules $In_{10}N_4$

In similarity to the virtual synthesis procedure used to develop the pre-designed $In_{10}As_3N$ and $In_{10}As_2N_2$ molecules, the pre-designed $In_{10}N_4$ one was derived from the $In_{10}As_4$ cluster of Chapter 3 by replacement of all arsenic atoms with nitrogen ones and subsequent minimization of the total energy of the cluster so obtained. In the process of this optimization, the positions of the centers of mass of the cluster's atoms were fixed (that is, spatial constraints were applied). The corresponding vacuum cluster was derived from the pre-designed one upon lifting the spatial constraints (the cluster's atoms were allowed to "move") followed by unconditional total energy minimization. Ground state electronic properties of these molecules are collected in Tables 6.5 and 6.6, and their MEP and MOs surfaces illustrated in Figs. 6.13 to 6.16.

Replacement of all arsenic atoms with nitrogen ones resulted in the deep ground state energy mimima of about -1921 H (Table 6.5) consistent with an earlier discovered tendency that a replacement of As atom with N one in tetrahedral In-As molecules brings the total energy down by about 3.4 H. OTEs of these pre-designed ROHF triplet and RHF singlet $In_{10}N_4$ (Table 6.6) are larger than ROHF/RHF OTEs of their pre-designed and vacuum tetrahedral counterparts containing As atoms. Further refinement of the OTE data by application of CI, MCSCF and MP2 approximation is due, and is expected to lessen the ROHF/RHF OTE values of the $In_{10}N_4$ molecules. Similar

TABLE 6.5 Ground state energies of the tetrahedral $In_{10}N_4$ molecules.

Molecule	ROHF ground state energy, Hartree	MCSCF ground state energy, Hartree	CAS (number of electrons × number of MOs)
Pre-designed $In_{10}N_4$	-1921.097652447	Not available	ROHF triplet
Vacuum $In_{10}N_4$	-1921.438151092	Not available	RHF singlet

TABLE 6.6 Direct optical transition energies (OTEs) of the tetrahedral $In_{10}N_4$ molecules.

Molecule	ROHF OTE, eV	MCSCF OTE, eV	CAS (number of electrons × number of MOs)
Pre-designed $In_{10}N_4$	4.1171	Not available	ROHF triplet
Vacuum $In_{10}N_4$	7.0396	Not available	RHF singlet

to other tetrahedral N-containing molecules, $In_{10}N_4$ ones are polar with the dipole moment of the pre-designed molecule being 0.9328 D and that of the vacuum molecule 4.1886 D. Thus, the dipole moment of the vacuum $In_{10}N_4$ molecule (whose ground state energy is the lowest among the studied tetrahedral molecules with N atoms) is over 4 times larger than that of the corresponding pre-designed molecule. This finding is in line with a similar phenomenon in the case of $In_{10}As_2N_2$ molecules. Thus, one is lead to a conclusion that in fcc-derived In-As molecules with high nitrogen content

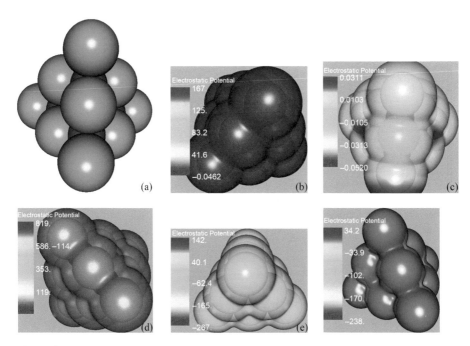

Fig. 6.13 The pre-designed ROHF triplet $In_{10}N_4$. Structure: (a). MEP surfaces (color scheme) corresponding to the CDD isovalues (b) 0.001, (c) 0.002, (d) 0.005, (e) 0.01 and (f) 0.03. MEP minimum and maximum values: −6.93221 and 1.35981, respectively (in arbitrary units). All dimensions are to scale. Indium atoms are blue, N deep blue.

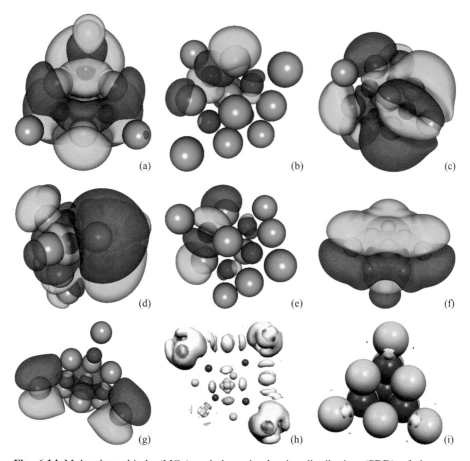

Fig. 6.14 Molecular orbitals (MOs) and the spin density distribution (SDD) of the pre-designed ROHF triplet $In_{10}N_4$. Isosurfaces of negative (semi-transparent red) and positive (semi-transparent green) components of MOs in the HOMO – LUMO region: (a) HOMO 115, isovalue 0.001; (b) HOMO 115, isovalue 0.01, (c) HOMO 116, isovalue 0.0005 (d) HOMO116, isovalue 0.001, (e) HOMO 116, isovalue 0.01, (f) LUMO 117, isovalue 0.001, and (g) LUMO 117, isovalue 0.01. SDD isosurfaces (yellow) corresponding to the isovalues (h) 0.001 and (i) 0.01, respectively. Indium atoms are blue, N deep blue. Atomic dimensions are reduced to show the structure.

electron charge re-distribution due to nitrogen atoms that act as electron charge thinks is the major process governing decline in the total energy of the molecules by about 3.4 H per N atom with inclusion of additional nitrogen atoms. This process offsets an increase in the total energy of the molecules due to increasing polarity regardless of the geometry of the molecules. [Notably, such molecules are not necessarily pyramidal. In particular, the structure of the vacuum $In_{10}As_2N_2$ (Figs. 6.11a to 6.11c) and $In_{10}N_4$ (Figs. 6.15a to 6.15c) molecules is only remotely reminds the tetrahedral pyramids of their parent molecules.]

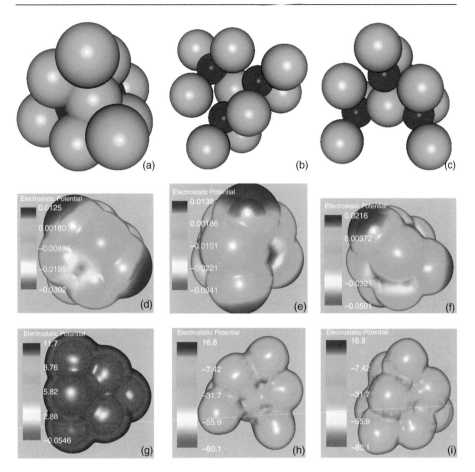

Fig. 6.15 The vacuum ROHF singlet $In_{10}N_4$. Structure: (a) to (c). MEP surfaces (color scheme) corresponding to the CDD isovalues (d) 0.0005, (e) 0.001, (f) 0.07, (g) 0.01, (h) 0.05 and (i) 0.05. MEP minimum and maximum values: −4.35883 and 1.69910, respectively (in arbitrary units). A "belt" of negative MEP due to electron charge accumulation by nitrogen atoms is prominent in (d) to (f). In (b), (c), (h) and (i) atomic dimensions are reduced. In (a), (d), and (e) to (g) all dimensions are to scale. Indium atoms are blue, N deep blue.

The MEP surfaces of the $In_{10}N_4$ molecules (Figs. 6.13 and 6.15) provide further insight into the structure of the delocalized electron charge "belt" facilitated by nitrogen atoms. In the case of the pre-designed molecule, delocalized electron charge is pushed further beyond the molecular surface (Figs. 6.13b to 6.13i) than that of its vacuum counterpart (Figs. 6.15d to 6.15g). For example, in the case of the CDD isovalue 0.001 the MEP surface of the pre-designed molecule is negative (Fig. 6.13b), while the corresponding MEP surface of the vacuum molecule (Fig. 6.15e) exhibits significant regions of positive MEP values that indicate electron charge deficit. Only in the immediate vicinity of the atomic "spheres" (as defined by their covalent radii) MEP surfaces of the vacuum molecule (Fig. 6.15g) become slightly negative. For this

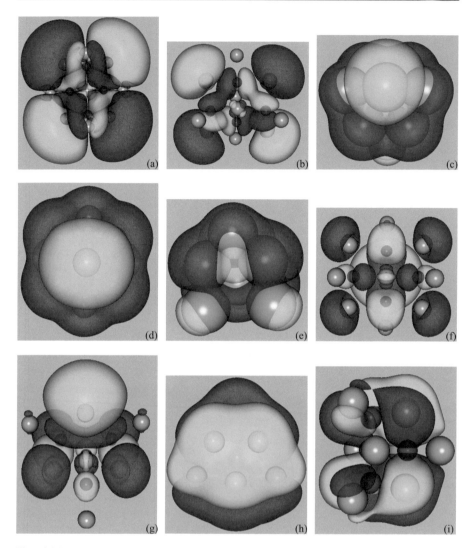

Fig. 6.16 Molecular orbitals (MOs) of the vacuum ROHF singlet $In_{10}N_4$. Isosurfaces of negative (semi-transparent red) and positive (semi-transparent green) components of MOs in the HOMO – LUMO region: (a) HOMO 114, isovalue 0.001; (b) HOMO 114, isovalue 0.01, (c) HOMO 115, isovalue 0.001 (d) HOMO115, isovalue 0.001, (e) HOMO 115, isovalue 0.001, (f) HOMO115, isovalue 0.01, and (g) HOMO115, isovalue 0.01, (h) LUMO116, isovalue 0.001, and (i) LUMO 116, isovalue 0.01. Indium atoms are blue, N deep blue. Atomic dimensions are reduced to show the structure.

CDD isovalue (0.01), the negative values of MEP of the pre-designed molecule are about two orders of magnitude higher (Fig. 6.13e). The "belt" of delocalized electronic charge is more distinct in the case of the vacuum molecule, as it reaches into the space outside of the molecular surface (Figs. 6.15d to 6.15f) further than that of the pre-designed molecule (Fig. 6.13c). Moreover, the delocalized electronic charge is more uniformly distributed over the "belt" of the vacuum molecule, with the major charge kept inside of the molecular volume closer to the nitrogen atoms (yellow and red regions in Figs. 6.15d to 6.15f).

Similar to the case of other fcc-derived N-containing molecules discussed in this chapter, the electronic charge re-distribution from molecular volumes to their surfaces provides for the development of non-standard ligand bonding breaking the octet rule. For example, the most significant MOs in the homo-LUMO region of the pre-designed $In_{10}N_4$ molecule include the major contributions as follows. (1) HOMO 115 (Figs. 6.14a and 6.14b) includes leading contributions from: $2p$ AO of one N atom, $4d$ AOs of 3 In atoms, and $2p$ AOs of the rest of N atoms. (2) The proper HOMO 116 (Figs. 6.14c to 6.14e) includes leading contributions from: $4d$ AOs of 2 In atoms, $2p$ AO of one N atom, and $2p$ AOs of 2 other N atoms. (3) LUMO 117 (Figs. 6.14f and 6.14g) includes leading contributions from: $2p$ AOs of 3 N atoms, $4d$ AOs of 3 In atoms and $5p$ AOs of 3 other In atoms resulting in the development of a strong 3 N + 3 In bond.

The most significant MOs of the vacuum $In_{10}N_4$ molecule exhibit somewhat high contributions from $5p$ AOs of In atoms. Thus, (1) HOMO 114 (Figs. 6.16a and 6.16b) contains large portions of $5p$ AOs of 4 In atoms, $2p$ AOs of 2 N atoms, and $4d$ AOs of 4 other In atoms. (2) The proper HOMO 115 (Figs. 6.16c to 6.16g) of this molecule includes large parts of $5p$ AO of one In atom, $2p$ AOs of 2 N atoms, $4d$ AOs of 4 other In atoms, $2p$ AOs of 2 remaining N atoms, and much smaller contributions of $4d$ AOs of the remaining 5 In atoms. (3) LUMO 116 (Figs. 6.16h and 6.16i) of this molecule is the only MO that is built of major contributions due to $5p$ AOs of 5 In atoms and $2p$ AOs of 4 N atoms. The $4d - 5p$, $4d - 4d$ and $5p - 5p$ In ligand bonding mediated by $2p$ arsenic and nitrogen AOs (with each nitrogen atoms exhibiting a tendency to form bonds with 3 In atoms) continues to define remarkable stability of these molecules.

Similar to other fcc-derived N-containing molecules, magnetic properties of $In_{10}N_4$ ones are of little interest for applications, unless the molecules are doped with "magnetic" atoms. The pre-designed molecule is a "ferromagnetic" triplet with its 2 uncompensated z-components of the spin magnetic moment contributed by spins localized near 2 In atoms (Figs. 6.14h and 6.14i). The vacuum molecule is an "antiferromagnetic" singlet with zero total magnetic moment.

6.6 Hexagonal molecules In_6N_6

Wurtzite bulk lattices offer numerous opportunities to simplify virtual design of QDs by using hexagonal symmetry elements for the development of pre-designed cluster structures that model atomic clusters nucleation in quantum confinement offered by such lattices or surfaces. It also is easy to introduce lattice defects in such structures to study effects of structural defects on optoelectronic and magnetic properties of such quantum-confined QDs. In this section two of such QDs are discussed. One of them is the pre-designed In_6N_6 cluster (Fig. 6.17a) derived from the "ideal" hexagonal prism, which is the major symmetry element of a wurtzite lattice with parameters

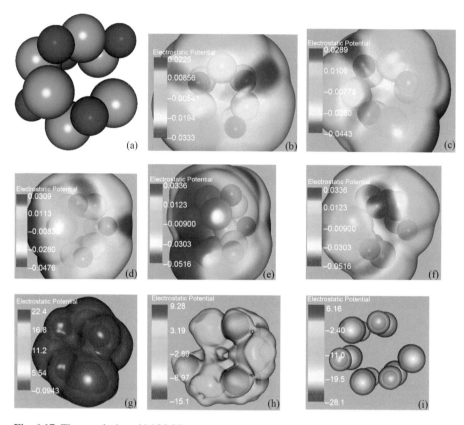

Fig. 6.17 The pre-designed MCSCF pentet In_6N_6. Structure: (a). MEP surfaces (color scheme) corresponding to the CDD isovalues (b) 0.001, (c) 0.005, (d) 0.007, (e) 0.01, (f) 0.01, (g) 0.05, (h) 0.1, and (i) 0.2. MEP minimum and maximum values: -3.41457 and 5.32946, respectively (in arbitrary units). Atomic dimensions are reduced. Indium atoms are blue, N deep blue.

$a = 3.5366$ Å, $c = 5.7009$ Å and $u = 0.3769$ Å obtained from experimental data. A structural defect in the form of two In and N atoms positioned one over another and simultaneously displaced from their positions in the "ideal" hexagonal prism was introduced into the structure of this cluster. Simultaneous displacement of these atoms opened the corresponding hexagonal rings of the original "ideal" prism. Interatomic distances between the dislocated pair of atoms and one of the two pairs of neighbouring In and N atoms were kept unchanged. This defect models the simplest dislocation of 2 atoms in a wurtzite lattice. The HF, CI, CASSCF and MCSCF methods were used to minimize the total energy of this defective cluster conditionally, that is, in the presence of spatial constraints applied to the centers of mass of the cluster's atoms. The result is the pre-designed In_6N_6 molecule of Figs. 6.17 and 6.18 that realizes a local minimum of the total energy of such a defective prismatic structure composed of In and N atoms. The corresponding vacuum In_6N_6 molecule (Figs. 6.19 and 6.20)

Quantum Dots of Indium Nitrides 267

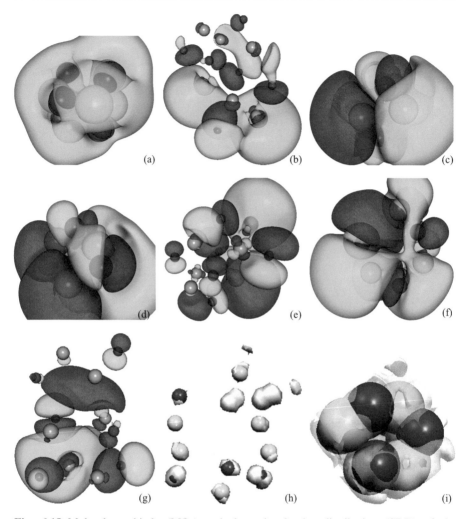

Fig. 6.18 Molecular orbitals (MOs) and the spin density distribution (SDD) of the pre-designed ROHF pentet In_6N_6. Isosurfaces of negative (semi-transparent red) and positive (semi-transparent green) components of MOs in the HOMO – LUMO region: (a) HOMO 79, isovalue 0.002; (b) HOMO 79, isovalue 0.02, (c) HOMO 80, isovalue 0.001, (d) HOMO 80, isovalue 0.001, (e) HOMO 80, isovalue 0.01, (f) LUMO 81, isovalue 0.002, and (g) LUMO 80, isovalue 0.02. SDD isosurfaces (yellow and transparent yellow) corresponding to the isovalues (h) 0.01 and (i) 0.005, respectively. Indium atoms are blue, N deep blue. Save for (i), atomic dimensions are reduced to show the structure. In (i) atomic dimensions are somewhat enlarged.

was obtained upon lifting the position constraints applied to the atoms of the pre-designed molecule, and subsequent total energy minimization using HF, CI, CASSCF and MCSCF approximations. As expected, the vacuum molecule is the perfect hexagonal prism (Fig. 6.19a) that realizes the global minimum of the perfect hexagonal

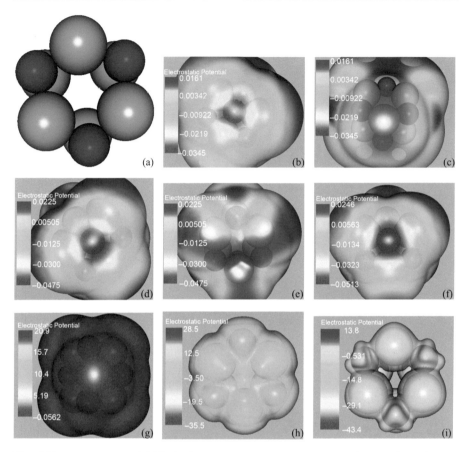

Fig. 6.19 The vacuum MCSCF singlet In_6N_6. Structure: (a). MEP surfaces (color scheme) corresponding to the CDD isovalues (b) 0.001, (c) 0.001, (d) 0.005, (e) 0.005, (f) 0.007, (g) 0.01, (h) 0.05, and (i) 0.1. MEP minimum and maximum values: −5.96534 and 10.60720, respectively (in arbitrary units). Atomic dimensions are reduced. Indium atoms are blue, N deep blue.

prism composed of 6 In and and 6 N atoms according to experimental parameters of a realistic wurtzite InN bulk lattice. Electronic properties of these molecules are encompassed in Tables 6.7 and 6.8.

The HF ground state energies of both wurtzite-derived molecules (Table 6.7) is over 700 H higher than those of the zincblende-derived molecules (Tables 6.1, 6.3 and 6.5). Thus, from a thermodynamic standpoint, the wurtzite-derived molecules are significantly less stable than fcc-derived ones. Moreover, the ground state energy of the pre-designed In_6N_6 molecule (that features a significant dislocation structural defect involving 2 atoms out of the total 12 atoms of this molecule) is only about 0.4 H higher than that of the vacuum molecule (see MCSCF data of Table 6.7). That is, the ground state energies of these molecules are within the error brackets of calculations

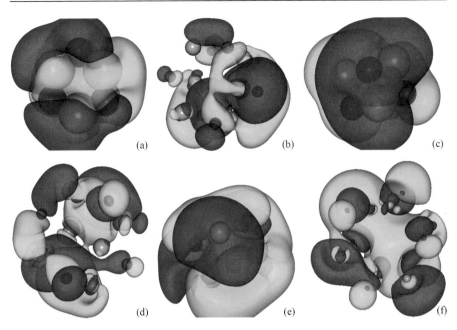

Fig. 6.20 Molecular orbitals (MOs) and the spin density distribution (SDD) of the vacuum MCSCF singlet In_6N_6. Isosurfaces of negative (semi-transparent red) and positive (semi-transparent green) components of MOs in the HOMO – LUMO region: (a) HOMO 77, isovalue 0.001; (b) HOMO 77, isovalue 0.01, (c) HOMO 78, isovalue 0.001, (d) HOMO 78, isovalue 0.01, (e) LUMO 79, isovalue 0.001, and (f) LUMO 79, isovalue 0.01. Indium atoms are blue, N deep blue. Save for (a) and (c), atomic dimensions are reduced to show the structure. In (a) and (c) all dimensions are to scale.

from each other. This fact has a profound consequence experienced by experimentalists: synthesis of wurtzite InN lattices with low concentration of structural defects is an extremely demanding task. Moreover, OTE of the pre-designed molecule is about 4 times smaller than that of the vacuum one, indicating that the pre-designed molecule realizes a local minimum of the total energy of the prismatic In_6N_6 cluster. Yet another

TABLE 6.7 Ground state energies of the wurtzite-derived In_6N_6 molecules.

Molecule	ROHF ground state energy, Hartree	MCSCF ground state energy, Hartree	CAS (number of electrons × number of MOs)
Pre-designed In_6N_6	−1186.855939	−1186.937738796	8 × 9
Vacuum In_6N_6	−1187.166724	−1187.271610806 −1187.163804381	8 × 7 12 × 10

TABLE 6.8 Direct optical transition energies (OTEs) of the wurtzite-derived In_6N_6 molecules.

Molecule	ROHF OTE, eV	MCSCF OTE, eV	CAS (number of electrons × number of MOs)
Pre-designed In_6N_6	1.6735	1.7440	8 × 9
Vacuum In_6N_6	6.3185	9.7495	8 × 7
		6.9780	12 × 10

tendency also is confirmed: in the case of nitrogen-containing molecules composed of In and As atoms the use of a larger CAS leads the lower computational values of OTE. Thus, 12 × 10 CAS MCSCF calculations (Table 6.8) produced the OTE value of the vacuum In_6N_6 molecule that is almost 3 eV lesser than that obtained by 8 × 7 CAS MCSCF calculations. Interestingly, the 8 × 9 CAS MCSCF OTE of the pre-designed molecule (1.7440 eV; Table 6.8) almost coincides with a recent experimental evaluation [55] of the fundamental band gap in hexagonal InN. This result indicates that the experimental value is largely overestimated, because from fundamental considerations small crystalline nanostructures are expected to possess OTEs significantly larger than those of their parent bulk lattices.

Similar to the fcc-derived N-containing molecules, the prismatic molecules are polar, with the dipole moment of the defective pre-designed one being 3.1388 D and much larger than that of the vacuum prismatic molecule (0.5650 D). This finding is further illustrated by MEP surfaces of these molecules (Figs. 6.17b to 6.17i and 6.19b to 6.19i). MEP surfaces of the pre-designed molecule confirm the total loss of hexagonal symmetry caused by the structural defect, while MEP surfaces of the vacuum molecule demonstrate its perfect hexagonal symmetry. Both molecules have developed stability "belts" of delocalized electron charge of In atoms collected by nitrogen atoms illustrated in Figs. 6.17b to 6.17f and 6.19b to 6.19f (greenish, yellowish and reddish regions of the MEP surfaces). Thus, while the surfaces of these molecules are negatively charged, the charge is distributed over the surfaces non-uniformly (Figs. 6.17b, 6.17c, and 6.19b to 6.19f).

The charge re-distribution toward surfaces of these molecules facilitates a new type of bonding violating the octet rule and similar to that specific to the fcc-derived molecules discussed above. The major difference in ELS structure of the two wurtzite-derived molecules is related to the molecular symmetries. Thus, apart from spontaneous degeneration, MOs of the pre-designed molecule are non-degenerate. In the case of the vacuum In_6N_6 molecule, both the proper HOMO 78 and LUMO 79 are non-degenerate, while the majority of other MOs in the near HOMO-LUMO region are twice degenerate, E-type orbits specific to hexagonally symmetric molecules.

The only significant difference in the nature of bonding in the wurtzite-derived molecules compared to that of the fcc-derived ones is manifested by the absence of or very small contributions from $5p$ AOs of In atoms. Thus, MCSCF calculations reveal that bonding in the most significant MOs of the pre-designed molecule is defined by the following major contributions. (1) HOMO 79 (Figs. 6.18a and 6.18b) includes contributions from: $4d$ AOs of 2 In atoms, $2p$ AOs of all 4 N atoms, and $4d$ AOs of another 3 In atoms. (2) The proper HOMO 80 (Figs. 6.18c to 6.18e) includes contributions from: $4d$ AOs of 2 In atoms, $2p$ AOs of 3 N atoms, $4d$ AOs of the rest of In atoms and $2p$ AO of the remaining N atom. (3) LUMO 81 (Figs. 6.18f and 6.18g) includes contributions from: $4d$ AOs of 2 In atoms, $2p$ AOs of 4 N atoms and $4d$ AOs of the rest of In atoms. Despite of the structural defect, the stability "belt" of delocalized electron charge collected by nitrogen atoms and re-distributed from indium atoms via $4d$ In ligand bonding mediated by $2p$ nitrogen AOs ensures an impressive stability of the defective molecule. The MO signature of the vacuum molecule is very similar. Its (1) HOMO 77 (Figs. 6.20a and 6.20b) contains the major contributions from $2p$ AOs of 2 N atoms, $4d$ AOs of 4 In atoms, $2p$ AOs of the other 2 N atoms and $4d$ AOs of the remaining 2 In atoms. (2) The proper HOMO 78 (Figs. 6.20c and 6.20d) includes hybridized $2p$ AOs of 3 N atoms, $4d$ AOs of 3 In atoms, somewhat smaller contributions from $2p$ AOs of the 3 remaining N atoms and $4d$ AOs of 2 other In atoms, and a minor contributions from $5p$ AO of the remaining In atom. (3) LUMO 79 (Figs. 6.20e and 6.20f) of the vacuum molecule is built of $2p$ AOs of 3 N atoms, $4d$ AOs of 3 In atoms, somewhat smaller contributions of $2p$ AOs of the remaining 3 N atoms, $4d$ AOs of 2 other In atoms, and small contributions due to $5p$ AOs of 2 In atoms one of which also contributes via its $4d$ AO. Due to a leading role of $4d$ AOs of In atoms and $2p$ AOs of N atoms, MOs of the wurtzite-derived molecules in their respective HOMO-LUMO regions have complex structures of knots made of $4d$ In ligand bonding regions mediated by $2p$ AOs of N atoms. The $5p$ In ligand bonding is not outstanding, and further MP2 studies are necessary to ascertain whether it is present, indeed. This finding is consistent with results of a recent experimental study that did not observe any signs of $5p$ In – $2p$ N bonding in gallium and indium hydrazides [66]. Similar to the case of the fcc-derived molecules, nitrogen atoms in the wurtzite-derived molecules tend to form bonds with In atoms, but may also bond directly other nitrogen atoms. In addition, In atoms in these molecules tend to bond 3 N atoms directly, so the In ligand bonding is always mediated by N atoms. These results are consistent with known experimental findings (see, for example Ref. 66 and references therein). Notably, according to GGA DFT results of Ref. 67, in small M_nN_m clusters (where n and m run from 1 to 2, and M stands for Al, Ga or In) featuring 2 nitrogen atoms a dinitrogen molecule is formed, and in In_nN_m clusters the In – N bond is non-shared. Thus, GGA DFT findings contradict to the existing experimental facts, while the above HF/CI/MCSCF findings are consistent with experimental evidence.

The pre-designed In_6N_6 molecule is a "ferromagnetic" ROHF/CI/CASSCF/MCSCF triplet with its uncompensated spins carried by In atoms. Interestingly, due to the structural defect and significant indium electron charge re-distribution toward nitrogen atoms enhanced by the defect, 3 of nitrogen atoms provide noticeable contributions to SDD. The total uncompensated z-component of the spin magnetic moment of this molecule is 2 Bohr magnetons. The vacuum molecule is an "antiferromgnetic" RHF/CI/CASSCF/MCSCF singlet with all its electron spins compensated. Thus, magnetic properties of the wurtzite-derived In_6N_6 molecules are closer to those of the corresponding fcc-derived molecules with two nitrogen atoms.

Importantly, MCSCF OTE of the pre-designed defective In_6N_6 molecule (Table 6.8) almost coincides with the experimental re-evaluations [68] of the fundamental band gap of wurtzite InN, pointing out that the experimental value of Ref. 68 is overestimated. At the same time, the obtained MCSCF OTE of the pre-designed In_6N_6 molecule supports the experimental value of the fundamental band gap of bulk wurtzite InN reported in Ref. 55 and that of the InN nanobelts of Ref. 69.

6.7 Conclusions

Two groups of small indium nitride QDs have been studied [25,64,70,71] using symmetry considerations and the first-principle quantum many body theoretical methods, including HF, CI, CASSCF and MCSCF approximations. The first group includes 8 fcc-derived pyramidal-like molecules built of 10 In atoms, 3 or 2 As atoms and 1 or 2 N atoms. The pre-designed molecules are ROHF/CI/CASSCF/MCSCF triplets, and their structure resembles that of their parent pre-designed $In_{10}As_4$, vacuum $In_{10}As_3P$ and $In_{10}As_2P_2$ pyramidal clusters. Two of the three vacuum molecules, $In_{10}As_2N_2$ and $In_{10}N_4$, realizing the global minima of the total energy of the respective atomic clusters, are RHF/CI/CASSCF/MCSCF singlets. The structure of these molecules was derived from that of the tetrahedral symmetry element of the fcc InAs bulk lattice, where 1, 2 or 4 As atoms were substituted with nitrogen atoms. The pre-designed molecules model nucleation in quantum confinement or on surfaces by accounting for excluded volume effects and some polarization due to the confinement. The vacuum molecules model nucleation in the absence of any external fields or foreign atoms. Considering a significant proportion of nitrogen that replaced As in several pre-designed molecules, one is lead to a conclusion that nitrogen substitution defects are energetically favorable in the case of the fcc bulk InAs lattice, surfaces and nanostructures. This conclusion is strongly supported by a finding that such substitution is accompanied by a decrease in the total energy of a 14-atomic pyramidal cluster containing 10 In and 4 As atoms by about 3.4 H per each substituted As atom.

The second group of the studied molecules was derived from a hexagonal symmetry element of the wirtzite InN bulk lattice in a way similar to that of the fcc-derived molecules. It includes one pre-designed and one vacuum In_6N_6 molecules. The ground state energies of these molecules are significantly (about 730 H) larger than those of the fcc-derived molecules, but these potential energy wells still are very deep signifying that the wurtzite-derived molecules are also very stable. This conclusion is further supported by the finding that the pre-designed prism-like molecule is a ROHF/CI/CASSCF/MCSCF pentet, and the perfect prism of the vacuum molecule is a ROHF/CI/ CASSCF/MCSCF singlet. All major tendencies revealed by these first-principles studies are similar for both molecular groups.

The ground state of the pre-designed In_6N_6 molecule is somewhat special as it is the only ROHF/CI/CASSCF/MCSCF pentet among all of the studied nitrogen-containing molecules. The reason for this is that this molecule features a dislocation defect that leads to a higher excited state being the ground state of this molecule. The ground state energy of this molecule is much higher than that of the fcc-derived pre-designed

$In_{10}As_4$ molecule [65], so it is less stable. Indeed, there is 12 atoms total in this molecule, as opposed to 14 atoms of the pre-designed $In_{10}As_4$ molecule. In the wurtzite-derived molecule the total energy contribution per atom is about -99 H, while in the case of the fcc-derived molecule it is about -137 H per atom. Interestingly, the ground state energy of the vacuum RHF/CI/CASSCF/MCSCF In_6N_6 singlet is very close to that of the pre-designed defective ROHF/CI/CASSCF/MCSCF In_6N_6 pentet. Thus, when synthesized experimentally, both wurtzite-derived prismatic molecules may be less stable than the fcc-derived pyramidal $In_{10}N_4$ ones. As discussed below in this section, experimentally synthesized InAs and III-nitride QDs are as a rule, tetrahedral pyramids. Thus, quantum many body theoretical findings are in an excellent agreement with experimental evidence confirming that the wurtzite-derived prismatic InN molecules are less stable than their fcc-derived pyramidal siblings.

A physical mechanism behind predicted larger instability of wurtzite-derived molecules may be accumulation of too much of electron charge in the vicinity of the 6 nitrogen atoms and a subsequent increase in the electron Coulomb repulsion that competes with a decrease in the total energy of the molecule due to electron charge re-distribution toward nitrogen atoms and the molecular surface. At non-ideal experimental conditions any permutation of the electron charge density in such circumstances may cause a jump in the Coulomb repulsion, disturbing the energy balance and destabilizing the molecule.

Importantly, the discussed first-principle quantum many body-theoretical results confirm that nitrogen atoms serve as electron charge sink both in the case of fcc-derived molecules, and in the case of wurtzite-derived ones. These findings provide a valuable insight into physical mechanisms behind experimentally observed accumulation of nitrogen near surfaces of bulk InAsN, InGaAsN/GaAs, InAsN, InN and other III-nitrides observed experimentally [56–60,72]. Due to the fact that substitution of As atom with N one causes a decrease of about 3.4 H in the total energy of a symmetry element containing the substituted atom, the nitrogen concentration gradient should drive nitrogen atoms toward the surface of a III-nitride that is negatively charged due to symmetry breaking. Near the surface nitrogen atoms collect some of the excessive electronic charge bringing the surface energy, and thus the total energy of the system down and stabilizing the surface. This nitrogen diffusion process should come to an end once the number of nitrogen atoms near the surface reaches some optimum, because the further increase in the number of nitrogen atoms per symmetry element increases the total energy of that element. This theoretical prediction is supported by calculations of the ground state energy of the wurtzite-derived molecules with 6 nitrogen atoms whose ground state energies are about 730 H higher than those of the fcc-derived ones.

Each of the studied molecules is electroneutral, but its surface is negatively charged due to electron charge re-distribution from the volume of the molecule to its surface. The physico-chemical mechanism accompanying this re-distribution is formation of complex bonds containing large contributions from In ligand bonding and violating the standard octet rule. Up to 100 of AOs of the participating atoms contribute to such bonding and shape about a hundred of highly hybridized and delocalized, topmost occupied MOs of the studied molecules. Nitrogen atoms always contribute via their

$2p$ AOs forming up to 3 bonds with In atoms or with In and other N atoms. This facilitates $4d$- and $5p$- In ligand bonding involving groups of In and N atoms, such as (2 In + N) + In + (N + 2 In), where N atoms are placed on the top of small pyramids formed by 3 In atoms, and where one or more In atoms bond directly to 2 N atoms. The $4d$ In – $2p$ N, and to much lesser degree, $5p$ In – $2p$ N nitrogen-driven In ligand bonding facilitates accumulation of electron charge of In atoms in proximity of nitrogen atoms, and also pushes some of the electron charge toward the molecular surface stabilizing the entire molecule. This process, in which N atoms play a role of electron charge sinks, is revealed as "belts" of electron charge filling the molecular volume and "embracing" the molecular surface (roughly, in the regions neighbouring nitrogen atoms). The above results obtained by the first-principle many body quantum theoretical computations is in an excellent agreement with well-known experimental observations that demonstrate a charge-redistribution role of nitrogen atoms via hybridization of their AOs as a leading factor of bonding and stability of both inorganic and organic nitrogen-containing molecules [73,74].

The ELSs of the studied molecules are typical for such small molecules and are described in detail in Ref. 25. It reflects remaining or acquired symmetries of the molecules via the presence of degenerate MOs. In the HOMO regions of the ELSs there exist bunches of closely lying MOs. With an increase in the number of atoms in an atomic cluster such bunches of MOs give rise to sub-bands of the valence bands of the corresponding larger structures and bulk lattices. As expected, ROHF OTEs of these molecules are overestimated. CI/CASSCF/MCSCF calculations with larger CASs provide much more realistic data for ELSs of the studied molecules, with the OTE values somewhat below or slightly above 2 eV (Tables 6.2 and 6.4). Those OTE values are significantly smaller than the corresponding OTE values of $In_{10}As_4$ molecules [65,75] lying in the range from about 2.3 eV to about 3 eV. These theoretical results are in a very good agreement with experimental observations demonstrating that the fundamental band gap of InAsN nanostructures and bulk lattices decreases with an increase in their nitrogen content [56–58,76–80]. At the same time, CI/CASSCF/MCSCF OTE below 1 eV of the vacuum $In_{10}As_3N$ molecule indicates that such a tendency may not necessarily hold for all nitrogen-containing small InAs molecules. Moreover, RHF OTEs of the vacuum $In_{10}As_2N_2$ (Table 6.6) and $In_{10}As_4$ (Table 6.6) molecules exceed 6 eV and 7 eV, respectively. To clarify thus issue, further CI/CASSCF/MCSCF/MP2 calculations of ELSs and OTEs of these molecules, and MP2 treatment of the rest of the studied molecules, are necessary.

Comparison of ELS, CDD, MEP, MO and SDD surfaces of the studied fcc—and wurtzite-derived pre-designed molecules with those of the vacuum ones confirms that quantum confinement effects have a profound impact on electronic and magnetic properties of molecules formed in such confinement. Despite of weakness of constraints applied to atomic positions of the pre-designed molecules that forbid changes in the atomic positions (such changes usually are in the range of a few hundredths of Angstrom), all electronic properties of the pre-designed molecules are affected disproportionally. In particular, the RHF vacuum singlets, such as $In_{10}As_2N_2$ and $In_{10}N_4$, become ROHF/CI/CASSCF/MCSCF triplets if synthesized in model quantum confinement. But OTE values are affected even more dramatically decreasing from 1.5 to

3 times, when the corresponding atomic clusters nucleate in quantum confinement. These findings lead to a conclusion that QD nucleation in quantum confinement may serve as the major tool for manipulations with electronic and magnetic properties of such QDs. In particular, OTEs of the QDs may be manipulated in the limits of several electron Volts, and their magnetic nature (apparent ferro- or antiferromagnetism) and spin magnetic moments up to an order of magnitude. OTEs of the wurtzite-derived molecules are more sensitive than fcc-derived ones to excluded volume and polarization effects, and possible dislocations introduced due to quantum confinement in the discussed studies. Indeed, the magnitude of RHF OTE of the pre-designed wurtzite-derived molecule is over 5 eV lesser than that of its vacuum counterpart.

Despite a possibility to manipulate magnetic properties of the studied molecules using quantum confinement, such manipulations do not produce sufficiently large magnetic moments in the molecules nucleated in "natural" quantum confinement (that is, in confinements whose structure and composition are similar to those of larger systems that are "parental" to nucleating molecules, as happened in the cases studied here). The magnitude of SDD values specific to the studied fcc- and wurtzite-derived vacuum molecules is very small, because unconstrained nucleation of those molecules leads to compensation of almost all electron spin magnetic moments contributed by $4d$ AOs of In atoms through high delocalization and hybridization of MOs. To enhance magnetic properties of the pre-designed molecules using quantum confinement, care must be taken to choose such confinement different in symmetries and composition from tetrahedral and hexagonal ones typical for those of the studied molecules. Effects of such "foreign" confinement will inevitably produce multiplet ground states of the fcc- and wurtzite-type molecules nucleated in such confinement or on such surfaces. Certainly, if a large spin magnetic moment is desirable, molecules derived from diluted magnetic semiconductor – type structures and those of magnetic ozides (Ni-O, Co-O, etc) described in Chapters 5 and 8 of this book should be chosen.

During the recent two decades, numerous experimental methods were established to synthesize a huge variety of semiconductor compound QDs, including those of III-nitrides, on specially designed substrates of in bulk lattice confinement. Such self-assembled small QDs grown on surfaces or in confinement tend to have tetrahedral pyramid geometry, both when the natural processes drive the self-assembly [81–90], and when special procedures are developed to ensure that kind of geometry [91–95]. Thus, the first principle quantum many body-theoretical studies of this book and chapter provide a guide for experimentalists, especially because experimental data on electronic and magnetic properties of QDs with their linear dimensions in the range of 1 nm are not available. Very important is the fact that (i) no fitting parameters, assumptions or other adjustments have been used in the process of the reported HF, CI, CASSCF and MCSCF calculations; (ii) the same software package (GAMESS) was used to synthesize virtually all of the considered molecules, and (iii) the SBKJC standard basis set was used in all cases. Thus, no unintentional contributions related to changes in virtual synthesis setup, adjustments, software and basic sets were introduced into calculations. This permits unambiguous direct comparison of the calculated data for various individual molecules or their groups.

A very important theoretical finding is that in all small molecules of semiconductor compound atoms, including indium nitride QDs, studied by self-consisted, first-principle quantum many body-theoretical methods, electronic charge of transition metal atoms is re-distributed from the volumes to the surfaces of the molecules. Such re-distribution is accompanied by formation of $4d$ or $5p$ In ligand bonds mediated by arsenic and nitrogen $4p$ and $2p$ AOs, respectively. While each of the nitrogen atoms tends to form 3 bonds with In atoms in such molecules, other types of bonds are also formed. Such bonding violates the octet rule and develops in response of low dimensional systems to symmetry breaking. The ELS, OTE, CDD, MEP and SDD surfaces of such molecules are sensitive to all details of the molecular synthesis environment, if any, and conditions of the molecular synthesis. In addition to reaction to symmetry breaking, the pre-designed molecules formed on surfaces or in confinement are also stressed because of excluded volume and polarization effects of the confinement. Therefore, such molecules tend to have more of electron charge re-distributed to their surfaces than in the case of the corresponding vacuum molecules. This theoretical result provides a fundamental explanation of well-known experimental observations that surfaces of low-dimensional and bulk indium nitride systems are charged.

Comparison of the ground state energies of the studied indium nitride molecules suggests that substitution of an arsenic or phosphorus atom in the fcc-derived In-As or In-As-P molecules studied in Refs. 63, 65 and 75 with nitrogen atoms is thermodynamically beneficial, as it leads to a decrease in the ground state energy of the molecules by about 3.4 H per a nitrogen atom. This result also is indirectly confirmed by a decrease in OTE of the fcc-derived In-N and In-As-N molecules with increasing nitrogen content. The obtained theoretical values of HF and CI/CASSCF/MCSCF OTEs of the studied indium nitride molecules are smaller than those of the corresponding fcc-derived In-As and In-As-P molecules, which is in excellent agreement with the corresponding experimental data [56–58,76–80]. Indeed, it was established experimentally that in InAs-nitride nanostructures and alloys synthesized using MBE methods a real band gap energy shifts to the red with increasing nitrogen content.

The performed theoretical studies of effects caused by a simple dislocation defect in the pre-designed wurtzite-derived molecule In_6N_6 suggest that the band gaps of indium nitride lattices with numerous dislocation defects should be narrower that those with a few such defects by a factor from 2 to 4. High sensitivity of the wurtzite-derived structures to dislocation defects may be used in experimental materials synthesis to develop low-dimensional materials for near IR sensors with tunable band gaps. First-principle theoretical calculations of OTEs of small indium nitride QDs discussed in this chapter provide the upper limits for the band gaps of similar, but larger fcc and wurtzite systems, and may be used to evaluate accuracy of the corresponding experimental data.

Theoretical results discussed in this chapter reveal a role of nitrogen atoms in, and physical and chemical mechanisms of, stabilization of indium nitride systems with broken symmetries. Thus, from the first principles, it has been established that nitrogen atoms act as electron charge sinks enabling In ligand bonding and re-distribution of In electron charge in such systems from the volume to the surface, which is manifested by the development of "belts" of highly delocalized electron charge. This fundamental theoretical result explains mechanisms behind experimentally established [56–61, 73,74] charge delocalization and collectivization in both inorganic and organic nitrogen-containing systems of any dimensions from molecular to bulk.

References

[1] Chen, R.-S., Yang, T.-H., Chen, H.-Y., Chen, L.-C., Chen, K.-H., Yang, Y.-J., Su, C.-H., and Lin, C.-R. (2009). High-gain photoconductivity in semiconducting InN nanowires. *App. Phys. Lett.* **95**, 162112.

[2] Walukiewicz, W., Ager III, J. W., Yu, K. M., Liliental-Weber, Z., Wu, J., Li, S. X., Jones, R. F., and Denlinger, J. D. (2006). Structure and electronic properties of InN and In-rich group III-nitride alloys. *J. Phys. D: Appl. Phys.* **39**, R83–R99.

[3] O'Leary, S. K., Foutz, B. E., Shur, M. S., and Eastman, L. F. (2006). Potential performance of indium-nitride-based devices. *Appl. Phys. Lett.* **88**, 152113.

[4] Asbeck, P. M., Yu, E. T., Lau, S. S., Sullivan, G. J., Van Hove, J., and Redwing, J. M. (1997). Piezoelectric charge densities in AlGaN/GaN HFETs. *Electron. Lett.* **33**, 1230–1231.

[5] Khaselev, O., and Turner, J. A. (1998). A Monolithic Photovoltaic-Photoelectrochemical Device for Hydrogen Production via Water Splitting. *Science* **280**, 425–427.

[6] Nagahama, S.-I., Yanamoto, T., Sano, M., and Mukai, T. (2001). Wavelength dependence of InGaAs laser diode characteristics. *Jpn. J. Appl. Phys.* (Part I) **40**, 3075–3081.

[7] Nakamura, S., Pearton, S., and Fasol, G. (2000). "The Blue Laser Diode: The Complete Story". Springer, Berlin.

[8] Van Hove, J. M., Hickman, R., Klaassen, J. J., and Chow, P. P. (1997). Ultraviolet-sensitive, visible-blind GaN photodiodes fabricated by molecular beam epitaxy. *Appl. Phys. Lett.* **70**, 2282.

[9] Cimalla, V., Pradarutti, B., Matthaus, G., Brukner, C., Riehemann, S., Notni, G., Nolte, S., Tunnermann, A., Lebedev, V., and Ambacher, O. (2007). High efficient terahertz emission from InN surfaces. *Phys. Status Solidi* (b) **244**, 1829–1833.

[10] Wang, K. R., Lin, S. J., Tu, W., Chen, M., Chen, Q.-Y., Chen T.-H., Chen, M.-L., Seo, H.-W., Tai, N.-H., Chang, S.-C., Lo, I., Wang, D.-P., and Chu, W.-K. (2008). InN nanotips as excellent field emitters. *Appl. Phys. Lett.* **92**, 123105.

[11] Lu, Y. S., Ho, C. L., Yeh, J. A., Lin, H. W., and Gwo, S. (2008). Anion detection using ultrathin InN ion selective field effect transistors. *Appl. Phys. Lett.* **92**, 212102.

[12] Bhuiyan, A. G., Hashimoto, A., and Yamamoto, A. (2003). Indium nitride (InN): a review of growth, characterization, and properties. *J. Appl. Phys.* **94**, 2779.

[13] Tang, T., Han, S., Jin, W., Lin, X., Li, C., Zhang, D., Zhou, C., Chen, B., Han, J., and Meyyapan, M. (2004). Synthesis and characterization of single-crystal indium nitride nanowires. *J. Mater. Res.* **19**, 423–426.

[14] Chang, C. Y., Chi, G. C., Wang, W. M., Chen, L. C., Chen, K. H., Ren, F., and Pearton, S. J. (2005). Transport properties of InN nanowires. *Appl. Phys. Lett.* **87**, 093112.

[15] Cheng, G., Stern, E., Turner-Evans, D., and Reed, M. A. (2005). Electronic properties of InN nanowires. *Appl. Phys. Lett.* **87**, 253103.

[16] Calleja, E., Grandal, J., Sanchez-Garcia, M. L., Niebelschutz, M., Cimalla, V., and Ambacher, O. (2007). Evidence of electron accumulation at nonpolar surfaces of InN nanocolumns. *Appl. Phys. Lett.* **90**, 262110.

[17] Shen, C. H., Chen, H. Y., Lin, H. W., Gwo, S., Klochikhin, A. A., and V. Yu. Davydov. (2006). Near-infrared photoluminescence from vertical InN nanorod arrays grown on silicon: effects of surface electron accumulation layer. *Appl. Phys. Lett.* **88**, 253104.

[18] Stoica, T., Meijers, R. J., Calarco, R., Richter, T., Sutter, E., and Luth, H. (2006). Photoluminescence and intrinsic properties of MBE-grown InN nanowires. *Nano Lett.* **6**, 1541–1547.
[19] Othonos, A., Zervos, M., and Pervolaraki, M. (2009). Ultrafast carrier relaxation in InN nanowires grown by reactive vapor transport. *Nano. Res. Lett.* **4**, 122–129.
[20] Schlager, J. B., Bertness, K. A., Blanchard, P. T., Robins, L. H., Roshko, A., and Sanford, N. A. (2008). Steady-state and time-resolved photoluminescence from relaxed GaN nanowires grown by catalyst-free molecular-beam epitaxy. *J. Appl. Phys.* **103**, 124309.
[21] Song, D. Y., Holtz, M. E., Chandolu, A., Bernussi, A., Nikishin, S. A., Holtz, M. W., and Gherasoiu, I. (2008). Effect of stress and free-carrier concentration on photoluminescence in InN. *Appl. Phys. Lett.* **92**, 121913.
[22] Li, S. X., Yu, K. M., Wu, J., Jones, R. E., Walukiewicz, W., Ager III, J. W, Shan, W., Haller, E. E., Lu, H., and Schaff, W. J. (2005). Fermi-level stabilization energy in group III nitrides. *Phys. Rev. B* **71**, 161201 (R).
[23] Naik, V. M., Naik, R., Haddad, D. B., Thakur, J. S., and Auner, G. W. (2005). Room-temperature photoluminescence and resonance-enhanced Raman scattering in highly degenerate InN films. *Appl. Phys. Lett.* **86**, 201913.
[24] Crawley, D., Nikolic, K., and Rorshaw, M. (2004). "3-D Nanoelectronic Computer Architecture and Implementation." Taylor & Francis/CRC Press, London.
[25] Pozhar, L. A. (2010). Small InAsN and InN clusters: electronic properties and nitrogen stability belt. *Eur. Phys. J. D* **57**, 343–354.
[26] Lieb, E. H. (1985). Density functionals for Coulomb systems. *In*: "Density Functional Methods in Physics. NATO Advanced Science Institute, Series B: Physics." (R. M. Dreizler and J. da Providencia, Eds.), Vol. 123, pp. 31–80. Springer, NY.
[27] AlZahrani, A. Z., Srivastava, G. P., Gard, R., and Migliorato, M. A. (2009). An ab initio study of electronic and structural properties of Mn in a GaAs environment. *J. Phys.: Condens. Matter* **21**, 485504.
[28] Perdew, J. P., Burke, K., and Ernzerhof, M. (1996). Generalized gradient approximation made simple. *Phys. Rev. Lett.* **77**, 3865–3868.
[29] Hafner, J. J. (2008). Ab initio simulations of materials using VASP: Density-functional theory and beyond. *J. Comput. Chem.* **29**, 2044–2078.
[30] Galicka, M., Buczko, R., and Kacman, P. (2011). Structure-dependent magnetism in Mn-doped III-V nanowires. *Nano Lett.* **11**, 3319–3323.
[31] Schuch, N., and Verstraete, F. (2007). Interacting electrons, density functional theory and quantum merlin Arthur. arXiv:0712.0483v1.
[32] Schuch, N., and Verstraete, F. (2009). Computational complexity of interacting electrons and fundamental limitations of density functional theory. *Nature Phys.* **5**, 732–735; doi:10.1038/nphys1370.
[33] Qteish, A., Al-Sharif, A. I., Fuchs, M., Scheffler, M., Boeck, S., and Neugebauer, J. (2005). Exact-exchange calculations of the electronic structure of AlN, GaN and InN. *Comp. Phys. Comm.* **169**, 28–31.
[34] Rinke, P., Winkelnkemper, M., Qteish, A., Bimberg, D., Neugebauer, J., and Scheffler, M. (2008). Consistent set of band parameters for the group-III nitrides AlN, GaN and InN. *Phys. Rev.* **B 77**, 075202.
[35] Kotani, T., van Schilfgaarde, M., and Faleev, S. V. (2007). Quasiparticle self-consistent GW method: a basis for the independent-particle approximation. *Phys. Rev. B* **76**, 165106.
[36] Christensen, N. E., Gorczyca, I., Laskowsky, R., Svane, A., Albers, R. C., Chantis, A. N., Kotani, T., and van Schilfgaarde, M. (2009). Electronic and optical properties of III-nitrides under pressure. *Phys. Status Solidi B* **246**, 570–575.

[37] Kaminska, A., Franssen, G., Suski, T., Gorczyca, I., Christensen, N. E., Svane, A., Suchocki, A., Lu, H., Schaff, W. J., Dimakis, E., and Georgakilas, A. (2007). Role of conduction-band filling in the dependence of InN photoluminescence on hydrostatic pressure. *Phys. Rev. B* **76**, 075203.
[38] Zhang, W.-Q., Sun, J.-M., Zhao, G.-F., and Zhi, L.-L. (2008). The structural and electronic properties of In_nN (n=1 – 13) clusters. *J. Chem. Phys.* **129**, 064310.
[39] Zhang, S., and Chen, N. (2005). Lattice inversion for interatomic potentials in AlN, GaN and InN. *Chem. Phys.* **309**, 309–321.
[40] Kandalam, A. K., Blanco, M. A., and Pandey, R. (2002). Theoretical study of Al_nN_n, Ga_nN_n and In_nN_n (n=4, 5, 6) clusters. *J. Phys. Chem. B* **106**, 1945–1953.
[41] Cardelino, B. H., Moore, C. E., Cardelino, C. A., Frazier, D. O., and Bachlmann, K. J. (2001). Theoretical study of indium compounds of interest for organometallic chemical vapor deposition. *J. Phys. Chem. A* **105**, 849–868.
[42] Singh, A. K., Briere, T. M., Kumar, V., and Kawazoe, Y. (2003). Magnetism in transition-metal-doped silicon nanotubes. *Phys. Rev. Lett.* **91**, 146802.
[43] Kumar, V., and Kawazoe, Y. (2003). Hydrogenated silicon fullerenes: effects of H on on the stability of metal-encapsulated silicon clusters. *Phys. Rev. Lett.* **90**, 055502.
[44] See: Lu, J., and Nagase, S. (2003). Structural and electronic properties of metal-encapsulated silicon clusters in a large size range. *Phys. Rev. Lett.* **90**, 115506, and references therein.
[45] Bryant, G. W., and Jaskolski, W. (2005). Surface effects on capped and uncapped nanocrystalls. *J. Phys. Chem. B* **109**, 19650–19656.
[46] Xie, R-H, Bryant, G. W., Zhao, J., Kar, T., and Smith Jr., V. H. (2005). Tunable optical properties of icosahedral, dodecahedral and tetrahedral clusters. *Phys. Rev. B* **71**, 125422.
[47] Schmidt, M. W., Baldridge, K. K., Boatz, J. A., Elbert, S. T., Gordon, M. S., Jensen, J. H., Koseki, S., Matsunaga, N., Nguyen, K. A., Su, S., Windus, T. L., Dupuis, M., and Montgomery Jr., J. A. General atomic and molecular electronic structure system. (1993). *J. Comput. Chem.* **14**, 1347–1363. http://www.msg.ameslab.gov/GAMESS.
[48] Chang, H. H., Lai, M. Y., Wei, J. H., Wei, C. M., and Wang, Y. L. (2004). Structure determination of surface magic clusters. *Phys. Rev. Lett.* **92**, 066103.
[49] Jia, J.-F., Liu, X., Wang, J.-Z., Liu, J.-L., Wang, X. S., Xue, Q.-K., Li, Z.-Q., Zhang, Z., and Zhang, S. B. (2002). Fabrication and structural analysis of Al, Ga and In nanocluster crystals. *Phys. Rev. B* **66**, 165412.
[50] Lai, M. Y., and Wang, Y. L. (1999). Self-organized two-dimensional lattice of magic clusters. *Phys. Rev. B* **64**, 241404.
[51] Biswas, R., and Li, Y.-P. (1999). Hydrogen flip model for light-induced changes of amorphous silicon. *Phys. Rev. Lett.* **82**, 2512–2515.
[52] Zhang, S. B., and Chadi, D. J. (1990). Stability of DX centers in $Al_xGa_{1-x}As$ alloys. *Phys. Rev. B* **42**, 7174–7177.
[53] Pchelyakov, O. P., Bolkhovityanov, Yu. B., Dvurechenskii, A. V., Sokolov, L. V., Nikiforov, A. I., Yakimov, A. I., and Voigtlander, B. (2000). Silicon-germanium nanostructures with quantum dots: formation mechanism and electrical properties. *Semiconductors* **34**, 1229–1247.
[54] Jeganathan, K., Shimizu, M., Ide, T., and Okumura, H. (2003). The effect of gallium gallium adsorbate on SiC (0001) surface for GaN by MBE. *Phys. Status Solidi* (b) **240**, 326–329.
[55] Davydov, V. Yu., Klochikhin, A. A., Seisyan, R. P., Entsev, V. V., Ivanov, S. V., Bechstedt, F., Furthmüller, J., Harima, H., Mudryi, A. V., Aderhold, J., Semchinova, O., Graul, J. (2002). Absorption and emission of hexagonal InN. Evidence of narrow fundamental band gap. *Phys. Stat. Solidi* (b) **229**, R1–R3.

[56] Kuroda, M., Katayama, R., Nishio, S., Onabe, K., and Shiraki, Y. (2003). Hall elect measurement of InAsN alloy films grown directly on GaAs (001) substrates by RF-MBE. *Phys. Stat. Solidi C* **0**, 2765–2768.
[57] Suemune, I., Sasikala, G., Kumano, H., Uesugi, K., Nabetani, Y., Matsumoto, T., Maeng, J.-T., and Seong, T. Y. (2006). Role of nitrogen precursor supplies on InAs quantum dot surfaces in their emission wavelength. *Jpn. J. Appl. Phys.* **45**, L529–L532.
[58] Soshnikov, I. P., LedentsovN. N., Volovik B. V., Kovsh, A., Maleev, N. A., Mikhrin, S. S., Gorbenko, O. M., Passenberg, W., Kuenzel, A., Grote, N., Ustinov, V. M., Kirmse, H., Neuman, W., Werner, P., Zakharov, N. D., Bimberg, D., and Zh. I. Alferov. (2001). Nitrogen-activated phase separation in InGaAsN/GaAs heterostructures grown by MBE. *In*: "9th Int. Symp. "Nanostructures: Physics and Technology, June 18–22, 2001." NT.24 Ioffe Institute, St Petersburg, Russia.
[59] Tanaka, I., Kamiya, I., Sakaki, H., and Fujimoto, M. (2000). Surface potential measurement of self-assembled InAs dots by scanning Maxwell stress microscopy. *Physica E* **7**, 373–376.
[60] Vajpeyi, A. P., Ajagunna, A. O., Tsiakatouras, G., Adikimenakis, A., Iliopoulos, E., Tsagaraki, K., Androulidaki, M., and Georgakilas, A. (2009). Spontaneous growth of III-nitride nanowires on Si by molecular beam epitaxy. *Microelectronic Eng.* **86**, 812–815.
[61] Bhatta, R. P., Thoms, B. D., Alevl, M., and Dietz, N. (2007). Surface electron accumulation in indium nitride layers grown by high pressure chemical vapor deposition. *Surf. Sci.* **601**, L120–L123.
[62] Boldyrev, A. I., and Wang, L.-S. (2001). Beyond classical stoichiometry: experiment and theory. *J. Phys. Chem. A* **105**, 10759–10775.
[63] Pozhar, L. A., and Mitchel, W. C. (2009). Virtual Synthesis of Electronic Nanomaterials: Fundamentals and Prospects. *In*: "Toward Functional Nanomaterials, Lecture Notes in Nanoscale Science and Technology" (Z. Wang, A. Waag and G. Salamo Ed.), Vol. 5, pp. 423–474. Springer, NY; see also http://arxiv.org/ftp/cond-mat/papers/0502/0502476.pdf).
[64] Pozhar, L. A., and Mitchel, W. C. (2007). Collectivization of electronic spin distributions and magneto-electronic properties of small atomic clusters of Ga and In with As, V and Mn. *IEEE Trans. Magnet.* **43**, 3037–3039.
[65] Pozhar, L. A., Yeates, A. T., Szmulowicz, F., and Mitchel, W. C. (2006). Virtual Synthesis of Artificial Molecules of In, Ga and As with Pre-Designed Electronic Properties Using a Self-Consistent Field Method. *Phys. Rev. B* **74**, 085306. See also Virtual J. Nanoscale Sci & Technol. **14**, No. 8 (2006): http://www.vjnano.org.
[66] Luo, B., Cramer, C. J., and Gladfelter, W. L. (2003). Gallium and indium hydrazides. Molecular and electronic structure of In[N(SiMe$_3$)NMe$_2$]$_3$ and related compounds. *Inorg. Chem.* **42**, 3431–3438.
[67] Costales, A., Kandalam, A. K., Pendas, A. M., Blanco, M. A., Recio, J. M., Pandey, R. (2000). The first principles study of polyatomic clusters of AlN, GaN, and InN. 2. Chemical bonding. *J. Phys. Chem. B* **104**, 4368–4374.
[68] Shubina, T. V., Ivanov, S. V., Jmerik, V. N., Solnyshkov, D. D., Vekshin, V. A., Kop'ev, P.S. (2004). Mie resonances, infrared emission, and the band gap in InN. *Phys. Rev. Lett.* **92**, 117407.
[69] See: Hu, M.-S., Wang, W. M., Chen, T. T., Hong, L. S., Chen, C. W., Chen, C. C., Chen, Y. F., Chen, K. H., and Chen, L. C. (2006). Sharp infrared emission from single-crystalline indium nitride nanobelts prepared using guided stream thermal chemical vapor deposition. *Adv. Functional Mater.* **16**, 537–541, and references therein.
[70] Pozhar, L. A. (2011). Opto-Electronic Properties of Small InAsN and InN Quantum Dots, *Phys. Stat. Solidi C* **8**, 2142–2144.

[71] Pozhar, L. A., and Mitchel, W. C. (2007). Opto-electronic properties and stability of artificial In-N molecules. *Mater. Res. Soc. Proc.* **955**, 39–44.
[72] Segura-Ruiz, J., Garro, N., Cantarero, A., Ilikawa, F., Denker, C., Malindretos, J., and Rizzi, A. (2009). Optical properties of InN nanocolumns: electron accumulation at InN non-polar surfaces and dependence on the growth conditions. *Phys. Stat. Solidi C* **6**, S553–S556.
[73] Jencks, W.P. (1987). "Catalysis in Chemistry and Enzymology". Courier Dover, New York.
[74] Breneman, C. M., and Weber, L. W. (1996). Charge and energy redistribution in sulfonamides undergoing conformational changes. Hybridization as a controlling influence over conformer stability. *Can. J. Chem.* **74**, 1271–1282.
[75] Pozhar, L. A., Yeates, A. T., Szmulowicz, F., and Mitchel, W. C. (2005). Small atomic clusters as prototypes of sub-nanoscale heterostructure units with pre-designed electronic properties. *Europhys. Lett.* **71**, 380–386.
[76] Talwar, D. N. (2008). Assessing the preferential chemical bonding of nitrogen in novel diluted III-As-N alloys. *In:* "Dilute III-V Nitride Semiconductors and Materials Systems: Physics and Technology. Springer Series in Materials Science" (A. Erol , Ed.), Vol. 105, pp. 222–245. Springer, Berlin.
[77] Alexandre, F., Gouardes, E., Gauthier-Lafaye, O., Buoadma, N., Vuong, A., and Therdez, B. (2002). Nitride-based long wavelength lasers on GaAs substrates. *J. Mater. Sci.: Materials in Electronics* **13**, 633–642.
[78] Hang, D. R., Huang, C. F., Hung, W. K., Chang, Y. H., Chen, J. C., Yang, H. C., Chen, Y. F., Shih, D. K., and Lin, H. H. (2002). Shubnikov-de Haas oscillations of two-dimensional electron gas in InAsN/InGaAs single quantum well. *Semicond. Sci. Technol.* **17**, 999–1003.
[79] Hsu, C.-C., Hsu, R.-Q., Wu, Y.-H., Chi, T.-W., Chiang, C.-H., Chen, J.-F., and Chang, M.-N. (2008). Analysis of InAsN quantum dots by transmission electron microscopy and photoluminescence. *Ultramicroscopy* **108**, 1495–1499.
[80] Wang, J.-S., and Lin, H.-H. (1999). Growth and postgrowth rapid thermal annealing of InAsN/InGaAs single quantum well on InP grown by gas source molecular beam epitaxy. *J. Vac. Sci. Technol. B* **17**, 1997–2000.
[81] Fukui, T., Ando, S., Tokura, Y., and Toriyama, T. (1991). GaAs tetrahedral quantum dot structures fabricated using selective area metalorganic chemical vapor deposition. *Appl. Phys. Lett.* **58**, 2018–2020.
[82] Fukui, T., and Ando, S. (1992). GaAs tetrahedral quantum dots grown by selective area MOCVD. *Superlatt. Microstruct.* **12**, 141–144.
[83] Nagamune, Y., Nishioka, M., Tsukamoto, S., and Arakava, Y. (1994). GaAs quantum dots with lateral dimension of 25 nm fabricated by selective metalorganic chemical vapor deposition. *Appl. Phys. Lett.* **64**, 2495–2497.
[84] Rajkumar, K. C., Kaviani, K., Chen, J., Chen, P., and Madhukar, A. (1992). Nanofeatures on GaAs (111)B via photolithography. *Appl. Phys. Lett.* **60**, 850–852.
[85] Madhukar, A., Rajkumar, K. C., and Chen, P. (1993). *In situ* approach to realization of three-dimensional confined structures via substrate encoded size inducing epitaxy on non-polar patterned substrates. *Appl. Phys. Lett.* **62**, 1547–1549.
[86] Rajkumar, K. C., Madhukar, A., Rammohan, K., Rich, D. H., Chen, P., and Chen, L. (1993). Optically active three-dimensionally confined structures realized via molecular beam epitaxial growth on planar GaAs (111) B. *Appl. Phys. Lett.* **63**, 2905–2908.
[87] Rajkumar, K. C., Madhukar, A., Chen, P., Konkar, A., Chen, L., Rammohan, K., and Rich, D. H. (1994). Realization of three-dimensionally confined structures via one-step *in situ* molecular beam epitaxy on appropriately patterned GaAs (111) B and GaAs (001). *J. Vac. Sci. Technol. B* **12**, 1071–1074.

[88] Williams, R. L., Aers, G. C., Poole, P. J., Levebvre, J., Chithrani, D., and Lamontagne, P. (2001). Controlling the self-assembly of InAs/InP quantum dots. *J. Cryst. Growth* **223**, 321–331.
[89] Levebvre, J., Poole, P. J., Aers, G. C., Chithrani, D., and Williams, R. L. (2002). Tunable emission from InAs quantum dots on InP nanotemplates. *J. Vac. Sci. Technol.* B **20**, 2173–2176.
[90] Chithrani, D., Williams, R. L., Levebvre, J., Poole, P. J., and Aers, G. C. (2004). Optical spectroscopy of single, site-selected InAs/InP self-assembled quantum dots. *Appl. Phys. Lett.* **84**, 978–980.
[91] Kapon, E., Pelucchi, E., Watanabe, S., Malko, A., Baier, M. H., Leifer, K., Dwir, B., Michelini, F., and Dupertuis, M.-A. Site- and energy-controlled pyramidal quantum dot heterostructures. (2004). Physica *E* **25**, 288–297.
[92] Michelini, F., Dupertuis, M.-A., and Kapon, E. (2004). Effects of the one-dimensional quantum barriers in pyramidal quantum dots. *Appl. Phys. Lett.* **84**, 4086–4088.
[93] Watanabe, S., Pelucchi, E., Dwir, B., Baier, M. H., Leifer, K., and Kapon, E. (2004). Dense uniform arrays of site-controlled quantum dots grown in inverted pyramids. *Appl. Phys. Lett.* **84**, 2907–2909.
[94] Baier, M. H., Watanabe, S., Pelucchi, E., and Kapon, E. (2004). High uniformity of site-controlled pyramidal quantum dots grown on prepatterned substrates. *Appl. Phys. Lett.* **84**, 1943–1945.
[95] Pelucchi, E., Watanabe, S., Leifer, K., Dwir, B., and Kapon, E. (2003). Site-controlled quantum dots grown in inverted pyramids for photonic crystal applications. *Physica E* **23**, 476–481.

Nickel Oxide Quantum Dots and Nanopolymer Quantum Wires

7.1 Introduction

Since 70's of the last century, nickel oxide nanoclusters and nanostructures have enjoyed an impressive variety of applications. Thus, small Ni_xO_y anions with x ranging from 1 to 12 and $y = 0, 1, 2$ produced by laser vaporization were thoroughly studied experimentally in conjunction with heterogeneous catalysis, formation of catalytic active sites and chemisorption [1–3]. Novel colloidal composites achieved through growth of Ni/NiO core-shell nanoparticles were shown to incorporate multiple functionalities, including optical and magnetic manipulations, thermal response and molecular trapping [4]. Photocatalytic hydrogen production and solar energy conversion systems were enhanced by the use of nickel oxide nanoclasters [5]. However, synthesis of advanced nanomaterials for sensor, electronics and spintronics applications [6–16] is among the major fields greatly benefitted from exceptionally useful magnetic and electronic properties of nickel oxide nanoclusters and nanostructures, especially those fabricated in quantum confinement or on surfaces [17,18]. Recently, electron spin states of exchange-biased, core–shell Ni-O nanoclusters attracted significant interest as prospective physical carriers of information qubits [19], because such states can be precisely controlled. It has been understood that the exchange bias phenomenon observed in Ni-O core-shell nanoclusters [20] may be used as a natural means of such control paramount for spintronics and electron spin-based quantum information processing.

Further advances in spintronics, quantum electronics, quantum computing and quantum information processing require a dramatic decrease in size of nanoclusters to only a few atoms. At the same time, recent experiments [19,21] indicate that some useful effects, such as exchange bias, may be lost in small, sub-critical size nanoclusters. Thus, first-principle understanding of structure - property relationships for and charge/spin localization in promising small Ni-O QDWs is extremely important both from fundamental and practical standpoints. This chapter is focused on electronic and magnetic properties of small Ni-O structures, in view of their utmost importance for nanoelectronics and emerging quantum information processing-based technologies.

Experimental assessment of synthesis conditions and properties of small Ni-O clusters is at the limits of capabilities of contemporary experimental techniques [17,18,22]. Moreover, understanding and interpretation of the available and emerging experimental results must be based on *ab initio* theoretical studies to reveal physical and chemical mechanisms behind observed properties of such systems. The existing first-principle theoretical studies of electronic and magnetic properties of small Ni-O clusters have used both density functional theory (DFT) and many body quantum theoretical methods (see, for example, Refs. 23–25). In available DFT-based theoretical studies Ni-O clusters are usually envisioned as elements of a bulk Ni-O lattice, and modeled either as surrounded by "point charges" [26], or as a periodic slab [2], to ensure that the electronic and magnetic properties of the clusters match those of the bulk Ni-O lattice. However, there is no fundamental justification behind expectations that the structure or/and properties of small systems composed of only few atoms should mimic those of the corresponding bulk systems. Moreover, even modern versions of DFT are non-variational theories, and thus cannot provide unambiguous computational results when applied to quantum mechanical systems (see a discussion and references in Ref. 27). Thus, in this chapter electronic and magnetic properties of a large family of small Ni-O clusters are studied using first-principle, many body quantum theoretical methods and their computational realizations, and in particular the virtual synthesis method [27–31] discussed and used in previous chapters of this book. These methods do not require introduction of heuristic adjustments, such as case-specific exchange-correlation functionals in DT-based methods. The virtual synthesis approach exploits several approximations of increasing accuracy, including restricted and restricted open shell Hartree-Fock (RHF, ROHF or HF, respectively), configuration interaction (CI), complete active space (CAS) self-consistent field (SCF), and multiconfiguration SCF (MCSCF) approximations, as realized by GAMESS software package [32,33] with the SBKJC standard basis set [34].

The "standard" optimization procedure of Refs. 32 and 33 has been applied to model nucleation of small Ni-O atomic clusters in the absence of any external fields exerted on the cluster atoms by foreign atoms or fields. Within the adopted approximations, this procedure converges to the equilibrium geometry and global minimum of the total energy of the optimized clusters in the absence of any external field and foreign atoms, that is, to the ground state of the corresponding molecules. Molecules synthesized in this fashion are called below vacuum molecules (*i.e.*, molecules synthesized virtually in physical "vacuum"). Recently, this standard optimization procedure has been modified [27–31,35] to incorporate the major effects of quantum confinement and surfaces on molecular synthesis without detailed atomistic modeling of environment surrounding an optimized atomic cluster (see a detailed discussion in Chapter 2). In this approach, excluded volume effects, and to a significant degree, polarization effects, are taken into account through spatial constraints applied to optimized systems. Such a modified procedure is necessary to bypass hardware and software restrictions that reduce applications of the many body quantum field theoretical methods to quantum systems of about 100 many-electron atoms. In the modified optimization procedure, symmetry elements of bulk lattices and other structures are used to pre-design clusters' geometry and composition, and to keep the centers of mass of the cluster atoms constrained to specified positions in space. Subsequently, the standard total energy minimization is applied to such spatially constrained clusters to obtain their ground state energy and electronic structure of the constrained molecules (called below pre-designed molecules). This energy minimization process converges to a local minimum of the total energy of an optimized cluster that corresponds to spatial constraints applied to the cluster atoms, thus modeling effects of quantum confinement or surface on properties of the molecule synthesized in such confinement.

In calculations presented and discussed below, experimental values of 1.24 Å and 0.66 Å were used as initial covalent radii of Ni and O atoms. The virtual synthesis method was applied to configure a large number of non-stoichiometric, vacuum and pre-designed Ni-O molecules composed of 3 to 46 atoms. Understanding electronic and magnetic properties of such molecules is necessary to predict and control magnetism of advanced nickel oxide nanomaterials, and exchange bias effects at nanoscale.

7.2 Molecules derived from Ni_2O cluster

The first two molecules (Table 7.1) of this study are derived from an atomic cluster composed of one O and two Ni atoms. They are the linear vacuum MCSCF septet Ni_2O (Fig. 7.1) and predesigned MCSCF triplet Ni_2O (Fig. 7.2), both of which have a deep ground state energy minimum.

Being a septet, the vacuum molecule is synthesized at the limits of applicability of the used approximations, and may not be stable at experimental conditions. Both molecules are "ferromagnetic", with their uncompensated electron spins aligned in the same direction and localized on Ni atoms, and the electron charge accumulated in the vicinity of O atoms (Figs. 7.1b and 7.2b). The highest occupied molecular orbits (HOMOs) of these molecules are composed of $3d$ orbits of Ni atoms ($3d_{xy}$ and $3d_{xz}$, in the case of the triplet, Fig. 7.2c), with only minor contributions from $2p_x$ and $2p_z$ orbits of O atoms.

The electronic energy level structure (ELS) of these molecules is depicted in Fig. 7.3. Unoccupied electron energy levels in conduction bands of these molecules form numerous, closely located sub-bands, reflecting enhanced metallicity of the molecules. Valence bands of the molecules contain distinct sub-bands separated by wide gaps. The ELS structure of the linear Ni_2O isomers supports a possibility of optical absorption due to $p \rightarrow d$ transitions, and is consistent with well-known experimental data [36] on optical absorption in a wide energy range for bulk NiO systems. The HF, CI and CASSCF/MCSCF data obtained using SBKJC basis and discussed above predict a complicated spectrum of both linear Ni_2O molecules in the region from about 2.5 to 4 eV, similar to that predicted for NiO molecules in the 1 to 3 eV region by the use of CI calculations with a different basis functions in Ref. 23. Interestingly enough, HF OTE of the vacuum linear Ni_2O molecule is about 3.5 eV (Table 7.1) that correlates reasonably close with CI OTE of the NiO molecule of Ref. 23, while MCSCF OTE of this molecule seem to be too large. Thus, it seems that HF calculations using SBKJC basis are as accurate as CI calculations with the basis of Ref. 23, and subsequent MCSCF calculations do not improve the HF results.

TABLE 7.1 Ground state energies and OTEs of small Ni-O molecules.

No.	Molecule	RHF/ROHF ground state energy, Hartree	MCSCF ground state energy, Hartree	RHF/ROHF OTE, eV	MCSCF OTE, eV	RHF/ROHF spin multiplicity
1.	Linear Ni_2O	−352.635956	−352.641378	3.5130	10.9036	vacuum septet
2.	Linear Ni_2O	−352.430358	−352.450052	1.2708	4.9838	predesigned triplet
3.	Triangular Ni_2O	−352.406852	−352.441908	1.9456	6.5443	vacuum singlet
4.	Triangular Ni_2O	−352.380775	−352.416856	6.1253	6.2695	predesigned singlet
5.	Square Ni_2O_2	−368.161288	−368.169562	0.0735	6.9607	vacuum pentet
6.	Square Ni_2O_2	−368.107922	−368.112991	8.7838	8.0164	predesigned singlet
7.	Modified Ni_2O_2	−368.175174	−368.216201	3.6354	2.7429	vacuum septet
8.	Modified Ni_2O_2	−367.719865	−368.067697	8.5797	2.3266	predesigned singlet
9.	Pyramidal Ni_4O	−689.301239	−689.385034	0.2177	0.1633	vacuum triplet
10.	Pyramidal Ni_4O	−689.350295	−689.378840	0.2176	0.8599	predesigned pentet
11.	Octahedral Ni_4O_2	−705.020224	−705.164661	0.4573	0.6504	predesigned septet
12.	Octahedral Ni_4O_2	−704.865259	−704.895966	1.1592	2.8273	vacuum septet
13.	Prismatic Ni_6O_6	−1104.943626	−1104.976724	3.9783	14.9390	predesigned pentet
14.	Prismatic Ni_6O_6	−1104.806148	−1105.045861	10.9635	14.3833	vacuum pentet
15.	QD Ni_7O_6	−1272.671838	−1272.528182	0.0381	3.0994	predesigned pentet
16.	QD Ni_7O_6	−1273.167164	−1273.256344	0.0952	0.4027	vacuum triplet
17.	QW Ni_8O_6	−1440.716675	−1440.911084	0.1360	0.0109	predesigned triplet
18.	QW Ni_8O_6	−1441.561976	−1441.654850	6.4110	1.0776	vacuum singlet

The pre-designed linear Ni_2O triplet is a stable molecule that is expected to be easily synthesized experimentally in quantum confinement. While the HF absolute values of its lowest unoccupied molecular orbital (LUMO) and highest occupied molecular orbital (HOMO) energy levels are not very accurate (which is a known feature of the HF approximation), the HF OTE of this molecule (about 1.27 eV) seems reasonable

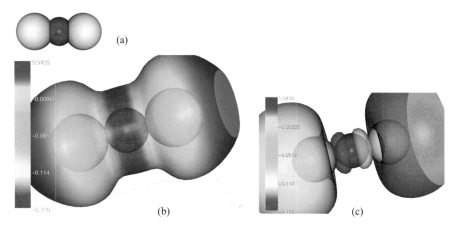

Fig. 7.1 Linear vacuum MCSCF septet Ni_2O: (a) structure; (b) surface of the molecular electrostatic potential [MEP; color scheme shown in the figure] corresponding to the fraction (cut) 0.01 of the maximum value (not shown) of the electron charge density distribution (CDD). (c) Isosurface of the positive (blue) and negative (reddish) parts of the highest occupied molecular orbit (HOMO) corresponding to the isovalue 0.01. Ni and O atoms are yellow and red spheres, respectively. Electron charge (c) is accumulated in the vicinity of the oxygen atom (yellowish to reddish regions). Atomic dimensions in (a) are to scale, and in (b) and (c) reduced. Other dimensions in the figures are to scale.

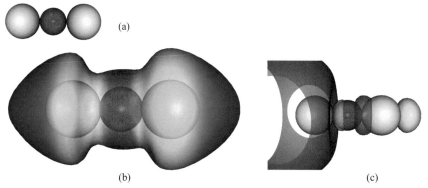

Fig. 7.2 Linear pre-designed MCSCF triplet Ni_2O: (a) structure; (b) surface of the molecular electrostatic potential [MEP; values are changing from negative (red) to positive (blue)] corresponding to the fraction (cut) 0.01 of the maximum value (not shown) of the electron charge density distribution (CDD); (c) isosurface of the positive (blue) and negative (reddish) parts of the highest occupied molecular orbit (HOMO) corresponding to the isovalue 0.02. Ni and O atoms are yellow and red spheres, respectively. Electron charge (c) is accumulated in the vicinity of the oxygen atom (yellowish to reddish regions). All dimensions are to scale.

in comparison with the data of Ref. 23 that show a number of local energy minima for NiO molecule with CI optical transition energies in 1 to 2 eV range. Due to nonstoichiometry, the Ni_2O triplet is much more metallic than NiO molecule of Ref. 23, and thus its OTE may be smaller than that of NiO molecule. However, this assumption is not confirmed by CI/MCSCF OTE (about 4.98 eV) of the linear Ni_2O triplet. The

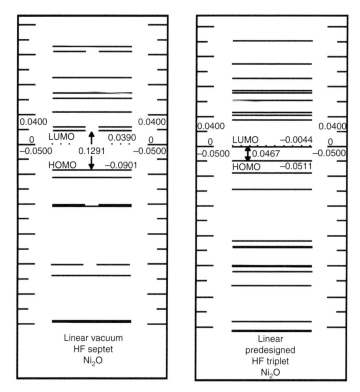

Fig. 7.3 HF electronic energy level structure of the vacuum (left) and pre-designed (right) linear Ni_2O molecules in the HOMO-LUMO region. The scales in the unoccupied and occupied MO regions of the figures are slightly different. The electronic energies and OTEs are in Hartree units (H).

major reason for a large value of CI/MCSCF OTE of this molecule is loss of symmetry of its MOs (Fig. 7.2) in comparison with those of the vacuum Ni_2O molecule. Indeed, spatial constraints applied to the center of mass of Ni and O atoms of the pre-designed linear Ni_2O molecule stress the molecule and destroy symmetry of its MOs compared to those of the vacuum Ni_2O molecule. The latter possesses doubly-degenerated occupied and unoccupied MOs, while the former does not (Fig. 7.2). The value of MCSCF OTE of the pre-designed linear Ni_2O triplet also correlates very well with the fact that bulk NiO is a semiconductor with a wide gap from about 3.6 to 4.0 eV (see, for example, a discussion and references in Ref. 37). Thus, the pre-designed molecule Ni_2O seems likely to have OTE larger than 4.0 eV.

While the total charge of the synthesized Ni_2O molecules is zero, non-stoichiometric arrangement leads to delocalization of the electron charge inside of these molecules (greenish to red regions in Figs. 7.1b and 7.2b). The Ni-containing "ends" of the molecules (greenish to blue regions in Figs. 7.1b and 7.2b) are deficient in electron charge, while "inner" regions containing O atom are slightly negative. Such electron charge

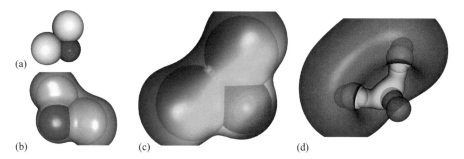

Fig. 7.4 Triangular vacuum MCSCF singlet Ni_2O: (a) structure; (b) isosurface of the electron charge density distribution (CDD) corresponding to the fraction (cut) 0.015 of its maximum value (not shown); (c) surface of the molecular electrostatic potential [MEP; values are varying from negative (red) to positive (blue)] corresponding to the fraction (cut) 0.02 of the maximum value (not shown) of CDD; (d) isosurfaces of the positive (blue) and negative (red) parts of the highest occupied molecular orbit (HOMO) corresponding to the isovalue 0.02. Ni and O atoms are represented by yellow and red spheres, respectively. Electron charge is accumulated (c) in the vicinity of the oxygen atom (yellowish to reddish regions). Atomic dimensions in (a), (b) and (c) are to scale, and in (d) reduced. Other dimensions in the figures are to scale.

and spin re-distributions are typical for non-stoichiometric molecules, and manifest violation of the standard octet rule [38,27–31,35].

There are at least two more spatial isomers of the studied linear Ni_2O molecules: triangular vacuum and pre-designed singlets Ni_2O (Table 7.1; Figs. 7.4 and 7.5). Deep potential wells of the HF and CI/MCSCF ground states of all four Ni_2O molecules

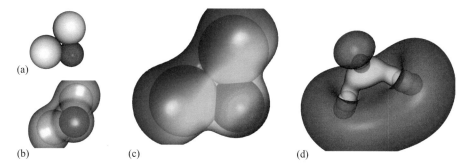

Fig. 7.5 Triangular pre-designed MCSCF singlet Ni_2O: (a) structure; (b) isosurface of the electron charge density distribution (CDD) corresponding to the fraction (cut) 0.015 of its maximum value (not shown); (c) surface of the molecular electrostatic potential [MEP; values are varying from negative (red) to positive (blue)] corresponding to the fraction (cut) 0.02 of the maximum value (not shown) of CDD; (d) isosurfaces of the positive (blue) and negative (red) parts of the highest occupied molecular orbit (HOMO) corresponding to the isovalue 0.02. Ni and O atoms are represented by yellow and red spheres, respectively. Electron charge is accumulated (c) in the vicinity of the oxygen atom (yellowish to reddish regions). Atomic dimensions in (a), (b) and (c) are slightly enlarged, and in (d) reduced. Other dimensions in the figures are to scale.

correspond to the ground state energies lying within calculation error brackets from each other. The triangular vacuum singlet Ni_2O has its HF ground state energy very close to that of the linear pre-designed triplet Ni_2O. Thus, one can conclude that the Ni-O-Ni angle of the triangular vacuum molecule may change in response to synthesis conditions to form a family of triangular molecules, including a degenerate one, which is the linear pre-designed molecule studied above. CI/MCSCF studies confirm, that the MCSCF OTEs of the triangular vacuum Ni_2O molecule and those of its pre-designed linear isomer differ more than the corresponding values obtained in the HF approximation. This may signify that both triangular vacuum and linear pre-designed molecules are stable and can be synthesized experimentally. At the same time, high excited ground state of the linear vacuum Ni_2O molecule means that the molecule may not be stable. The mere existence of the pre-designed triangular singlet Ni_2O demonstrates, yet again, that the Ni-O bond is both stretchable and flexible, and thus may accommodate the existence of various triangular conformations of Ni_2O cluster, depending on confinement conditions. In conjunction with the bulk NiO system, these properties of the Ni-O bond provide for bulk NiO being a wide band gap semiconductor.

The lengths of the Ni-O and Ni-Ni bonds (1.738 Å and 2.373 Å, respectively) in the triangular vacuum singlet (Fig. 7.4) are smaller than those in the predesigned one (1.900 Å and 2.480 Å, respectively, Fig. 7.5). Thus, it becomes obvious that the triangular shape serves to help the only O atom accommodate "extra" electron charge of Ni atoms by means of increasing the outside body angle Ni-O-Ni while still keeping Ni atoms as far from each other as possible. Accumulation of electron charge in the vicinity of O atom outside the Ni_2O triangle (Fig. 7.4c) is further demonstrated by the significant dipole moment of this molecule reaching 2.2251 D. In the pre-designed triangular molecule separation of Ni atoms from each other and the oxygen atom is enforced by spatial constraints, and is larger than that in the vacuum triangular molecule. This leads to the corresponding increase in the ground state energy (Table 7.1), and the dipole moment value (3.3601 D) of the pre-designed molecule.

In the linear Ni_2O molecules uncompensated electron spins of $3d$ electrons of Ni atoms are parallel to each other, so the molecules are "ferromagnetic" MCSCF multiplets. In contrast, both triangular MCSCF Ni_2O singlets are "antiferromagnetic," with their antiparallel electron spins localized on Ni atoms, and zero total magnetic moment. ELS patterns of the triangular molecules (Fig. 7.6) are very similar to those of the linear ones. They feature closely lying sub-bands of unoccupied electron energy levels that indicate enhanced metallicity, and well-developed "semiconductor"-type structure in the occupied MO regions. The symmetry of the triangular vacuum singlet is somewhat affected when the spatial constraints are applied to the center of mass of its atoms. This produces the corresponding pre-designed singlet with ELS that has less degenerate MOs in the HOMO-LUMO region.

In correspondence with known properties of the HF approximation, it works better for the triangular vacuum singlet, which realizes the global minimum of total energy of the Ni_2O cluster. In this case HOMO and LUMO energies both have correct signs, and the OTE is close to 2 eV (Table 7.1). In the case of the local minima of the total energy of the Ni_2O cluster, which are realized by the rest of the virtually synthesized

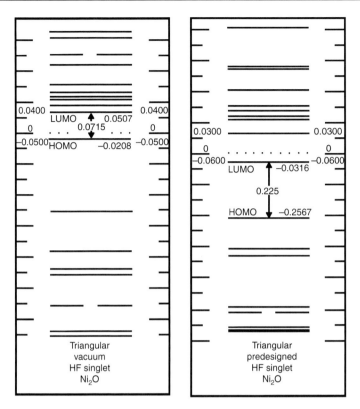

Fig. 7.6 HF electronic energy level structure of the vacuum (left) and pre-designed (right) triangular Ni_2O molecules in the HOMO-LUMO region. The scales in the unoccupied and occupied MO regions of the figures are slightly different. The electronic energies and OTEs are in Hartree units (H).

Ni_2O spatial isomers, the electron level energies in the HOMO-LUMO regions are calculated less accurately. In particular, the HOMO and LUMO energy levels seem to be too close in the case of the linear vacuum septet. For the pre-designed triangular molecules the HF LUMO energy levels lie slightly below zero. Similar to the case of the linear Ni_2O molecules, the MCSCF approximation does not improve HF OTE values of the triangular Ni_2O molecules. Considering that metallicity of all Ni_2O molecules is enhanced in comparison with that of Ni-O molecule of Ref. 23, it seems highly unlikely that the OTEs of these molecules exceed 5 eV, let alone exceeding 6 and 10 eV predicted by MCSCF calculations. This conclusion correlates with findings of Ref. 23, where the use of CI (although with a less accurate basis than SBKJC one) also did not improve the OTE data obtained for NiO molecule in the HF approximation. Thus, it may happen that MCSCF approximation used with the SBKJC basis does not work well for non-stoichiometric "metallic" oxides of transition metals.

Magnetic properties of the triangular Ni_2O molecules studied here by quantum many body-theoretical means are similar to those of a triangular molecule of Ref. 22

obtained using a DFT-based method, and supported by some experimental results also reported in Ref. 22. Both molecules (that of Fig. 7.4 and that of Ref. 22) possess antiferromagnetic arrangement of electron spins localized on Ni atoms. Unfortunately, the ground state and OTE data for the molecule of Ref. 22 are not available, so any quantitative comparison is not possible at this time.

It is also interesting to compare properties of Ni_2O molecules obtained by the virtual synthesis method here with those of a vacuum NiO dimer of Ref. 25 synthesized using a similar method, the spin-unrestricted HF approximation. The length of the Ni-O bond (1.738 Å) in the triangular vacuum singlet (Fig. 7.4) is comparable with the length of the Ni-O bond (1.784 Å) in this dimer, while both are significantly shorter than the Ni-O bond length of 2.084 Å in the case when NiO molecule were "embedded" into a model crystal of NiO studied in Ref. 25. [The bond length of 2.084 Å is also known from earlier experiments for bulk NiO (see Ref. 39 and references therein).] This fact corroborates results obtained and discussed above for the pre-designed triangular Ni_2O molecule of Fig. 7.5, with its Ni-O bond of 1.900 Å, and confirms that such a molecule synthesized in quantum confinement may, indeed, possess a longer Ni-O bond. [Note, that the Ni-O bond length of 1.900 Å was calculated from experimental data on covalent radii of Ni and O atoms in 12 coordinated metals [40].] The flexible and easily stretchable Ni-O bond ensures synthesis and stabilization of such molecules at various nucleation conditions. In the process of adjustment, electron spin alignment changes dramatically from the ferromagnetic to antiferromagnetic type, and *vice versa*, while the ground state energy changes within 0.2 H, that is, within calculation error brackets.

In conclusion, it's important to note that all Ni_2O molecules of this study have hybridized HOMOs composed largely of $3d$ orbits of Ni atoms, with smaller contributions due to $2p$ orbits of O atoms (Figs. 7.1c, 7.2c, 7.4d and 7.5d), similar to that of the Ni-O dimer of Walch and Goddard III [23]. With an increase in the ratio of the numbers of oxygen to nickel atoms in Ni - O molecules hybridization of their MOs in the HOMO-LUMO region increases, in correspondence with well-known experimental and theoretical evidence suggesting that wide gap semiconductors possess highly hybridized MOs in this region.

7.3 Molecules derived from Ni_2O_2 cluster

As demonstrated by studies presented in section 7.2, Ni_2O cluster can easily adjust to nucleation conditions adopting different spatial forms to minimize its total energy, and thus producing a number of spatial isomers. In the process of such adjustment spin alignment of the cluster atoms changes dramatically (from the ferromagnetic to antiferromagnetic type), while HF OTEs change from about 1 to over 6 eV (the corresponding change in MCSCF OTE values is up to about 2 times, see Table 7.1). This structural flexibility is further revealed in studies of this section, and is a common property of few-atomic Ni-O clusters.

Yet another interesting property of small Ni-O molecules is demonstrated via comparison of properties of several isomers produced by Ni_2O_2 cluster (Table 7.1;

Figs. 7.7 to 7.10). In particular, while three of four virtually synthesized Ni_2O_2 molecules are very stable and possess large OTEs, the predesigned molecules have their OTEs much larger than that of the corresponding vacuum molecules. This manifests a stabilizing role of environment in the process of synthesis of small Ni-O molecules.

The pre-designed Ni_2O_2 molecule (Fig. 7.8) was obtained by positioning the centers of mass of Ni and O atoms in the vertices of a square with the side length equal to the sum of covalent radii of Ni and O atoms, and subsequent conditional minimization of the cluster's total energy, while the centers of mass of the oxygen and nickel atoms

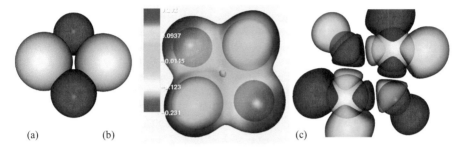

Fig. 7.7 Square vacuum MCSCF pentet Ni_2O_2: (a) structure; (b) surface of the molecular electrostatic potential [MEP; color scheme is shown in the figure] corresponding to the fraction (cut) 0.05 of the maximum value (not shown) of CDD; (c) isosurfaces of the positive (blue) and negative (red) parts of the highest occupied molecular orbit (HOMO) corresponding to the isovalue 0.01. Ni and O atoms are represented by yellow and red spheres, respectively. Electron charge is accumulated (b) in the vicinity of the oxygen atoms (yellowish to reddish regions on the outer side of the atoms). Atomic dimensions in (a) are to scale, and in (b) and (c) somewhat reduced. Other dimensions in the figures are to scale.

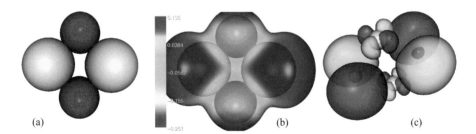

Fig. 7.8 Square pre-designed MCSCF singlet Ni_2O_2: (a) structure; (b) surface of the molecular electrostatic potential [MEP; color scheme is shown in the figure] corresponding to the fraction (cut) 0.02 of the maximum value (not shown) of CDD; (c) isosurfaces of the positive (blue) and negative (reddish) parts of the highest occupied molecular orbit (HOMO) corresponding to the isovalue 0.01. Ni and O atoms are represented by yellow and red spheres, respectively. Electron charge is accumulated (b) in the vicinity of the oxygen atoms (yellowish to reddish regions on the outer side of the atoms). Atomic dimensions in (a) and (b) are to scale, and in (c) reduced. Other dimensions in the figures are to scale.

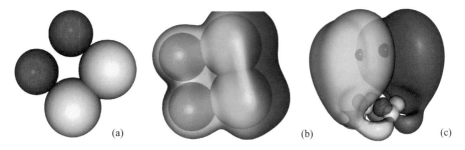

Fig. 7.9 Modified pre-designed MCSCF singlet Ni_2O_2: (a) structure; (b) surface of the molecular electrostatic potential [MEP; values are varying from negative (reddish) to positive (blue)] corresponding to the fraction (cut) 0.02 of the maximum value (not shown) of CDD; (c) isosurfaces of the positive (blue) and negative (reddish) parts of the highest occupied molecular orbit (HOMO) corresponding to the isovalue 0.02. Ni and O atoms are represented by yellow and red spheres, respectively. Electron charge is accumulated (b) between the oxygen atoms (yellowish to reddish regions somewhat "outside" of the atoms). Atomic dimensions in (a) and (b) are to scale, and in (c) reduced. Other dimensions in the figures are to scale.

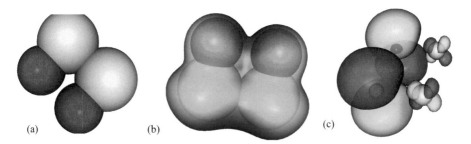

Fig. 7.10 Modified vacuum MCSCF singlet Ni_2O_2: (a) structure; (b) surface of the molecular electrostatic potential [MEP; values are varying from negative (red) to positive (blue)] corresponding to the fraction (cut) 0.02 of the maximum value (not shown) of CDD; (c) isosurfaces of the positive (blue) and negative (reddish) parts of the highest occupied molecular orbit (HOMO) corresponding to the isovalue 0.02. Ni and O atoms are represented by yellow and red spheres, respectively. Electron charge is accumulated (b) near the oxygen atoms (yellowish to red regions somewhat "outside" of the oxygen atoms). Atomic dimensions in (a) and (b) are to scale, and in (c) reduced. Other dimensions in the figures are to scale.

were kept in their initial positions. The corresponding vacuum molecule (Fig. 7.7) was synthesized using unconditional total energy minimization (that is, the atoms originally placed in the vertices of the square in the case of the pre-designed molecule were allowed moving, thus finally assuming their optimal positions corresponding to the global minimum of the total energy of the cluster).

Amazingly, it occurs that if one interchanges two neighboring nickel and oxygen atoms in the pre-designed Ni_2O_2 molecule, so the two atoms of Ni become positioned next to each other, it is still possible to minimize the total energy of the cluster so configured to obtain the modified pre-designed Ni_2O_2 molecule of Fig. 7.9 (see also Table 7.1, No. 8). Moreover, the modified pre-designed molecule so obtained is a very stable, "antiferromagnetic" singlet with a large HF OTE of over 8 eV. The corresponding modified vacuum molecule is synthesized by application of the unconditional energy optimization to the modified pre-designed molecule. It appears to be a "ferromagnetic septet", that may be more stable than the modified pre-designed molecule, because its ground state potential well is deeper (Table 7.1).

In the modified vacuum molecule (Fig. 7.10a) oxygen atoms are closer to the corresponding Ni atoms than those in the corresponding pre-designed molecule (Fig. 7.9a). In both modified molecules electron charge is accumulated in a region between the oxygen atoms somewhat outside of a space occupied by the atoms, in contrast to that of the square molecules. Thus, all modified molecules have significant dipole moments.

The ground states of all Ni_2O_2 molecules feature deep potential wells of about −368 H in depth, with the well depths lying within the computational error brackets from each other (Table 7.1). At the same time, HF OTEs of these molecules are very different ranging from about 0.07 eV for the square vacuum pentet to over 8 eV for both pre-designed molecules. With an exception of the square vacuum Ni_2O_2 molecule, HF approximation results for these molecules seem more realistic than the corresponding MCSCF results, and the latter are consistent with MCSCF results for Ni_2O molecules of section 7.2. While MCSCF ground state energies are consistent with HF ones, MCSCF OTEs seem too large to improve the values of HF OTEs. In the case of the square vacuum Ni_2O_2 molecule MCSCF OTE value is consistent with those of the rest of the molecules, while HF one is too small, and differs over an order of magnitude from the corresponding HF results for other Ni_2O_2 molecules.

Ni_2O_2 molecules are stoichiometric, so the classic octet rule is expected to work for them. Indeed, HOMO of the square vacuum pentet is almost equally proportioned mixture of $3d$ and $2p$ atomic orbits of Ni and O atoms, respectively (Fig. 7.7c). For the modified vacuum septet (Fig. 7.10c), modified pre-designed singlet (Fig. 7.9c), and the square pre-designed singlet (Fig. 7.8c) contributions from $2p$ orbits of the oxygen atoms prevail. These results are consistent with the MO configuration analysis of Ref. 23 for the stoichiometric molecule NiO.

The ELSs of the modified Ni_2O_2 molecules are illustrated in Fig. 7.11. Location of the LUMO energy levels slightly below zero energy level is consistent with known properties of the HF approximation and indicates that this approximation must be improved to obtain more realistic MOs and ELSs. LUMO regions feature many closely lying energy levels typical for "metallic" conduction bands of small clusters, while HOMO regions are composed of bands (solid slabs) of energy levels separated by wide spaces typical for "valence" bands of small atomic clusters and molecules.

The obtained results highlight a stabilizing role of quantum confinement in the process of molecular synthesis, point out at opportunities offered by manipulations with the structure and composition of quantum confinement, and hint at possible physical mechanisms governing the exchange bias development or loss. Indeed, both pre-designed Ni_2O_2 molecules are stable singlets, while the corresponding vacuum molecules may be less stable pentet and septet. Moreover, both pre-designed molecules are

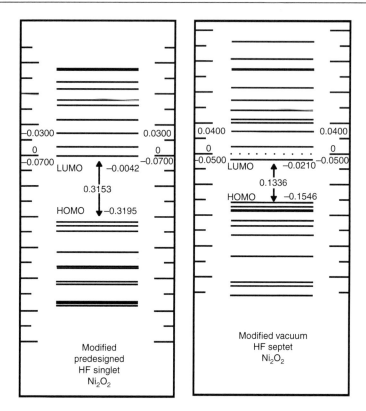

Fig. 7.11 HF electronic energy level structure of modified pre-designed (left) and vacuum (right) Ni$_2$O$_2$ molecules in the HOMO-LUMO region in the HF approximation. The scales in the unoccupied and occupied MO regions of the figures are slightly different. The electronic energies and OTEs are in Hartree units (H).

"antiferromagnetic", while vacuum ones are "ferromagnetic" multiplets. Certainly, a few-atomic cluster may be self-stabilized at conditions where there is no "foreign" atoms within about 3 to 4 Lennard-Jones atomic diameters from the centers of mass of cluster atoms. However, the resulting vacuum molecules may be less stable than the corresponding (pre-designed) molecules synthesized from the same original atomic cluster in an appropriate quantum confinement. This process may include a dramatic change of spin alignment that is crucial for the development or loss of exchange bias. As discussed in the following sections, the existence of stable Ni-O spatial isomers in quantum confinement is possible due to a stretchable and flexible Ni-O bond.

7.4 Quantum dots derived from larger Ni-O clusters

Adding more nickel and oxygen atoms and exploiting other crystalline symmetries allows derivation of several larger Ni-O clusters. (Table 7.1; Figs. 7.12 to 7.16).

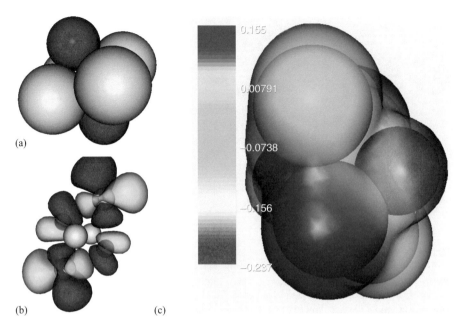

Fig. 7.12 Octahedral pre-designed MCSCF septet Ni_4O_2: (a) structure; (b) isosurfaces of the positive (blue) and negative (reddish) parts of the highest occupied molecular orbit (HOMO) corresponding to the isovalue 0.005; (c) surface of the molecular electrostatic potential [MEP; values are varying from negative (red) to positive (blue)] corresponding to the fraction (cut) 0.02 of the maximum value (not shown) of CDD. Ni and O atoms are represented by yellow and red spheres, respectively. Electron charge is accumulated (c) near the oxygen atoms (yellowish to reddish regions surrounding the oxygen atoms). Atomic dimensions in (a) are to scale, and in (b) and (c) somewhat reduced. Other dimensions in the figures are to scale.

The octahedral pre-designed septet Ni_4O_2 of Fig. 7.12 is obtained upon conditional optimization of a cluster composed of 4 Ni atoms whose centers of mass are positioned in the vertices of a square with its side length equal to two covalent radii of Ni atoms, and two oxygen atoms on the tops of two pyramids having the square as a common foundation (Fig. 7.12a). The pyramid side length is equal to the sum of covalent radii of Ni and O atoms.

Despite of the octahedral symmetry of this molecule borrowed from bulk NiO, this molecule is a "ferromagnetic" septet that indicates a result at the edge of applicability of the HF approximation, although its ground state has a deep potential well (Table 7.1). The HF OTE of this molecule is about 0.46 eV, and its MCSCF OTE is a bit larger (about 0.65 eV). Both values are well below the bulk NiO value, and hint again that this molecule may be unstable. The corresponding vacuum molecule also is a "ferromagnetic" septet with a parallelogram of Ni atoms in its base (as opposed to the square of Ni atoms being the base of the pre-designed molecule), and O atoms in the verges of the pyramids (Fig. 7.13a) shifted with regard to each other. Its ground

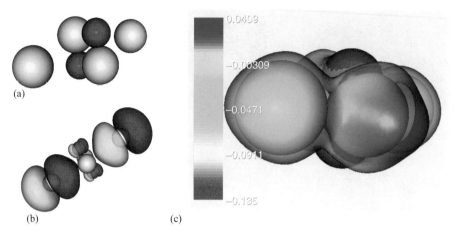

Fig. 7.13 Octahedral vacuum MCSCF septet Ni_4O_2: (a) structure; (b) isosurfaces of the positive (blue) and negative (reddish) parts of the highest occupied molecular orbit (HOMO) corresponding to the isovalue 0.01; (c) surface of the molecular electrostatic potential [MEP; values are varying from negative (red) to positive (blue)] corresponding to the fraction (cut) 0.01 of the maximum value (not shown) of CDD. Ni and O atoms are represented by yellow and red spheres, respectively. Electron charge is accumulated (c) near the oxygen atoms (yellowish to reddish regions surrounding the oxygen atoms). Atomic dimensions in (a) and (b) are reduced, and in (c) to scale. Other dimensions in the figures are to scale.

state potential well (Table 7.1) is slightly less deep than that of the pre-designed molecule, but its HF OTE and MCSCF OTE values are much larger, and are similar to the CI OTE of NiO molecule of Ref. 23. Electron charge in both Ni_4O_2 molecules is accumulated near oxygen atoms. The vacuum Ni_4O_2 molecule is almost flat, with its O atoms close to the plane of Ni-based parallelogram (Fig. 7.13a, c). This tendency for vacuum molecules to become almost two-dimensional structures persists for all larger vacuum Ni-O molecules (with an exception of Ni_4O and Ni_8O_6) of Table 7.1, and continues for NiO quantum wires (nanopolymers) discussed in the following section.

Comparison of properties of the octahedral Ni_4O_2 molecules with those of the pyramidal pre-designed and vacuum molecules Ni_4O helps understand whether or not the octahedral molecules are stable, and thus may be observed experimentally. The pyramidal pre-designed Ni_4O pentet (Table 7.1; Fig. 7.14a and 7.14b) studied here is simply the same structure as that of the pre-designed Ni_4O_2 (Fig. 7.12a) with one oxygen atom removed, that is, a pyramid with a square foundation built of Ni atoms. The corresponding vacuum Ni_4O triplet (Table 7.1) is obtained by unconditional optimization of the pre-designed structure. The structure of the pyramidal vacuum triplet (Fig. 7.14c) occurs dramatically different from that of the pre-designed molecule (Fig. 7.14a). The oxygen atom of this molecule moves closer to Ni atoms which arrange themselves into a spatial structure and destroy the former square foundation of the pre-designed pyramid. Thus, the final structure of this molecule occurs very similar to that of CH_4. Both Ni_4O molecules have almost identical HF and MCSCF ground state energies, and their HF OTEs coincide to the third digit after the dot. One

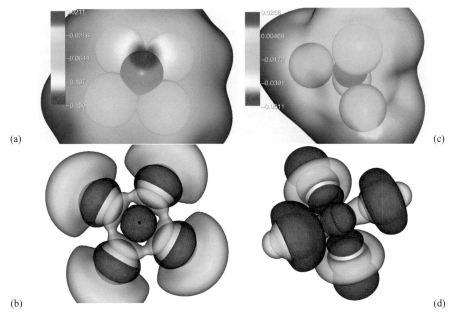

Fig. 7.14 Pyramidal molecules Ni_4O. Surfaces of the molecular electrostatic potential [MEP; values are varying from negative (red) to positive (blue)] corresponding to the fraction (cut) 0.01 of the maximum value (not shown) of CDD for (a) pre-designed MCSCF pentet and (c) vacuum MCSCF triplet. Isosurfaces of the positive (blue) and negative (reddish) parts of the highest occupied molecular orbits (HOMOs) corresponding to the isovalue 0.01 for (b) pre-designed MCSCF pentet and (d) vacuum MCSCF triplet. Ni and O atoms are represented by yellow and red spheres, respectively. In the pre-designed molecule electron charge is accumulated (a) near the oxygen atom (yellowish to reddish regions surrounding the oxygen atom). Atomic dimensions in (a) and (c) are to scale, and in (b) and (d) reduced. Other dimensions in the figures are to scale.

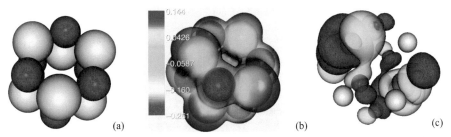

Fig. 7.15 Prismatic pre-designed MCSCF pentet Ni_6O_6: (a) structure; (b) surface of the molecular electrostatic potential [MEP; values are varying from negative (red) to positive (blue)] corresponding to the fraction (cut) 0.03 of the maximum value (not shown) of CDD; (c) isosurfaces of the positive (blue) and negative (red) parts of the highest occupied molecular orbit (HOMO) corresponding to the isovalue 0.03. Ni and O atoms are represented by yellow and red spheres, respectively. Electron charge is accumulated (b) near the oxygen atoms (yellowish to reddish somewhat "outside" of the oxygen atoms). Atomic dimensions in (a) and (b) are to scale, and in (c) reduced. Other dimensions in the figures are to scale.

would think that they are the same molecule, but the structures and MCSCF OTE values of these molecules are dramatically different. Thus, one concludes that there exists the *minimum minimorum* of the total energy of the cluster Ni_4O corresponding to the vacuum triplet, and a local minimum of the total energy (corresponding to the pre-designed pentet) lying very closely to the global one. The OTEs of Ni_4O molecules are rather small signifying their metallicity.

Once another oxygen atom is added to these pyramids, metallicity of the emerging octahedral pre-designed and vacuum distorted Ni_4O_2 molecules drops significantly, and their HF OTEs increase. Notably, MCSCF OTE of the pre-designed octahedral Ni_4O_2 septet decreases comparatively to OTE of the pre-designed pyramidal Ni_4O pentet. The major structural and energy characteristics revealed by this comparison of the two groups of molecules are consistent with theoretical expectations. Moreover, Ni_4O molecules are HF triplet and pentet, and therefore are obtained within a range of applicability of the HF approximation. This suggests that the octahedral molecules, despite of being obtained at the verge of the HF approximation, may be stable.

Many bulk semiconductors are wurtzite lattice structures containing a hexagonal prism as a symmetry element. Thus, in this work a hexagonal prism Ni_6O_6 cluster was optimized using conditional total energy minimization to obtain the corresponding pre-designed molecule. The pre-designed structure (Fig. 7.15a) is built of two "ideal" hexagons populated by alternating Ni and O atoms in their verges. One of the hexagons in this structure is turned by $\pi/3$ relatively to the rotation symmetry axis of the structure, to make O atoms of this hexagon facing Ni atoms of the other hexagon. Similar to the prismatic molecule of Ref. 22, the pre-designed Ni_6O_6 molecule is "ferromagnetic". It possesses a deep ground state energy minimum of about -1105 H, reasonable HF OTE, and a huge MCSCF OTE [further Møller-Plesset (MP2) studies are planned to ascertain the OTE value of this molecule and that of its vacuum counterpart].

Uncompensated electron spins of this molecule are localized on Ni atoms and produce the total magnetic moment of 4 μ_B, as opposed to 12 μ_B of the DFT-derived molecule of Ref. 22. Due to stoichiometry, MEP surfaces corresponding to isovalues smaller than 0.01 of the pre-designed molecule (not shown in Fig. 7.15) reveal that the molecule is only slightly electronegative, with the electron charge evenly smoothed over its "surface". Closer to the atoms, for isovalues 0.03 and larger, MEP surfaces of this molecule retain a general property of small Ni-O molecules: electron charge is accumulated near oxygen atoms (Fig. 7.15b). A highly hybridized HOMO (Fig. 7.15c) of this molecule consists primarily of $3d$ atomic orbits of Ni atoms and $2p$ atomic orbits of O atoms. Significant hybridization of MOs in the HOMO-LUMO region indicates [20] that a molecule should possesses a large OTE, similar to that of insulators. Indeed, the obtained CASSCF/MCSCF OTE values for both Ni_6O_6 molecules are the largest OTEs of this study, and most likely, too large to be realistic. Such MEP, HOMO and OTE properties manifest "semiconductor" nature of the pre-designed prismatic Ni_6O_6 molecule, which is the smallest "semiconductor" structure built of Ni and O atoms.

The prismatic vacuum Ni_6O_6 pentet (Table 7.1; Fig. 7.16) is obtained via unconditional total energy minimization of the pre-designed Ni_6O_6 cluster of Fig. 7.15a. The structure of this molecule resembles a somewhat distorted hexagonal prism where oxygen atoms moved slightly outside, and Ni atoms moved slightly inside, respectively,

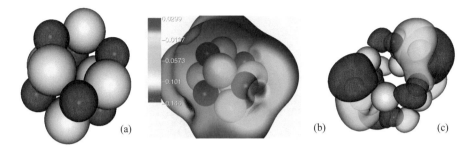

Fig. 7.16 Prismatic vacuum MCSCF pentet Ni_6O_6: (a) structure; (b) surface of the molecular electrostatic potential [MEP; values are varying from negative (red) to positive (blue)] corresponding to the fraction (cut) 0.01 of the maximum value (not shown) of CDD; (c) isosurfaces of the positive (blue) and negative (reddish) parts of the highest occupied molecular orbit (HOMO) corresponding to the isovalue 0.005. Ni and O atoms are represented by yellow and red spheres, respectively. Electron charge is accumulated (b) near the oxygen atoms (yellow to reddish regions somewhat "outside" the oxygen atoms). Atomic dimensions in (a) and (b) are to scale, and in (c) reduced. Other dimensions in the figures are to scale.

of their position in the ideal pre-designed hexagon. Both HF and MCSCF values of the ground state energy of this molecule are close to the corresponding values of its pre-designed counterpart, but its HF OTE of about 11 eV is much larger than that of the pre-designed molecule (in fact, it is the largest HF OTE of this study). MCSCF OTE values of both Ni_6O_6 molecules are comparable and both are too large, signifying that MCSCF approximation does not improve HF OTE values in this case. The electron charge of the prismatic vacuum molecule is not as evenly smoothed over the molecular "surface" as that of the pre-designed one (Fig. 7.16b), and is somewhat accumulated on the outside parts of regions containing oxygen atoms. HOMO of this molecule still is highly hybridized and similar to that of the pre-designed molecule. Other properties of the prismatic vacuum Ni_6O_6 pentet (Table 7.1, Fig. 7.16) also are similar to those of the pre-designed one.

The disk-like, predesigned "ferromagnetic" pentet Ni_7O_6 (Table 7.1, Fig. 7.17) is yet another structure that realizes one of the smallest nickel oxide quantum dots (QDs). It is built of Ni atoms with their centers of mass in the verges and the center of a closed packed hexagon, and O atoms occupying three tetrahedral holes on each side of the hexagonal structure (Fig. 7.17a). The corresponding vacuum triplet (Table 7.1, Fig. 7.18) is obtained from the pre-designed molecule using unconditional total energy minimization procedure. This vacuum molecule is almost entirely flat (Fig. 7.18a) and somewhat twisted around its rotation symmetry axis orthogonal to the plane of this molecule shown in Fig. 7.18a. Its HF and MCSCF ground state energy values are slightly lower than those of the pre-designed molecule, and its small HF OTE (Table 7.1) is still about 3 times larger than that of the pre-designed molecule. As usual in the case of Ni-O molecules, MCSCF OTEs of the disk-like pentet and triplet are sharply different from the corresponding quantities calculated in the HF

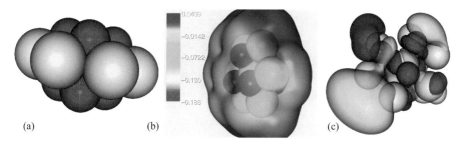

Fig. 7.17 Disk-like pre-designed MCSCF pentet Ni_7O_6: (a) structure; (b) surface of the molecular electrostatic potential [MEP; values are varying from negative (red) to positive (blue)] corresponding to the fraction (cut) 0.01 of the maximum value (not shown) of CDD; (c) isosurfaces of the positive (blue) and negative (reddish) parts of the highest occupied molecular orbit (HOMO) corresponding to the isovalue 0.005. Ni and O atoms are represented by yellow and red spheres, respectively. Electron charge is accumulated (b) near the oxygen atoms (yellowish to reddish regions somewhat "outside" of the oxygen atoms). Atomic dimensions in (a) and (b) are to scale, and in (c) reduced. Other dimensions in the figures are to scale.

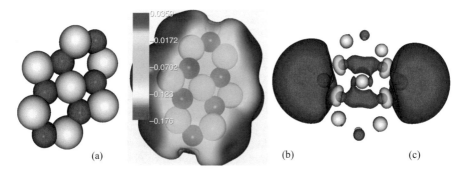

Fig. 7.18 Flat, disk-like vacuum MCSCF triplet Ni_7O_6: (a) structure; (b) surface of the molecular electrostatic potential [MEP; values are varying from negative (red) to positive (blue)] corresponding to the fraction (cut) 0.01 of the maximum value (not shown) of CDD; (c) isosurfaces of the positive (blue) and negative (reddish) parts of the highest occupied molecular orbit (HOMO) corresponding to the isovalue 0.01. Ni and O atoms are represented by yellow and red spheres, respectively. Electron charge is primarily accumulated (b) near two oxygen atoms at the sharper "ends" of the molecule (yellowish to reddish regions somewhat "outside" of the oxygen atoms). Atomic dimensions in (a) and (b) are to scale, and in (c) reduced. Other dimensions in the figures are to scale.

approximation. The MCSCF OTE value of the vacuum triplet is about 0.4 eV, and is closer to its HF OTE value than the corresponding values in the case of the pre-designed pentet. This OTE value hints at semi-metallic nature of the vacuum Ni_7O_6 molecule, and may explain extraordinary chemical reactivity of experimentally observed Ni_7O_6 clusters [1], which is about 4 times larger than that of other Ni-O clusters.

HOMO of the pre-designed pentet (Fig. 7.17c) is highly hybridized, which is a fact in support of its large MCSCF OTE value, as opposed to the small OTE value obtained in the HF approximation. Both $3d$ atomic orbits of several Ni atoms, and $2p$ atomic orbits of all oxygen atoms almost equally contribute to this HOMO. In the case of the vacuum molecule, HOMO is only slightly hybridized and consists primarily of large contributions from $3d$ atomic orbits of two Ni atoms in the "ends" of the 3-atomic Ni "line" in the center of the structure, and much smaller contributions from $2p$ atomic orbits of 4 "central" oxygen atoms (Figs. 7.18a and 7.18c). This HOMO structure supports expectation of high reactivity of this molecule and explains small MCSCF and HF OTE values (Table 7.1) specific to this molecule.

The largest of quantum dot-like, small Ni-O molecules studied here are the ball-like predesigned triplet Ni_8O_6 (Table 7.1; Fig. 7.19) and the corresponding football-like vacuum singlet Ni_8O_6 (Table 7.1; Fig. 7.20). They are obtained from the prismatic predesigned Ni_6O_6 structure (Fig. 7.15a) by addition of Ni atoms to the centers of each hexagon, and subsequent application of constrained and unconstrained total energy minimization procedures, respectively. The pre-designed Ni_8O_6 molecule recreates the smallest "cylindrical" core-shell quantum wire (QW) where a "string" of two Ni atoms, with their centers of mass on the rotation symmetry axis of the structure (Fig. 7.20a), represents a core, is surrounded by nickel oxide shell composed of alternating Ni and O atoms with their centers of mass in the verges of two hexagons. This molecule is a "ferromagnetic" triplet with a deep ground state energy minimum (Table 7.1) and a very small MCSCF OTE of about 0.01 eV manifesting metallicity of this QW (the HF OTE of this molecule is also small: 0.1360 eV). The corresponding "antiferromagnetic" vacuum Ni_8O_6 singlet is a significantly deformed version of the predesigned one (Figs. 7.20a and 7.20b), with its MCSCF OTE of about 1 eV

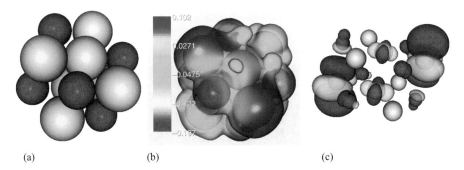

(a)   (b)   (c)

Fig. 7.19 Ball-like pre-designed MCSCF triplet Ni_8O_6: (a) structure; (b) surface of the molecular electrostatic potential [MEP; values are varying from negative (red) to positive (blue)] corresponding to the fraction (cut) 0.02 of the maximum value (not shown) of CDD; (c) isosurfaces of the positive (blue) and negative (reddish) parts of the highest occupied molecular orbit (HOMO) corresponding to the isovalue 0.01. Ni and O atoms are represented by yellow and red spheres, respectively. Electron charge is primarily accumulated (b) near the oxygen atoms (yellowish to reddish regions somewhat "outside" of the oxygen atoms). Atomic dimensions are reduced. Other dimensions in the figures are to scale.

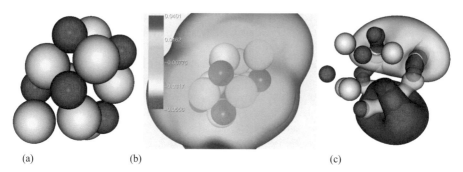

(a) (b) (c)

Fig. 7.20 Football-like vacuum MCSCF singlet Ni_8O_6: (a) structure; (b) surface of the molecular electrostatic potential [MEP; values are varying from negative (red) to positive (blue)] corresponding to the fraction (cut) 0.005 of the maximum value (not shown) of CDD; (c) isosurfaces of the positive (blue) and negative (reddish) parts of the highest occupied molecular orbit (HOMO) corresponding to the isovalue 0.01. Ni and O atoms are represented by yellow and red spheres, respectively. Electron charge is almost evenly distributed with some accumulation (b) near the oxygen atoms (yellowish and reddish regions somewhat "outside" of the oxygen atoms). Atomic dimensions in (a) and (b) are to scale, and in (c) reduced. Other dimensions in the figures are to scale.

indicating "semiconductor" nature of the vacuum QW. Its HF OTE is large: over 6.4 eV (Table 7.1). The shape of this vacuum molecule indicates that when quantum confinement constraints are lifted, the two Ni atoms of the "string" of the ball-like pre-designed molecule (Fig. 7.19) move apart, and the atoms in verges of the hexagons tend to "mix" around those two Ni atoms. The result is a football-like shape of the vacuum molecule that is strikingly similar to that of much larger (about 1 μm), experimentally synthesized QWs of Ref. 42 (Fig. 7.21).

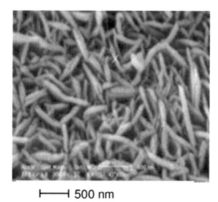

⊢——⊣ 500 nm

Fig. 7.21 SEM image of calcinated Ni-O powder synthesized at 400°C by a complexation-precipitation method using ammonium hydroxide as the complexation agent (from Ref. 42).

HOMO of the ball-like pre-designed QW Ni_8O_6 (Fig. 7.19c) is slightly hybridized and consists predominantly of contributions from $3d$ atomic orbits of 4 Ni atoms with only a small contribution from $2p$ atomic orbits of 3 oxygen atoms. This is consistent with the OTE data and indicates that the molecule is highly metallic. In contrast, HOMO of the football-like vacuum QW Ni_8O_6 (Fig. 7.20c) is a highly hybridized MO that consists of $3d$ atomic orbits of 2 Ni atoms, and $2p$ atomic orbits of 2 oxygen atoms. This is consistent with large HF and reasonable MCSCF OTE values that signify "semiconductor" nature of this molecule.

A variety of structural solutions for small Ni-O QDs and QWs has been expected [1], and is due to Ni - O exchange interactions that produce a flexible and stretchable Ni-O bond. The RHF/ROHF ground state energies of the studied systems are close to those calculated using the CASSCF/MCSCF approximation (Table 7.1), while the MCSCF OTEs differ significantly from HF ones, and in many cases seem to be too large. In such cases, more accurate approximations, and possibly a basis larger than SBKJC one, are necessary to ascertain OTE values of the studied molecules.

The obtained results reveal major tendencies and physical mechanisms that allow understanding experimentally observed properties of larger Ni-O systems, including large QDs, QWs and thin films. Thus, the above results indicate that as the number of Ni and O atoms grow, Ni-O molecules undergo remarkable transformation of their structure that accommodate "antiferromagnetic" and "ferromagnetic" alignment of uncompensated electron spins. About a half of the studied 8 very small molecules composed of 3 to 4 atoms (including triangular pre-designed and vacuum Ni_2O molecules, pre-designed square Ni_2O_2 and modified pre-designed Ni_2O_2) are "antiferromagnetic" singlets, while the rest of the small molecules are "ferromagnetic" spin multiplets. With further growth in the number of atoms, only "ferromagnetic" spin multiplets were virtually synthesized, until the football-like vacuum Ni_8O_6 QW, that again is a singlet. Results discussed in the following section show that father proportional increase in the numbers of Ni and O atoms in the molecules $Ni_{2+x}O_x$ with x even and varying from 8 to 22 produces "antiferromagnetic" pre-designed singlets (save for only one exception) that are almost one-dimensional nanopolymer QWs up to about 6 nm in length.

In the process of nucleation and synthesis in quantum confinement or on surfaces the structural transformation described above is expected to lead to the development or loss of exchange bias, provided the confinement/surface has appropriate spin alignment of its atoms. For example, the distorted octahedral vacuum septet Ni_4O_2 may be physically absorbed at a Ni-O surface edge built of $Ni_{18}O_{16}$ singlet QW. The absorbed molecule interacts only with a portion of the surface that is about 3.5 Lennard-Jones atomic diameters wider in linear dimensions than the molecule's linear dimensions (the rest of the surface adds only a small contribution to this interaction). Given a broad energy plateau near the minimum of the total energy of Ni-O clusters, at some thermodynamic conditions (pressure, temperature) the spin alignment of the octahedral molecule should not change, and thus the exchange bias interaction with the substrate should develop. When thermodynamic conditions change and a deeper minimum of the total energy would be reachable, the spin alignment of the octahedral molecule may change, and it may join $Ni_{18}O_{16}$ molecule to polymerize into a pre-designed $Ni_{22}O_{18}$ singlet (QW), so that the exchange bias effects disappear.

All small vacuum molecules of Table 7.1 tend to be as flat as possible. This tendency and an ability to self-assemble in a preferential direction lead to the development of almost one-dimensional Ni-O nanopolymer QWs discussed in the following section.

7.5 Ni-O nanopolymer quantum wires

Attempts to synthesize virtually non-stoichiometric Ni-O QWs longer than Ni_8O_6 of Table 7.1 and featuring hexagonal, pentagonal, square or triangular arrangement of alternating Ni and O atoms in the verges of the polygons forming QW "shells" and an axial string of Ni atoms, were not successful. Instead, it was found that there exists a tendency to "self-assembly" of Ni-O molecules beginning with the pre-designed octahedral molecule Ni_4O_2 (No. 11 in Table 7.1). In particular, addition of (i) three Ni atoms to each of the two rows of Ni atoms in the foundation of this molecule (Fig. 7.12a) and (ii) the corresponding three O atoms in the orthogonal plane to develop two "strings" of O atoms placed into octahedral holes above and below every Ni-based square as shown in Fig. 7.22a, and subsequent conditional minimization of the total energy of the obtained structure, lead to a pre-designed HF singlet $Ni_{10}O_8$ of Fig. 7.22.

Direct evaluations using Table 7.1 data show that such "polymerization" mechanism is energetically favourable and may produce quasi one-dimensional QWs possessing remarkably deep ground state potential wells and large OTEs. Indeed, comparing HF ground state energy data for the pre-designed Ni_7O_6 molecule (No. 15 in Table 7.1) with that of the pre-designed Ni_8O_6 triplet (No. 17 in Table 7.1) one can find out that addition of one Ni atom to a pre-designed Ni-O structure leads to a decrease

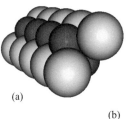
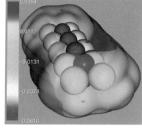
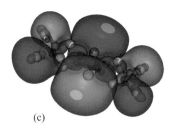

Fig. 7.22 Pre-designed HF singlet $Ni_{10}O_8$: (a) structure; (b) surface of the molecular electrostatic potential [MEP; values are varying from negative (red) to positive (blue)] corresponding to the fraction (cut) 0.005 of the maximum value (not shown) of CDD; (c) isosurfaces of the positive (blue) and negative (violet) parts of the highest occupied molecular orbit (HOMO) corresponding to the isovalue 0.005. Ni and O atoms are represented by yellow and red spheres, respectively. Electron charge is almost evenly distributed around the "ends" of the molecule (b) (light green regions). Atomic dimensions in (a) and (b) are to scale, and in (c) reduced. Other dimensions in the figures are to scale.

in the ground state energy by roughly 168 H. [The same result can be obtained comparing the ground state energies of any appropriate pair of molecules, such as the pre-designed molecules Ni_6O_6 and Ni_8O_6 (No. 13 and No. 17 in Table 7.1, respectively).] Further comparison of the ground state energies of the pre-designed octahedral Ni_4O_2 molecule (No. 11 in Table 7.1) and the pre-designed pyramidal pentet Ni_4O (No. 10 in Table 7.1) produces similar evaluation in the case of addition of one O atom to a pre-designed Ni-O structure: 16 H. Again, the latter evaluation can be checked by comparison of the ground state data for other appropriate pairs of molecules, such as the pre-designed molecules Ni_2O and Ni_2O_2 (No. 4 and No. 6 in Table 7.1, respectively), producing the same value of 16 H. Using these evaluations data one can predict the ground state energy of the pre-designed $Ni_{10}O_8$ HF singlet of Fig. 7.22. Indeed, taking the ground state of the pre-designed QW Ni_8O_6 as the calculation basis, and subtracting from that value the change of 368 H in the total energy due to addition of two Ni and two O atoms, one can predicts that the ground state energy of the $Ni_{10}O_8$ HF singlet is about -1809 H. Amazingly, this evaluation lies within calculation error brackets of the HF and MCSCF approximations (!) from the accurate HF value of −1808.2547340189 H (Table 7.2).

The "polymerization" mechanism discussed above was used to pre-design and optimize a range of QWs of Table 7.2. Beginning with QW $Ni_{10}O_8$, each subsequent QW is obtained from its predecessor by addition of two Ni and two O atoms along the length of the wire. All pre-designed QWs so polymerized and containing from 10 to 24 Ni atoms and from 8 to 22 O atoms, respectively, are "antiferromagnetic" HF singlets with deep energy minima and OTEs about 5 eV. These QWs are 4.98 Å in width, 2.19 Å in height, and of varying length from 2.49 nm (24.9 Å) to 5.96 nm (59.6 Å). Given that QW length is over an order of magnitude larger than their width and height

TABLE 7.2 RHF ground state energies and OTEs of small Ni-O nanopolymer quantum wires (QWs).

No.	Molecule (RHF singlet)	RHF/ROHF ground state energy, Hartree	RHF/ROHF OTE, eV	Dipole moment (Debye)	QW length (nm)
1.	$Ni_{10}O_8$	−1808.2547340189	5.3418	0.011406	2.48
2.	$Ni_{12}O_{10}$	−2176.1394863181	5.0450	2.408197	2.98
3.	$Ni_{14}O_{12}$	−2544.0484383246	5.3362	0.006731	3.47
4.	$Ni_{16}O_{14}$	−2911.9347598008	4.9089	3.392921	3.97
5.	$Ni_{18}O_{16}$	−3279.8419509998	5.3389	0.014099	4.46
6.	$Ni_{20}O_{18}$	−3647.7299132522	4.9334	4.005419	4.96
7.	$Ni_{20}O_{18}$ (triplet)	−3647.8951586947	0.7646	13.418265	4.96
8.	$Ni_{20}O_{18}$ (vacuum singlet)	−3647.9462432375	5.0069	4.399016	4.96
9.	$Ni_{24}O_{22}$	−4383.5248289489	4.9634	4.530491	5.95

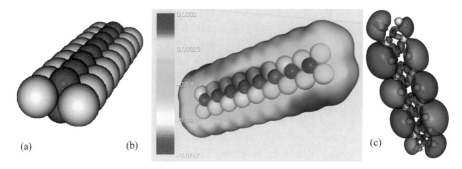

Fig. 7.23 Pre-designed RHF singlet $Ni_{18}O_{16}$: (a) structure; (b) surface of the molecular electrostatic potential [MEP; values are varying from negative (red) to positive (blue)] corresponding to the fraction (cut) 0.005 of the maximum value (not shown) of CDD; (c) isosurfaces of the positive (blue) and negative (violet) parts of the highest occupied molecular orbit (HOMO) corresponding to the isovalue 0.005. Ni and O atoms are represented by yellow and red spheres, respectively. Electron charge is almost evenly distributed along the "planes" of the molecule (b) (greenish regions). Atomic dimensions in (a) and (b) are to scale, and in (c) reduced. Other dimensions in the figures are to scale.

in all cases, the nanowires are quasi one-dimensional structures from applications' standpoint. Two examples of such Ni-O polymer nanowires are depicted in Figs. 7.23 and 7.24 below.

All HF singlet molecules of Table 7.2 have their electronic charge smoothly distributed over the molecular "surfaces" with some accumulation over the "ends". The first two smaller molecules of Table 7.2 do not possess any other charge accumulation regions (see, for example, Fig. 7.22b). As the molecular length increases, electron charge begins to accumulate in the central regions of the molecular "plains" (see, for example, Fig. 7.23b). With further increase in the molecular length the central region of charge accumulation splits into two and then three "spots" (Fig. 7.24c, reddish "spots" in the middle and closer to the "ends" of the molecular "plains") on each side of the molecule located symmetrically with regard to the rotational symmetry axes of the molecules.

HOMOs of all Ni-O nanopolymer QWs are highly hybridized (Figs. 7.22c, 23c and 24b), with somewhat larger contributions from $3d$ atomic orbits of Ni atoms and lesser ones from $2p$ atomic orbits of O atoms. This HOMO structure agrees very well with large OTE values for all synthesized HF singlets of Table 7.2 and signifies semiconductor nature of the QWs.

A very important observation follows from the results obtained in the case of conditional and unconditional total energy minimization of $Ni_{20}O_{18}$ cluster. It occurs that there exist at least two pre-designed isomers, the HF singlet and triplet of Fig. 7.25 (see also Table 7.2, Nos. 6 and 7, respectively) whose ground state energy values differ by only about 0.16 H (that is, lying within the calculation error brackets from each other, with the triplet's ground state energy lesser than that of the singlet). However, OTE values of the singlet and triplet differ almost by an order of magnitude: the singlet's OTE value is about 5 eV, and the triplet's one is about 0.7 eV. To clarify which

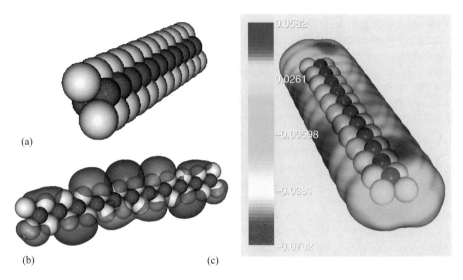

Fig. 7.24 Pre-designed RHF singlet $Ni_{24}O_{22}$: (a) structure; (b) isosurfaces of the positive (blue) and negative (violet) parts of the highest occupied molecular orbit (HOMO) corresponding to the isovalue 0.005; (c) surface of the molecular electrostatic potential [MEP; values are varying from negative (red) to positive (blue)] corresponding to the fraction (cut) 0.005 of the maximum value (not shown) of CDD. Ni and O atoms are represented by yellow and red spheres, respectively. Electron charge is almost evenly distributed (c) along the "planes" (greenish regions) of the molecule, with some accumulations in the center and near the "ends" (yellowish regions). All dimensions are to scale.

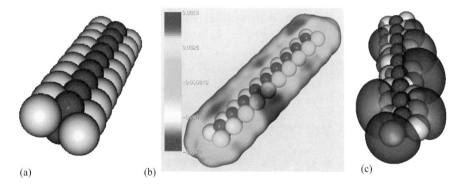

Fig. 7.25 Pre-designed HF triplet $Ni_{20}O_{18}$: (a) structure; (b) surface of the molecular electrostatic potential [MEP; values are varying from negative (red) to positive (blue)] corresponding to the fraction (cut) 0.005 of the maximum value (not shown) of CDD; (c) isosurfaces of the positive (blue) and negative (violet) parts of the highest occupied molecular orbit (HOMO) corresponding to the isovalue 0.005. Ni and O atoms are represented by yellow and red spheres, respectively. Electron charge is almost evenly distributed (b) over the "planes" of the molecule (yellowish regions), with some accumulations in the center and near the "ends" (brighter yellow regions). All dimensions are to scale.

of this molecules is stable, and thus may be synthesized in quantum confinement experimentally, the corresponding vacuum molecule was optimized (No. 8 in Table 7.2). The vacuum molecule occurred a HF singlet with its ground state energy value lying about 0.15 H below that of the pre-designed triplet, and its OTE value slightly larger than that of the singlet.

Comparative analysis of the obtained data for $Ni_{20}O_{18}$ QWs against those for QWs of Table 7.2 with the number of Ni atoms from 10 to 18 and O atoms from 8 to 16 leads to a conclusion that "antiferromagnetic" HF singlet states realize total energy minima for the pre-designed molecules of Table 7.2 from $Ni_{10}O_8$ to $Ni_{18}O_{16}$. Further increase in the number of Ni and O atoms produces QWs possessing almost flat total energy surfaces near the minima of the total energy corresponding to their ground states. This indicates a possibility of "polymerization" routes other than simply enlargement of the structures along one direction. Indeed, current studies of larger pre-designed QWs, $Ni_{26}O_{24}$ and $Ni_{30}O_{26}$, show that these QWs may have several isomers with their ground state energies lying within calculation error brackets from each other and that of the corresponding vacuum molecules.

The above results and discussion show that further addition of Ni and O atoms to $Ni_{18}O_{16}$ molecule to obtain $Ni_{20}O_{18}$ one may or may not results in a dramatic change in uncompensated electron spin alignment from "antiferromagnetic" one in $Ni_{18}O_{16}$ molecule to "ferromagnetic" in $Ni_{20}O_{18}$ and larger molecules. Projecting this results on a process of sorption of Ni and O atoms on Ni-O nanocluster surfaces one can conclude that in the process of cluster growth two types of larger clusters may emerge: those with antiferromagnetic and ferromagnetic spin alignment. Obviously, thermodynamic conditions must play a crucial role selecting a thermodynamicaly preferential type of spin alignment at a given range of the cluster sizes that produces a minimum of the total energy of the cluster. At the next step of materials development, such clusters may be deposited on a surface that may have one of the above types of spin alignments near a location where a cluster joins that surface. In the case when an "antiferromagnetic" cluster is deposited onto a "ferromagnetic" area on a surface, or *vice versa*, exchange bias interaction between the cluster and the surface will be significant.

7.6 Discussion and conclusions

A family of small Ni-O molecules, including the smallest oxidized core-shell QDs and QWs, have been synthesized using several progressively more accurate, many body quantum field theoretical methods. The obtained results capture the major magnetic and electronic properties of the studied molecules pointing toward possible physical mechanisms behind the exchange bias development and its loss. Thus, very small triangular Ni-O molecules in confinement or on surfaces are "antiferromagnetic" singlets, and somewhat larger molecules are primarily "ferromagnetic" multiplets that tend to become triplets and singlets as the numbers of Ni and O atoms proportionally increase up to 8 Ni and 6 O atoms. The next in the family of virtually synthesized molecules, the HF singlet $Ni_{10}O_8$ (No. 1 in Table 7.2), and all of the larger molecules of Table 7.2, have been obtained from the pre-designed octahedral molecule Ni_4O_2 (No. 11 in Table 7.1) by polymerization and occur to be quasi-one-dimensional QW singlets up $Ni_{20}O_{18}$. At this point it was found that the triplet $Ni_{20}O_{18}$ has its ground state energy slightly lesser than the corresponding singlet. While the difference in the ground state energies is within error brackets of the calculations, the mere fact signifies that when the small non-stoichiometric Ni-O QWs reach a "critical" length equal to about 18 Ni atomic diameters, there may emerge some other polymerization mechanism that would lead to QW growth in two other dimensions enabling wider, quasi-two-dimensional non-stoichiometric Ni-O quantum ribbons and football-like shapes similar to those of experimental Ni-O QWs of Fig. 7.21 and the vacuum Ni_8O_6 molecule (Fig. 7.20), and possibly large Ni-O QDs. Nevertheless, even larger quasi one-dimensional QW singlets, such as $Ni_{24}O_{22}$ (No. 9 in Table 7.2) and currently studied $Ni_{22}O_{20}$, $Ni_{28}O_{26}$, and $Ni_{30}O_{28}$ do exist.

The ground state energy of the vacuum singlets of Table 7.1 is lower than that of the corresponding "ferromagnetic" triplets and pentets. The opposite observation holds for predesigned singlets: their ground state energies are higher than those of the corresponding vacuum spin multiplets. Thus, projecting these results onto an experimental situation where small Ni-O molecules are deposited on a surface, one can predict that in the presence of interface, re-structuring of such predesigned "antiferromagnetic" singlets to "ferromagnetic" Ni-O molecules may be energetically favorable and easily realizable due to flexibility of the Ni-O bond. Alternatively, reconstruction of larger Ni-O multiplets to vacuum singlets (see Ni_8O_6 molecules, Table 7.1) with almost the same ground state energy is also possible (see Ni_8O_6 molecules, Table 7.1). Such processes may take place at antiferromagnet-ferromagnet interface between small core-shell Ni/Ni-O clusters up to some critical cluster size and a confinement or surface, leading to a decrease in the "surface" energy of the system, and also to a loss or the development of exchange bias. Examples of larger pre-designed singlets of Table 7.2 demonstrate a possible change in spin alignment of surface atoms in a vicinity of Ni and O atoms and small molecules absorbed by the surface. In particular, adding more Ni and O atoms to a surface may happen via polymerization mechanism described in the previous section. If spins of surface atoms are aligned "ferromagnetically", then such adsorption may lead to polymerization of fragments of the surface to produce large pre-designed "antiferromagnetic" singlets of Table 7.2 or similar, so that the total energy of the system decreases.

Properties of Ni-O molecules theoretically predicted here are in a general agreement with available experimental [1,20,22] and computational [13,22,26] data. It is also interesting to compare the obtained results and tendencies with experimental and computational findings for larger Ni-O systems (Table 7.3). The majority of systems considered experimentally or evaluated by DFT methods are much larger than those considered in this work, and are polycrystalline or amorphous, as opposed to single-crystal-type systems of this work. Experimental data of Table 7.3 were obtained at temperatures of about 300° C and high, while quantum field theoretical methods used in this work allow accurate calculations at temperatures near the absolute zero. Moreover, synthesis conditions in each of the cases of Table 7.3 were very different, and their details were not completely described in the corresponding publications, so

TABLE 7.3 Band gap width or OTE of Ni-O thin films and nanostructures: experiment and calculations.

No.	System	Preparation method	Temperature	Band gap width or OTE, eV
1.	1 μm Ni-O films, Ref. 42.	Spray pyrolysis	350° C	3.6
2.	200 to 500 Å Ni-O films, Ref. 43.	Spray pyrolysis	300° C	3.58 to 3.4
3.	Ni-O nanowire arrays, Ref. 44.	Optical absorption	320° C	3.74
4.	60 nm to 120 nm Ni-O films, Ref. 45.	Grown from solutions	320° C	3.25
5.	360 to 1000 μm Ni-O films, Ref. 46.	Optical absorption	300° C	4.4
6.	9 nm Ni-O nanoparticles, Ref. 47.	Thermal deposition	450° C	3.56
8.	Nanostructured Ni-O, Ref. 48.	Anodic deposition	320° C	3.55
9.	0.2 to 3 μm porous films of columnar 10 nm Ni-O grains, Ref. 49.	Sputtering	320° C	3.2 to 3.5
10.	Ni-O_6 clusters, Ref. 49.	Sputtering	320° C	4
11.	Submicron NiOx films, Ref. 50.	Spray pyrolysis	330 to 420° C	3.61 to 4
12.	NiO (100) and (111) surfaces, Ref. 51.	Thermal deposition	320° C	3.8
13.	Gradient-corrected DFT, Ref. 52.	Calculations, bulk NiO	N/A	5.93

it is impossible to ensure that theoretical models used in this chapter are applicable to mimic synthesis of the experimentally studied systems of Table 7.3.

However, the major tendencies, and values of quantities calculated in this chapter, can be compared to those reavealed by experimental data of Table 7.3. Thus, one can observe that OTEs of the molecules virtually synthesized in this work are larger than those of much larger structures evaluated from experimental data of Table 7.3, which is a correct tendency. The theoretical OTEs of the majority of nanopolymer Ni-O QWs of Table 7.2 lie in the range from about 4.9 to 5.33 eV, and therefore are in a very good correspondence with experimental data for much larger systems of Table 7.3, for which band gaps are expected to be smaller both because of larger system size, and because the measurements were made at high temperatures. [Nanoparticles of 9 nm in linear dimensions studied in Ref. 47 (No. 6 in Table 7.3) still are almost twice as large as the largest of systems of Table 7.2, and they were studied at high temperatures.] The only experimental system comparable in size with those studied theoretically in this work is Ni-O_6 clusters of Ref. 49 synthesized at 320° C. However, the nature of this oxide molecule is very different from that of "metallic" oxides of Table 7.1.

From experimental data of Table 7.3 it follows that synthesis conditions define the structure and properties of the synthesized systems. This observation is in a good agreement with theoretical data of Tables 7.1 and 7.3 proving that the smallest changes in synthesis conditions may significantly affect the structure of the synthesized molecules, and thus their electronic and magnetic properties. Notably, the theoretical OTEs of this work (Table 7.2) obtained using quantum field theoretical methods are much more consistent with experimental results of Table 7.3 than the DFT-based band gap of 5.93 eV calculated for bulk NiO in Ref. 52 that largely overestimates not only the band gap of bulk NiO, but also those of small Ni-O nanostructures and molecules.

The major conclusion derived from quantum many body-theoretical results obtained and discussed in this chapter are as follows.

1. Physical and chemical properties of nanoscale Ni-O QDs and QWs depend dramatically on details of the system structure and composition. In particular: (i) OTE of such systems can be changed within an order of magnitude by manipulations with the structure and composition of QDs/QWs and quantum confinement, and synthesis conditions; (ii) the octet rule does not hold for non-stoichiometric systems because (iii) electron charge is re-distributed to stabilize a non-stoichiometric molecule or a molecule synthesized in quantum confinement.

2. Some small Ni-O molecules and polymer nanowires up to 6 nm in linear dimensions exist as "ferromagnetic" multiplets with uncompensated electron spins parallel and localized on Ni atoms, while the majority of such studied systems are "antiferromagnetic" singlets with antiparallel electron spins. Depending on a particular structure, the ground state energy of "antiferromagnetic" singlets may be higher or lower than that of similar or larger "ferromagnetic" triplets, pentets and septets.

3. Re-structuring of such "antiferromagnetic" singlets to "ferromagnetic" Ni-O molecules and *vice versa* can be energetically favorable and accommodated by the highly flexible/stretchable Ni-O bond. At some thermodynamic conditions, such reversible re-structuring may lead to a decrease in the "surface" energy of the antiferromagnet - ferromagnet interface, and thus to the development or loss of exchange bias.

4. There exists a large class of "antiferromagnetic" Ni-O QWs up to 6 nm in length that are quasi one-dimensional polymers with dipole moments varying within an order of magnitude. Such nanowires can be used to create highly ordered nanoheterostractures with pre-designed dipole moment and electron spin distributions.

References

[1] Vann, W. D., Wagner, R. L., and Castleman Jr., A. W. (1998). Gas-phase reactions of nickel and nickel-rich oxide cluster anions with nitric oxide. 2. The addition of nitric oxide, oxidation of nickel clusters, and the formation of nitrogen oxide anions. *J. Phys. Chem. A* **102**, 8804–8811.

[2] Ferrari, A. M., and Pisani, C. (2008). Reactivity of non stoichiometric Ni_3O_4 phase supported at the Pd(100) surface: interaction with Au and other transition metal atoms. *Phys. Chem. Chem. Phys.* **10**, 1463–1470.

[3] Xu, X., Lü, X., Wang, N. Q., and Zhang, Q. E. (1995). Charge-consistency modeling of CO/NiO (100) chemisorption system. *Chem. Phys. Lett.* **235**, 541–547.

[4] Sánchez-Iglesias, A., Grzelczak, M., Rodriguez-González, B., Guardia-Girós, P., Pastoriza-Santos, I., Pérez-Juste, J., Prato, M., and Liz-Marzán, L. M. (2009). Synthesis of multifunctional composite microgels via in situ Ni growth on pNIPAM-coated Au nanoparticles. *ACS Nano* **3**, 3184–3190.

[5] Lin, H.-Y., Lee, T.-H., and Sie, C.-Y. (2008). Photocatalytic hydrogen production with nickel oxide intercalated $K_4Nb_6O_{17}$ under visible light irradiation. *Int. J. Hydrogen Energy* **33**, 4055–4063.

[6] Volkov, V. V., Wang, Z. L., and Zou, B. S. (2001). Carrier recombination in clusters of NiO. *Chem. Phys. Lett.* **337**, 117–124.

[7] Abiade, J. T., Miao, G. X., Gupta, A., Gapud, A. A., and Kumar, D. (2008). Corrigendum to "Structural and magnetic properties of self-assembled nickel nanoparticles in yttria stabilized zirconia matrix". *Thin Solid Films* **516**, 8763–8767.

[8] Miller, J. S., and Drillon M. (Eds.) (2002). "Magnetism: Molecules to Materials III. Nanosized Magnetic Materials." Wiley InterScience, New York.

[9] Nakamura, R., Lee, J.-G., Mori, H., and Nakajima, H. (2008). Oxidation behaviour of Ni nanoparticles and formation process of hollow NiO. *Philosophical Magazine* **88**, 257–264.

[10] Caruge, J.-M., Halpert, J. E., Bulović, V., and Bawendi, M. G. (2006). NiO as an inorganic hole-transporting layer in quantum-dot light-emitting devices. *Nano Lett.* **6**, 2991–2994.

[11] Roche, B., Voisin, B., Jehl, X., Wacquez, R., Sanquer, M., Vinet, M., Deshpande, V., and Previtali, B. (2012). A tunable, dual mode field-effect or single electron transistor. *arXiv*: 1201.3760v1 [cond-mat.mes-hall].

[12] Boris, A. V., Matiks, Y., Benckiser, E., Frano, A., Popovich, P., Hinkov, V., Wochner, P., Castro-Colin, M., Detemple, E., Malik, V. K., Bernhard, C., Prokscha, T., Suter, A., Salman, Z., Morenzoni, E., Cristiani, G., Habermeier, H.-U., and Keimer, B. (2011). Dimensionality control of electronic phase transition in nickel-oxide superlattices. *Science* **332**, 937–940.

[13] Irwin, M. D., Buchholz, D. B., Hains, A. W., Chang, R. P. H., and Marks, T. J. (2008). p-Type semiconducting nickel oxide as an efficiency-enhancing anode interfacial layer in polymer bulk-heterojunction solar cells. *Proc. Nat Acad. Sci.* **105**, 2783–2787.

[14] Nachman, M., Cojocaru, L. N., and Ribco, L. V. (2006). Electrical properties of nonstoichiometric nickel oxide. *Phys. Stat. Solidi* (b) **8**, 773–783.

[15] Allouti, F., Manceron, L., and Alikhani, M. E. (2009). On the performance of the hybrid TPSS meta-GGA functional to study the singlet open-shell structures: A combined theoretical and experimental investigation of Ni_2O_2 molecule. *J. Mol. Structure. Theochem* **903**, 4–10.

[16] Nogués, J., Sort, J., Langlais, V., Skumryev, V., Surinach, S., Muñoz, J. S., and Baró, M. D. (2005). Exchange bias in nanostructures. *Phys. Reports* **422**, 65–117.

[17] Chakhalian, J., Millis, A. J., and Rondinelli, J. (2012). Whither the oxide interface. *Nature Materials* **11**, 92–94.

[18] Hwang, H. Y., Iwasa, Y., Kawasaki, M., Keimer, B., Nagaosa, N., and Tokura, Y. (2012). Emergent phenomena at oxide interfaces. *Nature Materials* **11**, 103–113.

[19a] Koppens, F. H. L., Buizert, C., Tielrooij, K. J., Vink, I. T., Nowack, K. C., Meunier, T., Kouwenhoven, L. P., and Vandersypen, L. M. K. (2006). Driven coherent oscillation of a single electron spin in a quantum dot. *Nature* **442**, 766–771.

[19b] Loth, S., Baumann, S., Lutz, C. P., Eigler, D. M., Heinrich, A. J. (2012). Bistability in atomic-scale antiferromagnets. *Science* **335**, 196–199.

[20] Kodama, R. H., Makhlouf, S. A., and Berkowitz, A. E. (1997). Finite size effects in antiferromagnetic NiO nanoparticles. *Phys. Rev. Lett.* **79**, 1393–1396.
[21] Dobrynin, A. N., Temst, K., Lievens, K., Margueritat, J., Gonzalo, J., Afonso, C. N., Piscopiello, E., and Van Tandeloo, G. (2007). Observation of Co/CoO nanoparticles below the critical size for exchange bias. *J. Appl. Phys.* **101**, 113913.
[22] Yi, J. B., Ding, J., Feng, Y. P., Peng, G. W., Chow, G. M., Kawazoe, Y., Liu, B. H., Yin, J. H., and Thongmee, S. (2007). Size-dependent magnetism and spin-glass behavior of amorphous NiO bulk, clusters and nanocrystals: experiment and first-principle calculations. *Phys. Rev. B* **76**, 224402.
[23] Walch, S. P., and Goddard III, W. A. (1978). Electronic states of NiO molecule. *J. Amer. Chem. Soc.* **100**, 1338–1348.
[24] Fujimori, A., and Minami, F. (1984). Valence-band photoemission and optical absorption in nickel compounds. *Phys. Rev. B* **30**, 957–971.
[25] Xu, X., Nakatsuji, H., Ehara, M., Lu, X., Wang, N. Q., and Zhang, Q. E. (1998). Cluster modeling of metal oxides: the influence of the surrounding point charges on the embedded cluster. *Chem. Phys. Lett.* **292**, 282–288.
[26] Kadossov, E. B., Kaskell, K. J., and Langell, M. A. (2007). Effect of surrounding point charges on the density functional calculations of Ni_xO_x clusters (x=4-12). *J. Comput. Chem.* **28**, 1240–1251.
[27] Pozhar, L. A. (2010). Small InAsN and InN clusters: electronic properties and nitrogen stability belt. *Eur. Phys. J. D* **57**, 343–354.
[28] Pozhar, L. A., Yeates, A. T., Szmulowicz, F., and Mitchel, W. C. (2005). Small atomic clusters as prototypes for sub-nanoscale heterostructure units with pre-designed charge transport properties. *Eur. Phys. Lett.* **71**, 380–386.
[29] Pozhar, L. A., Yeates, A. T., Szmulowicz, F., and Mitchel, W. C. (2006). Virtual synthesis of artificial molecules of In, Ga and As with pre-designed electronic properties using a self-consistent field method. *Phys. Rev. B* **74**, 085306.
[30] Pozhar, L. A., and Mitchel, W. C. (2009). Virtual synthesis of electronic nanomaterials: fundamentals and prospects. *In*: "Toward Functional Nanomaterials, Lecture Notes in Nanoscale Science and Technology" (Z. Wang, Ed.), Vol. 5, pp. 423–474. Springer, New York.
[31] Pozhar, L. A., and Mavromichalis, C. (2010). Spin alignment in, and electronic and magnetic properties of small Co-O molecules. *J. Appl. Phys.* **107**, 09D708.
[32] Schmidt, M. W., Baldridge, K. K., Boatz, J.A., Elbert, S.T., Gordon, M.S., Jensen, J.H., Koseki, S., Matsunaga, N., Nguyen, K.A., Su, S., Windus, T.L., Dupuis, M., Montgomery, J.A. (1993). General Atomic and Molecular Electronic Structure System. *J. Comput. Chem.* **14**, 1347–1363. http:// www.msg.ameslab.gov/GAMESS
[33] Gordon, M. S., and Schmidt, M. W. (2005). Advances in electronic structure theory: GAMESS a decade later. *In* "Theory and Applications of Computational Chemistry: the First Forty Years" (C. E. Dykstra, G. Frenking, K. S. Kim, G. E. Scuseria, Eds.), pp. 1167–1189. Elsevier, Amsterdam. http:// www.msg.ameslab.gov/GAMESS
[34] Stevens, W. J., Krauss, M., Basch, H., and Jasien, P. (1992). Relativistic compact effective potentials and efficient, shared-exponent basis sets for the third-, fourth-, and fifth-row atoms. *Can. J. Chem.* **70**, 612–630.
[35] Pozhar, L. A., and Mitchel, W. C. (2007). Collectivization of electronic spin distributions and magneto-electronic properties of small atomic clusters of Ga and In with As, V and Mn *IEEE Trans. Magn.* **43**, 3037–3039.
[36] Powell, R. J., and Spicer, W. E. (1970). Optical properties of NiO and CoO. *Phys. Rev. B* **2**, 2182–2193.

[37] Sasi, B., and Gopchandran, K. G. (2007). Nanostructured mesoporous nickel oxide thin films. *Nanotechnology* **18**, 115613.
[38] Boldyrev, A. I., Wang, L.-S. (2001). Beyond classical stoichiometry: experiment and theory. *J. Phys. Chem. A* **105**, 10759–10775.
[39] Wyckoff, R. W. G. (1963). "Crystal Structures". Wiley, New York.
[40] Kittel., C. (2005). "Introduction to Solid State Physics", 8th ed. Wiley, New York, p.71.
[41] Kashani Moltagh, M. M., Youzbashi, A. A., and Sabaghadeh, L. (2011). Synthesis and characterization of nickel hydroxide/oxide nanoparticles by the complexation-precipitation method. *Int. J. Phys. Sci.*, **6**, 1471–1476.
[42] Mahmoud, A. A., Akl, A. A., Kamal, H., and Abdel-Hady, K. (2002). Opto-structural, electrical and electrochromic properties of crystalline nickel oxide thin films prepared by a spray pyrolysis. *Physica B* **311**, 366–375.
[43] Patil, P. S., and Kadam, L. D. (2002). Preparation and characterization of spray-pyrolyzed nickel oxide (NiO) thin films. *Appl. Surf. Sci.* **199**, 211–221.
[44] Lin, Y., Xie, T., Cheng, B., Geng, B., and Zhang, L. (2003). Ordered nickel oxide nanowire arrays and their optical absorption properties. *Chem. Phys. Lett.* **380**, 521–525.
[45] Varkey, A. J., and Fort, A. F. (1993). Solution growth technique for deposition of nickel oxide thin films. *Thin Solid Films* **235**, 47–50.
[46] Doyle, W. P., and Lonergau, G. A. (1958). Optical absorption in the divalent oxides of cobalt and nickel. *Discuss. Faraday Soc.* **26**, 27–33.
[47] Wang, X., Song, J., Gao, L., Jin, J., Zheng, H., and Zhang, Z. (2005). Optical and electrochemical properties of NiO via thermal decomposition of nickel oxalate nanofibers (NiC_2O_4). *Nanotechnology* **16**, 37.
[48] Boschloo, G. and Hagfeldt, A. (2001). Spectroelectrochemistry of nanostructured NiO. *J. Phys. Chem. B* **105**, 3039–3034.
[49] Wruck, D. A., and Rubin, M. (1993). Structure and electronic properties of electrochromic NiO films. *J. Electrochem. Soc.* **140**, 1096–1104.
[50] Desai, J. D., Min, S. K., Jung, K. D., and Joo, O.-S. (2006). Spray pyrolytic synthesis of large area NiO_x films from aqueous nickel acetate solutions. *Appl. Surface Sci.* **253**, 1781–1786.
[51] Cappus, D., Xu, C., Ehrlich, D., Dillmann, B., Ventrice, C. A., Jr., Al Shamery, K., Kuhlenbeck, H., Freud, H.-J. (1993). Hydroxyl groups on oxide surfaces: NiO (100), NiO (111) and Cr_2O_3 (111). *Chem. Phys.* **177**, 533–546.
[52] Bredow, T., and Gerson, A. R. (2000). Effect of exchange and correlation on bulk properties of MgO, NiO and CoO. *Phys. Rev. B* **61**, 5194–5201.

Quantum Dots of Indium Nitrides with Special Magneto-Optic Properties

8.1 Introduction

Indium nitrides, Ga-rich indium nitride thin films and quantum dots (QDs) possess exceptionally useful optical and optoelectronic properties. In particular, the bandgap of such systems is tunable from infrared, to visible and to near-ultraviolet bands of electromagnetic radiation spectrum [1] by manipulations with In, Ga and especially nitrogen content, QD size and shape, and confinement structure (see Chapter 6 for references and further details). Synthesis of InGaN structures with concentration of indium below 30% that possess the bandgaps from green to near-ultraviolet (or from 550 to 400 nm in wavelength and 2.3 to 3.1 eV in radiation energy, respectively) provided solutions to the so-called "green valley of death" problem [2,3] that once severely limited quantum efficiency of light emitting diodes (LED) generating light in the green part of the radiation spectrum which is extremely important for human vision. Rich electronic properties of indium nitrides, and in particular, their sensitivity to nitrogen content, are further enhanced in indium nitride nanostructures that are routinely used in field effect transistors, solar blind and other photodetectors, solar cells and laser diodes [4–7].

Higher electron drift velocity [8], Hall mobility [9] and week spin-orbit coupling [10] in bulk InN raises an interest to InN-based nanosystems doped with Ni or Co atoms as promising systems for spintronic and quantum information processing applications where the use of diluted magnetic semiconductor (DMS) systems is limited by fast dephasing of electron spins and electron charge delocalization effects. Indeed, Ni and Co dopant atoms introduce into InN-based nanosystems uncompensated electron spins localized on Ni or Co atoms that can be effectively manipulated with by the use of external magnetic fields or currents. Equally crucial for such applications, Ni or Co doping of InN-based nanosystems is expected not to significantly inhibit mobility of electrons and collective excitations, or affect week spin-orbit coupling responsible for long electron decoherence time and diffusion length that are crucial for spintronics and quantum information processing applications. Moreover, manipulations with electron spins using external magnetic fields or currents is more complicated in the case of DMS nanosystems with Mn dopant atoms than in the case of Ni- or Co- doped indium nitride nanosystems, because in the former case uncompensated electron spins are localized primarily in the vicinity of In atoms (see Chapter 5 for further details).

Another simple option to enhance magnetic properties of InN-based nanosystem is to dope them with vanadium atoms. In this case, vanadium-doped InN-based DMS receive uncompensated electron spins localized on V atoms, similar to the case of Ni- or Co-doped InN-based nanosystems. However, doping of InN with vanadium atoms is extremely difficult to achieve and control in practice. Thus, introduction of localized uncompensated electron spins into InN-based nanosystems seems much more practical in the case of Ni or Co dopants.

In this chapter the structure, electronic, opto-electronic, magnetic and magneto-optical properties of InN-based QDs doped with Ni or Co atoms are discussed in detail from the first-principle, quantum many-body theoretical standpoint. In addition to already mentioned novel applications for doped indium nitride nanosystems in spintronics and quantum information processing device development, it is important to note that such nanosystems may become valuable as a class of active magneto-optic materials for the development of unidirectional optical components, including isolators and circulators, for existing optical communication systems. Currently, yttrium iron garnet (YIG, or $Y_3F_5O_{12}$) and its modifications, such as bismuth-substituted YIG, and several other nonreciprocal materials [terbium gallium garnet ($Tb_3Ga_5O_{12}$), terbium aluminum garnet ($Tb_3Al_5O_{12}$), terbium-doped borosilicate glass] and birefringent crystals ($CaCo_3$, TiO_2, YVO_4, BaB_2O_4, and $LiNbO_3$), are used in beam displacers, isolators, circulators, prism polarizers, compensators and other optical components utilizing the polarization splitting [11] phenomenon. It is unlikely, that such materials could be integrated in spintronics and quantum information processing optical communication systems of the future. The majority of such materials used in contemporary optical communication components do not possess uncompensated electron spins, and those of them that do, have extremely poor electron mobility. Thus, InN-based nanosystems doped with Ni and Co atoms seem to be uniquely positioned to replace active magneto-optic materials of contemporary optical communication components in devices of similar functionality designed for quantum information and spintronics communication systems of the future.

While there exists a large volume of experimental data and some recent first-principle (DFT) results concerning electronic and optical properties of bulk indium nitride and InN-based nanostructures (see Chapter 6 for details and references), very little information is available on such properties of Ni- or Co- doped bulk indium nitrides and the corresponding nanosystems. This chapter is focused on discussion of what seems to be the first such results obtained in Refs. 12 and 13.

8.2 Virtual synthesis procedure for small indium nitride QDs doped with Ni or Co atoms

Similar to other studies of semiconductor compound nanosystems in this book, several small non-stoichiometric QDs of indium nitrides doped with Ni or Co atoms have been synthesized virtually [12,13] using first-principle, many body quantum theoretical methods. These methods include several approximations of increasing accuracy and predictive capacity (described in Chapter 2 in detail) that have been used to perform the total energy minimization of all of the doped indium nitride systems utilizing GAMESS software package [14–16] with the SBKJC standard basis set [17,18]. Thus, Hartree-Fock (HF) approximation permitted to obtain the equilibrium structures, molecular orbitals (MOs) and electronic properties of the QDs that were further used as input for the next order approximation procedure called the configuration interaction (CI) approximation. The resulting CI MOs served as input to obtain more accurate equilibrium electronic properties within the complete active space (CAS) self-consistent field (SCF) and multiconfiguration SCF (MCSCF) approximations. The output of MCSCF procedure provides electronic properties and MOs of the studied molecules where configuration interactions, static and (partially) dynamic correlations are accounted for. Modifications [19,20] of standard applications of HF, CI, CAS and MCSCF methods have been used to model quantum confinement effects. In particular, a conditional total energy minimization procedure permitted to mimic QD stabilization in quantum confinement. In this case, coordinates of the centers of mass of the constitutive atoms are fixed during the total energy minimizations procedure in all approximations, reflecting so-called excluded volume effects of quantum confinement and to a degree, polarization effects of quantum confinement or other external electromagnetic fields on properties of molecules synthesized under such conditions. The resulting models of quantum-confined QDs are called below pre-designed or constrained molecules. To obtain the corresponding unconstrained (or vacuum) molecules, the spatial constraints applied to atomic positions in the pre-designed molecules are lifted, and the HF, CI, CASSCF and MCSCF total energy minimization procedures are applied again. Such models of "vacuum" molecules reflect nucleation of chemically bonded structures (that from a chemical standpoint are molecules) out of groups of independent atoms in the absence of any constraints or external fields. The conditional total energy minimization procedures [19,20] allow alleviation of limiting effects of contemporary supercomputer hardware on full-scale quantum system modeling that reduce such modeling to systems composed of several tenths of many-electron atoms.

Manipulations with synthesis conditions of nanoscale systems reveal that in many cases quantum confinement effects are as strong as composition effects, and may even offset the latter, because both classes of effects result in breaking natural symmetry of molecules that are revealed when such molecules nucleate in the absence of any foreign atoms or outside fields. In particular, such symmetry breaking brings about the first singlet excitations otherwise excluded in pre-designed tetrahedral symmetry systems of this book. In their turn, the latter excitations lead to enhanced generation of the second-order harmonics in non-centro-symmetric InAs- and InN-based systems [21] that significantly widens horizons for applications of such systems.

All molecules of this chapter have been derived from already studied systems: the unconstrained $In_{10}As_2N_2$ molecule and pre-designed $In_{10}N_4$ molecule of Chapter 6. In their turn, the latter molecules were derived from $In_{10}As_4$ molecule of Chapter 3. Being synthesized virtually without any constraints applied, $In_{10}As_2N_2$ molecule has a distorted pyramidal shape, while the pre-designed $In_{10}N_4$ molecule is an almost perfect pyramid. Such a "family-like" approach to synthesis of new molecules allows direct comparison of their properties to those of their "parent" and "sibling" molecules, thus allowing detailed studies of a role of dopant atoms in stabilization of the molecules so synthesized, and also effects of change in composition on electronic, magnetic and optical properties of these molecules. In derivation of such molecules from their parent $In_{10}As_4$ (Chapter 3), initial values of the covalent radii of In and As atoms (1.44 Å and 1.18 Å, respectively) have been adopted from experiment for convenience.

8.3 Ni-doped molecules derived from unconstrained $In_{10}As_2N_2$ molecule

Four Ni-doped indium nitride molecules discussed in this section have been derived by replacement of one or two As atoms by Ni ones in the unconstrained molecule $In_{10}As_2N_2$ of Chapter 5 (without changing interatomic distances), and subsequent conditional or unconditional minimization of the total energy of the doped QDs so constructed. Ground state electronic properties of these molecules, their unconstrained parent $In_{10}As_2N_2$, and pre-designed $In_{10}As_4$ parents are collected in Table 8.1.

TABLE 8.1 Ground state data for Ni- and Co- containing molecules derived from unconstrained $In_{10}As_2N_2$ molecule of Chapter 6.

Molecule	RHF/ROHF ground state energy, Hartree	MCSCF ground state energy, Hartree	RHF/ROHF direct optical transition energy, eV	MCSCF or CI direct optical transition energy, eV	Spin multiplicity, RHF/ROHF or CI/MCSCF
$In_{10}As_4$	-1907.035049	-1907.077256	3.4286	2.9964	pre-designed triplet
$In_{10}As_2N_2$	-1914.312056	not available	6.2831	not available	vacuum singlet
$In_{10}AsNiN_2$**	-2076.451685	-2076.466091	1.5075	2.9945	predesigned triplet
$In_{10}AsNiN_2$*	-2076.486655	not available	1.2327	not available	vacuum triplet
$In_{10}Ni_2N_2$**	-2238.866104	not available	0.3293	not available	predesigned septet
$In_{10}Ni_2N_2$*	-2238.636824	-2238.782366	3.6926	0.8289	vacuum singlet
In_9CoAsN_3**	-1873.277601	-1873.356717	0.7048	2.6994	predesigned triplet
In_9CoN_4**	-1876.983518	-1877.090751	4.7647	5.9430	predesigned singlet

* Vacuum molecules.
** Pre-designed molecules derived from the vacuum molecule $In_{10}As_2N_2$.

Immediate comparison of the RHF/ROHF ground state energies of the parent molecule $In_{10}As_2N_2$ and Ni-containing ones reveals that substitution of As atoms by Ni ones brings the ground state energy down by about 162 H per Ni atom, both in the pre-designed and vacuum molecules. Notably, the RHF/ROHF ground state energies of the pre-designed Ni-containing molecules lie within 1 H from those of the corresponding vacuum molecules of the same composition. In contrast, replacement of two As atoms by N ones in the pre-designed $In_{10}As_4$ molecule with subsequent unconstrained total energy minimization, brings about the vacuum molecule $In_{10}As_2N_2$ of Chapter 6 (that is, the parent molecule of the Ni-containing ones) whose RHF ground state energy is lower only by about 3.5 H per N atom than that of the pre-designed $In_{10}As_4$ molecule. More accurate values (CI, CASSCF and MCSCF) of the ground state energies are lower only by about 1 H, at most, than those provided by RHF/ROHF approximations for all systems considered of this chapter. Thus, the ground state data of Table 8.1 reveal a remarkable fact that substitution of As atoms by Ni ones in tetrahedral InAs structures is extremely favorable energetically. Moreover, such substitution brings down the direct optical transition energy (OTE), pointing toward possible enhanced mobility of electrons in nanostructures composed of the Ni-containing molecules compared to those built of $In_{10}As_2N_2$ units. The structure and electronic properties of these Ni-containing molecules derived from their vacuum $In_{10}As_2N_2$ parent are further illustrated in Figs. 8.1 to 8.4.

Despite of the tremendous decrease in the ground state energy, all of the Ni-doped molecules possess the structure (Figs. 8.1a, 8.1b, 8.2a, 8.2b, 8.3a, 8.3b, 8.4a and 8.4b) very similar to that of their distorted pyramidal parent. This is, of course, obvious in the case of the pre-designed molecules in Figs. 8.1 and 8.3, where As atoms were replaced by Ni ones under the existing constraints applied to all atomic positions. However, in the case of the vacuum molecules in Figs. 8.2 and 8.4, retention of the major features of the structure of the parent molecule indicates that that structure accommodates a wide range of spatial constraints or their absence.

Physical mechanisms providing for such structural integrity are revealed by analysis of electron charge re-distribution and bonding in the Ni-containing molecules. In the case of the molecules with one Ni atom, molecular electrostatic potential (MEP) surfaces corresponding to small CDD isovalues (Figs. 8.1c, 8.2c and 8.2d) indicate that outside regions close to molecular "surfaces" (as defined by the covalent radii of In atoms framing the molecules) are electroneutral, or even slightly electropositive. Furthermore, MEP surfaces corresponding to increasingly larger absolute values of CDD become increasingly negative (Figs. 8.1d to 8.1f, and 8.2e and 8.2f). Given that the total charge of the molecules is zero, these findings indicate that electron charge of these molecules is contained in the bulk of the molecules. Comparison of MEP surfaces corresponding to rather large CDD isovalues (Figs. 8.1e and 8.2e, and 8.1f and 8.2f) reveals that electron charge of the unconstrained $In_{10}AsNiN_2$ molecule (Fig. 8.2) is more "tightly packed" inside the molecular volume than that of the pre-designed $In_{10}AsNiN_2$ one. Indeed, in Figs. 8.2e and 8.2f one can observe a well-developed, delocalized electron charge "stability belt", which is provided for by N and Ni atoms [22]. This "belt" is somewhat more prominent than a similar charge accumulation region of the corresponding pre-designed molecule (Figs. 8.1e and 8.1f). Tightening of the charge accumulation region

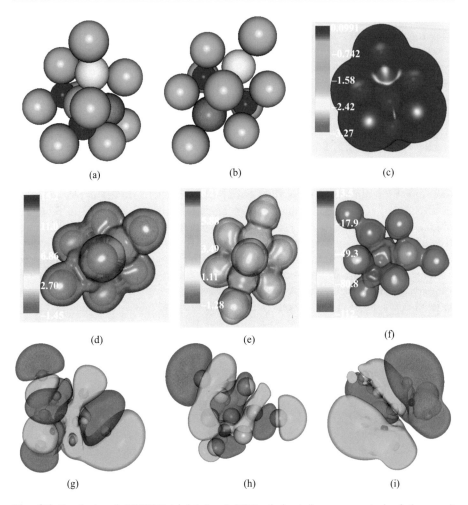

Fig. 8.1 Pre-designed MCSCF triplet $In_{10}AsNiN_2$ derived from unconstrained (vacuum) molecule $In_{10}As_2N_2$ of Chapter 6. Structure: (a) and (b). MEP surfaces (color codding schemes are shown in the figures) corresponding to the isovalues (c) 0.005, (d) 0.05, (e) 0.07, and (f) 0.1 of the charge density distribution (in arbitrary units). Isosurfaces of positive (green) and negative (red) parts of the highest occupied (HO) and lowest unoccupied (LU) molecular orbits (MOs): (g) HOMO 121, (h) HOMO 122 and (i) LUMO 123 corresponding to the isovalue 0.001 (in arbitrary units). Indium atoms are blue, As brown, Ni yellow and N dark blue. Atomic dimensions are reduced; all other dimensions are to scale.

in the unconstrained molecule is achieved by a slight optimization of the molecule's structure (where atoms were "moved" in the process of the unconditional total energy minimization) that caused a significant decrease of the electron repulsion energy by about 2000 H. At the same time, other contributions to the total energy were adjusted accordingly, resulting in the very similar values of the ground state energy (Table 8.1) of these molecules. Notably, the huge decrease in the electron-electron repulsion energy

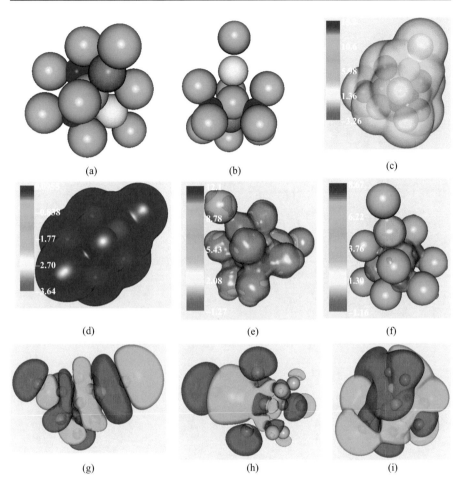

Fig. 8.2 Unconstrained (vacuum) ROHF triplet $In_{10}AsNiN_2$. Structure: (a) and (b). MEP surfaces (color codding schemes are shown in the figures) corresponding to the isovalues (c) 0.005, (d) 0.004, (e) 0.08, and (f) 0.1 of the charge density distribution (in arbitrary units). Isosurfaces of positive (green) and negative (red) parts of the highest occupied (HO) and lowest unoccupied (LU) molecular orbits (MOs): (g) HOMO 121, (h) HOMO 122 and (i) LUMO 123 corresponding to the isovalue 0.001 (in arbitrary units). Indium atoms are blue, As brown, Ni yellow and N dark blue. Atomic dimensions are reduced; all other dimensions are to scale.

of the unconstrained molecule compared to that of the pre-designed one is facilitated by very small changes (in the range of several hundredths of Angstrom) in the atomic positions of atoms in the unconstrained molecule comparatively to those in the pre-designed one. At the same time, nuclear repulsion energy remains almost unchanged indicating that the structure of the parent $In_{10}As_2N_2$ and pre-designed $In_{10}AsNiN_2$ molecules is extremely robust with respect to small atomic displacements. Similar findings and conclusions also apply to the majority of molecular structures derived from tetrahedral pyramids of zincblende InN and GaN bulk lattices, and other high symmetry structures

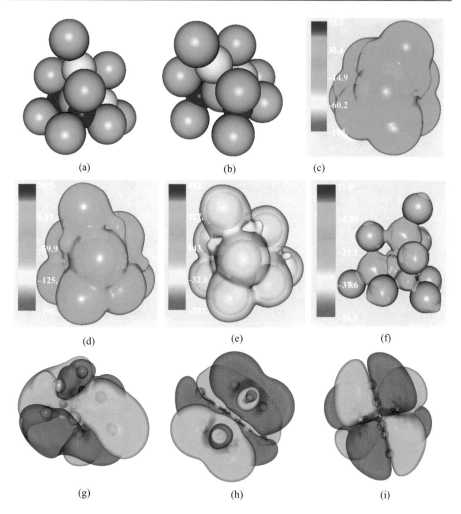

Fig. 8.3 Pre-designed ROHF septet $In_{10}Ni_2N_2$ derived from unconstrained (vacuum) molecule $In_{10}As_2N_2$ of Chapter 6. Structure: (a) and (b). MEP isosurfaces (color codding schemes are shown in the figures) corresponding to the isovalues (c) 0.01, (d) 0.03, (e) 0.04, and (f) 0.1 of the charge density distribution (in arbitrary units). Isosurfaces of positive (green) and negative (red) parts of the highest occupied (HO) and lowest unoccupied (LU) molecular orbits (MOs): (g) HOMO 130, (h) HOMO 131 and (i) LUMO 132 corresponding to the isovalue 0.001 (in arbitrary units). Indium atoms are blue, Ni yellow and N dark blue. Atomic dimensions are reduced; all other dimensions are to scale.

discussed in this book. These observations simply confirm that high symmetry structures are robust with respect to low intensity vibrations of the structure atoms, which is a universally known experimental fact.

However, even such small atomic displacements significantly affect electronic, magnetic and optical properties of the structures through symmetry breaking, and lead

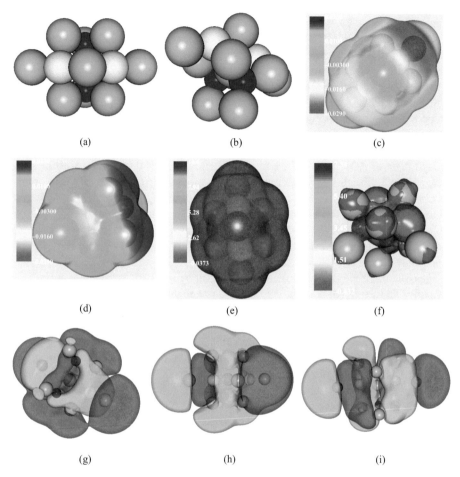

Fig. 8.4 Unconstrained (vacuum) MCSCF singlet $In_{10}Ni_2N_2$. Structure: (a) and (b). MEP surfaces (color codding schemes are shown in the figures) corresponding to the isovalues (c) 0.005, (d) 0.005, (e) 0.01, and (f) 0.1 of the charge density distribution (in arbitrary units). Isosurfaces of positive (green) and negative (red) parts of the highest occupied (HO) and lowest unoccupied (LU) molecular orbits (MOs): (g) HOMO 127, (h) HOMO 128 and (i) LUMO 129 corresponding to the isovalues 0.001, 0.002 and 0.001, respectively (in arbitrary units). Indium atoms are blue, Ni yellow and N dark blue. Atomic dimensions are reduced; all other dimensions are to scale.

to changes in the nature of bonding in such molecules. In the considered case of molecules with one Ni atom of Table 8.1 molecular orbits (MOs) in the highest occupied (HO) – lowest unoccupied (LU) MO regions of these molecules are highly hybridized. Predictably, the topmost HOMO 122 and HOMO 123 of the pre-designed molecule (Figs. 8.1g and 8.1h) feature two major contributions from the $5p$ atomic orbit (AO) of an In atom closest to the Ni atom, and a $3d$ AO of the Ni atom itself. These two contributions are hybridized with numerous smaller contributions from $5p$ AOs of almost all In atoms, $4p$ AOs of arsenic and $2p$ AOs of nitrogen. Similar structure is also typical for MOs in HOMO-LUMO region of the unconstrained $In_{10}AsNiN_2$ molecule (Figs. 8.2g and 8.2h), as well as other Ni-containing molecules of this chapter. Each particular shape of such an MO is defined by proportion in which the contributing AOs of the upper shell electrons of the constitutive atoms are mixed. Abundance of contributions from p-type AOs of many or all constitutive atoms provides for strong π-type bonding in the Ni-containing molecules, while contributions from $3d$ AOs of Ni complicate the shape of their MOs and introduce elements of antibonding. In some cases of such HOMOs, and in many cases of LUMOs, indium atoms also contribute through their $4d$ AOs. Such are the cases of LUMO 123 of the pre-designed $In_{10}As$-NiN_2 molecule (Fig. 8.1i), where noticeable contributions from $4d$ AOs of 3 In atoms are observed, and LUMO 123 of the vacuum $In_{10}AsNiN_2$ molecule (Fig. 8.2i) that features contributions from $4d$ AOs of 2 In atoms. These $4d$ AO contributions further complicate the shape of the LUMOs, and are responsible for the development of antibonding between two strongly π-bonded parts of LUMO 123 of the pre-designed molecule (Fig. 8.1i).

The overall π-type In ligand bonding mediated by nitrogen and arsenic atoms is a distinctive feature of MOs in the HOMO-LUMO regions of all Ni-containing molecules of this chapter. The bond length specific to various contributing atomic pairs varies in a wide range. Thus, in the case of the molecules with one Ni atom, the length of the strongest In-In ligand bond (developed in the unconstrained molecule) is 3.488 Å, while many much weaker In-In ligand bonds with the bonding length running from 5.537 Å to 3.290 Å also develop. The bondlengths of the strongest π-type In-N bonds in the pre-designed $In_{10}AsNiN_2$ molecule are 2.322 Å and 2.154 Å. Somewhat less strong In-N bonds with the bondlengths of 2.130 Å, 2.154 Å and 3.208 Å also develop in this molecule, where each of 6 In atoms bond one of 2 N atoms. In the vacuum molecule with one Ni atom the In-In bond lengths are somewhat longer, In-As bond length somewhat shorter, and In-N bond lengths somewhat closer to each other in the absolute value (2.164 Å, 2.209 Å and 2.203 Å) than those in the corresponding pre-designed molecule. In both molecules with one Ni atom there exist direct and strong In-Ni bonds with 4 typical bondlength values the shortest of which is 3.015 Å (in both molecules) and the longest 3.533 Å (in the vacuum molecule).

Interestingly, in each of the $In_{10}AsNiN_2$ molecules the bondlengths of strongest individual In-As bonds are larger than the sum of the covalent radii of In and As atoms (2.64 Å) in the zincblende InAs bulk lattice. This fact indicates that (i) the In-As interatomic distance in bulk zincblende lattices is defined by collective interaction effects of a large number of electrons of many atoms, rather than by a local force equilibrium only, and (ii) near surfaces of InAs bulk structures that interatomic distance may increase. Notably, bondlengths of several individual In-Ni bonds in each of these molecules are shorter or equal to those of the corresponding In-As bonds. Thus, replacement of an As atom with a Ni one does not perturb significantly integrity of the overall π-type bonding in these molecules, despite generally antibonding nature of $3d$ AO contribution of the Ni atom. This explains remarkable integrity of the structure of these molecules inherent from that of their unconstrained $In_{10}As_2N_2$ parent. As transpires from analysis of bonding in the other two molecules with 2 Ni atoms derived from this parent, their structural integrity is also explained by dominating π-type bonding in these molecules.

Replacement of another As atom by Ni one in the vacuum $In_{10}As_2N_2$ parent, and conditional minimization of the total energy of the QD so obtained in the presence of position constraints applied to the centers of mass of its atoms, produces the pre-designed $In_{10}Ni_2N_2$ septet of Table 8.1 with the structure (Figs. 8.3a and 8.3b) inherent from the parent molecule. The corresponding vacuum singlet $In_{10}As_2N_2$ (Figs. 8.4a and 8.4b) has been derived from the pre-designed one by lifting the spatial constrains applied to atomic positions of the pre-designed molecule and subsequent unconditional minimization of the total energy. The structure of this molecule is very similar to, but somewhat more symmetric than that of its pre-designed sibling (Figs. 8.3a and 8.3b). This observation further confirms structural integrity of Ni-containing molecules. All Ni-containing molecules derived from the vacuum $In_{10}As_2N_2$ parent inherit its structure, as a result of the similar nature of π-type bonding in all Ni-containing molecules of this section.

Comparing the ground state energies of Ni-containing molecules with one and two Ni atoms (Table 8.1), one immediately observes that the molecules with 2 Ni atoms have their ground state energies lower by almost exactly 162 H that those of the molecules with one Ni atom. That is, a decrease in the RHF/ROHF ground state energy per Ni atom achieved by replacement of an As atom with a Ni one is the same for all 4 molecules of these section. Where available, more accurate MCSCF calculations of the ground state energy of these molecules improve the ground state energy values only slightly and have no effect on the above observation. However, MCSCF OTE and RHF/ROHF OTE values of these molecules differ significantly, confirming to a general expectation that MCSCF calculations that partially include dynamic correlations are necessary to obtain reasonable values of OTEs of Ni-containing molecules.

Furthermore, comparison of the OTE values of Ni-containing molecules listed in Table 8.1 readily reveals that the molecules with two Ni atoms are more metallic than those with one Ni atom. The pre-designed $In_{10}Ni_2N_2$ molecule is the most metallic of all 4 such molecules. While its MCSCF OTE value is not yet available, it is expected to be smaller than that of its vacuum counterpart. The reason for this is that in the case of unconstrained total energy minimization the *minimum minimorum* of the total energy of an atomic cluster is reached. In the case of conditional minimization of the total energy, only a local energy minimum corresponding to a particular set of constraints is achieved. In the majority of studied cases (such as those discussed in Chapters 3 to 8, for examples), the equilibrium state of an atomic cluster (that is, a molecule composed of the cluster's atoms) corresponding to the global minimum of the total energy of the cluster is also characterized by a large OTE value.

Analysis of MEP surfaces further confirms significant metallicity of the molecules with two Ni atoms. Indeed, in the case of smaller CDD isovalues, up to 0.01, the MEP surfaces are outside of the molecular volumes and are negative signifying that some electron charge is pushed outside the molecules (Figs. 8.3c to 8.3e, and 8.4c to 8.4e). Compared to MEP isosurfaces of the molecules with one Ni atom (Figs. 8.1c to 8.1e and 8.2c to 8.2e) for similar CDD isovalues, negative MEP isosurfaces of the molecules with 2 Ni atoms are positioned farther from the "surfaces" of the molecules. The

Ni-N charge accumulation regions inside these molecules include larger portions of the molecular bulks (Figs. 8.3e, 8.3f and 8.4f). The MEP surfaces of the unconstrained $In_{10}Ni_2N_2$ molecule corresponding to very small CDD values show a distinct dipole setup, where electron charge is accumulated at the face of the molecular structure closest to the two Ni atoms (Fig. 8.4c), while the opposite face closest to the two N atoms (Fig. 8.4d) is electron charge deficient. In the case of the pre-designed $In_{10}Ni_2N_2$ molecule (Fig. 8.3c) such a dipolar picture does not form. Instead, similar to the case of "diluted magnetic semiconductor" QDs of Chapter 5, a rather uniform shell of charge deficit (hole) is observed. The hole embraces the "surface" of this molecule on the outside and inside of the molecular bulk, and the cores of In atoms (Fig. 8.4d).

The charge distribution differences reflected by MEP isosurfaces of the $In_{10}Ni_2N_2$ molecules can be traced to the nature of bonding in these molecules. Molecular orbits of the pre-designed molecule in the HOMO-LUMO region, such as HOMO 130 and HOMO 131 (Figs. 8.3g and 8.3h) are defined by almost equal contributions from $3d$ AOs of both Ni atoms, $2p$ AOs of both N atoms, $5p$ AOs of 5 In atoms, and somewhat smaller contributions of $4d$ AOs of 2 In atoms. Contributions from $5p$ AOs of the remaining 5 In atoms are much smaller, although non-zero. Thus, HOMOs of the pre-designed $In_{10}Ni_2N_2$ molecule in the HOMO-LUMO region have distinct antibonding features (Fig. 8.3h) due to large contributions from d-type AOs of both Ni and 2 In atoms. The p-type AO contributions provide for large π-type In-In ligand and In-N bonding parts in these HOMOs. The LUMOs (Fig. 8.3i) of these molecules have similar properties.

While the type of AOs contributing to MOs of the vacuum $In_{10}Ni_2N_2$ molecule in the HOMO-LUMO region is the same as that of the pre-designed molecule, proportions in which those contributions are hybridized in such MOs are very different. For example, HOMOs 127 and 128 of this molecule are formed largely by hybridization of $5p$ AOs of 6 In atoms and $2p$ AOs of both N atoms. Contributions from $5p$ AOs of the other 4 In atoms are also significant, while d-type AOs of Ni and one In atoms do not contribute much. This setup provides for highly delocalized π-type bond that includes large In-In ligand and In-N parts (Figs. 8.4g and 8.4h). In the case of LUMO 129 of this molecule (Fig. 8.4i), increased contributions from $3d$ AOs of Ni atoms (at the expense of decreased contributions from $5p$ AOs of several In atoms) give rise to antibonding regions of these MOs.

The lengths of In-N and In-Ni bonds in the pre-designed and vacuum $In_{10}Ni_2N_2$ molecules are close in value to those in the corresponding $In_{10}AsNiN_2$ molecules, while the In-In bond lengths are somewhat larger. This is reflected by continuing integrity of the structure of the $In_{10}Ni_2N_2$ molecules that is very similar to that of the $In_{10}AsNiN_2$ molecules and the parent of the family, the vacuum $In_{10}As_2N_2$ molecule. Interestingly, the two Ni atoms in each of $In_{10}Ni_2N_2$ molecules develop Ni-Ni ligand bonds of the length 4.134 Å in the pre-designed molecule and 4.306 Å in the vacuum molecule. Similar to the case of the $In_{10}AsNiN_2$ molecules, both Ni atoms in the $In_{10}As_2N_2$ molecules orchestrate re-distribution and delocalization of electron charge of In atoms toward inner parts of the molecules and closer to nitrogen atoms. In both types of molecules, Ni atoms are an indispensable player responsible for the charge stabilization "belts" of these molecules. Bunches of closely lying MOs of such molecules came to being largely due to strong hybridization of In and Ni AOs occupied with electrons shared between In, Ni and N atoms. These stacks of MOs are physically lying in the inner regions of the molecules forming the charge stabilization "belts" that provide for remarkable stability of all Ni-containing molecules. Due to larger number of Ni AOs, the number of such MOs in the Ni-containing molecules is significantly larger than that in the case of indium nitride molecules of Chapter 6. Thus, Ni atoms are responsible for very deep potential wells of the ground states of the Ni-containing indium nitride molecules.

In the case of indium nitride molecules of Chapter 6 that do not contain Ni atoms, a smaller number of AOs in constitutive N and As atoms, and less complicated hybridization of those AOs, do not permit the development of as many closely lying MOs in inner molecular regions, as in the Ni-containing indium nitride molecules of these chapter, leading to less deep potential wells of the ground states of the molecules of Chapter 6.

Three of the four Ni-containind molecules discussed above are "ferromagnetic" multiplets with their uncompensated spins aligned and localized near Ni and In atoms. In particular, the total value of the z-component of the uncompensated electron spin (μ_z) of the pre-designed triplet $In_{10}AsNiN_2$ is 1 μ_B (where μ_B denotes Bohr's magneton) with both uncompensated electron spins localized near the Ni atom. The vacuum $In_{10}AsNiN_2$ molecule is also a "ferromagnetic" triplet with the same μ_z, but with its two uncompensated electron spins localized near an In atom. Neither of these indium nitride molecules with one Ni atom develops a region of electron charge deficit (holes). Therefore, charge carriers in materials that may be synthesized of such nanostructures are likely to be electrons.

The pre-designed $In_{10}Ni_2N_2$ molecule is the only molecule in this group of 4 molecules that possess a relatively large magnetic moment. It is a "ferromagnetic" septet ($\mu_z = 6$ μ_B) with its uncompensated electron spins localized near both In and Ni atoms. Thus, 2 such spins are localized near Ni (13) (where 13 indicates the number of this atom in the structure $In_{10}Ni_2N_2$), two other spin lie near Ni (12) and In (4) atoms, respectively, and most interesting, two remaining spins are localized at almost equal distances from In (5) and In (6) atoms. This molecule, being rather metallic, also houses a charge deficit region. Thus, materials synthesized of such molecules may be of significant interest both for contemporary electronics, and for future spintronics and quantum information processing applications. Such materials possess electron and hole charge carriers, relatively large uncompensated magnetic moment and significant electron and hole mobility. In all these respects, such materials may be superior to the existing DMS films, and may also be superior to DMS nanostructured materials synthesized of components discussed in Chapter 5.

In contrast, the vacuum $In_{10}Ni_2N_2$ molecule is the only "antiferromagnetic" singlet in this group of 4 Ni-containing indium nitride molecules. Its uncompensated magnetic moment is zero, so the molecule is not of interest for applications emphasizing magnetic properties. However, metallicity of this molecule indicates that materials synthesized from such components may have enhanced electron mobility while possessing a significant OTE.

All 4 considered indium nitride molecules containing Ni atoms possess significant dipole moments, the largest of which (5.2010 D) is a property of the vacuum

$In_{10}AsNiN_2$ triplet, and the smallest (3.5474 D) one is exhibited by the pre-designed $In_{10}Ni_2N_2$ septet. Two other molecules, the pre-designed $In_{10}AsNi_2N_2$ triplet and the vacuum $In_{10}Ni_2N_2$ singlet, have the dipole moments 4.5995 D and 4.3682 D, respectively. It seems that the smallest value of the dipole moment of the pre-designed $In_{10}Ni_2N_2$ molecule is achieved at the expense of a significant increase in its uncompensated magnetic moment. This effect of electron charge delocalization and re-distributions due to structural and polarization influence of model quantum confinement on systems synthesized in such confinement may open entirely new horizons for synthesis of ferromagnetic nanosystems with enhanced piezoelectric properties.

Comparison of electronic, magnetic and optical properties of the pre-designed molecules and those of their unconstrained (or vacuum) counterparts once again indicates a tremendous role played by synthesis conditions affecting the structure, and thus the properties of synthesized entities. In the case of $In_{10}AsNiN_2$ molecules, effects of model quantum confinement amount to a very small change of about 0.04 H in the ground state energy, a change of about 0.33 eV in the OTE, and do not affect magnetic properties. In contrast, in the case of $In_{10}Ni_2N_2$ molecules effects of model quantum confinement are tremendous. In particular, the pre-designed molecule virtually synthesized in such confinement possesses a large uncompensated magnetic moment and a region of electron charge deficit (hole), while the corresponding molecule synthesized in "free space" does not exhibit any of these properties. At the same time, RHF OTE of the vacuum molecule is an order of magnitude larger than that of the corresponding pre-designed molecule, while the ground state energy is almost the same. The more complex electronic structure of atoms settling into a molecule at conditions provided by quantum confinement, the stronger influence such quantum confinement exerts on electronic, magnetic and optical properties of the synthesized molecules.

8.4 Ni-doped molecules derived from the pre-designed $In_{10}N_4$ molecule

To better understand structure-property relations, an almost ideally pyramidal, pre-designed molecule $In_{10}N_4$ of Chapter 3 has been chosen as a parent of another family of Ni-doped indium nitrides. This parent molecule has its structure inherent from its own "parent", the pre-designed pyramidal molecule of $In_{10}As_4$, which is a symmetry element of the zincblende InAs bulk lattice (see Chapter 6 for further details). The pre-designed $In_{10}N_4$ molecule has coordinates of the centers of mass of its atoms exactly coinciding with those of the pre-designed $In_{10}As_4$ molecule, but As atoms were replaced there by N ones.

Replacement of one or two nitrogen atoms by nickel ones and subsequent conditional or unconditional minimization of the total energy of the $In_{10}N_4$ QDs so derived produced the three Ni-containing molecules in their ground states (Table 8.2).

The pre-designed $In_{10}N_3Ni$ septet (Fig. 8.5a) has the structure of its pre-designed $In_{10}N_4$ (and $In_{10}As_4$) parents. This provides an opportunity to evaluate a change in

TABLE 8.2 Ground state data for molecules derived from the pre-designed $In_{10}N_4$ molecule of Chapter 6.

Molecule	RHF/ROHF ground state energy, Hartree	MCSCF or CI ground state energy, Hartree	RHF/ROHF direct optical transition energy, eV	MCSCF or CI direct optical transition energy, eV	Spin multiplicity, RHF/ROHF or CI/MCSCF
$In_{10}N_4$	-1921.097652	not available	4.1171	not available	pre-designed triplet
$In_{10}N_3Ni$	-2079.647165	-2079.668420	0.5252	2.3837	predesigned septet
$In_{10}N_3Ni*$	-2079.884152	not available	2.0299	not available	vacuum triplet
$In_{10}N_2Ni_2$	-2238.427818	-2238.430916	1.1891	0.3374 (CI)	predesigned pentet
In_9CoN_4*	-1877.290887	not available	6.6505	not available	vacuum singlet

* Vacuum molecules.

electronic properties of the three exactly same pyramidal structures framed by ten In atoms that is caused by replacement of all As atoms in $In_{10}As_4$ pyramid by N ones, and then by replacement of one of N atoms by a nickel atom. After each replacement, of course, the total energy of the derived structures was minimized while positions of their atoms were constrained to the same points in space.

As already discussed in Chapters 3 and 6, such pyramidal structure is very robust with respect to a change in the nature of the atoms "framed" in the "bodies" of these structures. However, the electronic, magnetic and optical properties of the three pyramidal molecules differ significantly from $In_{10}As_4$ to $In_{10}N_4$ molecules, and dramatically from these two molecules to their pre-designed $In_{10}N_3Ni$ derivative. First, the ground state energy of the $In_{10}As_4$ and $In_{10}N_4$ molecules differ only by about 7 H (Tables 8.1 and 8.2) indicating that replacement of As atoms brings a decrease in the total energy of about 3.5 H per nitrogen atom. At the same time, replacement of one N atom with Ni atom in this pyramidal structure facilitates a decrease in the total energy by about 158 H. This decrease in the total energy of the $In_{10}N_4$ pyramidal structure after replacement of a nitrogen atom with Ni one is only about 4 H smaller than the loss of the total energy of about 162 H achieved by replacement of one As atom with Ni one in distorted pyramids of $In_{10}AsNiN_2$ molecules (Table 8.1). This, of course, is due to higher symmetry of the almost perfect pyramidal structure of the pre-designed $In_{10}N_3Ni$ molecule comparatively to those of the distorted $In_{10}AsNiN_2$ pyramids. Correspondingly, the ground state energy of the pre-designed $In_{10}N_3Ni$ molecule is about 3H lower than that of the $In_{10}AsNiN_2$ molecules. At the same time, ROHF OTE of this molecule is almost 3 times smaller, and its MCSCF OTE is about 0.6 H smaller than those of the pre-designed $In_{10}AsNiN_2$ pyramids. However, the largest difference in properties of the pre-designed $In_{10}N_3Ni$ molecule of Table 8.2 and both $In_{10}AsNiN_2$

molecules of Table 8.1 is brought about by their different chemical nature: the pre-designed $In_{10}N_3Ni$ molecule is a "ferromagnetic" septet with its large z-component of the uncompensated spin magnetic moment equal to 6 μ_B, while the other two molecules are triplets with $\mu_z = 1$ μ_B. Having in mind possible technological applications in spintronics and quantum information processing, nanostructured films built of such pyramidal $In_{10}N_3Ni$ elements seem more easy and convenient to synthesize, and they

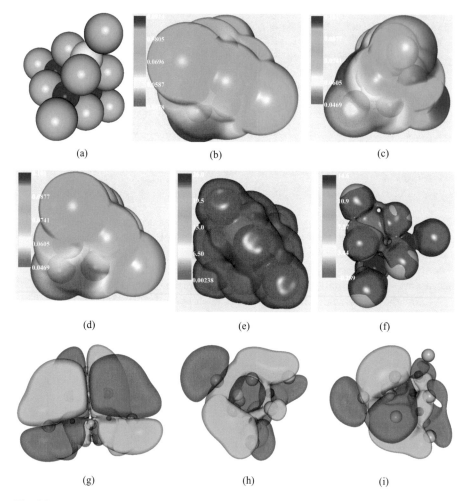

Fig. 8.5 Pre-designed MCSCF septet $In_{10}N_3Ni$ derived from the pre-designed molecule $In_{10}N_4$ of Chapter 6. Structure: (a). MEP surfaces (color codding schemes are shown in the figures) corresponding to the isovalues (b) 0.001, (c) 0.005, (d) 0.005, (e) 0.03, and (f) 0.5 of the charge density distribution (in arbitrary units). Isosurfaces of positive (green) and negative (red) parts of the highest occupied (HO) and lowest unoccupied (LU) molecular orbits (MOs): (g) HOMO 123, (h) HOMO 124 and (i) LUMO 125 corresponding to the isovalue 0.001 (in arbitrary units). Indium atoms are blue, Ni yellow and N dark blue. Atomic dimensions are reduced; all other dimensions are to scale.

also may produce up to 6 times larger magnetization than those synthesized of distorted pyramidal $In_{10}AsNiN_2$ elements.

The vacuum molecule $In_{10}N_3Ni$ is derived from its pre-designed sibling by lifting the spatial constraints applied to the latter molecule's atoms to allow atomic motion, while minimizing the total energy. The resulting shape of the vacuum molecule (Figs. 8.6a and 8.6b) differs significantly from the nearly perfect pyramid of the pre-designed molecule and reminds the distorted pyramids of the $InAsNiN_2$ molecules in Figs. 8.1 and 8.2. The reason for this is that a nickel atom is much larger than a nitrogen one. While

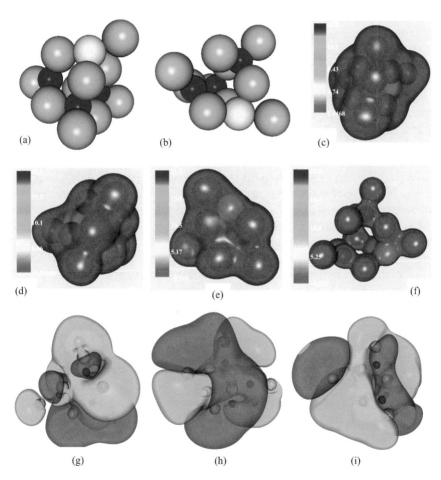

Fig. 8.6 Unconstrained (vacuum) ROHF triplet $In_{10}N_3Ni$. Structure: (a) and (b). MEP surfaces (color codding schemes are shown in the figures) corresponding to the isovalues (c) 0.001, (d) 0.0035, (e) 0.005, and (f) 0.03 of the charge density distribution (in arbitrary units). Isosurfaces of positive (green) and negative (red) parts of the highest occupied (HO) and lowest unoccupied (LU) molecular orbits (MOs): (g) HOMO 121, (h) HOMO 122 and (i) LUMO 123 corresponding to the isovalue 0.001 (in arbitrary units). Indium atoms are blue, Ni yellow and N dark blue. Atomic dimensions are reduced; all other dimensions are to scale.

the pre-designed pyramid of $In_{10}N_4$ molecule is somewhat loose (because the nitrogen atoms occupy much less space than As atoms they replaced in the $In_{10}As_4$ pyramid), replacement of one N atom by Ni one significantly affects electron charge of In atoms closest to the Ni one. As a result, there is a local stress in the vicinity of the Ni atom in the pre-designed $In_{10}Ni_3N$ molecule demonstrated by its MEP surfaces (Figs. 8.5b to 8.5d) and a large value of its dipole moment (8.6935 D). When atoms of this stressed molecule are allowed to move by elimination of confinement (that is, by lifting position constraints), the structure is re-shaped, as In atoms move further from the Ni one. The In atom in the vertex of the pyramid closest to the Ni atom moves out of the ideal vertex position and becomes somewhat hanging over a face of the perfect pyramid giving the resulting structure a shape reminiscent of clown's cap (Fig. 8.6a). These adjustments are sufficient to provide for uniform re-distribution of electron charge of In atoms (as demonstrated by MEP surfaces of this molecule for smaller CDD isovalues, Figs. 8.6c to 8.6e), and to bring down the dipole moment of this molecule to 2.8940 D, which is the smallest value among those specific to all Ni-containing molecules of this chapter.

This charge re-distribution is facilitated by significant changes in bonding of the molecule's atoms. While in the pre-designed $In_{10}N_3Ni$ molecule only 5 In atoms are involved into ligand bonding with equal values of the bondlengths (4.284 Å), in the vacuum $In_{10}N_3Ni$ molecule 7 In atoms bond each other directly, with bondlengths significantly shorter - from 3.110 Å to 4.143 Å. In both molecules all of In atoms bond N atoms, but the In-N bondlengths in the vacuum molecule are shorter running from 2.102 Å to 2.399 Å in value, while in the pre-designed molecule such bondlengths fall in the range from 2.220 Å to 2.934 Å. In the pre-designed molecule Ni atom bonds directly 3 nearest In atoms with bonds of 2.220 Å, 2.717 Å and 2.934 Å in length. At the same time, in the vacuum molecule the Ni atom bonds 4 In atoms directly, and the bondlengths of these bonds are closer to each other in value: 2.521 Å, 2.594 Å, 2.599 Å and 2.694 Å. Similar to many other Ni-containing molecules of these chapter, the nickel atom in $In_{10}N_3Ni$ molecules does not bond nitrogen atoms.

Differences in bonding of the two $In_{10}N_3Ni$ molecules are facilitated by differences in the nature of hybridization of MOs in their respective HOMO-LUMO regions. Thus, topmost MOs of the pre-designed molecule in its HOMO-LUMO region features large contributions from $4d$ AOs of 3 In atoms and $3d$ AOs of Ni, and significantly smaller contributions from $2p$ AOs of nitrogen atoms and $5p$ AOs of the remaining 7 In atoms (Figs. 8.5g to 8.5i). These contributions provide for significant antibonding parts of these MOs. In contrast, contributions to MOs in the HOMO-LUMO region of the vacuum molecule from In and Ni AOs of d-type are small, so the nature of bonding in this molecule is defined by In and N AOs of p-type (Figs. 8.6g to 8.6i). In both molecules, MOs from the respective HONO-LUMO regions feature significant contributions from p-type ligand bonding of In atoms.

Effects of quantum confinement modeled by application of spatial constraints to atomic positions in the pre-designed molecule are accountable for remarkable deviation of its optical and magnetic properties from those of the vacuum molecule. For example, ROHF OTE of the pre-designed molecule is about 4 times smaller, and its MCSCF OTE is about 0.35 H larger, than ROHF OTE of the vacuum molecule. Both molecules are "ferromagnetic" multiplets with their uncompensated electron spins parallel. The pre-designed molecule is a septet ($\mu_z = 6\ \mu_B$), while the vacuum one is a triplet ($\mu_z = 1$ μ_B). Two of the six uncompensated electron spins of the pre-designed molecule are localized in the vicinity of 3 In atoms [In (2), In (9), and In (10)], three other such spins are localized near nitrogen atoms – one spin per atom, and one spin is localized on Ni atom. Appearance of 3 electron spins in the body of this molecule near nitrogen atoms is a remarkable phenomenon highlighting a role of the molecule's charge stability "belt" (Fig. 8.5f) provided for by N and Ni atoms and similar to those discussed in section 3 of this chapter. The nitrogen-nickel charge stability "belt" of the vacuum molecule (Fig. 8.6f) plays a similar role: one of the molecule's uncompensated electron spins is localized near a nitrogen atom, while the other such spin is positioned between this nitrogen atom and the nearest indium one. Similar to the case of some DMS nanosystems of Chapter 5, none of the two uncompensated spins of this molecule are located near the Ni atom.

The pre-designed $In_{10}N_2Ni_2$ molecule of Table 8.2 is derived from the pre-designed $In_{10}N_4$ one by replacement of 2 nitrogen atoms with two nickel ones and subsequent conditional minimization of the total energy of the cluster so derived. Comparison of properties of this molecule (Fig. 8.7) with those of its pre-designed sibling $In_{10}N_3Ni$ and parent $In_{10}N_4$ of the same pyramidal shape (Fig. 8.7a) with all atoms located in the same positions allows elucidation of chemical composition effects without interference of structural ones.

The ground state energy of the pre-designed $In_{10}N_2Ni_2$ pentet is almost 317 H less (Table 8.2) than that of its patent pyramid $In_{10}N_4$ and about 158.66 H less than that of the pre-designed pyramid with one Ni atom. This finding confirms a previously reached conclusion that in such pyramidal molecules replacement of a nitrogen atom with a Ni one is the major single factor behind a sharp decrease in the total energy.

The ROHF OTE value of the pre-designed $In_{10}N_2Ni_2$ pentet lies between those of the pre-designed $In_{10}N_4$ and $In_{10}N_3Ni$ molecules (Table 8.2). Most likely, all three of these ROHF OTE values are not very accurate due to the fact that the RHF/ROHF approximation significantly overestimates energy differences (Chapter 2). Indeed, MCSCF OTE of 0.3374 H of this molecule seems much more reasonable, as it reflects a significant increase in metallicity of this molecule comparatively to that of its pre-designed $In_{10}N_3Ni$ sibling. The dipole moment 3.9279 D of the $In_{10}N_2Ni_2$ pentet is smaller than that of the pre-designed $In_{10}N_3Ni$ septet confirming again its increased stability with respect to that of the latter molecule.

This conclusion is further supported by the nature of MEP surfaces of the pre-designed $In_{10}N_2Ni_2$ pentet for smaller CDD isovalues (Figs. 8.7b and 8.7c) that reflect more uniform charge distribution outside the surface of this molecule than that of the pre-designed $In_{10}N_3Ni$ septet (Figs. 8.5b to 8.5d). Due to increased metallicity, MEP surfaces of $In_{10}N_2Ni_2$ pentet is also uniform for CDD isovalues of about 0.01 (in arbitrary units; Fig. 8.7d), while such uniformity in the case of the $In_{10}N_3Ni$ septet is reached only for larger CDD isovalues (Fig. 8.5e) starting from 0.03 (in arbitrary units). Further on, the N-Ni charge stability "belt" of the $In_{10}N_2Ni_2$ pentet (Figs. 8.7e and 8.7f) is much thicker and reveals itself in the wider range of the CDD isovalues than that of the $In_{10}N_3Ni$ septet (Fig. 8.5f).

The nature of bonding in these two pre-designed molecules is similar. Predictably, MOs of these molecules in their respective HOMO-LUMO regions are results of complicated hybridization of *d*- type AOs of In and Ni atoms with *p*-type AOs of In

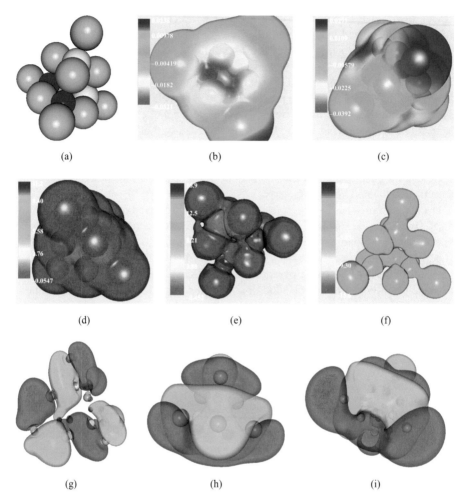

Fig. 8.7 Pre-designed ROHF pentet $In_{10}N_2Ni_2$ derived from the pre-designed molecule $In_{10}N_4$ of Chapter 6. Structure: (a). MEP surfaces (color codding schemes are shown in the figures) corresponding to the isovalues (b) 0.001, (c) 0.003, (d) 0.01, (e) 0.05, and (f) 0.07 of the charge density distribution (in arbitrary units). Isosurfaces of positive (green) and negative (red) parts of the highest occupied (HO) and lowest unoccupied (LU) molecular orbits (MOs): (g) HOMO 129, (h) HOMO 130 and (i) LUMO 131 corresponding to the isovalues 0.01, 0.001 and 0.001, respectively (in arbitrary units). Indium atoms are blue, Ni yellow and N dark blue. Atomic dimensions are reduced; all other dimensions are to scale.

and N atoms (Figs. 8.5g to 8.5i and 8.7g to 8.7i). However, in the case of $In_{10}N_2Ni_2$ pentet, d-type contributions to such MOs are much smaller than dominating p-type AO contributions that in many cases include a huge In-In ligand π-bond. As revealed by the shape of the HOMO 130 and LUMO 131 (Figs. 8.7h and 8.7i), large overall π-type bonds embracing all atoms are formed in the $In_{10}N_2Ni_2$ pentet due to this phenomenon. At the same time, larger d-type AO contributions result in relatively large antibonding parts of the HOMO 124 and LUMO 125 of the $In_{10}N_3Ni$ septet. The bondlength 4.284 Å of the strongest In-In bonds in the $In_{10}N_2Ni_2$ pentet is the same as that in the case of the $In_{10}N_3Ni$ septet. However, 3 types of less strong In-In bonds with the bondlengths 4.277 Å, 4.282 Å and 4.291 Å are also present in the pentet, while no other In-In bonds are present in the septet. The bondlengths of In-N bonds in both molecules are essentially the same. The In-Ni bonding, however, is very different in these molecules. Thus, the $In_{10}N_2Ni_2$ pentet has 10 strong In-Ni bonds with their bondlengths in the range from 2.220 Å to 3.997 Å, and 2 weaker In-Ni bonds of 6.547 Å in length. The $In_{10}N_3Ni$ septet, however, has only 3 strong In-Ni bonds of 2.220 Å, 2.717 Å and 2.934 Å in length. With acquisition of an additional Ni atom instead of N atom the electronic structure of the $In_{10}N_2Ni_2$ pentet acquires an opportunity to develop many more MOs to facilitate electron charge redistribution toward the body of the molecule (and thus stabilize the molecule) than the electronic structure of the $In_{10}N_3Ni$ septet with only one Ni atom.

Interestingly, the ground state energy of the pre-designed $In_{10}N_2Ni_2$ pentet almost coincides with those of both pre-designed and vacuum distorted $In_{10}N_2Ni_2$ pyramids of section 8.3 derived from the significantly distorted vacuum $In_{10}As_2N_2$ pyramid. The ROHF OTEs of the pre-designed $In_{10}N_2Ni_2$ pentet (Table 8.2) and the pre-designed $In_{10}Ni_2N_2$ septet of section 8.3 (Table 8.1) are also close in value, while their MOs are similar in nature. The reason for this is that the electronic and optical properties of these molecules are defined by the same Ni content responsible for a steep decrease by over 300 H in the ground state energy of and numerous MOs available for electron charge re-distribution in these molecules compared to those of their respective $In_{10}N_4$ and $In_{10}As_2N_2$ "parents". [Structure effects contribute to the total energy of these molecules less than several tens of Hartree.] This observation perfectly correlates with the facts that replacement of an As atom with Ni one translates into a loss of 162 H in the ground state energy per Ni atom in the molecules of Table 8.1 derived from the vacuum $In_{10}As_2N_2$. Similarly, replacement of a nitrogen atom with a nickel one brings about a loss of over 158 H per Ni atom in the ground state energy of the molecules of Table 8.2 derived from the pre-designed $In_{10}N_4$ molecule.

The pre-designed $In_{10}Ni_2N_2$ molecule is a "ferromagnetic" pentet with its z-component of the total electron spin magnetic moment equal to 4 μ_B. Two of the uncompensated electron spins are localized between near In atoms and the other two on Ni atoms. While the total magnetic moment of this molecule is relatively small, it still is comparable with, and in many cases even larger than those of the In-As-Mn and In-As-V molecules of Chapter 5 modeling components of DMS nanosystems. Therefore, nanostructures synthesized of tetrahedral $In_{10}N_2Ni_2$ components may also be of interest for applications in spintronics and quantum information processing device development.

8.5 Co-doped In-As-N and In-N molecules

In contrast to Ni-doped molecules where Ni atoms replaced As or N atoms in the bulk of the parent molecules, a Co atom in Co-doped molecules has replaced one of In atoms in the frame of the parent structures, the vacuum $In_{10}As_2N_2$ or the pre-designed $In_{10}N_4$ molecules.

Thus, to synthesize virtually the pre-designed In_9CoAsN_3 triplet of Table 8.1, one of In atoms in the frame of the distorted pyramid of the vacuum $In_{10}As_2N_2$ molecule was replaced with a cobalt atom, and one of As atoms in the bulk of that molecule was replaced with a nitrogen atom. The total energy of the QD so obtained was minimized while spatial constraints were applied (the atomic positions were kept fixed). Thus, the emerged pre-designed In_9CoAsN_3 triplet retains the shape (Fig. 8.8a) of its parent molecule, and therefore, all differences in electronic properties of these and similar molecules can be traced directly to differences in their composition and positions of the dopant atoms.

In sharp contrast to the case of Ni-containing molecules, all molecules with Co atoms have their ground state energies larger than those of their corresponding parent molecules or Ni-containing siblings. The ground state energy of the pre-designed In_9CoAsN_3 triplet (Table 8.1) is about 41 H larger than that of its vacuum $In_{10}As_2N_2$ parent, and over 200 H larger than those of any of Ni-containing molecules with one Ni atom in Tables 8.1 and 8.2. While this finding indicates somewhat lesser stability of this Co-containing molecule compared to its parent and the Ni-containing siblings, the Co-containing pre-designed triplet still has a deep ground state energy minimum, and thus is very stable. Both ROHF OTE and MCSCF OTE of this molecule is significantly smaller (Table 8.1) than those of its closest sibling, the pre-designed $In_{10}AsNiN_2$ molecule, but both molecules are "ferromagnetic" triplets. Because Co and N atoms have 35 electrons total, as opposed to the total electron count of 82 of As and In atoms they substituted, the "surface" of the pre-designed In_9CoAsN_3 triplet is slightly electropositive, as all electron charge is kept within its bulk (Figs. 8.8b to 8.8e). Due to the three nitrogen atoms, and somewhat to As one, the charge stability "belt" of this molecule (Fig. 8.8e) is well developed, although Co atom contributes insignificantly to this charge stabilization.

Molecular orbits of this molecule in the HOMO-LUMO region are complex shapes built of $3d$ Co AOs, $2p$ N AOs, small $4d$ contribution of In AOs, and $4p$ and $5p$ AOs of As and In atoms, respectively, hybridized in various proportions (Figs. 8.8f to 8.8i). Because of dominating p-type contributions, these MOs are primarily bonding with large π-bonding parts of In ligand bonding; antibonding contributions are small. All atoms participate in this bonding. In particular, Co atom directly bonds all 3 nitrogen atoms with bonds of 2.131 Å in length. It also bonds 2 In atoms: the bondlengths of these bonds are 4.571 Å and 3.723 Å. Direct In-N bonding is the most developed one in this molecule, with In-N bondlengths lying in the range from 2.660 Å to 3.218 Å. The bondlengh of the two strongest In-As bonds is 2.130 Å that almost exactly coincides with the length of the strongest Co-As bond, which is 2.131 Å. No direct N-N or N-As bonds has been found.

Despite the larger spin magnetic moment of the Co atom, both In_9CoAsN_3 and $In_{10}AsNiN_2$ pre-designed molecules are "ferromagnetic" triplets with the same $\mu_z=1$ μ_B. Thus, no gain in the total magnetic moment of the constrained Co-containing molecule is achieved comparatively to that of the Ni-containing one. Similar to the pre-designed $In_{10}AsNiN_2$ molecule, where two uncompensated electron spins are localized near the Ni atom, both uncompensated electron spins of the Co-containing molecule reside near Co atom.

The only gain from using Co atom instead of Ni atom in the Co-containing derivative of the pre-designed $In_{10}As_2Ni_2$ molecule is a decrease of about 0.3 H in MCSCF

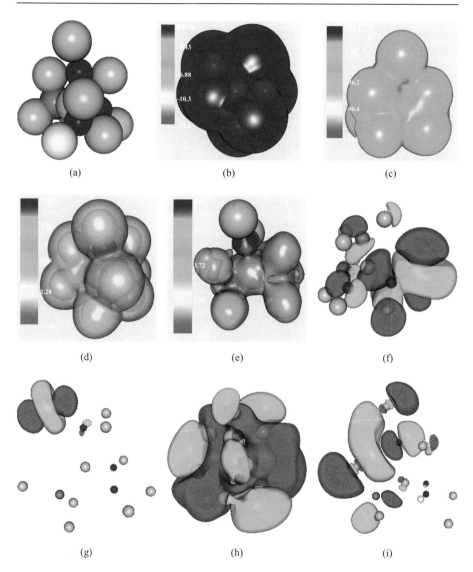

Fig. 8.8 Pre-designed MCSCF triplet In_9CoAsN_3 derived from unconstrained (vacuum) molecule $In_{10}As_2N_2$ of Chapter 6. Structure: (a). MEP surfaces (color codding schemes are shown in the figures) corresponding to the isovalues (b) 0.001, (c) 0.02, (d) 0.03 and (e) 0.08 of the charge density distribution (in arbitrary units). Isosurfaces of positive (green) and negative (red) parts of the highest occupied (HO) and lowest unoccupied (LU) molecular orbits (MOs): (f) HOMO 114, (g) HOMO 114, (h) LUMO 115 and (i) LUMO 115 corresponding to the isovalues 0.001, 0.02, 0.001 and 0.02, respectively (in arbitrary units). Indium atoms are blue, As brown, Co yellow and N dark blue. Atomic dimensions are reduced; all other dimensions are to scale.

OTE and somewhat larger dipole moment of the Co-containing molecule (5.8313 D, as opposed to 4.5995 D of the Ni-containing one). From this example one can see again, that quantum confinement may have remarkable effects on the properties of molecules synthesized in it. For example, in the case of the Co-containing pre-designed triplet the model quantum confinement almost entirely "averaged out" a composition effect due to replacement of In and As atoms with Co and N ones, respectively. Due to the effects of model quantum confinement, a difference in magnetic and optical properties of the pre-designed Co-containing molecule and its Ni-containing sibling is similar to that of the pre-designed and vacuum $In_{10}AsNiN_2$ molecules. This finding indicates that nanostructured materials with In_9CoAsN_3 as "building blocks" may not offer advantages for applications interested in magnetic properties, but may be useful when tuning of OTE or electric properties is necessitated by device applications.

The pre-designed molecule In_9CoN_4 of Table 8.1 was derived from the same pre-designed $In_{10}As_2N_2$ parent by replacement of one In atom with a Co one and simultaneous replacement of both As atoms with N ones, and subsequent minimization of the total energy keeping atomic position fixed. The latter spatial constraints ensure that the shape of this molecule (Fig. 8.9a) is exactly the same as that of its parent, the pre-designed molecule $In_{10}As_2N_2$, and siblings, the pre-designed In_9CoAsN_3 and $In_{10}AsNiN_2$ molecules. This is done to elucidate effects of nitrogen content in such structures.

Replacement of another As atom with a nitrogen atom brings the ground state energy of the pre-designed In_9CoN_4 molecule down by about 3.6 H compared to that of the pre-designed In_9CoAsN_3 one (Table 8.1). This value is very close to the energy loss per N atom achieved by replacement of As atoms in the tetrahedral pyramid of $In_{10}As_4$ and other structures derived from it, which is about 3.5 H (for further discussion, see Chapter 6 and the ground state energy data in the first rows of Tables 8.1 and 8.2). At the same time, MCSCF OTE value of the pre-designed In_9CoN_4 molecule is about 2.2 times larger than that of the pre-designed In_9CoAsN_3 molecule. This increase in OTE is a sign of exceptional stability of the pre-designed In_9CoN_4 molecule gained due to acquisition of yet another nitrogen atom instead of As one. The dipole moment of this molecule is large (7.0246 d) due to a significant charge re-distribution toward the bulk of the molecule facilitated by the nitrogen atoms, and asymmetry of the entire CDD of this molecule due to the Co atom in its frame. Remarkably, this molecule is an "antiferromagnetic" singlet; that is, the uncompensated magnetic moment of the Co atom was entirely compensated through electron charge re-distribution mechanism giving rise to the only pre-designed singlet among all Ni- and Co- containing molecules discussed in this chapter. This observation points toward a conclusion that an increase in the number of nitrogen atoms in the "body" of zincblende InAs-derived molecules translates into a decrease in the uncompensated electron spin magnetic moment of the molecules.

Analysis of MEP surfaces of the pre-designed In_9CoN_4 molecule confirms that farther from the "surface" of this molecule there is no electron charge (Fig. 8.9b), and a region near the "surface" is highly electropositive (Fig. 8.9d). This latter region embraces the "surface" (including the cores of In atoms) both on outer and inner sides of it and forms a tetrahedral-like electron charge deficit hole of about 1 nm in linear dimensions. This hole is similar to the holes of DMS-like nanosystems of Chapter 5, but is better defined in space and does not embrace any uncompensated polarized electron spins, because the pre-designed In_9CoN_4 molecule is an "antiferromagnetic" singlet. Electron charge of this molecule is contained inside of its bulk (Figs. 8.9d), and a significant portion of it forms a strong charge stability "belt" (Fig. 8.9e) supported by nitrogen atoms.

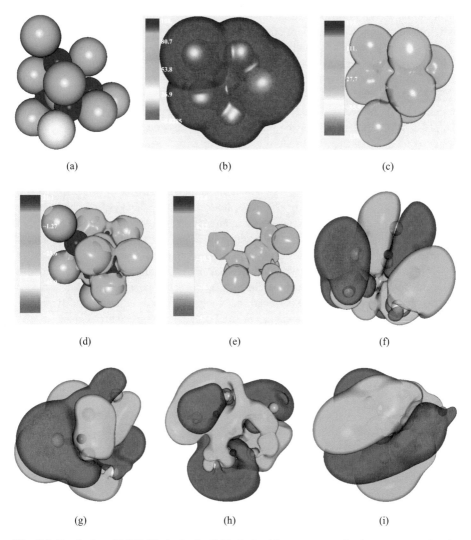

Fig. 8.9 Pre-designed MCSCF singlet In_9CoN_4 derived from unconstrained (vacuum) molecule $In_{10}As_2N_2$ of Chapter 6. Structure: (a). MEP surfaces (color codding schemes are shown in the figures) corresponding to the isovalues (b) 0.0005, (c) 0.04, (d) 0.07 and (e) 0.1 of the charge density distribution (in arbitrary units). Isosurfaces of positive (green) and negative (red) parts of the highest occupied (HO) and lowest unoccupied (LU) molecular orbits (MOs): (f) HOMO 112, (g) HOMO 113, (h) and (i) LUMO 114 corresponding to the isovalue 0.001 (in arbitrary units). Indium atoms are blue, Co yellow and N dark blue. Atomic dimensions are reduced; all other dimensions are to scale.

Bonding in the pre-designed In_9CoN_4 molecule is defined by dominating p-type AO contributions of almost all participating atoms, save for the Co one, which contributes through its $3d$ AOs. One of such $3d$ Co AO contributions is enhanced by a contribution from a $4d$ AO of one of In atoms in the case of HOMO 112 (Fig. 8.9f), thus providing for significant antibonding parts of this HOMO. In contrast, $4d$ AOs of In atoms do not contribute to HOMO 113 (Fig. 8.9g) and LUMO 114 (Figs. 8.9h and 8.9i) of this molecule, where contributions from p-type AOs of all the atoms, but Co one, define distinct sandwich-like structures of strong π-bonds of these MOs. The In-In, In-N, In-Co and Co-N bonds in this molecule are similar in length to those in the pre-designed In_9CoAsN_3 molecule.

The only unconstrained molecule In_9CoN_4 (Table 8.2) was derived from the pre-designed $In_{10}N_4$ molecule directly, bypassing indirect options, such as its derivation from the pre-designed In_9CoAsN_3 or the pre-designed In_9CoN_4 molecules. Thus, in the pre-designed $In_{10}N_4$ molecule an In atom in the pyramidal "frame" was replaced by a Co atom, and the structure was optimized without any constraints applied to positions of its atoms. Interestingly, the shape of this molecule (Figs. 8.10a and 8.10b) is closer to that of the distorted unconstrained $In_{10}Ni_2N_2$ tetrahedral pyramid than to that of the unconstrained $In_{10}N_3Ni$ molecule (Figs. 8.5a, b). The major reason for this is a loss of tetrahedral symmetry bearing on electronic structure when an indium atom in the almost perfect pyramid of the parent molecule $In_{10}N_4$ is replaced with a Co atom. A loss of tetrahedral symmetry of the parent molecule due to replacement of an "inner" nitrogen atom with a Ni one is not as significant, because it is somewhat alleviated by re-distribution of electron charge from the surface to the body of the structure. Only replacement of two nitrogen atoms with two Ni ones in the "body" of the pre-designed $In_{10}N_4$ structure leads to tetrahedral symmetry loss comparable to that caused by replacement of only one In atom with a Co one.

The RHF ground state energy of the unconstrained In_9CoN_4 molecule (Table 8.2) is about 0.3 H lower than that of the pre-designed In_9CoN_4 molecule of Table 8.1 derived from the vacuum $In_{10}As_2N_2$ parent, and only about 0.2 H lower than the MCSCF ground state energy of the pre-designed In_9CoN_4 molecule. This finding implicates that the global minimum of the total energy of the In_9CoN_4 atomic cluster is indeed realized as the ground state of the vacuum In_9CoN_4 molecule of Table 8.2. The major reason for such a conclusion follows from the fact that the pre-designed In_9CoN_4 molecule of Table 8.1 (which itself is not tetrahedrally symmetric) was derived from a different parent molecule in the process of different manipulations with composition of that molecule, than those relevant to the case of the unconstrained In_9CoN_4 molecule. Yet the difference in the ground state energies of these molecules falls in the same value range as those of other pairs of the pre-designed and vacuum molecules featuring the same composition and studied in this chapter. In similarity to OTE properties of other such pairs, the RHF OTE values of these molecules are very different, with RHF OTE of the vacuum molecule being about 2 eV larger than that of its pre-designed sibling, while being only by about 0.6 eV larger than the MCSCF OTE value of the sibling. Both molecules are "antiferromagnetic" singlets with zero uncompensated electron spin magnetic moment. Owning to better CDD uniformity, the dipole moment of the vacuum In_9CoN_4 molecule (6.9564 D) is smaller than that of its pre-designed In_9CoN_4 sibling of Table 8.1, which is also typical for pre-designed/unconstrained molecule pairs.

Quantum Dots of Indium Nitrides with Special Magneto-Optic Properties 343

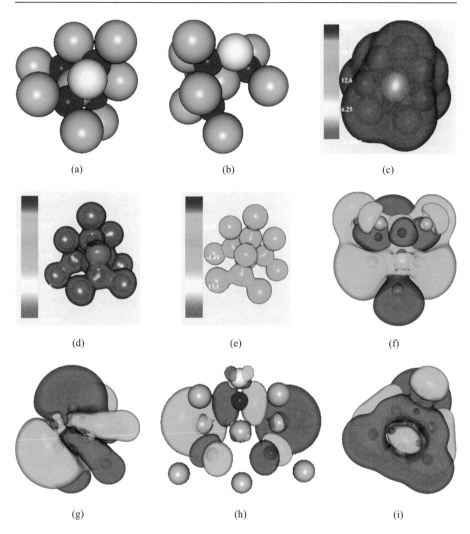

Fig. 8.10 Unconstrained (vacuum) RHF singlet In_9CoN_4 derived from the pre-designed molecule $In_{10}N_4$ of Chapter 6. Structure: (a) and (b). MEP surfaces (color codding schemes are shown in the figures) corresponding to the isovalues (c) 0.001, (d) 0.03 and (e) 0.05 of the charge density distribution (in arbitrary units). Isosurfaces of positive (green) and negative (red) parts of the highest occupied (HO) and lowest unoccupied (LU) molecular orbits (MOs): (f) HOMO 112, (g) and (h) HOMO 113, and (i) LUMO 114 corresponding to the isovalues 0.002, 0.001, 0.02 and 0.002, respectively (in arbitrary units). Indium atoms are blue, Co yellow and N dark blue. Atomic dimensions are reduced; all other dimensions are to scale.

In contrast to the pre-designed In_9CoN_4 molecule of Table 8.1, the vacuum In_9CoN_4 molecule of Table 8.2 does not possess an electron charge deficit region (or a hole), as demonstrated by MEP surfaces of this molecule. Thus, for small CDD isovalues its MEP surfaces outside the molecular "surface" correspond to slightly negative CDD values (Fig. 8.10c), and with an increase in CDD absolute values MEP surfaces of this molecule become more and more negative as they "move" into the bulk of the molecule (Figs. 8.10d and 8.10c). Electron charge of this molecule is re-distributed uniformly, so the charge stability "belt" in this molecule (Fig. 8.10e) is not as prominent as it is in the case of its pre-designed counterpart (Fig. 8.9e). Molecular orbits of the vacuum In_9CoN_4 molecule are hybridized of AOs of the same type as those of its pre-designed counterpart of Table 8.2, where contributions from p-type AOs primarily define the shape of MOs (see, for example, its HOMO 112 in Fig. 8.10f). However, a contribution of a $3d$ AO of the Co atom and a $4d$ AO of one of In atoms in the case of HOMO 113 of the vacuum molecule (Figs. 8.10g and 8.10h) are relatively large, so this MO features some antibonding parts. In contrast, LUMO 114 of this molecule includes only a small contribution from a $3d$ AO of Co, and no contributions from $4d$ AOs of In atoms, and thus is defined by hybridization of p-type AOs of In and N atoms.

As reflected by uniformity of its MEP surfaces, bonding in the vacuum In_9CoN_4 molecule is very reach. In particular, there exists 11 π-type In-N bonds contributing to the overall bonding in this molecule with bondlengths in the range from 2.108 Å to 2.491 Å. Notably, there is only one strong π-type In-In ligand bond of 3.437 Å in length, while all other In-In bondins are mediated by nitrogen atoms. In contrast, the Co atom in this molecule bonds 4 In atoms (the bondlengths run from 3.295 Å to 3.895 Å), while it bonds only 3 closest nitrogen atoms (the bondlengths are 2.038 Å, 2.099 Å and 2.157 Å).

All 3 Co-containing molecules of this section are of little interest for applications relying on outstanding magnetic properties. At the same time, these molecules are great examples of a crucial role of quantum confinement effects that may define properties of synthesized entities in a way that alleviate effects of composition and enhance effects of structure. Most importantly, magnetic properties of the studied Co-containing molecules demonstrate that the use of a few atoms with the large uncompensated electron spin magnetic moment does not necessarily translates into the large magnetic moment of the synthesized structure. In fact, among the three Co-containing molecules of this section only one is a "ferromagnetic" triplet with its modest $\mu_z = 1$ μ_B, and two other are "antiferromagnetic" singlets. From a standpoint of magnetic properties, nothing is gained using a Co atom instead of Ni one in the studied structures.

8.6 Conclusions

Electronic, optical and magnetic properties of 10 indium nitride nanosystems containing Ni or Co atoms have been studied in this chapter using the first-principle, many body quantum field-theoretical methods. The results show that the properties of these molecules are comparable to those of DMS-like nanosystems of Chapter 5. However, from technological standpoint, Ni-containing indium nitride systems are easier to synthesize and manipulate with than the Mn- and V-containing nanosystems of Chapter 5. The use of Co atoms as providers of uncompensated electron spins in the studied systems does not offer any advantages comparatively to the use of Ni atoms for the same purpose. The obtained results provide an insight into fundamental physical and chemical mechanisms defining electronic, optical and magnetic properties of Ni- and Co-containing nanoscale indium nitrides. Several major conclusions below may help develop effective strategies for experimental synthesis of novel indium nitride – based nanomaterials with rich electronic, optical and magnetic properties challenging those of DMS-like nanostructures.

1. All small indium nitride nanostructures containing Ni or Co atoms and discussed in this chapter exhibit electronic, magnetic and optical properties largely different from those of bulk systems, and very similar to those of molecules. Their ELSs are comprised of bundles of well-separated energy levels, CDDs and MEP surfaces are non-uniform for larger CDD isovalues, MOs are products of strong hybridization of upper shell occupied AOs of their constitutive atoms, and the total magnetization is defined by a few uncompensated electron spins. Such strongly correlated systems cannot be properly described in the framework of theoretical approaches developed for bulk indium nitrides or other bulk systems and require the first-principle, quantum many body-theoretical investigation.

2. Coulomb interaction-driven re-distribution of electron charge of In atoms toward inner parts of the studied molecules containing nitrogen and/or nickel atoms is the major physical mechanism providing for exceptional stability of these molecules. Substitution of As atoms with Ni ones in the parent indium nitride molecules promotes further already significant stability of the parent systems for the following reasons. (i) A decrease in the number of electrons of Ni atoms compared to that of As ones leads to a decrease in Coulomb repulsion between electrons of In atoms in the scaffold and electrons of N and Ni atoms in the body of the molecules. (ii) Compared to that of As atoms, ELS of Ni atoms possess more unoccupied $3d$, $4s$ and $4p$ levels providing for more options for the development of low lying hybridized MOs built of contributions from topmost AOs of almost all constitutive atoms. This permits to accommodate a larger portion of electron charge of In atoms in the body of the molecules. (iii) At the same time, proton charge of nickel nuclei still is large enough to ensure significant nucleus-electron Coulomb attraction stabilizing that portion of electron charge of In atoms in the body of the molecules. The considerations of items (ii) and (iii) also explain an increase in molecular stability in the case of replacement of N atoms with Ni ones in the parent indium nitrides molecules, despite of a larger number of electrons in Ni atoms than that of N atoms.

3. Using a Co atom instead of one of the In atoms in the frames of the parent indium arsenide molecules leads to some increase in the ground state energy, and thus somewhat decreased stability of the Co-containing molecules compared to their parent structures and siblings with one Ni atom. The major reason for this is a strong impact of a "foreign" ELS of the Co atom on tetrahedral symmetry of the

In-defined ELS of the molecular scaffolds. While a Co atom also possesses many unoccupied $3d$, $4s$ and $4p$ AOs that contribute to the overall In electron charge re-distribution, this re-distribution is not directed toward inner regions of the molecules and primarily interferes with ligand In – In providing for antibonding mediated by the Co atom. The overall electron charge distribution in this case is less uniform than in the case of indium nitride molecules with one Ni atom positioned inside their bulks.

4. Similar to electronic properties, magnetic properties of the studied molecules are defined by Coulomb interactions driving electron charge re-distribution in these molecules to ascertain the minimum values of the total energy. Depending on composition (that is, on a particular ELS structure and nucleus charge of the constitutive atoms) and synthesis conditions, such re-distribution may or may not lead to compensation of uncompensated electron spins introduced by Ni or Co atoms. Thus, localization of uncompensated spins in the bodies of such molecules (near N atoms) or near In atoms instead of being localized on Ni or Co atoms is a direct consequence of electron charge re-distribution. Similarly, compensation of electron spins introduced by Ni or Co atoms in the case of the vacuum singlet $In_{10}Ni_2N_2$, and both vacuum (Table 8.1) and pre-designed (Table 8.2) singlets In_9CoN_4, further demonstrates the leading role of electron charge collectivization and re-distribution caused by interplay of direct Coulomb interaction effects and "integrated" Coulomb interaction effects introduced by quantum confinement.

5. In the majority of the studied cases, the uncompensated electron spin magnetic moment of the virtually synthesized Ni- or Co- containing molecules is insignificant. Three of the ten such molecules are singlets, 4 triplets, one pentet and two septets. Interestingly, the pre-designed $In_{10}N_2Ni_2$ pentet (Table 8.2) and two septets – the pre-designed $In_{10}Ni_2N_2$ and $In_{10}N_3Ni$ molecules (Table 8.1) – are Ni-containing pre-designed molecules, while two of 3 Co-containing molecules are singlets. While not suitable for applications relying on the use of magnetic properties, the latter singlets can be of interest for quantum computing realized using antiferromagnetic QDs [23].

The three higher multiplets containing Ni atoms may have a bright future in applications currently using DMS nanomaterials. The $In_{10}N_2Ni_2$ pentet and $In_{10}Ni_2N_2$ septet contain 2 Ni atoms each and possess MCSCF OTEs that are rather small. Materials synthesized of such structural units may be too metallic for applications in spintronics or quantum information processing, because of possible decrease in the electron spin diffusion length and decoherence time.

The pre-designed $In_{10}N_3Ni$ septet is the easiest to synthesize, and undoubtedly exists naturally on surfaces and in voids of bulk InN materials exposed to Ni contamination. It is an exceptionally stable molecule with a deep ground state potential well, a large OTE of 2.3837 eV (Table 8.2) and a π-type bonding HOMO (Fig. 8.5h). The latter points toward aromaticity [24] of this molecule, and thus indicates that larger structures may be synthesized of $In_{10}N_3Ni$ structural elements. Such structures may become indispensable materials for spintronics and quantum information processing devices that require large magnetization, higher electron spin mobility and large electron spin decoherence times.

6. All three Ni-containing molecules possessing the largest uncompensated electron spin magnetic moments are pre-designed using model quantum confinement. The unconstrained siblings of the two pre-designed $In_{10}Ni_2N_2$ and $In_{10}N_3Ni$ septets (Tables 8.1 and 8.2, respectively) are $In_{10}Ni_2N_2$ singlet and $In_{10}N_3Ni$ triplet, respectively. Therefore, in these cases the model confinement effects are the major leading mechanism behind electron charge re-distribution in these molecules that results in their large spin magnetic moments. This conclusion is also supported by an observation that only one or two uncompensated electron spins in these molecules reside near Ni atoms, while the rest of the spins are localized near In or N atoms.

   In contrast, model quantum confinement effects do not lead to the appearance of any uncompensated electron spin magnetic moment in the pre-designed In_9CoN_4 molecule. Similar to its vacuum sibling, it remains in its singlet state. These and numerous other findings imply that effects of quantum confinement on electronic, optical and magnetic properties of entities synthesized in such confinement are multifaceted. In some cases such effects may entirely change the properties (for example, alleviating effects of structure and composition), and in other cases they may be negligibly small. Yet in all cases, such effects may be used efficiently to manipulate synthesis of nanosystems with desirable electronic, magnetic and optical properties.

7. Two of the pre-designed molecules – already mentioned $In_{10}Ni_2N_2$ septet and In_9CoN_4 singlet of Table 8.1 - house one well-developed electron charge deficit region (hole) each. The holes have a tetrahedral shape, are of about 1 nm in linear dimensions and better defined spatially than similar holes found in DMS-like nanosystems of Chapter 5. This observation indicates that quantum confinement may have very subtle effects on electron charge distributions of systems synthesized in such confinement. Thus, in the above case of the two In_9CoN_4 singlets, the ground state energy of the pre-designed molecule (Table 8.1) almost coincides with that of its vacuum sibling (Table 8.2), but their ROHF OTEs differ by about 2 eV, and their uncompensated magnetic moments are zero. The pre-designed molecule develops a well-defined electron charge deficit hole, while the vacuum one does not. Such effects of quantum confinement have been used experimentally to enhance hole conductivity in some nanostructures (see corresponding discussions in Chapter 5). They may also be used to manipulate a mode of conductance in nanosystems possessing both hole and electron conductance modes.

8. Ability of quantum confinement to affect spin multiplicity of entities synthesized in such confinement may be useful in studies of the development and loss of exchange bias in nanostructures [25,12,13]. This phenomenon requires further investigation and may be utilized to synthesize nanomaterials for magnetic memory systems and other devices for spintronics.

9. While there exist a large array of experimental and some theoretical studies of electronic, optical and magnetic properties of indium nitride nanosystems, no such reports concerning properties of indium nitride nanosystems doped with Ni or Co atoms have been found in literature. The only available density functional

theory-based, computational investigation [26] of Ni-doped indium nitrides deals with charge states of Ni substitution impurities in (In, Ni)N bulk systems. Unfortunately, the results of this study cannot be directly compared to those discussed in this chapter for obvious reasons.

References

[1] Wu, J. (2009). When group-III nitrides go infrared: new properties and perspectives. *J. Appl. Phys.* **106**, 011101.
[2] Fuhrmann, D., Netzel, C., Rossow, U., Hangleiter, A., Ade, G., and Hinze, P. (2006). Optimization scheme for the quantum efficiency og GaInN-based green-light-emitting diodes. *Appl. Phys. Lett.* **88**, 071105.
[3] Huang, C.-Y., Yan, Q., Zhao, Y., Fujito, K., Feezell, D., Van de Walle, C. G., Speck, J. S., DenBaars, S. P., and Nakamura, S. (2011). Influence of Mg-doped barriers on semipolar multiple-quantum-well green light-emitting diodes. *Appl. Phys. Lett.* **99**, 141114.
[4] Van der Walle, C. G., Lyons, J. L., and Janotti, A. (2010). Controlling the conductivity of InN. *Phys. Status Solidi A* **207**, 1024–1036.
[5] Ji, L. W., Su, Y. K., Chang, S. J., Liu, S. H., Wang, C. K., Tsai, S. T., Fang, T. H., Wu, L. W., and Xu, Q. K. (2003). InGaN quantum dot photodetectors. *Solid-State Electronics* **47**, 1753–1756.
[6] Shubina, T. V. (2010). Plasmonics effects and optical properties of InN composites with In nanoparticles. *Phys. Status Solidi A* **207**, 1054–1061.
[7] Sharma, T. K., and Towe, E. (2010). Application-oriented nitride substrates: The key to long-wavelength nitride lasers beyond 500 nm. *J. Appl. Phys.* **107**, 024516.
[8] Theodoropolpu, N., Hebard, A. F, Overberg, M. E., Abernathy, C. R., Pearton, S. J., Chu, S. N., and Wilson, R. G. (2001). Magnetic and structural properties of Mn-implanted GaN. *Appl. Phys. Lett.* **78**, 3475–3477.
[9] Foutz, B. E., O'Leary, S. K., Shur, M. S., and Eastman, L. F. (1999). Transient electron transport in wurtzite GaN, InN and AlN. *J. Appl. Phys.* **85**, 7727–7734.
[10] Davydov, V. Yu., Klochikhin, A. A., Emtsev, V., Smirnov, A. N., Goncharuk, I. N., Sakharov, A. V., Kurdykov, D. A., Baidakova, M. V., Vekshin, V. A., and Ivanov, S. (2003). Photoluminescence and Raman study of hexagonal InN and In-rich InGaN alloys. *Phys. Status Solidi B*, **240**, 425–428.
[11] Eldada, L. (2004). Optical communication components. *Rev. Scientific Instruments* **75**, 575.
[12] Pozhar, L. A. (2011). Magneto-optic properties of InAsN and InN quantum dots doped with Co or Ni atoms. *J. Appl. Phys.* **109**, 07C303.
[13] Pozhar, L. A. (2011). Electronic and magneto-optic properties of Co- and Ni-doped small quantum dots of indium nitrides. *Phys. Status Solidi C* **8**, 2261–2263.
[14] Schmidt, M. W., Baldridge, K. K., Boatz, J. A., Elbert, S.T., Koseki, S., Matsunaga, N., Nguyen, K.A., Su, S., Windus, T.L., Dupuis, M., and Montgomery, J.A. (1993). General Atomic and Molecular Electronic Structure System. *J. Comput. Chem.* **14**, 1347–1363.
[15] Roos, B. O. (1987). The complete active space self-consistent field method and its applications in electronic structure calculations. *Adv. Chem. Phys.* **69**, 399–445.
[16] www.msg.ameslab.gov/GAMESS.
[17] Stevens, W. J., Basch, H., Krauss, M., and Jasien, P. (1992). Relativistic compact effective potentials and efficient, shared exponent basis sets for the third-, fourth- and fifth-row atoms. *Can. J. Chem.* **70**, 612–630.

[18] Cundari, T. R., and Stevens, W. J. (1993). Effective core potential methods for the lanthanides. *J. Chem. Phys.* **98**, 5555–5565.

[19] Pozhar, L. A., Yeates, A. T., Szmulowicz, F., and Mitchel, W. C. (2006). Virtual synthesis of artificial molecules of In, Ga and As with pre-designed electronic properties using a self-consistent field method. *Phys. Rev. B* **74**, 085306. See also: (2006). *Virtual J. Nanoscale Sci & Technol.* 14, No. 8, http://www.vjnano.org.

[20] Pozhar, L. A., Yeates, A. T., Szmulowicz, F., and Mitchel, W. C. (2005). Small atomic clusters as prototypes for sub-nanoscale heterostructure units with pre-designed charge transport properties. *EuroPhys. Lett.* **71**, 380–386.

[21] Pisarev, R. V., Kaminski, B., Lafrentz, M., Pavlov, V. V., Yakovlev, D. R., and Bauer, M. (2010). Novel mechanisms of optical harmonics generation in semiconductors. *Phys. Status Solidi B* **247**, 1498–1504.

[22] Pozhar, L. A. (2010). Small InAsN and InN clusters: electronic properties and nitrogen stability belt. *Eur. Phys. J. D* **57**, 343–354.

[23] Meier, F., Levy, J., and Loss, D. (2003). Quantum computing with antiferromagnetic spin clusters. *Phys. Rev. B* **68**, 134417.

[24] Boldyrev, A. I., and Wang, L.-S. (2001). Beyond classical stoichiometry: experiment and theory. *J. Phys. Chem.* **105**, 10759–10775.

[25] Nogues, J., Sort, J., Langlais, V., Skumryev, V., Sarinach, S., Munoz, J. S., and Baro, M. D. (2005). Exchange bias in nanostructures. *Phys. Reports* **422**, 65–118.

[26] Usman, Z., Cao, C., Khan, M., Mahmood, T., and Niazi, A. R. (2013). Effect of Ni charge states on structural, electronic, magnetic, and optical properties of InN. *J. Phys. Chem. A* **117**, 5650–5654.

Appendix

Examples of Virtual Templates of Small Quantum Dots and Wires of Semiconductor Compound Elements

This Appendix contains examples of virtual templates of quantum dots and wires discussed in preceding chapters. The templates are output files (containing an extension .out) obtained after conditional and unconditional GAMESS optimization of the corresponding atomic clusters. After introductory information on the names of GAMESS' developers and contributors, the first lines of these files contain input information and initial coordinates of the atoms, such as in Example below.

```
                              EXAMPLE.
EXECUTION OF GAMESS BEGUN Mon Mar  1 15:15:10 2004

            ECHO OF THE FIRST FEW INPUT CARDS -
 INPUT CARD> $CONTRL SCFTYP=MCSCF RUNTYP=ENERGY ECP=SBKJC MPLEVL=0 $END
 INPUT CARD> $CONTRL ICHARG=0 MULT=3 COORD=UNIQUE NOSYM=1 $END
 INPUT CARD> $CONTRL INTTYP=HONDO ITOL=20 ICUT=9 MAXIT=200 $END
 INPUT CARD> $CONTRL EXETYP=RUN $END
 INPUT CARD> $SYSTEM TIMLIM=10000 MWORDS=1800 PARALL=.F. $END
 INPUT CARD> $GUESS GUESS=MOREAD NORB=124 $END
 INPUT CARD> $MCSCF CISTEP=GUGA FORS=.T. ACURCY=1.E-7 $END
 INPUT CARD> $MCSCF ENGTOL=1.E-12 MAXIT=200 FULLNR=.T. $END
 INPUT CARD> $DRT GROUP=C1 FORS=.T. NMCC=111 NDOC=3 NALP=2 $END
 INPUT CARD> $DRT NVAL=8 ISTSYM=1 $END
 INPUT CARD> $GUGDIA NSTATE=4 $END
 INPUT CARD> $DATA
 INPUT CARD>TITLE: MCSCF Triplet State Energy of GaAs bonded cell
 INPUT CARD>CN        1
 INPUT CARD>
 INPUT CARD>GA         31.0     0.0000000000     0.0000000000     0.0000000000
 INPUT CARD>    SBKJC   0
 INPUT CARD>    D       1
 INPUT CARD>    1             0.2070000000  1.00000000
 INPUT CARD>
 INPUT CARD>GA         31.0     5.6532500000     5.6532500000     0.0000000000
```

```
INPUT CARD>     SBKJC   0
INPUT CARD>     D       1
INPUT CARD>     1               0.2070000000  1.00000000
INPUT CARD>
INPUT CARD>GA           31.0    2.8266250000     2.8266250000     0.0000000000
INPUT CARD>     SBKJC   0
INPUT CARD>     D       1
INPUT CARD>     1               0.2070000000  1.00000000
INPUT CARD>
INPUT CARD>GA           31.0    5.6532500000     0.0000000000     5.6532500000
INPUT CARD>     SBKJC   0
INPUT CARD>     D       1
INPUT CARD>     1               0.2070000000  1.00000000
INPUT CARD>
INPUT CARD>GA           31.0    0.0000000000     5.6532500000     5.6532500000
INPUT CARD>     SBKJC   0
INPUT CARD>     D       1
INPUT CARD>     1               0.2070000000  1.00000000
INPUT CARD>
INPUT CARD>GA           31.0    2.8266250000     2.8266250000     5.6532500000
INPUT CARD>     SBKJC   0
INPUT CARD>     D       1
INPUT CARD>     1               0.2070000000  1.00000000
INPUT CARD>
INPUT CARD>GA           31.0    2.8266250000     0.0000000000     2.8266250000
INPUT CARD>     SBKJC   0
INPUT CARD>     D       1
INPUT CARD>     1               0.2070000000  1.00000000
INPUT CARD>

..... DONE SETTING UP THE RUN .....
1800000000 WORDS OF MEMORY AVAILABLE

     RUN TITLE
     ---------
TITLE: MCSCF Triplet State Energy of GaAs bonded cell

THE POINT GROUP OF THE MOLECULE IS CN
THE ORDER OF THE PRINCIPAL AXIS IS      1

ATOM       ATOMIC                    COORDINATES (BOHR)
           CHARGE        X                    Y                    Z
GA          31.0      0.0000000000       0.0000000000        0.0000000000
GA          31.0     10.6830934401      10.6830934401        0.0000000000
GA          31.0      5.3415467200       5.3415467200        0.0000000000
GA          31.0     10.6830934401       0.0000000000       10.6830934401
GA          31.0      0.0000000000      10.6830934401       10.6830934401
GA          31.0      5.3415467200       5.3415467200       10.6830934401
GA          31.0      5.3415467200       0.0000000000        5.3415467200
GA          31.0      0.0000000000       5.3415467200        5.3415467200
GA          31.0     10.6830934401       5.3415467200        5.3415467200
GA          31.0      5.3415467200      10.6830934401        5.3415467200
AS          33.0      2.6707733600       2.6707733600        2.6707733600
AS          33.0      2.6707733600       8.0123200801        8.0123200801
AS          33.0      8.0123200801       2.6707733600        8.0123200801
AS          33.0      8.0123200801       8.0123200801        2.6707733600
```

A filename (such as Ga10As4-MCSCF-M3-8x13-cont6.out) does not necessarily coincides with the title of a particular run (which is more descriptive), such as "INPUT CARD>TITLE: MCSCF Triplet State Energy of GaAs bonded cell" in the Example above. As a rule, filenames and run titles contain the major information on particular approximations and conditions used in the run, beginning with the composition of the cluster, the approximation used (HF or MCSCF), spin multiplicity (M1, M3, *etc.*) and the cluster structure. If a filename and/or title do not provide details on the origin of a cluster (such as the run title in Example), this means that the cluster is pre-designed, that is, optimized while spatial constraints were applied to the atomic coordinates. Alternatively, the filename and/or title that contain an abbreviation (such as VacGeom, Vac, etc.) hints that the cluster was optimized unconditionally, that is that the obtained molecule is a "vacuum" one. Further information in the filenames or run titles concerns details of the runs. For example, "8 × 13" in the filename describes a chosen CAS (see Chapter 2 for discussion of approximations used in calculations), and "cont6" signifies that there were 5 subsequent previous runs, and that the molecular orbits (MOs) received as an output of the 5-th run were used as input MOs for the 6-th (current) run. Thus, the filenames and run titles of the output files included in this Appendix are self-explanatory and should not create misunderstanding as to the nature of the corresponding templates. As a rule, virtual template files are of about 3 to 50 MB in size. Each chapter of the book is represented by two or more virtual templates (*i.e.*, by two GAMESS output files) included in the list below that can be downloaded from the book website http://booksite.elsevier.com/9780123969842/index.php or the website http://www.PermaNature.com.

Chapter 3.	Ga10As4-MCSCF-M3-8x13-cont6.out
	In10As4Own-VacGEOM-M3-MCSCF-8x8.out
Chapter 4.	In10As3POwn-VacGEOM-M3-MCSCF-8x10.out
	Ga10As3P-M3-MCSCF-6x9.out
Chapter 5.	[In10As4]Ga-VacGEOM-M3-MCSCF-8x10.out
	Ga10As2V2-HF-M7.out
Chapter 6.	In10As3NOwn-InAsPVacG-M3-MCSCF-12x11.out
	In6N6Own-M5-MCSCF-8x9.out
Chapter 7.	Ni4O2-Octa-M7-PreD-MCSCF-10x11.out
	Ni4O2-OctaM7Vac-MCSCF-10x10.out
	Ni24O22-Sheet-HF-M1-C3.out
	Ni14O12-Sheet-HF-M1-C3.out
Chapter 8.	In9CoN4OwnVac-PreDes-MCSCF-8x7.out
	In10AsNiN2OwnVac-GEOM-M3.out

To visualize the molecules, one has to convert the virtual template files from UNIX to DOS text format using one of available convertors (such as UNIX2DOS, that can be downloaded from elsewhere), change the file extension .out to .gam, and use Molekel software to open the files. Molekel is free software, and its newer versions can be downloaded from http://molekel.cscs.ch/wiki/pmwiki.php/Main/DownloadBinary

Unfortunately, the newer versions of Molekel are unable to visualize the spin density distributions (SDD), so readers are advised to use an older version, molekel4.3.win32.exe, to perform SDD visualization and analysis.

Readers who would like to use the virtual templates of the molecules described in this book for research purposes may request the corresponding .dat files from the author over e-mail LPozhar@yahoo.com. Such files contain MOs and spectra of the molecules that can serve as input for calculations of various properties of the molecules, as well as for design of other templates on the basis of the developed ones using GAMESS, NWChem or Molpro software packages.

Index

A

Ab initio electronic structure calculations, 103
  computational techniques, 94
Ab initio modules, 101
Ab initio numerical calculations, 104
Ab initio software packages, 101
  quantum chemistry software packages, 102
  and use, 101–103
Active electrons, 85
Ad hoc fitting, 192
Ad hoc schemes, 194
"Antiferromagnetic" cluster, 310
AOs. *See* Atomic orbits (AOs)
Arbitrary scalar function, 12
Atomic clusters, 168, 241
Atomic orbits (AOs), 78, 158, 201, 295, 324
  infinite number of, 78

B

Basis function sets, 98–101
Bogoliubov–Tiablikov's Green's functions, 9
Bohr radius, 69
Bohr's magneton, 199
Boltzmann constant, 9
Bond lengths, 172
Born–Oppenheimer approximation, 4, 74, 82, 96, 115
Bottom-up chemical growth methods, 147
Bottom-up experimental methods, 151

C

Cartesian Gaussian-type functions (GTFs), 99
Cartesian tensor, 11
Cartesian vectors, 13
CAS. *See* Complete active space (CAS)
CASPT2 method, 94
CASSCF. *See* Complete active space self-consistent field (CASSCF) method
CDDs. *See* Charge density distributions (CDDs)
Central-field assumption, 71

CGTFs. *See* Contracted *Gaussian*-type functions (CGTFs)
Charge carriers, 140
  number operator, 14
Charge conservation equation, 9
Charge density
  charge density correlations, 28
  expectation values of, 9–10
  fluctuation, 32
  induced, 16
  operators, 8, 14, 29
Charge density distributions (CDDs), 135, 197, 216, 241
  isovalues, 225, 327
Charge hole, 185
CIS. *See* Configuration interaction-singles (CIS) method
Classical valence theory, 140
Co atom, spin magnetic moment of, 338
Co-containing molecules, 344
Coherent potential approximation (CPA), 193
Column operators, 43
Complete active space (CAS), 154, 244, 283, 318
  second order perturbation theory, 94
Complete active space self-consistent field (CASSCF) method, 4, 85, 241
  variation procedures, 115
Complex energy, 37
Composition effect, 182
Condensed matter theory, 33
Conditional total energy minimization procedures, 117
Configuration interaction (CI), 82–88, 241, 283
  approximation, 318
  computations, types of, 84
  methods, 4, 94
  molecular wavefunction, 84, 85
Configuration interaction-singles (CIS) method, 88
Configuration state functions (CSF), 83, 87, 98

Conservation equations, 6
Continued fraction equation, 49
Continuity equation, 20
Contracted *Gaussian*-type functions
  (CGTFs), 100
 computational advantage of, 101
Contraction coefficients, 101
Controlled approximations, 4
Correlation functions, 3, 42, 43
  lower-order, 35
Coulomb blockade, 195
Coulomb effect, integrated quantum, 199
Coulomb integrals, 76
  energy integrals, 87
Coulomb interactions, 240, 346
  driven redistribution, 345
Coulomb operator, 76
Coulomb repulsion, 273
Coupled cluster (CC) method, 94
  amplitudes, 96
  approximations, 89–98
    errors of, 98
  calculations, 61
CPA. *See* Coherent potential
  approximation (CPA)
CSF. *See* Configuration state functions (CSF)
Current density, 20
  conservation equation, 21
  expectation values of, 9–10
  operator, 8, 10

D

Dangling bonds, 105
Davison method, 87
DCI. *See* Direct configuration interaction
  (DCI) method
Degenerate energy
  levels, perturbation theory treatment of,
    66–68
  perturbation-theoretical treatment of, 66
Density functional theory (DFT), 62, 114,
    193, 194, 239, 283, 291, 311, 318
  intrinsic deficiencies of, 240
  spin-polarized, 194
Density matrix, 78, 80, 82
DFT. *See* Density functional theory (DFT)
Dielectric permittivity, 23
  tensor, 14, 15, 23
  to STFT, 16

Diluted magnetic III–V semiconductor
    compounds
  $In_{10}As_3Mn$ molecules
    predesigned, 197–206
      highest occupied and lowest
        unoccupied molecular orbits, 203
      molecular electrostatic potential
        (MEP) of, 199
    vacuum, 197–206
      electron charge density distribution
        (CDD), 200
      electronic energy level structure
        (ELS) in, 202
      highest occupied and lowest
        unoccupied molecular orbits, 204
      molecular electrostatic potential
        (MEP), 200
      spin density distribution (SDD),
        isosurfaces of, 201
  $In_{10}As_2V_2$ molecules
    predesigned, 223
    vacuum, 223
  $In_{10}As_3V$ molecules
    predesigned, 206–215
      electron charge density distribution
        (CDD), 208
      highest occupied and lowest
        unoccupied molecular orbits, 212
      HOMO–LUMO region of, 209
      molecular electrostatic potential
        (MEP) of, 210
      spin density distribution (SDD),
        isosurfaces of, 214
      structure of, 207
    vacuum, 215
      electron charge density distribution
        (CDD), 209
      highest occupied and lowest
        unoccupied molecular orbits, 213
      HOMO–LUMO region, 211
      molecular electrostatic potential
        (MEP) of, 211
  $In_{10}As_3V_2$ molecules
    predesigned, molecular electrostatic
      potential (MEP) of, 224
    vacuum, molecular electrostatic
      potential (MEP) of, 224
  $Ga_{10}As_3V$ molecules
    with one vanadium atom, 214–222

predesigned
  highest occupied and lowest unoccupied molecular orbits, 219
  molecular electrostatic potential (MEP) of, 217
  spin density distribution (SDD), isosurfaces of, 222
vacuum
  highest occupied and lowest unoccupied molecular orbits, 220
  molecular electrostatic potential (MEP) of, 218
  spin density distribution (SDD), isosurfaces of, 222
$Ga_{10}As_3V_2$ molecules
  predesigned, molecular electrostatic potential (MEP) of, 225
  vacuum, molecular electrostatic potential (MEP) of, 225
$Ga_{10}As_2V_2$ molecules, predesigned and vacuum, 223
ground states of, 197
InAs/GaAs-based molecules
  RHF/ROHF, ground state data for, 201
  with two vanadium atoms, 222–228
  introduction, 191–195
  small quantum dots of, 191–231
  virtual synthesis of, 195–197
Diluted magnetic semiconductor (DMS) systems, 154, 191, 192, 231, 317
  applications of, 229
  films, 329
  magnetic properties of, 193, 194
  nanosystems, 230
Dipole moment, 168, 216, 226, 260, 270, 333, 340
Dirac's brackets, 75
Dirac's delta function, 10, 39
Direct configuration interaction (DCI) method, 87
Dislocation defect, 242
Dispersion relations, 39
DMS. *See* Diluted magnetic semiconductor (DMS) systems
Double excitation operator, 96
Dyson equation, 45, 193

E

ECF. *See* Exchange-correlation function (ECF)
Edge-magneto-plasmons, 150
EGFs. *See* Empirical Green's functions (EGFs)
Electric field intensity, 18, 23
Electron
  charge
    deficit, 330
    shells, 205
    delocalization, and bonding in molecules, 135–138
    density distribution (CDD), 119, 258
    probability density, 86
    stability, 175
  correlations, 83
    energy, 84
    nuclei, 82
  energies, 74
  interaction integrals, 92
  probability density, 62
  quantum numbers, 74
  repulsion
    energy, 163, 321
    integrals, 98
  shells, 195
  spin magnetic moment, 342
  systems, 68
ELS. *See* Energy level structure (ELS)
Empirical Green's functions (EGFs), 61
  analytical derivation of, 103
  analytical properties of, 53
  spectral density of, 41
  time derivatives of, 53
Energy level structure (ELS), 76, 105, 152, 194, 239, 285, 345
  HOMO regions of, 274
Energy minimization process, 244, 252, 284
Equations of motions, 35
Ergodic systems, 41
  thermodynamic properties of, 34
Exact-exchang (EXX) Kohn–Sham density functional theory, 240
Exchange-correlation function (ECF), 194, 239
Exchange integrals, 76
Exchange operators, 76
Excitation operators, 95, 96
  application of, 95

Expectation values, 9
Experimental synthesis techniques, 151
Exponential operator, 95
Extended variation method, 62–64

F

Fabry–Perot laser diodes
   InGaAsP, 148
fcc-derived molecules, 272
fcc-pyramidal scaffolds, 151
FCI. *See* Full configuration interaction
   (FCI) calculations
Fermi energy, 192
"Ferromagnetic" triplets, 158, 329
Ferromagnetism (FM), discovery of, 191
Feynman diagrams, 42
Field effect transistors, 239
First-principle quantum many-body
   theory, 241
First-principle quantum theory of solids, 192
Fock matrix, 81
   elements, 77
Fock operator, 79, 80, 89, 90
Four-center electron repulsion integrals, 100
Frozen-core (FC) approximation, 84, 93
Full configuration interaction (FCI)
   calculations, 84

G

$Ga_{10}As_4$ molecules, 129–133
   CASSCF triplets
      ground state interaction energies
         for, 132
      isosurfaces of MOs from
         HOMO–LUMO region
         of predesigned 6×9, 136
         of vacuum 8×8, 136
      predesigned 8×13, 129, 138
         MEP isosurfaces of, 130
      vacuum 8×8, 130, 138
         MEP isosurfaces of, 130
   SDD isosurfaces of
      predesigned 6×9 and vacuum 8×8, 133
      predesigned 8×10 and vacuum
         8×10, 134
      predesigned 8×13 and vacuum 8×8, 134
   triplet
      direct optical transition energies
         (OTEs) of, 131

      ELSs of vacuum and predesigned
         ROHF, 131
      ground state energies of, 131
$Ga_{10}As_3P$ molecules, 164
   MEPs of, 164
   triplets
      chemical properties of, 157
      pyramidal shape of, 158
$Ga_{10}As_2V_2$ molecules, ground-state energies
   of, 226
Gallium
   based molecules, 129, 159
   Ga–Ga bonds, 175
Gallium arsenide phosphides systems, 184
   Ga–As molecules with one and two
      phosphorus atoms, 153–162
   $Ga_{10}As_2P_2$ molecules
      ground state energies of, 154
      predesigned CAS 6×8 MCSCF triplet,
         160
      unconstrained HF singlet, 161
      vacuum and predesigned, ROHF
         electronic level structure of, 158
   $Ga_{10}As_3P$ molecules
      ground state energies of, 154
      triplet
         predesigned CAS 6×9 MCSCF, 155
         unconstrained CAS 8×8
            MCSCF, 156
      vacuum and predesigned, ROHF
         electronic level structure of, 153
   introduction, 147–151
   quantum confinement, composition effects
      of, 173–183
   small quantum dots of, 147–186
   virtual synthesis procedure, 151–153
Gallium arsenide (GaAs) systems, 113, 194
   based DMS materials, 214
   based DMS systems, 229
   based molecules, 222
      magnetic properties of, 226
   defined quantum confinement, 178
   lattice of, 218
      of fcc, 105
      zincblende, 206, 215
   phosphide
      materials, 147
      molecules, 163
   properties of, 113
   V DMS systems, 214

GAMESS software, 101, 196, 275
   optimization, 351
   package, 152, 196
(Ga, Mn)As crystalline nanowires
   MBE-based formation of, 191
Gaussian functions, 100
   d-type, 99
   f-type, 99
   p-type, 99
   s-type, 99
Gaussian software package, 101
Generalized continuous fraction method, 27, 52
Generalized gradient approximation (GGA), 194, 239
Generalized Hartree–Fock approximation (GHFA), 52
Generalized longitudinal sum rule, 13–14
GGA. *See* Generalized gradient approximation (GGA)
GHFA. *See* Generalized Hartree-Fock approximation (GHFA)
Graphical unitary group approach (GUGA), 87
Green death valley effect, 147
   problem, 317
Green's functions (GF), 107, 193
   based quantum-statistical mechanical methods, 5
   importance of, 33
Ground-state energy(ies), 62, 65, 91, 310, 311
   value approximations, 70
Ground-state wavefunction, 65
   Hartree–Fock (HF) wavefunction, 61
"Guess" density matrix, 81
"Guess" functions, 67
GUGA. *See* Graphical unitary group approach (GUGA)

H

Hall mobility, 317
Hall resistance, hysteresis of, 191
Hamiltonian, 7, 13, 31, 35, 36, 61, 62, 66, 70, 73, 87, 88
   eigenfunctions of, 63
   of equation, 23
   operator, 97
   representation of, 64
   systems, 43

Hartree–Fock (HF) approximations, 75, 196, 290, 297, 318
   based methods, 4, 76
Hartree–Fock equations, 89
Hartree–Fock–Roothaan equations, 77, 80, 98
Hartree–Fock self-consistent field method, 70–82
   Fock operator
      and calculation of physically meaningful quantities
      matrix elements of, 78–82
   for molecules, 73–77
Hartree–Fock wavefunction, 78
Hartree method, 72
Hartree orbits, 72
Hartree procedure, 72
Hartree self-consistent field (SCF) method, 70–73
   wavefunction, 89
Hartree wavefunction, 73
Hausdorff expansion, 97
Hessian matrix, 82
High-efficiency light harvesting, 149
Higher electron drift velocity, 317
Higher-order MP perturbation theory, 119
Highest occupied molecular orbits (HOMO), 119, 154, 162, 250, 285, 286
   HOMO–LUMO regions, 135, 168, 169, 328, 334, 338
Hilbert space, 29, 43, 51
Holes, 154
HOMO. *See* Highest occupied molecular orbits (HOMO)
Homogeneous systems, 26, 27
   quantum systems, 37
   thermodynamic limit for, 38
Hubbard potential, 194, 239
Hybrid bonding, 185
   pd-type, 230

I

ICs. *See* Integrated circuits (ICs)
Impurity band (IB) model, 193
$[In_{10}As_4]_{Ga}$ triplet
   "artificial" molecules, 125–129
   direct optical transition energies (OTEs), 127
   ground state energies of, 127
   interaction energies for, 127

[$In_{10}As_4$]$_{Ga}$ triplet *(cont.)*
  predesigned ROHF, 125
    ELSs of, 128
    MOs from HOMO–LUMO region
      of, 137
  vacuum
    8×10 CASSCF, 126, 137
    ROHF, ELSs of, 128
In_3As molecule
  ground and excited state energies of, 119
  pyramidal, 117
    in HOMO-LUMO region, 120
    MP2 In3As singlet, 120
  RHF electronic level structure of, 118
  transition energies of, 119
$In_{10}As_4$ molecules
  electronic properties of, 124
  ground-state data of, 125
  predesigned and vacuum, 121–125
$InAsNiN_2$ molecules, distorted pyramids
  of, 333
$In_{10}As_2N_2$ molecules
  HOMO–LUMO region of, 259
  magnetic properties of, 260
[$In_{10}As_3P$]$_{Ga}$ molecule
  predesigned, ROHF ground-state energy
    of, 174
$In_{10}As_2P_2$ molecule, 168
  bonding in, 172
$In_{10}As_3P$ molecules, 163
  SDD isosurfaces, 167
$In_{10}As_4$ triplets
  direct optical transition energies
    (OTEs), 122
  ELSs of vacuum and predesigned ROHF, 124
  ground state energies of, 122
  MEP isosurfaces of, 123
  pyramidal predesigned and vacuum
    CASSCF, 121
In_9CoN_4 molecule, 342
  unconstrained, RHF ground-state energy
    of, 342
Indium
  electrons, 163
  In–In bonds, 172, 182
    ligand bonds, 167
  In ligand
    atoms, 209
    bonding, π–type, 326

Indium arsenide (InAs), 113
  based DMS systems, 229
  based molecules, 212, 221
    magnetic properties of, 226
  In–As–Co/Ni systems, 106
  V systems, 206
  zincblende
    lattice, 244
    symmetry elements of, 243
Indium arsenide phosphide systems, 173, 184
  $In_{10}As_2P_2$ molecules
    ground state energies of, 163
    triplet
      predesigned CAS 8×10 MCSCF, 170
      unconstrained CAS 8×10
        MCSCF, 171
      vacuum and predesigned, ROHF
        electronic level structure of, 169
  $In_{10}As_3P$ molecules
    ground state energies of, 163, 174
    triplet
      predesigned ROHF, 165
      unconstrained CAS 8×10
        MCSCF, 166
      vacuum and predesigned, ROHF
        electronic level structure of, 164
  In–As molecules with one and two atoms
    of phosphorus, 162–173
  [$In_{10}As_2P_2$]$_{Ga}$ triplet
    predesigned CAS 8×9 MCSCF, 180
    unconstrained CAS 8×10 MCSCF, 181
  [$In_{10}As_3P$]$_{Ga}$ triplet
    predesigned ROHF triplet, 176
    unconstrained ROHF, 177
  introduction, 147–151
  quantum confinement, composition effects
    of, 173–183
  small quantum dots of, 147–186
  virtual synthesis procedure, 151–153
Indium nitride (InN), 317
  $In_{10}AsNiN_2$ molecules
    with special magneto-optic properties
      from predesigned ROHF septet
      $In_{10}Ni_2N_2$ derived, 324
      ROHF triplet structure, 323
  $In_{10}As_2N_2$ molecules
    direct optical transition energies
      (OTEs), 253
    ground state energies of, 252

predesigned MCSCF triplet, 255
　molecular orbitals (MOs), 256
　spin density distribution (SDD), 256
predesigned ROHF triplet
　molecular orbitals (MOs), 254
　spin density distribution (SDD), 254
　structure, 253
with special magneto-optic properties
　predesigned MCSCF triplet
　　$In_{10}AsNiN_2$, 322
　unconstrained, Ni/Co-containing
　　molecules from ground state data
　　for, 320
vacuum RHF singlet, molecular orbitals
　(MOs) of, 258
vacuum singlet structure, 257
$In_{10}AS_3N$ molecules
　direct optical transition energies
　　(OTEs), 244
　ground state energies of, 244
　predesigned MCSCF triplet, 249
　　molecular orbitals (MOs), 250
　　spin density distribution (SDD), 250
　predesigned ROHF triplet
　　molecular orbitals (MOs), 246
　　spin density distribution (SDD), 246
　　structure, 245
　vacuum MCSCF triplet
　　molecular orbitals (MOs), 248
　　spin density distribution (SDD), 248
　　structure, 247
based nanosystems, 318
　magnetic properties of, 318
conclusions, 272–276
In_9CoN_4 molecules
　singlet predesigned MCSCF
　　derived from unconstrained molecule
　　　$In_{10}As_2N_2$ of, 341
　singlet unconstrained RHF
　　derived from predesigned molecule
　　　$In_{10}N_4$, 343
　ground-state energies of, 276
　introduction, 239–241
　nanosystems, 344
　　magnetic properties of, 347
$In_{10}Ni_2N_2$ molecule, 337
　predesigned, 329
In_6N_6 molecules
　hexagonal molecules, 265–272

predesigned, 271
　ground state of, 272
　MCSCF pentet structure, 266
　ROHF pentet molecular orbitals
　　(MOs), 267
　ROHF pentet spin density distribution
　　(SDD), 267
　vacuum MCSCF singlet
　　molecular orbitals (MOs), 269
　　spin density distribution (SDD) of, 269
　　structure, 268
　wurtzite-derived molecules
　　direct optical transition energies
　　　(OTEs), 270
　　ground state energies of, 269
$In_{10}N_4$ molecules
　MEP surfaces of, 263
　predesigned molecule
　　ground state data for, 331
　　HOMO–LUMO region of, 265
　　Ni-doped molecules derived from,
　　　330–337
　pyramidal molecules, 260–265
　vacuum, MOs of, 265
pyramidal InAs-based molecules
　with one nitrogen atoms, 243–251
　with two nitrogen atoms, 252–260
quantum dots (QD), 239–276
semiconductors, 239
with special magneto-optic properties
　Co-doped In–As–N/In–N molecules,
　　337–344
　conclusions, 344–347
　introduction, 317–318
　$In_{10}Ni_2N_2$, structure, 325
　$In_{10}N_2Ni_2$, derived from the predesigned
　　molecule $In_{10}N_4$, 336
　$In_{10}N_3Ni$ molecules
　　septet predesigned MCSCF, derived
　　　from predesigned molecule
　　　$In_{10}N_4$, 332
　　structure, ROHF triplet, 333
　small quantum dots of, 317–347
　unconstrained $In_{10}As_2N_2$ molecule
　　Ni-doped molecules derived from,
　　　320–330
　　triplet In_9CoAsN_3 derived from, 339
　virtual synthesis procedure doped with
　　Ni or Co atoms, 318–320

Indium nitride (InN) *(cont.)*
  tetrahedral $In_{10}N_4$ molecules
    direct optical transition energies (OTEs) of, 261
    ground state energies of, 260
    predesigned ROHF triplet
      molecular orbitals (MOs), 262
      spin density distribution (SDD), 262
      structure, 261
    vacuum ROHF singlet
      molecular orbitals (MOs) of, 264
      structure, 263
  virtual synthesis of, 241–243
  wurtzite lattices, symmetry elements of, 243
InGaN structures, synthesis of, 317
INO. *See* Iterative natural-orbital (INO) method
Integrated circuits (ICs), 113
Integrated optoelectronic circuits (IOEC), 150
Interaction operator, 48
IOEC. *See* Integrated optoelectronic circuits (IOEC)
Isothermal susceptibility, 40
Iteration process, 78
Iterative natural-orbital (INO) method, 86

K

Kerr rotation spectroscopy
  time-resolved, 113
Kinetic energy operator, 7
KKR approach. *See* Korringa, Kohn, and Rostoker (KKR) approach
Kohn–Sham equation, 193
Korringa, Kohn, and Rostoker (KKR) approach, 193
KRF. *See* Kubo's relaxation functions (KRF)
Kronecker delta, 45, 76
Kronecker's symbols, 11, 45
Kubo charge density relaxation functions, 29
Kubo's relaxation functions (KRF), 29, 34
  of equation, 44

L

LCs. *See* Linear combinations (LCs)
LDA. *See* Local density approximation (LDA)
LEDs. *See* Light emitting diodes (LEDs)

Legendre polynomials, 71
Lennard–Jones atomic diameters, 305
Ligand bonds, 137, 185
  π-type, 137
Light emitting diodes (LEDs), 147
  quantum efficiency of, 317
  structure, 147
Linear combinations (LCs), 100
Linear response equation, 29
Linear response theory, 18
Linear response transport coefficients, 33
Lippmann-Schwinger equation, 193
Local density approximation (LDA), 194, 239
Localized surface plasmons (LSPs), 147
Longitudinal conductivity, 23
Lowest unoccupied molecular orbital (LUMO), 119, 154, 161, 250, 286
LSPs. *See* Localized surface plasmons (LSPs)
LUMO. *See* Lowest unoccupied molecular orbital (LUMO)

M

Magnetic moment, 24
Magnetic permeability, 24
Many-body perturbation theory (MBPT), 89
Many-body quantum theoretical methods, 94, 318
Many-particle system, 3
Matrix elements, 78, 80, 81, 91
Matrix multiplication operation, 53
Maximum-intensity frequencies, 88
Maxwell's equations, 7, 15, 17, 23
Maxwell's system of equations, 21
MBPT. *See* Many-body perturbation theory (MBPT)
MC-QDPT2 optimization, 118
MCSCF. *See* Multiconfiguration self-consistent field (MCSCF) approximations
MEPs. *See* Molecular electrostatic potentials (MEPs)
Mesoscale systems, 4
Mesoscopic systems, 195
Metal-organic chemical vapor deposition (MOCVD), 147
MnAs nanoparticles, 195
MOCVD. *See* Metal-organic chemical vapor deposition (MOCVD)

Index 363

Molecular electrostatic potentials (MEPs), 119, 152, 169, 197, 248, 300, 321
  isosurfaces, 327, 344
Molecular orbitals (MOs), 75, 201, 318, 324, 353
  occupied, 82
  properties of, 202
  structure of, 161, 259
  unoccupied, 82
Møller–Plesset (MP) perturbation theory, 89–94, 115
  calculations, disadvantage of, 93
  limitations of, 94
  MP2 theory, 92
    direct, 93
    localized, 93
    semidirect, 93
    studies, 300
Mori's projection operator approach, 5, 42
MOs. See Molecular orbitals (MOs)
MRCI. See Multireference configuration interaction (MRCI) method
Multiconfiguration quasidegenerate perturbation method (MC-QDP2), 117
Multiconfiguration self-consistent field (MCSCF) approximations, 4, 283, 290, 300, 318
  calculations, 87
  ground-state energy, 158, 295, 298
  method, 85, 115
Multielectron atomic clusters, 116
Multireference configuration interaction (MRCI) method, 86, 94

N

Nanopolymer quantum wires (QWs)
  Ni–O, OTEs of, 312
  predesigned Hartree–Fock (HF)
    $Ni_{10}O_8$ singlet
      molecular electrostatic potential (MEP), isosurface of, 306
      structure, 306
    $Ni_{18}O_{16}$ singlet
      molecular electrostatic potential (MEP), isosurface of, 308
      structure, 308
    $Ni_{24}O_{22}$ singlet
      isosurface of, 309
      structure, 309
    $Ni_{20}O_{18}$ triplet
      molecular electrostatic potential (MEP), isosurface of, 309
      structure, 309
  predesigned multiconfiguration self-consistent field (MCSCF)
    Ni_6O_6 pentet
      molecular electrostatic potential (MEP), isosurface of, 299
      structure, 299
    Ni_4O_2 septet
      molecular electrostatic potential (MEP), isosurface of, 297
      structure, 297
    Ni_2O singlet
      molecular electrostatic potential (MEP), isosurface of, 289
      structure, 289
    Ni_2O_2 singlet
      HF electronic energy level structure of, 296
      isosurface of molecular electrostatic potential, 293
      structure, 293
    Ni_2O triplet
      molecular electrostatic potential (MEP), isosurface of, 287
      structure, 287
    Ni_8O_6 triplet
      molecular electrostatic potential (MEP), isosurface of, 303
      structure, 303
  vacuum multiconfiguration self-consistent field (MCSCF)
    Ni_2O_2 pentet
      molecular electrostatic potential (MEP), isosurface of, 293
      structure, 293
    Ni_6O_6 pentet
      molecular electrostatic potential (MEP), isosurface of, 301
      structure, 301
    Ni_7O_6 pentet
      molecular electrostatic potential (MEP), isosurface of, 302
      structure, 302
    Ni_2O septet
      molecular electrostatic potential (MEP), isosurface of, 287
      structure, 287

Nanopolymer quantum wires (QWs) *(cont.)*
  Ni_4O_2 septet
    molecular electrostatic potential
      (MEP), isosurface of, 298
    structure, 298
  Ni_2O singlet
    molecular electrostatic potential
      (MEP), isosurface of, 289
    structure, 289
  Ni_2O_2 singlet
    HF electronic energy level structure
      of, 296
    isosurface of molecular electrostatic
      potential, 294
    structure, 294
  Ni_8O_6 singlet
    molecular electrostatic potential
      (MEP), isosurface of, 304
    structure, 304
  Ni_7O_6 triplet
    molecular electrostatic potential
      (MEP), isosurface of, 302
    structure, 302
Nanoscale systems, 93, 240
  DMS systems, ELS of, 206
  heterostructures, 192
  photovoltaic devices, 149
  synthesis conditions of, 319
Nanowire (NW) systems, 150
  properties of, 150
Natural orbitals method, 86
Near infrared band (NIR), 119
N-electron system
  in MP theory, 90
NGF. *See* Nonequilibrium Green's functions
  (NGF)-based methods
Nickel
  atoms, 344
    parallelogram of, 297
    containing molecules
      ground-state energies of, 327
      OTE values of, 327
    doped molecules, 337
  Ni–O bond
    properties of, 288, 289
  Ni–O molecules, 310
    properties of, 311
Nickel oxide
  nanoclusters/nanostructures, 283, 291
  nanomaterials, 285

Nickel oxide quantum dots, 283–313
  discussion and conclusions, 310–313
  introduction, 283–285
  Ni–O
    clusters
      calcinated powder synthesized, SEM
        image of, 304
      quantum dots derived from, 296–306
      ground state energies and OTEs
        of, 286
    nanoclusters, 283
    nanopolymer quantum wires, 306–310
      HOMOs of, 308
      RHF ground state energies and OTEs
        of, 307
    thin films and nanostructures
      band gap width/OTE of, 312
  Ni_2O
    cluster, molecules derived from,
      285–292
    predesigned MCSCF singlet
      molecular electrostatic potential
        (MEP), isosurface of, 289
      structure, 289
    predesigned MCSCF triplet
      molecular electrostatic potential
        (MEP), isosurface of, 287
      structure, 287
    properties of, 292
    spatial isomers of, 289
    vacuum MCSCF septet
      molecular electrostatic potential
        (MEP), isosurface of, 287
      structure, 287
    vacuum MCSCF singlet, 286
      molecular electrostatic potential
        (MEP), isosurface of, 289
      structure, 289
  Ni_2O_2
    cluster, molecules derived from,
      292–296
    ground states of, 295
    predesigned MCSCF singlet
      HF electronic energy level structure
        of, 296
      isosurface of molecular electrostatic
        potential, 293
      structure, 293
    vacuum MCSCF pentet

molecular electrostatic potential
(MEP), isosurface of, 293
structure, 293
vacuum MCSCF singlet
HF electronic energy level structure
of, 296
isosurface of molecular electrostatic
potential, 294
structure, 294
Ni_4O, 298
Ni_4O_2
octahedral, properties of, 298
predesigned MCSCF septet
molecular electrostatic potential
(MEP), isosurface of, 297
structure, 297
vacuum MCSCF septet
molecular electrostatic potential
(MEP), isosurface of, 298
structure, 298
Ni_6O_6
predesigned MCSCF pentet
molecular electrostatic potential
(MEP), isosurface of, 299
structure, 299
vacuum MCSCF pentet
molecular electrostatic potential
(MEP), isosurface of, 301
structure, 301
Ni_8O_6
predesigned MCSCF triplet
molecular electrostatic potential
(MEP), isosurface of, 303
structure, 303
vacuum MCSCF singlet
molecular electrostatic potential
(MEP), isosurface of, 304
structure, 304
$Ni_{10}O_8$, predesigned HF singlet
molecular electrostatic potential (MEP),
isosurface of, 306
structure, 306
$Ni_{18}O_{16}$, predesigned HF singlet
molecular electrostatic potential (MEP),
isosurface of, 308
structure, 308
$Ni_{24}O_{22}$, predesigned HF singlet
isosurface of, 309
structure, 309

$Ni_{20}O_{18}$, predesigned HF triplet
molecular electrostatic potential (MEP),
isosurface of, 309
structure, 309
Ni_7O_6, vacuum MCSCF
pentet
molecular electrostatic potential
(MEP), isosurface of, 302
structure, 302
triplet
molecular electrostatic potential
(MEP), isosurface of, 302
structure, 302
pyramidal molecules Ni_4O
molecular electrostatic potential (MEP),
isosurface of, 299
NIR. See Near infrared band (NIR)
Noether's theorems, 62
Noncentrosymmetric systems, 148
Nondegenerate perturbation theory, 64–65
Nonequilibrium Green's functions
(NGF)-based methods, 4
semiheuristic, 5
Nonlinear optical nanosystems, 148
Nonstoichiometric molecules, 167
Nonvariational theory, 194
Nuclear repulsion, 74
Number density operator, 8

O

Occupation numbers, 86
Octet rule, 201, 265, 288
One-electron operator, 75
One-electron spatial orbital, 68
Operator algebra theorem, 83
Optical transition energy (OTE), 118, 121,
152, 241, 285, 321
first-principle theoretical calculations
of, 276
value, 308
Optoelectronics, 150
OTE. See Optical transition energy (OTE)
Overlap integrals, 77

P

Pauli exclusion principle, 71
Perdew–Burke–Ernzerhof exchange-
correlation, 194, 239

Perturbation theory, 91
  based methods, 64
Photo-generated carriers, 149
Photonic devices, 147
Planck constant, 33
Polarized (P) basis sets, 99
Polymerization mechanism, 306
Predesigned Ni-containing molecules
  RHF/ROHF ground-state energies of, 321
Projection operator methods, 3, 52, 103
Pyramidal InAs-based molecules
  with one nitrogen atom, 243–251
  with two nitrogen atoms, 252–260

Q

QDs. *See* Quantum dots (QDs) systems
QDW. *See* Quantum dots and wires (QDWs)
QICs. *See* Quantum integrated circuits (QICs)
QSM. *See* Quantum statistical mechanics (QSM)
Quantum charge transport theory, 6
Quantum chemical calculations, 98
Quantum chemistry software packages
  GAISSIAN, 31
  GAMESS, 31
  Molpro, 31
  NWChem, 31
Quantum conductivities, 26
Quantum confinement, 3, 106, 116, 139, 154, 173, 178, 199, 204, 241, 313, 334, 338
  composition effect of, 172, 185, 241, 347
  nucleation and synthesis in, 305
  stabilization role of, 295
Quantum confinement model, polarization influence of, 329
Quantum dots and wires (QDWs)
  elements, 114
  magnetic properties of, 133
Quantum dots (QDs) systems, 4, 113, 147, 239, 301, 317
  14-atomic, 151
  excitons in, 113
  properties of, 150
    electronic, 274, 318
    magnetic, 274
  quantum-confined, 318
  virtual design of, 265
  virtual templates of, 351

Quantum field-theoretical approximations, 152
Quantum field theory, 89
Quantum information processing devices, 239
Quantum integrated circuits (QICs), 113
Quantum Liouville (von Neumann's) equation, 9
Quantum many-body theoretical methods, 228
Quantum many-body theory-based optimization, 241
Quantum mechanical methods, 4
  of ground-state wavefunction calculation, 61
Quantum statistical mechanics (QSM), 195, 196
Quantum systems
  theory for, 63
  transport properties of, 29
Quantum transport coefficients, 27, 42
  explicit expressions for, 32
Quantum transport property tensors, 28
Quantum wells
  based devices, 149
    advantages, 149
  based solar cells, 149
  and wires, 147
Quantum wires (QWs)
  core-shell, 303
  geometry, 150
  nanopolymer. *See* Nanopolymer quantum wires (QWs)
  synthesis of, 191
  virtual templates of, 351
Quasiparticle self-consistent GW (QSGW), 240
QWs. *See* Quantum wires (QWs)

R

Radial factor, 71
Rayleigh–Schödinger perturbation theory, 65
Realistic quantum systems, 104
Reference function, 86
Resonant tunneling spectroscopy, 195
Restricted Hartree–Fock (RHF) method, 115
  optical transition energy (OTE), 330
  RHF-ROHF approximation method, 225
Restricted open-shell Hartree–Fock (ROHF) approximations, 93, 115, 152, 163, 168, 175, 196, 222
  dipole moment, 178

Index   367

electronic energy level structure (ELS), 122
ground-state energies, 163
optimization procedure, 121
ROHF/MCSCF triplets, 167
singlet, 226
RHF. *See* Restricted Hartree-Fock (RHF) method
R-matrix theory, 5
ROHF. *See* Restricted open-shell Hartree-Fock (ROHF) approximations

S

SBKJC standard basis set, 285, 318
Scalar dielectric permittivity, 17
Scalar function, 13
Scalar magnetic permeability, 25, 26
SCF. *See* Self-consistent field (SCF)
Schmidt's method, 80
Schrödinger equations, 66, 71, 73, 97, 115, 193, 196
  electronic, 74
  time-independent, 62
SDDs. *See* Spin density distributions (SDDs)
Self-consistent field (SCF)
  algorithm, 88
  approximations, 4, 283, 318
  ground-state wavefunction, 88
  MOs, calculations of, 85
Self-consistent method, 35
Self-interaction correction local density approximation (SIC-LDA), 194
Semiconductors, 204, 300
  elements, virtual templates, small quantum dots and wires of, 351–354
  nanosystems, 318
  nanowire, 147
  p-type, 239
Semiphenomenological models, 191, 193
Semiphenomenological theory, of conductance, 5
SIC-LDA. *See* Self-interaction correction local density approximation (SIC-LDA)
Simulated emission of radiation (SPASER concept), 147
Single-electron wavefunction, 69
Slater determinants, 69, 73, 78, 83, 84, 88, 90
  configuration state functions (CSF), 86
  molecular wavefunction, 83
  spin components of, 68–69
  variation modification of, 69–70
  wavefunctions, 91
Slater-type atomic orbitals (STOs), 98
  basis set
    double-zeta (DZ), 99
    triple-zeta (TZ), 99
Small quantum dots (QDs), 111–140
  optoelectronic and magnetic properties, 184
Small systems at equilibrium
  ab initio software packages and use, 101–103
  basis function sets, 98–101
  configuration interactions, 82–88
  coupled-cluster approximation, 94–98
  Hartree–Fock self-consistent field method, 70–82
  introduction, 61–62
  Møller–Plesset (MP) perturbation theory, 89–94
  quantum properties of, first-principles calculations, 61–107
  variational methods, 62–70
    degenerate energy levels, perturbation theory treatment of, 66–68
    extended variation method, 62–64
    nondegenerate perturbation theory, 64–65
    perturbation method, 64–65
    Slater determinants
      spin components of, 68–69
      variation modification of, 69–70
    variation theorem, 62–64
    wavefunctions, spin components of, 68–69
  virtual synthesis method, 103–107
Solar energy conversion systems, 283
Solid lattice, 183
Solid-state electronic structure theory, 4
Space–time Fourier transforms (STFTs), 10, 11, 15
  approximations for, 31
  of dielectric permittivity, 18
  of expectation values, 10–11
  of generalized susceptibility and microcurrent–microcurrent EGFs, 11–13
  identification of, 17
  linearity of, 13

Spatially homogeneous systems, 23
Spatially inhomogeneous quantum systems
  charge and spin transport in, 7–27
    charge and current densities, expectation values of, 9–10
    generalized longitudinal sum rule, 13–14
    induced magnetic moment and magnetic permeability, 24–26
    space–time Fourier transforms
      of expectation values, 10–11
      of generalized susceptibility and microcurrent–microcurrent EGFs, 11–13
    weak external electromagnetic field
      dielectric permittivity in, 14–18
      generalized susceptibility in, 18–22
      longitudinal quantum conductivity in, 23–24
      transversal conductivity in, 24–27
      transversal conductivity of, 26–27
    equilibrium Green's functions, calculation of, 29–32
    introduction, 3–7
    optical properties, tensor of refractive indices, 27–29
    transport properties of, 1–53
    Zubarev–Tsercovnikov's projection operator method, 33–53
Spatially inhomogeneous systems
  quantum transport properties of, 61
Spectral theorem, 39
  of equation, 40
Spin-constrained unrestricted Hartree–Fock (SUHF) approach, 93
Spin density, 183
Spin density distributions (SDDs), 133, 140, 162, 199, 241, 354
  isosurfaces, 158
  of molecules, 133–135
Split-valence basis sets, 99
Standard matrix algebra methods, 87
Standard valence theory, octet rule of, 199, 250
Statistical mechanics, 3
Steady-state statistical operator, 9
STFTs. *See* Space-time Fourier transforms (STFTs)
Stoner-like exchange parameter, 239

STOs. *See* Slater-type atomic orbitals (STOs)
SUHF. *See* Spin-constrained unrestricted Hartree-Fock (SUHF) approach
Supersymmetry-based techniques, 5

T

TCFs. *See* Time correlation functions (TCFs)
Tensor of refractive indices (TRI), 27
  nonlinearity of, 28
Terbium aluminum garnet ($Tb_3Al_5O_{12}$), 318
Terbium gallium garnet ($Tb_3Ga_5O_{12}$), 318
TFTs. *See* Time Fourier transforms (TFTs)
Thermodynamic limits, 3
Time correlation functions (TCFs), 35, 36
  properties of, 38
  time derivatives of, 53
Time-dependent projection operators, 51
Time Fourier transforms (TFTs), 37, 38, 41, 50
Traditional III-V semiconductor compounds
  "artificial" molecules [$In_{10}As_4$]$_{Ga}$, 125–129
  In_3As molecule
    ground and excited state energies of, 119
    pyramidal, 117
      in HOMO-LUMO region, 120
      MP2 In3As singlet, 120
    RHF electronic level structure of, 118
    transition energies of, 119
  $In_{10}As_4$ molecules
    predesigned and vacuum, 121–125
    triplets
      direct optical transition energies (OTEs), 122
      ELSs of vacuum and predesigned ROHF, 124
      ground state energies of, 122
      MEP isosurfaces of, 123
      pyramidal predesigned and vacuum CASSCF, 121
  In and As atoms, smallest 3D molecule of, 117–121
  electron charge delocalization and bonding in molecules, 135–138
  $Ga_{10}As_4$, 129–133
    CASSCF triplets
      ground state interaction energies for, 132

Index   369

predesigned 6×9, 133, 136
predesigned 8×10, 134
predesigned 8×13, 129, 130, 134, 138
vacuum 8×8, 130, 133, 134, 136, 138
vacuum 8×10, 134
direct optical transition energies (OTEs) of, 131
ground state energies of, 131
vacuum 8×8, 130
vacuum and predesigned ROHF, ELSs of, 131
$[In_{10}As_4]_{Ga}$ triplet
  direct optical transition energies (OTEs), 127
  ground state interaction energies for, 127
  MOs from HOMO–LUMO region of predesigned ROHF, 137
  predesigned ROHF, 124
    MOs from HOMO–LUMO region of, 137
    vacuum and predesigned (right) ROHF, ELSs of, 128
  vacuum 8×10 CASSCF, 126
    MOs from HOMO–LUMO region of, 137
introduction, 113–117
  virtual synthesis setup, 115–117
small quantum dots of, 111–140
spin density distributions of molecules, 133–135
Transport coefficients
  first-principles predictions of, 32
Transversal Kerr effect spectroscopy, 195
TRI. *See* Tensor of refractive indices (TRI)
Triangular Ni_2O molecules
  magnetic properties of, 291
Two-electron repulsion
  energy, 135
  integrals, 78

U

UMP-series methods, 93
Uncontrolled approximations, 4
Unrestricted Hartree–Fock (UHF) self-consistent field determinant, 93

V

Valance bands (VB), 193, 285
Valence shell electrons, 84
Variational methods, 62–70
  degenerate energy levels, perturbation theory treatment of, 66–68
  extended variation method, 62–64
  nondegenerate perturbation theory, 64–65
  perturbation method, 64–65
  Slater determinants
    spin components of, 68–69
    variation modification of, 69–70
  variation theorem, 62–64
  wavefunctions, spin components of, 68–69
Variation coefficients, 88
Variation perturbation method, 64–65
Variation theorem, 62–64, 70, 75
V-containing molecules, 206, 209
Virtual spin-orbits, 88, 91
Virtual synthesis method, 103–107, 126, 139, 153, 173, 205
  properties of, 106
Visualization software, 102
  Molekel, 102

W

Wavefunctions, spin components of, 68–69
Weak external electromagnetic field
  dielectric permittivity in, 14–18
  generalized susceptibility in, 18–22
  longitudinal quantum conductivity in, 23–24
  transversal conductivity in, 24–27
Wurtzite bulk lattices, 265
  InN bulk lattice, 272
  lattice structures, 300
Wurtzite-derived molecules, 272
  energy level structure (ELS), 270
  Hartree–Fock (HF) ground-state energies of, 268
  HOMO–LUMO regions, 270
  predesigned, dislocation defect in, 276
  predicted instability of, 273

X

XCIS method, 88

Y

YIG. *See* Yttrium iron garnet (YIG)
Yttrium iron garnet (YIG), 318

Z

Zener model, p–d, 191–193
"Zero-dimensional" nanostructures, 149
Zero-order wave function, 66, 67, 69
Zeroth-order approximation, 62, 70
Zincblende
   derived molecules, 197, 268
   GaAs bulk lattice, 173
   lattices, 326
Zubarev's equation, 18, 22, 27
Zubarev's expressions, 32
Zubarev–Tsercovnikov's method, 42, 43, 50
   advantage of, 51
   analytical methods, 51
   generalization of, 53
Zubarev-Tsercovnikov's transport coefficients, 6
Zubarev–Tserkovnikov projection, 27
Zubarev–Tserkovnikov's (ZT) projection operator method, 5, 33–53
   advantages of, 6, 51–52
   disadvantage of, 6, 51–52
   energy-dependent representation, 42–50
   prospects of applications, 53
   time-dependent representation, 50–51
   two-time temperature GFs used in statistical physics
   definitions and properties of, 33–42

Printed in the United States
By Bookmasters